ESSENTIAL CLASSICAL MECHANICS

Problems and Solutions

Other Related Titles from World Scientific

Essential Classical Mechanics
by Choonkyu Lee and Hyunsoo Min
ISBN: 978-981-3234-64-2

Fundamental Principles of Classical Mechanics: A Geometrical Perspective
by Kai S. Lam
ISBN: 978-981-4551-48-9

Principles of Physics: From Quantum Field Theory to Classical Mechanics (2nd Edition)
by Jun Ni
ISBN: 978-981-3227-09-5

ESSENTIAL CLASSICAL MECHANICS

Problems and Solutions

Choonkyu Lee
Seoul National University, South Korea

Hyunsoo Min
University of Seoul, South Korea

NEW JERSEY • LONDON • SINGAPORE • BEIJING • SHANGHAI • HONG KONG • TAIPEI • CHENNAI • TOKYO

Published by

World Scientific Publishing Co. Pte. Ltd.
5 Toh Tuck Link, Singapore 596224
USA office: 27 Warren Street, Suite 401-402, Hackensack, NJ 07601
UK office: 57 Shelton Street, Covent Garden, London WC2H 9HE

British Library Cataloguing-in-Publication Data
A catalogue record for this book is available from the British Library.

ESSENTIAL CLASSICAL MECHANICS
Problems and Solutions

ISBN 978-981-3270-05-3
ISBN 978-981-3270-97-8 (pbk)

For any available supplementary material, please visit
https://www.worldscientific.com/worldscibooks/10.1142/10985#t=suppl

Desk Editor: Ng Kah Fee

Typeset by Stallion Press
Email: enquiries@stallionpress.com

Printed in Singapore

Preface

This book consists of problems in *Essential Classical Mechanics* together with their solutions. The degree of difficulty with these problems varies from quite simple to very challenging; but none too easy, as all problems in physics demand some subtlety of intuition (in addition to a lot of practice). It is through problem-solving that students appreciate classical mechanics better and gain confidence in applying various techniques. Even if you are unable to solve some problems, you need not be discouraged. If you were learning to ride a bicycle, falling off the bicycle would be just a sign that you need more practice to get familiar with the machine. Same thing can be said for learning physics. Until you can solve problems, your understanding on the subject cannot be said to be sufficiently deep. For any given physics problem, there are usually more than one way of solving it. In this "Problems and Solutions" book, you will find one way of solving the problems, and it is our hope that this book can serve as a useful guide to students who are trying problems in classical mechanics.

This book has two parts, Part I with the summary and problems of each chapter in *Essential Classical Mechanics* and Part II with detailed solutions to all problems. The summaries are condensed, yet informative enough to contain all key concepts and important formulas of the given chapters. The notations are those explained in the textbook and we occasionally (especially in the "Solution" part) refer to equations and figures from the book. But, as for the statements of problems in Part I, we made some effort to make clear what are things that are being asked, independently of equations appearing in *Essential Classical Mechanics.* This is for the sake of some readers (e.g., graduate students) who just want to test and enhance their problem-solving ability in mechanics; for them, the summaries in Part I should serve as useful guidelines as to the scope of background knowledge required before they tackle specific problems in the book.

Choonkyu Lee
Department of Physics and Astronomy and
Center for Theoretical Physics
Seoul National University

Hyunsoo Min
Department of Physics
University of Seoul

Contents

Part I

Summary and Problems

1 In Three-Dimensional Space: Vector Description

SUMMARY

1.1 Vectors in 3-dimensional Euclidean space

- Physical quantities in mechanics are represented using notions like vectors and scalars defined in 3-dimensional Euclidean space.
- Choosing a set of Cartesian axes, one can represent a vector $\vec{A}$ by $\vec{A} = (A_x, A_y, A_z)$ or $\vec{A} = A_x\mathbf{i} + A_y\mathbf{j} + A_z\mathbf{k}$ and, if c is a scalar, $c\vec{A} = (cA_x, cA_y, cA_z)$.
- Addition (or subtraction) of two vectors $\vec{A}$ and $\vec{B}$ is defined by the parallelogram law or, in terms of their components, by $\vec{A} \pm \vec{B} = (A_x \pm B_x, A_y \pm B_y, A_z \pm B_z)$.
- If a different choice is made for the Cartesian axes, the components of any physical vector must transform in the same way as the coordinates (x, y, z) of the position vector $\vec{r}$.

1.2 The scalar and vector products

- The scalar (or dot) product of two vectors $\vec{A} = A_x\mathbf{i} + A_y\mathbf{j} + A_z\mathbf{k} \equiv A_i\mathbf{e}_i$ and $\vec{B} = B_x\mathbf{i} + B_y\mathbf{j} + B_z\mathbf{k} \equiv B_i\mathbf{e}_i$, having the angle θ between them, is given by

$$\begin{aligned}\vec{A} \cdot \vec{B} &= |\vec{A}||\vec{B}| \cos\theta \\ &= A_xB_x + A_yB_y + A_zB_z \quad (\equiv A_iB_i).\end{aligned}$$

- The vector (or cross) product $\vec{C} = \vec{A} \times \vec{B}$, as defined by the formula

$$\begin{aligned}\vec{A} \times \vec{B} &= (A_y B_z - A_z B_y)\mathbf{i} + (A_z B_x - A_x B_z)\mathbf{j} \\ &\quad + (A_x B_y - A_y B_x)\mathbf{k} \\ &\equiv (\epsilon_{ijk} A_j B_k)\mathbf{e}_i,\end{aligned}$$

represents the vector that points in the direction perpendicular to the plane containing $\vec{A}$ and $\vec{B}$, and has the magnitude equal to the area of the parallelogram with edges $\vec{A}$ and $\vec{B}$.
- Triple products:

$$\vec{A} \cdot (\vec{B} \times \vec{C}) = \epsilon_{ijk} A_i B_j C_k = \begin{vmatrix} A_1 & A_2 & A_3 \\ B_1 & B_2 & B_3 \\ C_1 & C_2 & C_3 \end{vmatrix},$$

$$\vec{A} \times (\vec{B} \times \vec{C}) = (\vec{A} \cdot \vec{C})\vec{B} - (\vec{A} \cdot \vec{B})\vec{C}.$$

1.3 Differentiation and integration of vector-valued functions

- If $\vec{r}(t)$ $(= x_i(t)\mathbf{e}_i)$ denotes the position of a particle at time t, the instantaneous velocity and acceleration of the particle are given by

$$\vec{v}(t) = \frac{d\vec{r}(t)}{dt} = \frac{dx_i(t)}{dt}\mathbf{e}_i,$$

$$\vec{a}(t) = \frac{d\vec{v}(t)}{dt} = \frac{d^2\vec{r}(t)}{dt^2} = \frac{d^2x_i(t)}{dt^2}\mathbf{e}_i.$$

For a particle moving in a circle on the xy-plane according to $\vec{r}(t) = \rho\cos\theta(t)\mathbf{i} + \rho\sin\theta(t)\mathbf{j}$, we have

$$\vec{v}(t) = \rho\dot{\theta}\mathbf{t}, \quad \vec{a}(t) = \rho\ddot{\theta}\mathbf{t} - \rho\dot{\theta}^2\mathbf{n},$$

where $\mathbf{t} = -\sin\theta\mathbf{i} + \cos\theta\mathbf{j}$ and $\mathbf{n} = \cos\theta\mathbf{i} + \sin\theta\mathbf{j}$ denote the unit tangent and unit normal to the path, respectively.

- Given $\vec{v}(t)$ $(= \frac{d\vec{r}(t)}{dt})$, we have $\vec{r} = \int \vec{v}\, dt$ or

$$\vec{r}(t) = \vec{r}(t_0) + \int_{t_0}^{t} \vec{v}(t')dt',$$

and, given $\vec{a}(t)$ $(= \frac{d\vec{v}(t)}{dt})$,

$$\vec{v}(t) = \vec{v}(t_0) + \int_{t_0}^{t} \vec{a}(t')dt',$$

$$\vec{r}(t) = \vec{r}(t_0) + \vec{v}(t_0)t + \int_{t_0}^{t} \left\{ \int_{t_0}^{t'} \vec{a}(t'')dt'' \right\} dt'.$$

For a projectile under the Earth's gravity, one can take the acceleration to be $\vec{a} = -g\mathbf{k}$ and so, if $\vec{v}_0 = \vec{v}(t=0)$ and $\vec{r}_0 = \vec{r}(t=0)$,

$$\vec{v}(t) = \vec{v}_0 - gt\mathbf{k},$$

$$\vec{r}(t) = \vec{r}_0 + \vec{v}_0 t - \frac{1}{2}gt^2\mathbf{k} \quad (t > 0).$$

1.4 Scalar and vector fields

- Physical quantities which are also functions of position can be represented by a scalar field $\phi(x, y, z)$ or a vector field $\vec{V}(x, y, z) = V_i(x, y, z)\mathbf{e}_i$, depending on the nature of distributed entities.
- The gradient of a scalar field ϕ is a vector field $\vec{\nabla}\phi(x_1, x_2, x_3) \equiv \frac{\partial \phi(x_1, x_2, x_3)}{\partial x_i}\mathbf{e}_i$. Then the difference between the values of a scalar field ϕ at $\vec{r} = (x, y, z)$ and $\vec{r} + d\vec{r} = (x + dx, y + dy, z + dz)$ can be expressed by the formula $d\phi = d\vec{r} \cdot \vec{\nabla}\phi$, and $\vec{\nabla}\phi\big|_{\vec{r}=\vec{r}_0}$ is always normal to the surface defined by the equation $\phi(\vec{r}) = \phi(\vec{r}_0)$ at the very point $\vec{r} = \vec{r}_0$.
- Given a curve $C\colon \vec{r} = \vec{r}(t)$ $(\alpha \le t \le \beta)$ in space and a vector field $\vec{V}(\vec{r}) = V_i(\vec{r})\mathbf{e}_i$, one can consider the line integral $\int_C \vec{V} \cdot d\vec{r} \equiv \int_\alpha^\beta \vec{V}(\vec{r}(t)) \cdot \frac{d\vec{r}(t)}{dt} dt$. If a vector field $\vec{V}(\vec{r})$ can be written as a gradient of a certain scalar field ϕ, i.e., $\vec{V}(\vec{r}) = \vec{\nabla}\phi(\vec{r})$, its line integral over an arbitrary closed path vanishes, and $\vec{V}(\vec{r})$ is then called a conservative vector field.

Problems

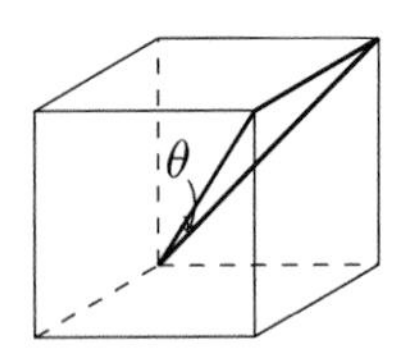

1-1 With the help of vector operations, find the cosine of the angle θ (see the figure) between a long diagonal and an adjacent face diagonal of a cube.

1-2 A plane in $\mathbb{E}^3$ is described by the equation

$$Ax + By + Cz + D = 0,$$

where x, y and z denote Cartesian coordinates.

(i) Find the unit vector $\mathbf{n}$ normal to this plane.
(ii) Let $P = (x_0, y_0, z_0)$ be any point not on this plane. Find the shortest distance from P to the plane.
(iii) Can you find the angle between the above plane and another plane defined by the equation $ax + by + cz + d = 0$?

1-3 Suppose a physical vector $\vec{A}$ in $\mathbb{E}^3$ can be expressed using an orthonormal triplet $(\mathbf{e}_1, \mathbf{e}_2, \mathbf{e}_3)$ as

$$\vec{A} = A_1\mathbf{e}_1 + A_2\mathbf{e}_2 + A_3\mathbf{e}_3 \equiv A_i\mathbf{e}_i.$$

Then let (A'_1, A'_2, A'_3) denote the components of this vector with respect to a new orthonormal triplet $(\mathbf{e}'_1, \mathbf{e}'_2, \mathbf{e}'_3)$ which is obtained from $(\mathbf{e}_1, \mathbf{e}_2, \mathbf{e}_3)$ through a counterclockwise rotation by an angle θ about the axis in the direction of $\mathbf{u} = \frac{1}{\sqrt{2}}(\mathbf{e}_1 + \mathbf{e}_2)$.

(i) As we write $A'_i = O_{ij}A_j$, find the appropriate 3×3 matrix $O = (O_{ij})$ explicitly.
(ii) Verify that the matrix thus obtained satisfies the conditions $OO^T = I = O^TO$ and $\det O = 1$.

1-4 Show that, for the totally antisymmetric symbol ϵ_{ijk} defined by $\epsilon_{123} = 1$ and $\epsilon_{ijk} = \epsilon_{jki} = \epsilon_{kij} = -\epsilon_{jik} = -\epsilon_{kji} = -\epsilon_{ikj}$, the following identities hold:

(i) $\epsilon_{ijk}\epsilon_{ljk} = 2\delta_{il}$ and $\epsilon_{ijk}\epsilon_{ijk} = 3!$;
(ii) $\epsilon_{ijk}\epsilon_{lmk} = \delta_{il}\delta_{jm} - \delta_{im}\delta_{jl}$.

Also, when A and B are two arbitrary 3×3 matrices, use the properties of the ϵ-symbol to justify following formulas:

(iii) $\epsilon_{lmn} A_{il} A_{jm} A_{kn} = \epsilon_{ijk} \det A$ and $\epsilon_{lmn} A_{li} A_{mj} A_{nk} = \epsilon_{ijk} \det A$;
(iv) $\det(AB) = \det A \cdot \det B$.

1-5 If three vectors $\vec{A}$, $\vec{B}$ and $\vec{C}$ in the 3-dimensional Euclidean space are linearly independent, any vector $\vec{D}$ may be written as suitable linear combination of those three, viz.,

$$\vec{D} = \alpha \vec{A} + \beta \vec{B} + \gamma \vec{C}.$$

Express the coefficients α, β and γ in terms of suitable geometrical quantities (e.g., scalar product, vector product) made out of vectors $\vec{A}$, $\vec{B}$, $\vec{C}$ and $\vec{D}$.

1-6 Consider a triangle with the length of its three sides equal to a, b and c. Then, with the help of the vector algebra, prove the sine law of trigonometry

$$\frac{a}{\sin \alpha} = \frac{b}{\sin \beta} = \frac{c}{\sin \gamma},$$

where α, β, γ denote the angles as shown in the figure.

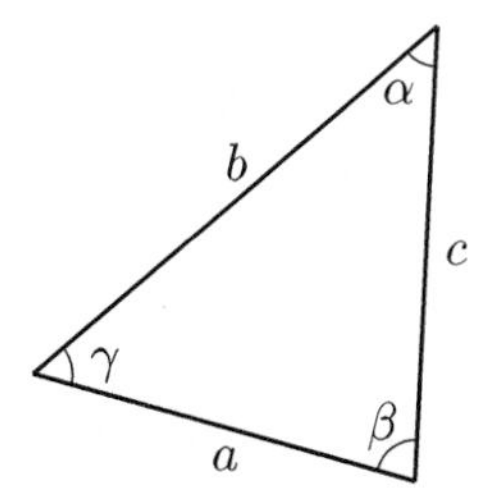

1-7 In $\mathbb{E}^3$, consider four vectors whose directions are outward perpendiculars to the four faces of a tetrahedron and whose lengths are equal to the areas of the faces they represent. Show that the sum of these four vectors is equal to the zero vector. Can you also generalize the discovery further to the case involving all faces of a *general* polyhedron?

1-8 Consider a particle which is constrained to move on the surface of a sphere (but its motion on the sphere can be quite arbitrary). Assuming that the sphere is at rest and has radius R, prove that the following relation always holds:

$$\mathbf{n} \cdot \vec{a}(t) = -\frac{\vec{v}(t) \cdot \vec{v}(t)}{R}.$$

Here, $\vec{v}(t)$ is the particle's velocity at time t, $\vec{a}(t)$ its acceleration, and $\mathbf{n}$ denotes the outward unit normal drawn in the direction of the particle position from the center of the sphere.

1-9 Suppose that a particle moves with a *uniform* speed along the trajectory as shown in the figure. At what point will the magnitude of the acceleration of the particle be a maximum? [Can you also support your answer by a mathematical argument?]

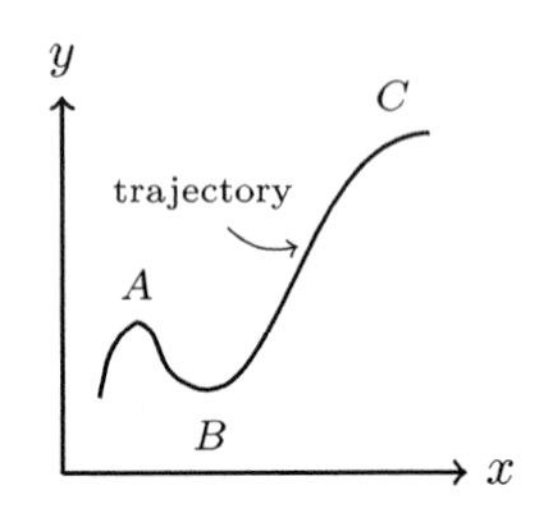

1-10 A wheel of radius b rolls along the ground with constant forward acceleration a_0. Show that, at any given instant, the magnitude of the acceleration of any point on the rim of the wheel is $\sqrt{a_0^2 + \frac{v^4}{b^2}}$ relative to the center of the wheel and is also $a_0\sqrt{2 + 2\cos\theta + \frac{v^4}{a_0^2 b^2} - \left(\frac{2v^2}{a_0 b}\right)\sin\theta}$ relative to the ground. Here v is the instantaneous forward speed, and θ defines the location of the point on the wheel, measured forward from the highest point. Which point has the greatest acceleration relative to the ground?

1-11 In Problem 1-10, suppose that $a_0 = 0$, i.e., the center of the wheel moves with a uniform velocity $\vec{v} = v\mathbf{i}$. In this case, find (and plot) the trajectory taken by a particular point P of the wheel as the wheel rolls. You may assume that at $t = 0$ the point P is at the highest position of the wheel. [This path is called a *cycloid*.]

1-12 A gun is mounted on a hill of height h above a level plain. Show that, if the air resistance is neglected, the greatest horizontal range for given muzzle velocity v_0 is obtained by firing at an angle of elevation θ such that

$$\csc^2\theta = 2\left(1 + \frac{gh}{v_0^2}\right).$$

1-13 Consider a ball thrown into air, from ground level, towards a wall of height h which is located at a distance L away. (See the figure.) When we want to have the ball pass over the wall, find the minimum speed V (by choosing the best angle for the throw) necessary for the ball initially.

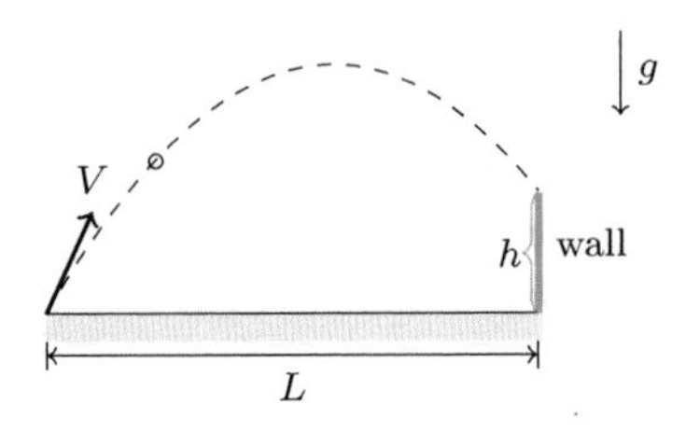

1-14 A fountain consists of a small hemispherical rose (sprayer) which lies on the surface of the water in a basin, as illustrated in the figure. The rose has many evenly distributed small holes in it, through which water spurts at the same speed in all directions. What would be the shape of the water "bell" formed by the jets?

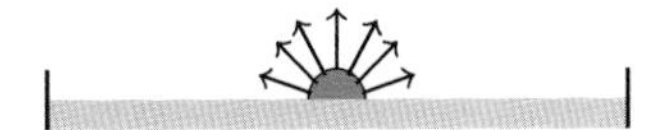

1-15 Answer the following questions (by using the properties of the gradient for a given function).

(i) Find the equation of the tangent plane to the surface $z = xy - 1$ (in $\mathbb{E}^3$) at the point $P = (2, 1, 1)$.

(ii) Consider the ellipse which is described by the equation $r_1 + r_2 =$ constant (see the figure). How is the unit normal $\mathbf{n}$ to the ellipse at point P given? Also show that $\overrightarrow{AP}$ and $\overrightarrow{BP}$ make equal angles with the tangent to the ellipse.

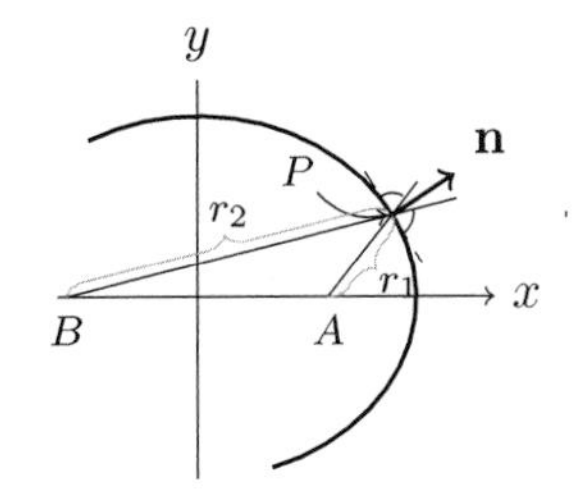

1-16 Consider a vector field given as $\vec{V}(\vec{r}) = \vec{A} \times \vec{r}$ (here, $\vec{A}$ is a constant vector) and two curves on the xy-plane — a straight path C_1 and a circular path C_2. Evaluate the line integrals $\int_{C_1} \vec{V}(\vec{r}) \cdot d\vec{r}$ and $\int_{C_2} \vec{V}(\vec{r}) \cdot d\vec{r}$. Note that both C_1 and C_2 start at $(a, 0)$ and end at $(-a, 0)$.

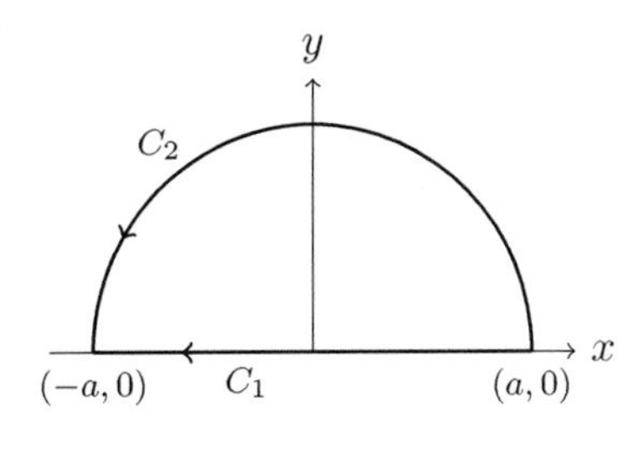

1-17 Evaluate the line integral

$$\oint_C (\mathbf{k} \times \vec{r}) \cdot d\vec{r}$$

for many different *closed* curves on the xy-plane. What do you find? Can you establish generally what you find by resorting to some simple geometrical arguments?

2 Evolution in Time: Basic Elements of Newtonian Mechanics

SUMMARY

2.1 Galileo–Newton spacetime and Newton's first law

- For nonrelativistic mechanics, one may adopt the Galileo–Newton spacetime model; the basic assumptions of which comprise the homogeneity of time, the (homogeneous and isotropic) Euclidean space $\mathbb{E}^3$ for the physical space, and the Galilean relativity, i.e., the principle that the world seen by all "unaccelerated" observers should be exactly alike.
- To each choice of origins in time and space and a set of three Cartesian axes in $\mathbb{E}^3$ corresponds a frame of reference in which the position and time of any event in spacetime may be represented using related time and space coordinates (t, x, y, z). It is assumed that there exists a special class of reference frames, called Newtonian inertial frames (attached to unaccelerated observers), in which an (every) isolated body, far removed from all other matter, would move along a geometrically-straight line with uniform coordinate velocity, i.e.,

$$\frac{dx(t)}{dt} = \text{const.}, \quad \frac{dy(t)}{dt} = \text{const.}, \quad \frac{dz(t)}{dt} = \text{const.}$$

This is Newton's first law, also called Galileo's law.

2.2 Symmetries of physical spacetime

- In the Galileo–Newton spacetime, fundamental laws of physics should assume the same form in all (Newtonian) inertial frames — i.e., the equations should be form-invariant under spacetime translation ($t \to t' = t + b$, $x_i \to x_i' = x_i + a_i$), time-independent rotation of spatial axes ($x_i \to x_i' = O_{ij}x_j$, where $O^{-1} = O^{\mathrm{T}}$ and $\det O = 1$), Galilean transformation ($t \to t' = t$, $x_i \to x_i' = x_i - u_i t$), and spatial inversion ($t \to t' = t$, $x_i \to x_i' = -x_i$).
- Given two observers, one in a frame R and the other in a frame R' which is related to R by a Galilean transformation $x_i' = x_i - u_i t$, an object with the velocity $\vec{v}(t) = \frac{dx_i(t)}{dt}\mathbf{e}_i$ according to the observer in R has the velocity

$$\vec{v}\,'(t) = \frac{dx_i'(t)}{dt} = \vec{v}(t) - \vec{u}$$

according to the observer in R'. Thus the direction of the velocity vector of an object in general depends on who is observing.
- When some particles have very large speed (not negligible compared to the speed of light), mechanical discussions must be based on the new spacetime model — that of Einstein. In this model, if (t, x, y, z) and (t', x', y', z') represent the spacetime coordinate readings of a given event by two inertial observers in uniform relative motion (say, along the x-axis), they are related not by a Galilean transformation but by a Lorentz boost transformation

$$t' = \frac{t - ux/c^2}{\sqrt{1 - (u/c)^2}}, \quad x' = \frac{x - ut}{\sqrt{1 - (u/c)^2}}, \quad y' = y, \quad z' = z.$$

Related velocity transformation laws, considered in Problem 2-4, are more complicated than the above Galilean transformation laws: here, "time duration" becomes also relative, i.e., observer-dependent. (Note that $\sqrt{1 - \vec{v}^{\,2}/c^2}\, dt = \sqrt{1 - \vec{v}\,'^2/c^2}\, dt' = d\tau$, $d\tau$ being particle's proper-time interval.)

2.3 Newton's second and third laws

- Newton's second law states that the acceleration of a particle in an inertial frame is proportional to the vector sum of all physical forces acting on the particle. Explicitly, in an isolated system of N bodies, the equation of motion of the ith body is given by

$$m_i \vec{a}_i \equiv m_i \frac{d^2 \vec{r}_i}{dt^2} = \sum_{j(\neq i)=1}^{N} \vec{F}_{ij} \,,$$

  where m_i is the proportionality constant called the mass of the ith body, and $\vec{F}_{ij}$ the force on the ith body due to the jth.
- Any physical two-body force $\vec{F}_{ij}$ is subject to Newton's third law: "action" and "reaction" are equal and opposite, i.e.,

$$\vec{F}_{ij} = -\vec{F}ji \quad (i \neq j).$$

- If the position of an extended body is taken as that of its center of mass, Newton's second and third laws apply to a system of composite objects as well.
- In Newtonian mechanics, the total momentum $\vec{P} = \sum_i \vec{p}_i$ (with $\vec{p}_i = m_i \vec{v}_i$) and total mass $M = \sum_i m_i$ in an isolated system are conserved.

2.4 Various types of forces in classical mechanics

- There are two fundamental classical forces, satisfying the symmetry principles of the Galileo–Newton spacetime, in the form of Newton's law of universal gravitation and the electrostatic/Coulomb force between electrically charged objects:

$$\vec{F}_{ij}^{(\text{gravi})} = -G m_i m_j \frac{\vec{r}_i - \vec{r}_j}{|\vec{r}_i - \vec{r}_j|^3}, \quad \vec{F}_{ij}^{(\text{elec})} = \frac{q_i q_j}{4\pi\epsilon_0} \frac{\vec{r}_i - \vec{r}_j}{|\vec{r}_i - \vec{r}_j|^3}.$$

  Both are central, inverse-square forces and obey Newton's third law.

- If one is mainly interested in the motion of a small body (with mass m) in the presence of other, much larger, bodies, one may represent the effect of the larger bodies on the small body in terms of "external forces" and consider the one-body equation of motion

$$m\frac{d^2\vec{r}(t)}{dt^2} = \vec{F}(\vec{r}(t), \vec{v}(t), t).$$

In this context, the symmetry principles of the Galileo–Newton spacetime need not manifest.

- In the presence of electric and magnetic fields, a moving charge is known to obey the Lorentz force law, i.e.,

$$\frac{d\vec{p}(t)}{dt} = q\big[\vec{E}(\vec{r}(t), t) + \vec{v}(t) \times \vec{B}(\vec{r}(t), t)\big],$$

where $\vec{p}(t)$ denotes the particle's momentum, i.e., $\vec{p} = m\vec{v}$. [In the relativistic case, one must instead use the relativistic momentum $\vec{p} = \frac{m_0\vec{v}}{\sqrt{1-\vec{v}^2/c^2}}$ where m_0 is the rest mass of the particle.]

- Newton's second and third laws may include various phenomenological forces such as velocity-dependent resistive forces, tension, normal forces, and friction.

Problems

2-1 Let $R = (t, x, y, z)$ define an inertial system, and $R' = (t', x', y', z')$ a non-inertial system which is related to R by

$$t' = t,\ x' = x\cos\omega t - y\sin\omega t,\ y' = x\sin\omega t + y\cos\omega t,\ z' = z,$$

where ω is a t-independent constant. When the motion of a certain small object is described with respect to R' by the equation

$$\frac{dx'(t)}{dt} = c_1,\ \frac{dy'(t)}{dt} = c_2,\ \frac{dz'(t)}{dt} = c_3\ \ (c_1,\ c_2,\ c_3:\ \text{constants}),$$

find its acceleration and trajectory as seen in the inertial system R.

2-2 Two ships are traveling parallel to each other in opposite directions with speeds v_1 and v_2. One ship fires on the other. At what angle φ should the gun be aimed at the target ship in order to make a hit, if the shot is fired at the instant when both vessels are on the straight line perpendicular to their courses? The shell velocity v_0 is assumed constant.

2-3 A boat crosses a river with a constant velocity v relative to the water, perpendicular to the current. The width of the river is d and the speed of the current is zero at the banks and increases linearly toward the center of the river, at which point its value is u. Find the trajectory of the boat, and the distance x_0 that it goes down along the current, from the point where it leaves one bank to the point where it reaches the other.

2-4 To derive the velocity transformation in special relativity, consider a particle with a trajectory $x^i = x^i(t)$ as seen by an inertial observer in $R = (t, x, y, z)$. Then, with respect to another inertial observer using the primed coordinates $R' = (t', x', y', z')$ which are specified by

$$t' = \frac{t - ux/c^2}{\sqrt{1 - \left(\frac{u}{c}\right)^2}}, \; x' = \frac{x - ut}{\sqrt{1 - \left(\frac{u}{c}\right)^2}}, \; y' = y, \; z' = z,$$

show that the velocity components of the particle with respect to R', i.e., $v'^i(t') = \frac{dx'^i(t')}{dt'}$, are related to the corresponding quantities in R, i.e., $v^i(t) = \frac{dx^i(t)}{dt}$, by

$$v'^1(t') = \frac{v^1(t) - u}{1 - uv^1(t)/c^2}, \qquad v'^2(t') = \frac{\sqrt{1 - \left(\frac{u}{c}\right)^2}\, v^2(t)}{1 - uv^1(t)/c^2},$$

$$v'^3(t') = \frac{\sqrt{1 - \left(\frac{u}{c}\right)^2}\, v^3(t)}{1 - uv^1(t)/c^2}.$$

Using this explicit formula, verify also that if the particle had the speed c with respect to R (i.e., $\sqrt{v^i v^i} = c$), then $\sqrt{v'^i v'^i}$ is again equal to c.

2-5 In the Minkowski space, it is convenient to represent the trajectory of a particle (also called the *world line*) in terms of the particle's *proper time* τ which is specified by the differential relation (here, $\vec{v}(t) = \frac{d\vec{x}(t)}{dt}$, i.e., the usual velocity of the particle at time t)

$$d\tau = \sqrt{1 - \frac{1}{c^2}\vec{v}^2(t)}\; dt.$$

(i) Show that $d\tau$ behaves as a Lorentz scalar, i.e., remains unchanged under Lorentz (boost) transformation.

(ii) Then, for a general trajectory of the particle, use the result of (i) to show that it is possible to introduce the so-called *four-velocity*

$$U^\mu = \left(U^0 \equiv \frac{c}{\sqrt{1-\frac{\vec{v}^2}{c^2}}},\ U^1 \equiv \frac{v_x}{\sqrt{1-\frac{\vec{v}^2}{c^2}}},\ U^2 \equiv \frac{v_y}{\sqrt{1-\frac{\vec{v}^2}{c^2}}},\ U^3 \equiv \frac{v_z}{\sqrt{1-\frac{\vec{v}^2}{c^2}}} \right)$$

which behaves, under Lorentz (boost) transformation, in the same way as the *four-coordinate* of the spacetime, i.e., $x^\mu = (x^0 \equiv ct, x^1 \equiv x, x^2 \equiv y, x^3 \equiv z)$.

2-6 The two components of a double star are observed to move — due to mutual gravitational attraction — in circles of radii r_1 and r_2 (on a single plane). What is the ratio of their masses?

2-7 Consider three (adjoined) freight cars of masses M_1, M_2 and M_3, which are pulled with force F by a locomotive. See the figure. Assuming that friction is negligible, find the forces acting on *each* freight car.

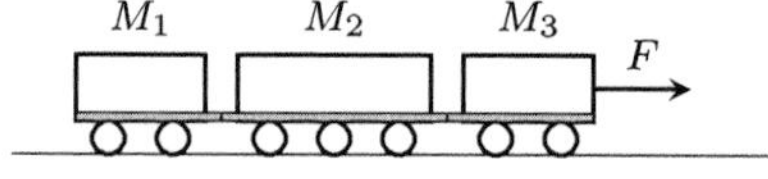

2-8 Consider a hanging rope, a *uniform* rope of mass m and length l, which hangs from the limb of a tall tree. When a man (with mass M) is at its end, find the tension at a distance x from the bottom of the rope.

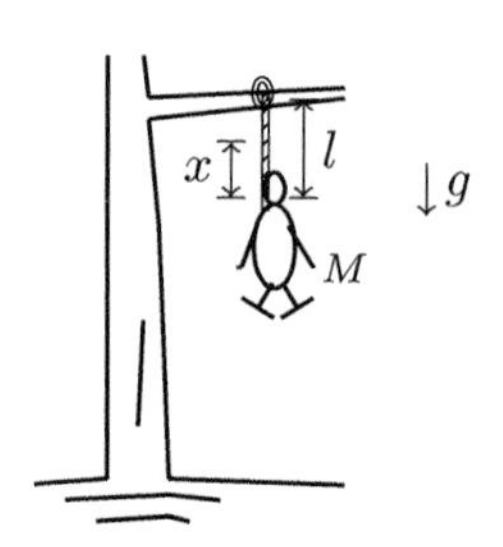

2-9 Consider two heavy objects of masses m_1 and m_2, which are connected to a very light (but inextensible) cable of length l and hung over a fixed smooth (i.e., frictionless) circular cylinder of radius a. The axis of the cylinder is horizontal.

(i) Draw a diagram showing all forces acting on the two masses.

(ii) Prove that the tensions the cable exerts on the two masses are of equal magnitude. What is the magnitude of the normal force exerted on the unit length of the cable by the cylinder?

(iii) Determine the position of equilibrium (i.e., the angles θ_1 and θ_2 in the figure when the two masses are at rest).

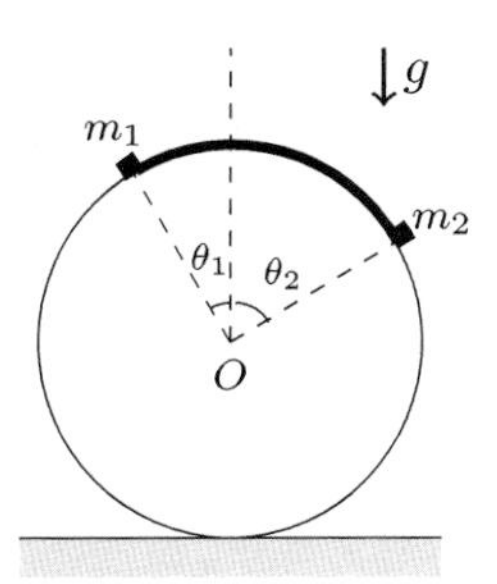

2-10 A uniform rope of mass m and length L, which is pivoted at one end and has a small stone of mass M attached at the other end, is whirling with uniform angular velocity ω. Neglecting the effect of gravity, find the tension in the rope at distance r from the pivot.

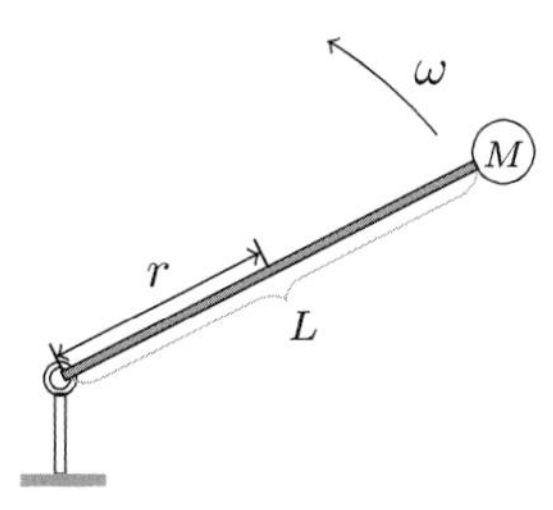

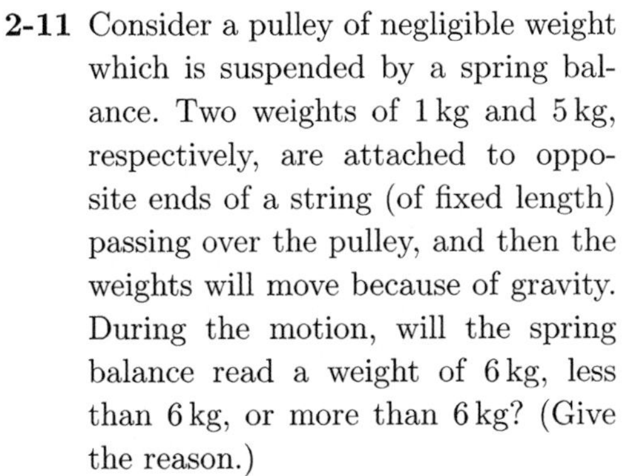

2-11 Consider a pulley of negligible weight which is suspended by a spring balance. Two weights of 1 kg and 5 kg, respectively, are attached to opposite ends of a string (of fixed length) passing over the pulley, and then the weights will move because of gravity. During the motion, will the spring balance read a weight of 6 kg, less than 6 kg, or more than 6 kg? (Give the reason.)

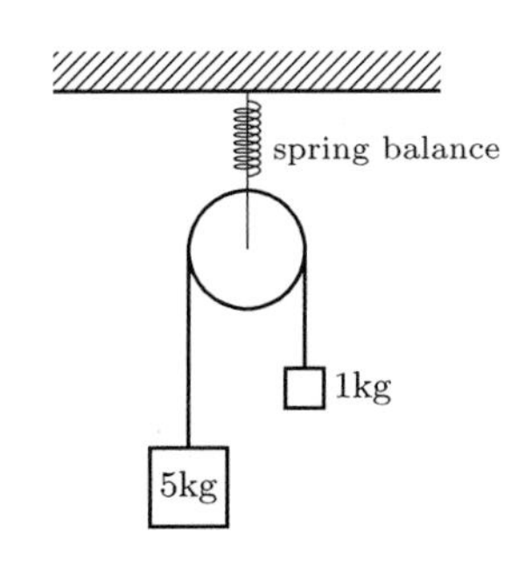

2-12 Let us consider a uniform cable, of mass λ per unit length and total length $2l$, hanging freely under the influence of its own gravity between two points A and B. The points A and B are at the same height and at a distance $2a$ (here $a < l$) apart. See the figure.

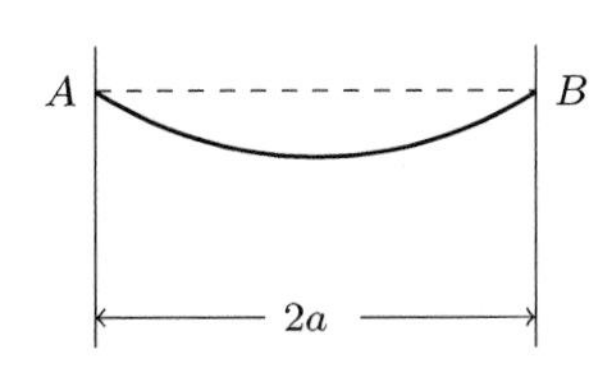

(i) Determine the curve of the cable.

(ii) At which point(s) in the cable does the maximum tension occur? Also, what is the magnitude of this maximum tension?

3 One-Dimensional Motion

SUMMARY

3.1 Linear motion of a particle

- To analyze the one-dimensional equation of motion

$$m\frac{d^2x(t)}{dt^2} \equiv \frac{dp(t)}{dt} = F(x(t), v(t), t),$$

it is useful to define the *impulse* and *work* due to the force F during the time interval (t_1, t_2) by

$$\Delta p = \int_{t_1}^{t_2} F(x(t), v(t), t)\, dt \quad (= mv(t_2) - mv(t_1)),$$

$$\begin{aligned} \Delta W &= \int_{t_1}^{t_2} v(t)\, F(x(t), v(t), t)\, dt \\ &\left(= \frac{1}{2}mv(t_2)^2 - \frac{1}{2}mv(t_1)^2\right) \\ &= \int_{x_1}^{x_2} F(x, v(x), t(x))\, dx, \end{aligned}$$

where $x_1 = x(t_1)$ and $x_2 = x(t_2)$.

- Given a force $F = F(t)$ (i.e., a function of t only), the particle's velocity and position as functions of time are given by

$$v(t) = v_0 + \frac{1}{m}\int_{t_0}^{t} F(t')\, dt',$$

$$x(t) = x_0 + v_0(t - t_0) + \frac{1}{m}\int_{t_0}^{t}\left\{\int_{t_0}^{t''} F(t')dt'\right\} dt''.$$

Given a conservative force $F = F(x)$ (i.e., a function of x only), the sum of the kinetic energy $T = \frac{1}{2}mv^2$ and the potential energy $V(x) = -\int^x F(x')\,dx'$ is constant in time, viz., we have the energy conservation law

$$\frac{1}{2}mv(t)^2 + V(x(t)) = E = \text{const.}$$

- In a conservative system with the potential energy function $V(x)$, a particle's motion may be determined by integrating

$$\dot{x}^2 = \frac{2}{m}\big[E - V(x)\big].$$

This gives rise to

$$t - t_0 = \pm\sqrt{\frac{m}{2}}\int_{x_0}^{x}\frac{dx'}{\sqrt{E - V(x')}} \qquad (x(t_0) = x_0).$$

- Given a force which is a function of velocity, i.e., $F = F(v)$, $v = v(t)$ can be found explicitly by integrating $m\frac{dv}{dt} = F(v)$, viz.,

$$\frac{1}{m}(t - t_0) = \int_{v_0}^{v}\frac{dv'}{F(v')}.$$

In the case of a body falling near the surface of the Earth, using this method with $F = -mg - \kappa v$ (here $\kappa > 0$ corresponds to the air resistivity) yields

$$v(t) = -\frac{mg}{\kappa}\left(1 - e^{-\frac{\kappa}{m}t}\right) \quad (\text{with } v(t=0) = 0)$$

and

$$x(t) = x_0 + \int_0^t v(t')dt' = x_0 + \frac{m^2 g}{\kappa^2}\left(1 - \frac{\kappa}{m}t - e^{-\frac{\kappa}{m}t}\right).$$

The limiting value of velocity, $v(t = \infty) = -\frac{mg}{\kappa}$, is called the terminal velocity. Analogous consideration can be employed to explain how a steady electric current sets in inside a conductor.

3.2 One-dimensional harmonic oscillator

- In a conservative system with a generic potential $V(x)$, the small-amplitude dynamics around its equilibrium point (taken to be at $x = 0$) can approximately be described in terms of a quadratic potential $\frac{1}{2}kx^2$, or equivalently a linear force $F = -kx$ where $k = V''(0)$. This leads to the linear equation of motion

$$m\ddot{x} + kx = 0$$

with the general solution

$$x(t) = \begin{cases} c_1 e^{\kappa t} + c_2 e^{-\kappa t} & \text{with } \kappa = \sqrt{-\frac{k}{m}}, \quad \text{if } k < 0 \\ d_1 \cos\omega t + d_2 \sin\omega t & \text{with } \omega = \sqrt{\frac{k}{m}}, \quad \text{if } k > 0. \end{cases}$$

The (simple) harmonic oscillator refers to the case with $k > 0$, i.e., when $x = 0$ corresponds to a potential minimum or stable equilibrium point.

- Alternative descriptions are possible for the above linear system. First, the above $k > 0$ solution can also be presented using (complex) exponential functions as

$$x(t) = Ce^{i\omega t} + C^* e^{-i\omega t} \quad \left(C = \frac{1}{2}(d_1 - id_2) \right)$$

or, writing $C = \frac{1}{2}Ae^{-i\theta_0}$ (A, θ_0: real), as

$$x(t) = \text{Re } z(t), \quad \text{with } z(t) = Ae^{i(\omega t - \theta_0)}.$$

Furthermore, if one introduces a phase-space vector $\mathbf{X} = \binom{x}{v}$, the above equation of motion takes the form

$$\dot{\mathbf{X}} = \begin{pmatrix} 0 & 1 \\ -\frac{k}{m} & 0 \end{pmatrix} \mathbf{X}$$

whose general solution is

$$\mathbf{X}(t) = \begin{cases} c_1 e^{\kappa t} \begin{pmatrix} 1 \\ \kappa \end{pmatrix} + c_2 e^{-\kappa t} \begin{pmatrix} 1 \\ -\kappa \end{pmatrix}, & \text{if } k < 0 \\ Ce^{i\omega t} \begin{pmatrix} 1 \\ i\omega \end{pmatrix} + C^* e^{-i\omega t} \begin{pmatrix} 1 \\ -i\omega \end{pmatrix}, & \text{if } k > 0\,. \end{cases}$$

These forms may be used to draw phase portraits for respective cases.

3.3 Damped harmonic oscillator

- For an oscillator which is subject to an additional damping force proportional to velocity, we have the equation of motion

$$m\ddot{x} + b\dot{x} + kx = 0$$

or

$$\ddot{x} + 2\gamma\dot{x} + \omega_0^2 x = 0 \qquad \left(\gamma \equiv \frac{b}{2m} > 0, \text{ and } \omega_0 \equiv \sqrt{\frac{k}{m}}\right).$$

Here, depending on the relative magnitudes between ω_0 and γ, the solution exhibits somewhat different characters. That is

(i) $\omega_0 > \gamma$ (underdamping): Oscillation with an exponentially decreasing amplitude results, with the explicit time dependence described by the form

$$x(t) = Ae^{-\gamma t}\cos(\omega t - \theta_0) \qquad \left(\omega = \sqrt{\omega_0^2 - \gamma^2}\right).$$

(ii) $\omega_0 < \gamma$ (overdamping): The displacement tends to zero exponentially (i.e., no oscillation), with the explicit time dependence described by the form

$$x(t) = c_1\, e^{-\gamma_+ t} + c_2\, e^{-\gamma_- t} \qquad \left(\gamma_\pm = \gamma \pm \sqrt{\gamma^2 - \omega_0^2}\right).$$

Note that the value $x = 0$ is possible for certain finite time t (> 0) only if $-\frac{c_1}{c_2} > 1$ or $\frac{c_1+c_2}{c_2} < 0$. If we assume that $x(t = 0) = x_0 = c_1 + c_2 > 0$, this condition becomes simply $c_2 < 0$. Since the initial velocity v_0 is given by

$$v_0 = -c_1\gamma_+ - c_2\gamma_- = -x_0\gamma_+ + c_2(\gamma_+ - \gamma_-),$$

this condition implies that $v_0 < -x_0\gamma_+$; if v_0 satisfies this condition, the oscillator *overshoots* (i.e., passes the equilibrium point).

(iii) $\omega_0 = \gamma$ (critical damping): With this choice the oscillator settles down at the equilibrium position in the shortest time, the time dependence being described by the form

$$x(t) = (C + Dt)e^{-\gamma t} \quad (C,\, D\text{: real constants}).$$

- In phase-space description, the flow of the phase point $\mathbf{X} = \binom{x}{v}$ is governed by the first-order differential equation

$$\dot{\mathbf{X}} = A\mathbf{X}, \quad A = \begin{pmatrix} 0 & 1 \\ -\omega_0^2 & -2\gamma \end{pmatrix}.$$

To solve this equation, one may then try the form

$$\mathbf{X}(t) = c_1 e^{\lambda_1 t}\mathbf{X}_{(1)} + c_2 e^{\lambda_2 t}\mathbf{X}_{(2)},$$

where $\mathbf{X}_{(1)}$, $\mathbf{X}_{(2)}$ denote eigenvectors of the above 2×2 matrix A, corresponding to its eigenvalues λ_1 and λ_2, respectively. This study may be utilized to obtain phase portraits for the system, which exhibits qualitatively different patterns depending on whether the oscillator is underdamped, overdamped, or critically damped.

3.4 Driven damped oscillator and resonance

- For a forced or driven damped oscillator described by the equation of motion

$$m\ddot{x} + b\dot{x} + kx = F(t)$$

($F(t)$ is the driving force), one can give the general solution by the form $x(t) = x_h(t) + x_s(t)$; here, $x_h(t)$ denotes the solution of the unforced problem (i.e., with $F(t) = 0$) and $x_s(t)$ is any particular solution in the presence of the driving force $F(t)$.

- With a trigonometric driving force $F(t) = F_1 \cos(\omega_1 t + \theta_1)$, the problem is aptly dealt with by considering the complexified equation of motion

$$m\ddot{z} + b\dot{z} + kz = Fe^{i\omega_1 t} \quad (F \equiv F_1 e^{i\theta_1})$$

under the identification $x(t) = \text{Re } z(t)$. By this method, one readily obtains a particular solution in the form

$$x_s(t) = \frac{(F_1/m)}{\sqrt{(\omega_0^2 - \omega_1^2)^2 + 4\gamma^2\omega_1^2}} \cos(\omega_1 t + \theta_1 - \alpha)$$

$$\left(\alpha = \tan^{-1} \frac{2\gamma\omega_1}{\omega_0^2 - \omega_1^2}\right).$$

This gives *steady-state term* of the solution, as for $t \gg \frac{1}{\gamma}$ the full displacement $x(t)$ approaches this steady-state expression regardless of the initial conditions.
- If the damping is small and the driving frequency ω_1 happens to be close to the oscillator frequency ω_0, the amplitude of the steady-state term becomes very large (of the order of $\frac{F_1}{b\omega_1}$) — the system is *in resonance.* The quality factor $Q = \frac{\omega_0}{2\gamma}$ provides a measure of the sharpness of the resonance peak.
- With a general periodic driving force (of period $T = \frac{2\pi}{\omega}$), one may resolve the force in Fourier series, viz.,

$$F(t) = \sum_{n=-\infty}^{\infty} F_n \, e^{in\omega t} \quad \left(F_n = \frac{1}{T}\int_0^T F(t) e^{-in\omega t} dt\right)$$

and apply the above method with each trigonometric force.
- With an arbitrary nonperiodic driving force, the Green's function method can be used. For instance, in the case of an underdamped oscillator subject to an arbitrary driving force $F(t)$, we have

$$x_s(t) = \int^t G(t - t') \, F(t') dt'$$

with the Green's function $G(t-t')$ given by (here, $\omega = \sqrt{\omega_0^2 - \gamma^2}$)

$$G(t-t') = \begin{cases} 0\,, & t < t' \\ \frac{1}{m\omega} e^{-\gamma(t-t')} \sin\omega(t-t'), & t > t' \end{cases} .$$

3.5 Nonlinear oscillations and chaos

- Nonlinear dynamical systems can exhibit very complex behaviors. Even for a conservative nonlinear system like that of a simple pendulum or an anharmonic oscillator, closed-form solutions to the corresponding equation of motion, viz.,

$$\ddot{\theta} + \frac{g}{l}\sin\theta = 0 \quad (\text{: simple pendulum}),$$
$$\ddot{x} + \omega_0^2 x + \beta x^3 = 0 \quad (\text{: anharmonic oscillator}),$$

are difficult to obtain (or involve less-familiar functions like elliptic functions). But, for small-amplitude oscillations about the equilibrium point (i.e., around $\theta = 0$ or $x = 0$), one may apply a systematic perturbation theory to obtain, in the case of the simple pendulum, a series solution

$$\theta(t) = \left(\theta_m + \frac{1}{192}\theta_m^3\right)\sin[\omega(t-t_0)] + \frac{1}{192}\theta_m^3 \sin[3\omega(t-t_0)] + O(\theta_m^5)$$

with

$$\omega = \sqrt{\frac{g}{l}}\left[1 - \frac{1}{16}\theta_m^2 + O(\theta_m^4)\right],$$

where θ_m is the maximum angle of oscillation. Notice that, in contrast to the case of a harmonic oscillator, the angular frequency ω and the oscillation period $T = \frac{2\pi}{\omega}$ now become dependent on the amplitude of oscillation (or on the total energy).

- If a nonlinear damped oscillator is subject to an oscillating driving force of strength F, its long-time behavior is crucially dependent on the strength of F. If F is not too large, it shows limit-cycle behaviors, i.e., approaches periodic solutions as $t \to \infty$; but, the period here may not always coincide with the frequency of the driving force as doubled period and some other periods may also occur. If the strength of F is increased further beyond some threshold value, the limit-cycle behavior is taken over by a chaotic motion.

- The principal property that characterizes a chaotic dynamical system is the "sensitive dependence on initial conditions" (the butterfly effect), which renders long-time prediction on such system impossible. For phase trajectories given in higher-dimensional phase spaces, these chaotic behaviors are indeed more common.

Problems

3-1 In Sec. 3.1 of the text, we have discussed the Euler's method of integrating the equation of motion (written in the first order form), which is based on dividing the given time interval into many small time intervals and using uniform approximation for the motion in each small time interval. In the limit of infinitesimal time step, the exact solution of the problem should follow. When a particle of mass m is subject to the force $F = -kx$ (k is a positive constant), find $x(t)$ satisfying the initial conditions $x(t = 0) = 0$ and $\dot{x}(t = 0) = v_0$ by this method.

3-2 Let us consider the case in which the particle moves in a fluctuating vertical current of air. Let $f'(t)$ be the *upward* velocity of the air at time t, and v the downward velocity of the particle relative to the ground so that $u = v + f'(t)$ becomes the (downward) relative velocity in air. Assuming the resistive force is proportional to the relative velocity, the equation of motion can be written in the form

$$m\frac{dv}{dt} = mg - \kappa(v + f'(t)) \quad (\kappa > 0).$$

(i) When the initial condition is such that $v = 0$ at $t = 0$, how is the velocity $v(t)$ at time $t > 0$ given?

(ii) Find the corresponding expression for $x(t)$, the traversed distance in time t.

(iii) When $f'(t)$ is equal to $C \cos pt$ (C, p are some constants), find $x(t)$ explicitly and discuss the nature of motion as $t \to \infty$.

3-3 A particle of mass m is projected vertically upwards in the atmosphere with a speed such that, if the air resistance were ignored, the particle would attain a maximum height h above the initial location. Assume the resistive force from the air to have the form $F = -kv^2$ (if $v > 0$), where k is a positive constant and v is the particle's velocity.

(i) When x denotes the vertical position of the particle, rewrite the equation of motion to take the form

$$\frac{dv}{dx} = \mathcal{F}(v).$$

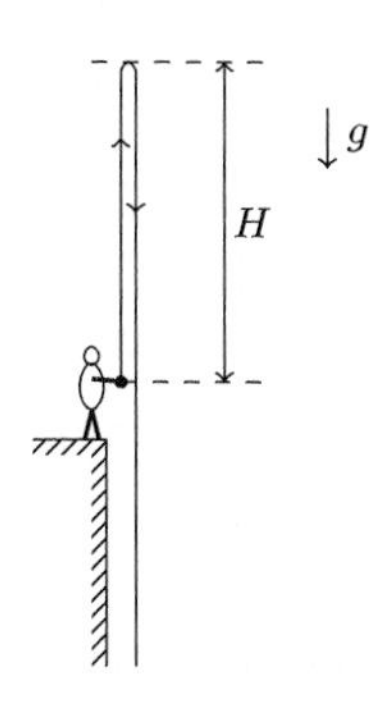

(ii) Solve the differential equation to find the actual maximum height H attained by the particle (see the figure). Also what is the terminal velocity here?

3-4 The potential energy function of a particle of mass m is $V(x) = -\frac{1}{2}c(x^2 - a^2)^2$, where a and c are positive constants. Sketch this function, and describe the possible types of motion for the following three cases:

(i) $E > 0$,
(ii) $E < -\frac{1}{2}ca^4$, and
(iii) $-\frac{1}{2}ca^4 < E < 0$.

3-5 The potential energy for the force between two atoms in a diatomic molecule has the approximate form

$$V(x) = -\frac{a}{x^6} + \frac{b}{x^{12}},$$

where x is the distance between the atoms and a, b are positive constants.

(i) Find the force as a function of x, and plot it.
(ii) Assuming that one of the atoms is very heavy and remains practically at rest while the other moves along a straight line, describe the possible motions.

(iii) Find the equilibrium distance and the period of small oscillations about the equilibrium position if the mass of the lighter atom is m.

3-6 Let the potential energy function of a particle of mass m be given by

$$V(x) = \frac{cx}{x^2 + a^2} \quad (c, a: \text{ positive constants}).$$

(i) Sketch the potential curve as a function of x.
(ii) Find the position of stable equilibrium, and the period of small oscillations about it.
(iii) Given that the particle starts from the point of stable equilibrium with velocity v_0, find the values of v_0 for which (a) it oscillates in a finite region, and (b) it escapes to $+\infty$.

3-7 For the equation of motion of a harmonic oscillator

$$m\ddot{x} + kx = 0,$$

obtain the general solution $x(t) = \sqrt{\frac{2E}{k}} \cos(\omega t - \theta_0)$ by making direct use of the formula (see (3.18) in the text), $t - t_0 = \pm\sqrt{\frac{m}{2}} \int_{x_0}^{x} \frac{dx'}{\sqrt{E - V(x')}}$.

3-8 In one dimension, consider a relativistic particle (of rest mass m_0) which is subject to a conservative force $F = F(x)$. The equation of motion is given by the form

$$\frac{d}{dt}\left(\frac{m_0 v}{\sqrt{1 - v^2/c^2}}\right) = F(x).$$

(i) Show that the energy conservation law for this system is described by the form

$$\frac{m_0 c^2}{\sqrt{1 - v^2/c^2}} + V(x) = E_{\text{rel}} \ (: \text{ const.}),$$

where $V'(x) = -F(x)$.
(ii) Based on this energy conservation law, discuss, for various potential shapes, possible allowed motions (depending on the value of E_{rel}) in a qualitative manner. [Consult Fig. 3-2 in the text, for the corresponding nonrelativistic discussion.]

3-9 A weight of mass m is vertically hung from the end of a spring which provides a restoring force equal to k times its extension. The weight is released from rest with the spring unextended. Find the position as a function of time (assuming negligible damping).

3-10 Consider the 1-dimensional system shown in the figure. Here the particle of mass m can move only vertically, in the presence of a uniform field of gravity. Set up the equation of motion for the particle and then solve it.

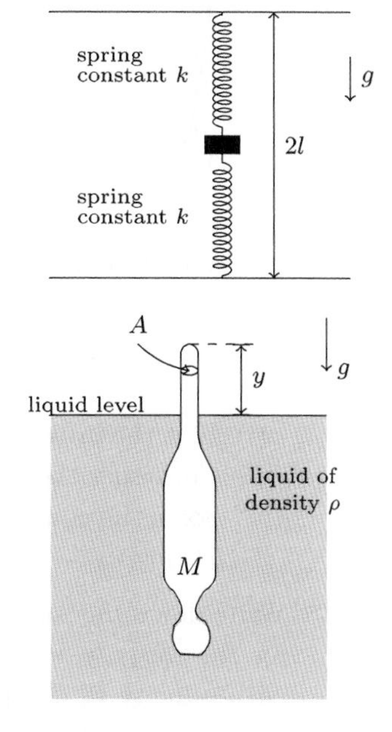

3-11 A hydrometer floats in a liquid of density ρ as shown in the figure. It is slightly displaced from its normal position of equilibrium and periodic motion about the equilibrium results. Show that the period of oscillations is given by

$$T = 2\pi\sqrt{\frac{M}{gA\rho}},$$

where M is the mass of the hydrometer and A is the cross sectional area.

3-12 A particle of mass m moves in the region $x > 0$ under the force $F = -m\omega_0^2(x - \frac{a^4}{x^3})$, where ω_0 and a are positive constants.

(i) Sketch the potential energy function.

(ii) Find the position of equilibrium, and the period of small oscillations about it.

(iii) The assumed force is somewhat special in that, as in the case of a harmonic oscillator, the period of oscillation is independent of the energy value (which can be arbitrarily large). Can you demonstrate this?

3-13 Consider two one-dimensional potentials $V_a(x)$ and $V_b(x)$, having the shapes given in the figure.

(i) Suggest plausible functional forms for $V_a(x)$ and $V_b(x)$, which describe the given shapes closely.

(ii) Considering the case that a particle of mass m moves in the given potential, provide for the respective potentials the corresponding phase portraits in the (x, v)-plane.

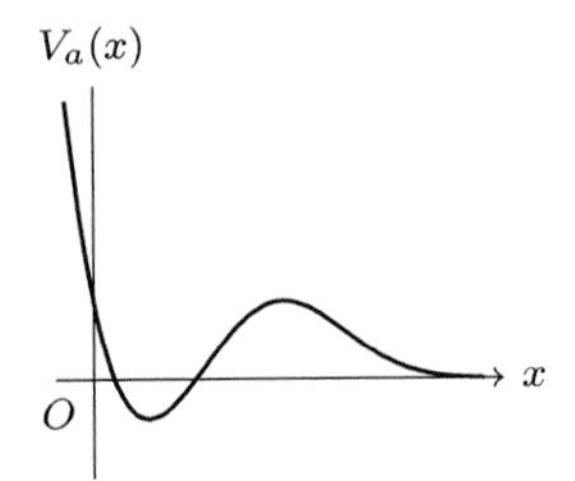

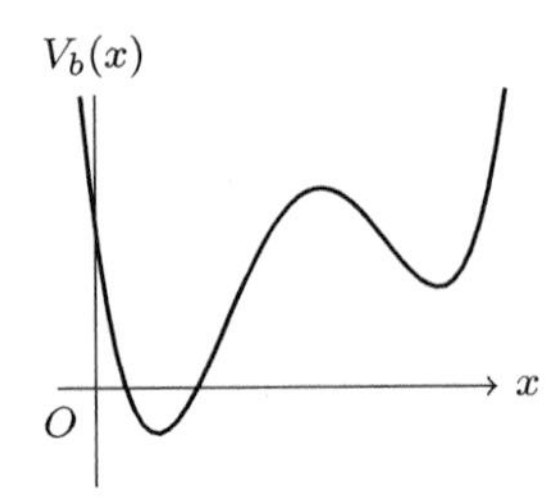

3-14 The potential energy function of a particle of mass m is given by

$$V(x) = \begin{cases} -\frac{m\omega^2}{2}(a^2 - x^2), & \text{for } |x| < a \\ 0, \quad \text{for } |x| > a. \end{cases}$$

(i) Plot the potential energy function.
(ii) Construct the phase portrait.
(iii) Solve the equation of motion using initial conditions $x(0) = -a$, $\dot{x}(0) = v_0$.

3-15 Write down the solution to the damped oscillator equation of motion for the case $\omega_0 > \gamma$, assuming that the oscillator starts from $x = 0$ with velocity v_0. Then show that as ω_0 is reduced to the critical value (i.e., $\omega_0 = \gamma$), the solution tends to the corresponding solution for the critically damped oscillator.

3-16 Repeat the calculation of Problem 3-9, now assuming that the system is critically damped. Given that the final position of equilibrium is 0.4 m below the point of release, find how close to the equilibrium position of the particle is after 1 sec.

3-17 A mass of 1000 kg drops from a height of 10 m on a platform of negligible mass. It is desired to design a spring and dashpot on which to mount the platform so that the platform will settle to a new equilibrium position 0.2 m below its original position as quickly as possible after the impact without overshooting.

(i) Find the spring constant k and the damping constant b of the dashpot. (Be sure to examine that your solution $x(t)$ satisfies the initial conditions and does not overshoot.)
(ii) When the values of k and b are as given in (i), find, to two significant figures, the time required for the platform to settle within 1 mm of its final position. Also, if the damping constant

happens to be larger than the value of (i) by 20 percent (but the same k value), what will be the corresponding time?

3-18 Given a body of mass m connected to a spring, the equation of motion with a damping term included can be taken as

$$m\ddot{x} + b\dot{x} + kx = 0 \quad (b, k > 0)$$

or

$$\ddot{x} + 2\gamma\dot{x} + \omega_0^2 x = 0 \quad \left(\gamma \equiv \frac{b}{2m}, \omega_0 \equiv \sqrt{\frac{k}{m}}\right).$$

Here, x denotes the displacement from the relaxed position of the spring (i.e., from the point O in the figure).

(i) The given body is displaced from the point O to a certain point $x = a$ (> 0) and is left there with zero initial velocity (i.e., $\dot{x}(t = 0) = 0$). Assuming that $\gamma < \omega_0$ (i.e., underdamping), find $x(t)$ describing the subsequent motion.

(ii) Now, for this oscillator, we wish to include in addition the effect of the sliding friction force, $\pm\mu mg$ (μ is the coefficient of the kinetic friction), existing between the body and the horizontal surface. Assuming the same initial conditions as in (i), find $x(t)$ for the first oscillation period, i.e., for $0 \le t \le T$ with $T = \frac{2\pi}{\omega}$ where $\omega = \sqrt{\omega_0^2 - \gamma^2}$. Under what situation does the body stop moving beyond a certain point? (It may be assumed that the coefficient of the static friction is also equal to μ here.)

(iii) In the case of (ii), explain in words the general character of motion for $t > T$ (when the body did not come to rest prior to the time $t = T$).

3-19 A column vector $\mathbf{X} = \binom{x}{v}$, representing phase points of the 1-D damped oscillator system, satisfies the vector equation of the form

$$\dot{\mathbf{X}} = A\mathbf{X}, \quad A = \begin{pmatrix} 0 & 1 \\ -\omega_0^2 & -2\gamma \end{pmatrix}.$$

With the help of the general solutions $\mathbf{X}(t)$ appropriate to the cases of overdamping (i.e., $\omega_0 < \gamma$) and critical damping (i.e., $\omega_0 = \gamma$), exhibit the shape of the corresponding phase portraits explicitly, taking some representative values for the ratio ω_0/γ (e.g., $\omega_0/\gamma = \frac{1}{3}, \frac{2}{3}, 1$).

3-20 If a damped harmonic oscillator is subject to a periodic driving force, the frequency of the driving force is picked up by the oscillator in its steady state. Suggest a few concrete physical examples in which this frequency-transferring mechanism plays a role.

3-21 A particle of mass m is attached to one end of a light spring of spring constant k. The other end of the spring is forced to undergo an oscillation, so that its position at time t is given by $A\sin\omega t$ (see the figure).

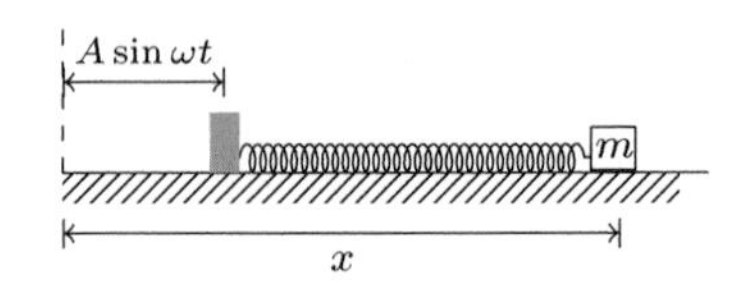

The mass is also subject to a resistive force equal in magnitude to b times its speed. Write the equation of motion of the particle and find its general solution. (Assume $\frac{b}{2m} < \sqrt{\frac{k}{m}}$.)

3-22 Consider a 1-dimensional harmonic oscillator which is subject to a periodic driving force $F(t) = F_1\cos(\omega_1 t + \theta_1)$. The equation of motion is then given by the form

$$m\ddot{x} + m\omega_0^2 x = F_1\cos(\omega_1 t + \theta_1).$$

(i) Find the general solution to this equation of motion, assuming that $\omega_1^2 \neq \omega_0^2$.
(ii) Also find the general solution when ω_1^2 is exactly equal to ω_0^2. (Why is this case so different from other cases? Explain.)

3-23 Solve the equation of motion of the damped harmonic oscillator (assume underdamping, i.e., $\omega_0 > \gamma$) driven by a damped harmonic force

$$F_{\text{ext}}(t) = F_0\, e^{-\alpha t}\cos\omega_1 t.$$

(Hint: Notice that $e^{-\alpha t}\cos\omega_1 t = \text{Re}\left[e^{(-\alpha+i\omega_1)t}\right]$.)

3-24 An oscillator with free oscillation period T is critically damped and subjected to a periodic force having the "saw-tooth" form

$$F(t) = c(t - nT), \quad \left(n - \frac{1}{2}\right)T < t < \left(n + \frac{1}{2}\right)T$$

for each integer n (c is a positive constant). Find the steady-state solution by the Fourier series method.

3-25 The equation describing an undamped harmonic oscillator with an arbitrary driving force, $\ddot{x} + \omega_0^2 x = \frac{1}{m}F(t)$, may be integrated directly by considering a complex variable

$$z = \dot{x} + i\omega_0 x.$$

(i) Show that the equation of motion can be rewritten as

$$\dot{z} - i\omega_0 z = \frac{1}{m} F(t).$$

(ii) Solve this (complex) linear differential equation to obtain the general solution for $x(t)$.

3-26 Suppose that a certain driving force $F(t)$, with $F(-\infty) = 0$ and $F(\infty) = F_0$ ($\neq 0$), acts upon an undamped harmonic oscillator. Determine the amplitude of oscillation as $t \to \infty$, assuming that it was at rest at $t = -\infty$. Also find the energy gained by the oscillator during the total time the driving force acts. [Hint: One may use the result of Problem 3-25.]

3-27 For an underdamped linear oscillator (with a general driving force $F(t)$)

$$m\ddot{x} + b\dot{x} + kx = F(t) \qquad \left(\gamma^2 < \omega_0^2, \text{ with } \gamma = \frac{b}{2m} \text{ and } \omega_0 = \sqrt{\frac{k}{m}}\right)$$

we have constructed in the text a special solution of the form

$$x_s(t) = \int_{t_0}^{t} \left[\frac{1}{m\omega} e^{-\gamma(t-t')} \sin\omega(t-t')\right] F(t')\,dt' \qquad \left(\omega = \sqrt{\omega_0^2 - \gamma^2}\right).$$

Verify directly (i.e., by evaluating $m\ddot{x}_s + b\dot{x}_s + kx_s$) that the given expression for $x_s(t)$ corresponds to a particular solution of the problem.

3-28 A driving force $F(t) = F_0(1 - e^{-at})$ (a, F_0 are some positive constants) acts on an oscillator, which is at rest at $t = 0$. Assume that the mass is m, the spring constant $k = 4ma^2$, and the damping constant $b = ma$. Find the motion by using the Green's function method.

3-29 Find the Green's function for an overdamped (i.e., $\gamma^2 > \omega_0^2$) oscillator. Use it to solve the problem of an overdamped oscillator that is initially at rest in its equilibrium position and is subjected from $t = 0$ to a driving force increasing linearly with time, i.e., $F(t) = ct$ for $t > 0$.

3-30 Consider a mildly damped linear oscillator, satisfying the equation of motion

$$m\ddot{x} = -b\dot{x} - m\omega_0^2 x + F(t),$$

where $\gamma \equiv \frac{b}{2m}$ is nonzero but much smaller than ω_0.

(i) Given the initial conditions $x(t=0) = 0$ and $\dot{x}(t=0) = v_0$, find the motion for $t > 0$ when $F(t)$, an impulsive force, is such that an impulse δp is given at certain instant $t = \tau\,(> 0)$.

(ii) Now suppose that an unending series of impulsive forces (of equal strength) are provided at constant time intervals, that is, impulse δp is given to the oscillator at every time $t = n\tau$ ($n = 1, 2, 3, \cdots$). Determine the motion of the oscillator for t large (i.e., $t \gg \frac{1}{\gamma}$ and $t \gg \tau$). Also discuss in an explicit manner under what conditions the phenomenon of resonance can occur here.

3-31 For the simple pendulum

$$\ddot{\theta} + \omega_0^2 \sin\theta = 0 \quad \left(\omega_0 = \sqrt{\frac{g}{l}}\right)$$

oscillating solutions can be described by the equation

$$\int_0^{\varphi} \frac{d\varphi'}{\sqrt{1 - a^2 \sin^2\varphi'}} = \omega_0(t - t_0),$$

where $a = \sin\frac{\theta_{\rm m}}{2}$ ($\theta_{\rm m}$ denotes the allowed maximum angle) and the variable φ is defined by the relation $\sin\varphi = \frac{1}{a}\sin\frac{\theta}{2}$. By studying the above solution for relatively small a, verify that it leads to the expansion

$$\theta(t) = \left(2a + \frac{3}{8}a^3\right)\sin\left[\omega(t - t_0)\right] + \frac{1}{24}a^3 \sin\left[3\omega(t - t_0)\right] + O(a^5),$$

where $\omega = \omega_0[1 - \frac{a^2}{4} + O(a^4)]$.

3-32 Given the nonlinear oscillator system (β is the perturbation parameter)

$$\ddot{x} + \omega_0^2 x + \beta x^2 = 0,$$

use the perturbation method to obtain the following approximate solution to $O(\beta^2)$:

$$x(t) = A_1\cos\omega t + \beta\left(\frac{A_1^2}{6\omega_0^2}\right)[-3 + \cos 2\omega t] + \beta^2\left(\frac{A_1^3}{48\omega_0^4}\right)\cos 3\omega t,$$

where $\omega = \omega_0 \left(1 - \beta^2 \frac{5A_1^2}{12\omega_0^4}\right)$.

3-33 Given a nonlinear oscillator system of the form (β is the perturbation parameter)

$$\ddot{x} = -\omega_0^2 x - \beta \frac{d}{dx} V_1(x),$$

the frequency shift may be found with the help of the perturbation method or by a direct evaluation of the integral for the oscillation period

$$T = 2 \int_{x_1(E)}^{x_2(E)} \frac{dx}{\sqrt{2(E - \frac{1}{2}\omega_0^2 x^2 - \beta V_1(x))}},$$

where x_1, x_2 denote appropriate turning points. For small β, use the second method to find the leading-order terms of the frequency shifts in the case of (i) $V_1(x) = \frac{1}{4}x^4$ and (ii) $V_1(x) = \frac{1}{3}x^3$. [Hint: The observation made in footnote 50 of this chapter may prove useful.] Do the results agree with those found by the perturbation method?

3-34 Let us consider the one-dimensional relativistic harmonic oscillator with the equation of motion (cf. Problem 3-8)

$$\frac{d}{dt}\left[\frac{m_0 \dot{x}}{\sqrt{1 - \dot{x}^2/c^2}}\right] = -kx \quad (k > 0).$$

When the given variable $x(t)$ oscillates with maximum amplitude x_0 (here assume that $|\dot{x}/c| \ll 1$ throughout the motion), find the leading *relativistic correction* to the angular frequency of oscillation ω. [Note that ω reduces to $\omega_0 = \sqrt{\frac{k}{m_0}}$ in the extreme nonrelativistic limit.]

3-35 Let us consider a (weakly) nonlinear oscillator in the presence of an oscillating driving force, governed by the equation

$$m\ddot{x} = -b\dot{x} - kx + \epsilon x^3 + f \cos \omega t \quad (b, k > 0).$$

For small enough ε (and f not too large), we would expect that the dynamics of this system be not much different from that of a linear, driven, damped oscillator. For instance, this will also have a steady-state solution which oscillates with the same period as the driving force; but, because of the small nonlinear term present in the equation of motion, corresponding *higher harmonic* terms may also enter the steady-state motion. Verify the last assertion by considering explicitly the approximate solution of the above equation, which has the following form

$$x(t) = A\cos(\omega t - \alpha) + B\cos[3(\omega t - \alpha)] + C\sin[3(\omega t - \alpha)].$$

4 Motion of a Particle in Two or Three Dimensions

SUMMARY

4.1 External force problems in three dimensions

- A 3-dimensional external force $\vec{F}$ is said to be conservative if it can be written in the form $\vec{F} = -\vec{\nabla}V(\vec{r})$ ($V(\vec{r})$ is an appropriate time-independent potential energy function), so that $\vec{\nabla} \times \vec{F} = 0$. For the related equation of motion $m\frac{d^2\vec{r}}{dt^2} = -\vec{\nabla}V(\vec{r})$, we then have the first integral in the form of the energy conservation law

$$\frac{1}{2}m\dot{\vec{r}}^2 + V(\vec{r}(t)) = \text{const.}$$

- Given a force $\vec{F}$ satisfying the condition $\vec{\nabla} \times \vec{F} = 0$, the corresponding potential is given by the line integral $V(\vec{r}) = -\int_{\vec{r}_0}^{\vec{r}} \vec{F} \cdot d\vec{r}$. Any time-independent central force, i.e., $\vec{F}(\vec{r}) = f(r)\frac{\vec{r}}{r}$ (with $r \equiv \sqrt{x^2 + y^2 + z^2}$) is conservative as it is derived from the radial potential $V(r) = -\int^r f(r')dr'$.
- Given a particle subject to an external force $\vec{F}$, the rate of change of its angular momentum $\vec{L} = \vec{r} \times \vec{p}$ is determined by the torque $\vec{\Gamma} = \vec{r} \times \vec{F}$ according to $\frac{d\vec{L}}{dt} = \vec{\Gamma}$.
- If there are only central forces, the particle's angular momentum is conserved, i.e., $\vec{L} = \vec{r}(t) \times \vec{p}(t) = \text{const.}$, which is the statement of the angular momentum conservation in its simplest form. This leads to Kepler's second law: the time rate of swept-out area A by the radius vector drawn from

the force center is constant, being equal to $\frac{dA}{dt} = \frac{1}{2}r^2\dot{\theta} = \frac{|\vec{L}|}{2m}$ (: const.).

4.2 Laws of motion in non-Cartesian bases

- Using cylindrical coordinates (ρ, φ, z) in terms of which one has $x = \rho\cos\varphi$ and $y = \rho\sin\varphi$, a particle's instantaneous velocity and acceleration are given by

$$\vec{v} = \dot{\rho}\hat{\rho} + \rho\dot{\varphi}\hat{\varphi} + \dot{z}\hat{z}, \quad \vec{a} = (\ddot{\rho} - \rho\dot{\varphi}^2)\hat{\rho} + (\rho\ddot{\varphi} + 2\dot{\rho}\dot{\varphi})\hat{\varphi} + \ddot{z}\hat{z},$$

and the kinetic energy becomes

$$T = \frac{1}{2}m(\dot{\rho}^2 + \rho^2\dot{\varphi}^2 + \dot{z}^2).$$

- Using spherical coordinates (r, θ, φ) related to Cartesian coordinates by $x = r\sin\theta\cos\varphi$, $y = r\sin\theta\sin\varphi$ and $z = r\cos\theta$, we can express various mechanical quantities by

$$\begin{aligned}
\vec{v} &= \dot{r}\,\hat{r} + r\dot{\theta}\,\hat{\theta} + (r\sin\theta)\dot{\varphi}\,\hat{\varphi},\\
\vec{a} &= (\ddot{r} - r\dot{\theta}^2 - r\dot{\varphi}^2\sin^2\theta)\hat{r} + (r\ddot{\theta} + 2\dot{r}\dot{\theta} - r\dot{\varphi}^2\sin\theta\cos\theta)\hat{\theta}\\
&\quad + \left\{\frac{d}{dt}(r\dot{\varphi}\sin\theta) + \dot{r}\dot{\varphi}\sin\theta + r\dot{\theta}\dot{\varphi}\cos\theta\right\}\hat{\varphi},\\
\vec{L} &\equiv \vec{r}\times\vec{p} = mr^2\dot{\theta}\,\hat{\varphi} - mr^2\sin\theta\dot{\varphi}\,\hat{\theta},\\
T &= \frac{1}{2}m(\dot{r}^2 + r^2\dot{\theta}^2 + r^2\sin^2\theta\dot{\varphi}^2).
\end{aligned}$$

If we use the angular-momentum square

$$\vec{L}^2 = m^2r^4(\dot{\theta}^2 + \dot{\varphi}^2\sin^2\theta),$$

the radial equation of motion can be put into the form

$$m\ddot{r} - \frac{\vec{L}^2}{mr^3} = F_r,$$

and the kinetic energy by the form $T = \frac{1}{2}m\dot{r}^2 + \frac{\vec{L}^2}{2mr^2}$.

- If a particle's motion is resolved in tangential and normal directions to its trajectory, we have

$$\vec{v} = \dot{s}\mathbf{t}, \qquad \vec{a} = \ddot{s}\mathbf{t} - \frac{\dot{s}^2}{\rho}\mathbf{n},$$

where s denotes the arc length along the path, $\mathbf{t}$ ($\equiv \frac{d\vec{r}}{ds}$) is the unit tangent and $\mathbf{n}$ ($= -\rho\frac{d\mathbf{t}}{ds}$, ρ being the radius of curvature) the unit normal to the path.

4.3 Isotropic and anisotropic oscillators

- If a particle moves under a central force $\vec{F} = f(r)\frac{\vec{r}}{r}$, the motion occurs in a plane due to the angular momentum conservation. On the plane of motion, particle's motion is governed using polar coordinates (r, θ) by

$$m\ddot{r} = f(r) + \frac{L^2}{mr^3}, \quad mr^2\dot{\theta} = L \quad (\text{: const.}).$$

The energy conservation equation becomes

$$\frac{1}{2}m\dot{r}^2 + V_{\text{eff}}(r) = E \quad (\text{: const.})$$

with the effective radial potential

$$V_{\text{eff}}(r) = -\int^r f(r')\, dr' + \frac{L^2}{2mr^2}.$$

- To obtain the orbit equation $r = r(\theta)$ in a central potential, it is convenient to use the variable $u = \frac{1}{r}$ since we have $\dot{r} = -\frac{L}{m}\frac{du}{d\theta}$. Then the orbit is determined by solving the equation

$$\frac{L^2}{2m}\left(\frac{du}{d\theta}\right)^2 + V_{\text{eff}}\left(\frac{1}{u}\right) = E.$$

- In the case of the isotropic oscillator with the force $\vec{F} = -k\vec{r}$ ($k > 0$), we have the effective radial potential $V_{\text{eff}}(r) = \frac{1}{2}kr^2 + \frac{L^2}{2mr^2}$. Then, from the orbit equation, we find general orbits described by the equation

$$\frac{1}{r^2} = \frac{mE}{L^2} + \sqrt{\frac{m^2E^2}{L^4} - \frac{mk}{L^2}}\cos\big[2(\theta - \theta_0)\big],$$

which corresponds to an ellipse with its center at the center of force. This is a periodic motion with period $T = 2\pi\sqrt{\frac{m}{k}}$. One can reach the same conclusion by studying the equations of motion in Cartesian coordinates.

- An anisotropic harmonic oscillator, described by the equation of motion

$$m\ddot{x} + k_x x = 0, \quad m\ddot{y} + k_y y = 0, \quad m\ddot{z} + k_z z = 0$$

$(k_x, k_y, k_z > 0)$, has fully periodic (or closed) orbits regardless of initial conditions only when the three angular frequencies $\omega_x \equiv \sqrt{\frac{k_x}{m}}$, $\omega_y \equiv \sqrt{\frac{k_y}{m}}$ and $\omega_z \equiv \sqrt{\frac{k_z}{m}}$ are commensurable.

- For 2-dimensional anisotropic oscillators, the nature of possible orbits can be studied on the basis of trajectory equations

$$x(t) = A_x \sin(\omega_x t + \theta_x), \quad y(t) = A_y \sin(\omega_y t + \theta_y).$$

This leads to Lissajous figures. If $\frac{\omega_y}{\omega_x} = \frac{m}{n}$ (: a rational number), the Lissajous figures correspond to closed algebraic curves; but if $\frac{\omega_y}{\omega_x}$ is an irrational number (or ω_x and ω_y are incommensurate), a single orbit — not a closed curve — fills the whole available region in the plane densely.

4.4 Kepler problem

- Given the inverse-square-law force $\vec{F} = \frac{k}{r^2}\hat{r}$, the corresponding potential energy function is $V(r) = \frac{k}{r}$. In the Kepler problem, we have $k = -GMm < 0$; but, in the electrostatic interaction with $k = \frac{qQ}{4\pi\epsilon_0}$, both positive and negative values are possible for k. For an object subject to this type of force, its motion can be completely specified based on

  (i) (radial energy equation) The radial motion, $r = r(t)$, can be found by studying

$$\frac{1}{2}m\dot{r}^2 + V_{\text{eff}}(r) = E, \quad \text{with } V_{\text{eff}} = \frac{L^2}{2mr^2} + \frac{k}{r}.$$

  (ii) (orbit equation) The actual orbit of a given object, $r = r(\theta)$, is derived by solving the differential equation

$$\frac{L^2}{2m}\left[\left(\frac{du}{d\theta}\right)^2 + u^2\right] + ku = E, \quad \text{for } u = \frac{1}{r}.$$

- With $k > 0$ (i.e., the repulsive case), we have only unbounded orbits with $E > 0$; but for $k < 0$ (i.e., attractive force), we can have bounded orbits (with $E < 0$) as well as unbounded orbits (with $E > 0$). Solving the orbit equation for detailed orbit forms leads to the equations of conic sections (with a focus at the force center); explicitly,

$$\text{for } k > 0: \quad r = \frac{L^2/m|k|}{e\cos(\theta-\theta_0)-1},$$
$$\text{for } k < 0: \quad r = \frac{L^2/m|k|}{e\cos(\theta-\theta_0)+1},$$

where $e = \sqrt{1+\frac{2EL^2}{mk^2}}$ is the eccentricity. Based on this, one finds that, for $k > 0$ (and so e greater than 1), the orbit becomes a hyperbola. For $k < 0$, on the other hand, one has different possibilities:

 (i) $e = 0$ (with $E = -\frac{mk^2}{2L^2}$): a circular orbit with radius $r = \frac{L^2}{m|k|}$.
 (ii) $e < 1$ (with $-\frac{mk^2}{2L^2} < E < 0$): a closed orbit having the form of an ellipse with semi-major axis $a = \frac{|k|}{2|E|}$ and semi-minor axis $b = \left(\frac{L^2}{2m|E|}\right)^{1/2}$. [This case is relevant for planetary orbits, with the orbital period T given by the formula (consistent with Kepler's third law) $T^2 = \frac{4\pi^2 ma^3}{|k|}$.]
 (iii) $e = 1$ (with $E = 0$, and L finite): a parabolic orbit which is taken by an object escaping to infinity with zero velocity.
 (iv) $e > 1$ (with $E > 0$): a hyperbolic orbit with semi-major axis $a = \frac{|k|}{2E}$, representing an unbounded motion.

- While this Newton-mechanics-based theory can give a highly accurate account for planetary orbits (and also for spaceship flights), the same cannot be said for the hydrogen atom, i.e., for an electron bound to the nucleus due to the inverse-square Coulomb force. In this case, the electron can take on for its energy only certain discrete values called "energy levels" and, for a fully satisfactory explanation of this fact, one must use the quantum theory. But, prior to having full quantum theory, N. Bohr was able to deduce these energy levels by applying an *ad hoc* "quantization rule" on the classical orbits — this is a precursor to the so-called semiclassical approximation method to quantum mechanics.

4.5 Rutherford scattering and the discovery of the atomic nucleus

- In a scattering experiment, one has a fixed (or heavy) target get bombarded by a uniform beam of particles and then measures with detecting apparatus the number of particles scattered in various directions. Here a useful notion is that of differential (scattering) cross section $\frac{d\sigma}{d\Omega}$. It is defined as the ratio

$$\frac{d\sigma}{d\Omega}(\theta, \varphi) = \frac{(dW/d\Omega)}{F},$$

where F is the incident flux (i.e., the number rate of incident particles per unit beam area), and $dW \equiv \left(\frac{dW}{d\Omega}\right)d\Omega$ can be identified with the number rate of scattered particles emerging into the given solid angle range $d\Omega = \sin\theta d\theta d\varphi$ as viewed from the target position.
- If one has detailed knowledge regarding the force acting between incident particles and the target, the differential cross section is a calculable quantity. For this, one needs to know the relation $b = b(\Theta)$ connecting the scattering angle Θ to the impact parameter b (for given incident energy $E = \frac{1}{2}mv_0^2$) under the force in question; with $b = b(\Theta)$ at hand,

$$\frac{d\sigma}{d\Omega} = \frac{b(\theta)}{\sin\theta}\left|\frac{db(\theta)}{d\theta}\right|.$$

In the case of scattering off a hard sphere target of radius R, one has $b = R\cos\frac{\Theta}{2}$ and hence an isotropic differential cross section in the form $\frac{d\sigma}{d\Omega} = \frac{1}{4}R^2$.
- In Rutherford's scattering experiment, the relevant differential cross section is that for scattering of particles of charge

q and mass m under the influence of the Coulomb force $\vec{F} = \frac{qQ}{4\pi\epsilon_0}\frac{1}{r^2}\hat{r}$ due to a fixed point charge Q at the origin. From the hyperbolic orbits of the inverse-square-law force, one here finds that the relation between the impact parameter and the scattering angle is $b = a\cot\frac{\Theta}{2}$ where $a = \frac{|qQ/4\pi\epsilon_0|}{mv_0^2}$. This leads to the Rutherford formula for scattering cross section

$$\frac{d\sigma}{d\Omega} = \frac{1}{4}\left(\frac{qQ}{4\pi\epsilon_0 m v_0^2}\right)^2 \frac{1}{\sin^4\frac{\theta}{2}}.$$

The experimental validity of this formula, especially at large angle, was the basis of Rutherford's conclusion for an almost point-like nucleus.

4.6 Systems with velocity-dependent forces

- Taking air resistance into account, the equation of motion for a projectile may be taken as

$$m\frac{d\vec{v}}{dt} = m\frac{d^2\vec{r}}{dt^2} = -R(v,z)\mathbf{t} - mg\mathbf{k},$$

where $-R(v,z)\mathbf{t}$, with $v = |\vec{v}|$ and $\mathbf{t} = \vec{v}/v$, represents the resistive force. If R is independent of the altitude z, one may use the relation $\frac{d\vec{v}}{dt} = v\frac{dv}{ds}\mathbf{t} - \frac{v^2}{\rho}\mathbf{n}$ (here, s denotes the arc length, $\mathbf{n}$ is the unit normal to the path, and ρ the radius of curvature) to have the equations

$$mv\frac{dv}{ds} = -R(v) - mg\sin\theta, \qquad m\frac{v^2}{\rho} = mg\cos\theta,$$

where θ denotes the inclination to the horizontal of the tangent to the path. Then, writing $R(v) = mg\phi(v)$ and using $\rho = -\frac{ds}{d\theta}$, find the solution to the equation of hodograph

$$\frac{1}{v}\frac{dv}{d\theta} = \frac{\phi(v) + \sin\theta}{\cos\theta}.$$

Once one has such solution $v = v(\theta)$ at hand, all desired information about the trajectory is given by quadrature.

- In the presence of a uniform magnetic field $\vec{B} = B\mathbf{k}$, a charged particle obeys the equation of motion $m\ddot{\vec{r}} = q\dot{\vec{r}} \times (B\mathbf{k})$ or, in component forms,

$$m\ddot{x} = qB\dot{y}, \quad m\ddot{y} = -qB\dot{x}, \quad m\ddot{z} = 0.$$

As a result, the particle moves along a circular helix with a constant speed; the angular frequency equals the cyclotron frequency $\omega = \frac{qB}{m}$, and the radius of the helix, ρ, is related to the transverse speed $v_\perp$ according to $\rho = \frac{mv_\perp}{qB}$. If (together with the above magnetic field) a uniform electric field $\vec{E} = E_x\mathbf{i} + E_z\mathbf{k}$ is also introduced, we obtain the equations of motion relevant to the discussion of the (classical) Hall effect

$$m\ddot{x} = qB\dot{y} + qE_x, \quad m\ddot{y} = -qB\dot{x}, \quad m\ddot{z} = qE_z.$$

The resulting motion then corresponds to a juxtaposition of a circular rotation with the cyclotron frequency and a uniform drift motion with velocity $\vec{v} = -\frac{E_x}{B}\mathbf{j}$ (i.e., in the direction of $\vec{E} \times \vec{B}$).

Problems

4-1 Determine which of the following forces are conservative and find the potential energy functions for the conservative forces:

(i) $F_x = 6abz^3y - 20bx^3y^2$, $F_y = 6abxz^3 - 10bx^4y$, $F_z = 18abxz^2y$,

(ii) $F_x = 18abyz^3 - 20bx^3y^2$, $F_y = 18abxz^3 - 10bx^4y$, $F_z = 6abxyz^2$,

(iii) $F_x = 2ax(z^3 + y^3)$, $F_y = 2ay(z^3 + y^3) + 3ay^2(x^2 + y^2)$, $F_z = 3az^2(x^2 + y^2)$,

(iv) $F_\rho = -3a\rho^2 \cos\varphi$, $F_\varphi = a\rho^2 \sin\varphi$, $F_z = 2az^2$,

(v) $F_r = -2ar \sin\theta \cos\varphi$, $F_\theta = -ar \cos\theta \cos\varphi$, $F_\varphi = ar \sin\varphi$.

Here, (ρ, φ, z) and (r, θ, φ) denote cylindrical and spherical coordinates, respectively.

4-2 Suppose we have a force field of the form

$$\vec{F} = -\frac{\vec{p}}{r^3} + c\frac{(\vec{p}\cdot\vec{r})\vec{r}}{r^5} \quad (r \equiv |\vec{r}| = \sqrt{x^2 + y^2 + z^2})$$

where $\vec{p}$ is a constant nonzero vector. Choose the value of c such that the resulting force may be a conservative one, and for that value of c find the corresponding potential energy function.

4-3 Consider a particle on the xy-plane, which is subject to a force field

$$\vec{F} = \gamma \frac{-y\mathbf{i} + x\mathbf{j}}{x^2 + y^2} \quad (x^2 + y^2 > 0)$$

where γ is a positive constant. Is the given force conservative? (How about the energy conservation law?) Write the equation of motion using polar coordinates and then, based on those equations, indicate noticeable features from the motion.

4-4 Given a particle of mass m and charge q moving in an electrostatic field $\vec{E}(\vec{r})$, we have the energy conservation law in the form

$$\frac{1}{2}m\dot{\vec{r}}^2 + q\phi(\vec{r}) = E \quad (:\ \text{const.}),$$

where $\phi(\vec{r})$ is the electrostatic potential constructed such that $-\vec{\nabla}\phi(\vec{r}) = \vec{E}(\vec{r})$. The equation $\phi(\vec{r}) = C$ (: const.) for each value of C defines a constant potential surface, and to visualize $\phi(\vec{r})$ one often employs so-called *contour lines* which are formed by the lines of intersection of the constant potential surfaces with a plane.

(i) When the electric field $\vec{E}(\vec{r})$ is that produced by two electric charges (of equal amount Q) located at positions $(-d, 0, 0)$ and $(d, 0, 0)$, draw contour lines in the xy-plane and in the drawing indicate the related electric field by using arrows.

(ii) The same question as (i), except that the electric field is now that due to two charges Q, $-Q$ positioned at points $(-d, 0, 0)$ and $(d, 0, 0)$, respectively.

4-5 A particle of mass m is attached to the end of a light string of length l. The other end of the string is passed through a small hole, and is slowly pulled through it. The particle is originally spinning around the hole with angular velocity ω. Find the angular velocity when the length of the string has been reduced to $\frac{1}{2}l$. Find also the tension in the string when its length is r, and verify that the increase in kinetic energy is equal to the work done by the force pulling the string through the hole. (Ignore gravity.)

4-6 A rod is constrained to rotate in a vertical plane with constant angular velocity ω about the fixed point O on the rod. A bead of mass m slides freely on the rod. See the figure.

(i) When r denotes the displacement of the bead along the rod from O, how is it given as a function of time t?

(ii) When the rod is horizontal, the bead passes the pivot O with a velocity v_0 along the rod. Show that the motion of the bead along the rod is simple harmonic if and only if $v_0 = g/2\omega$.

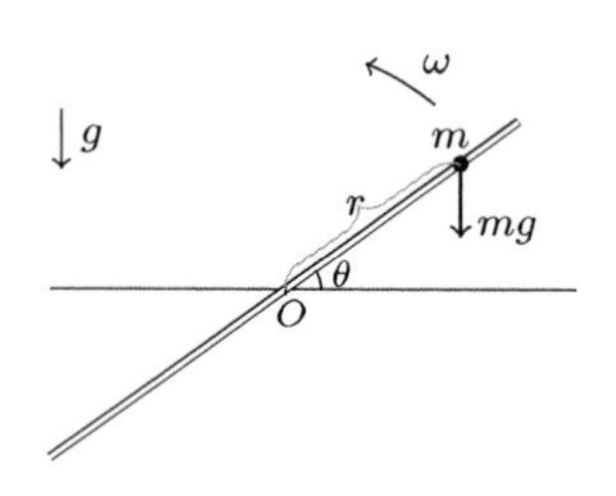

4-7 Consider a 3-dimensional pendulum where mass m hangs by a spring of the relaxed length l and spring constant k. (See the figure.)

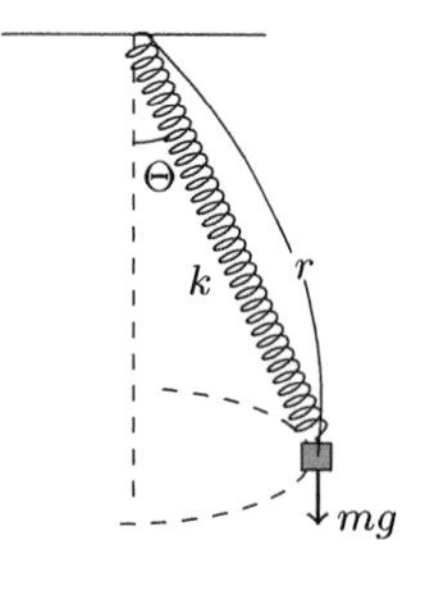

(i) Write the equation of motion using spherical coordinates (with the angle Θ in the figure as one of the coordinates).

(ii) Now, if the mass m is moving in a circular path of constant radius at certain constant angular frequency ω, find the angle Θ.

4-8 Consider a particle of mass m, moving in a central force field of the form $\vec{F} = -cr^n\hat{r}$ ($c > 0$; n can take either sign). Assume that the particle has angular momentum L $(= |\vec{L}|)$.

(i) What will be the value of the radius if the particle is moving in a *circular* orbit?

(ii) Plot the corresponding effective 1-dimensional radial potential $V_{\text{eff}}(r)$ for various values of n, and show that the circular orbits (considered in (i)) are unstable if $n < -3$.

4-9 A particle of mass m moves in a circle of radius R under the influence of an attractive central force

$$\vec{F} = -\frac{K}{r^2}e^{-r/a}\hat{r} \quad (K,\ a \text{ are positive constants}).$$

(i) Show that the circular orbit is stable if the radius R is less than a.

(ii) Find the frequency of small radial oscillations about this circular motion.

4-10 Consider a particle of mass m that moves under the influence of gravity along a frictionless wire (on a vertical plane) whose shape is described by a function of given form $z = f(x)$. See the figure.

(i) Show that the particle satisfies the equation of motion in the form

$$\frac{d^2 s}{dt^2} = -g\frac{dz}{ds},$$

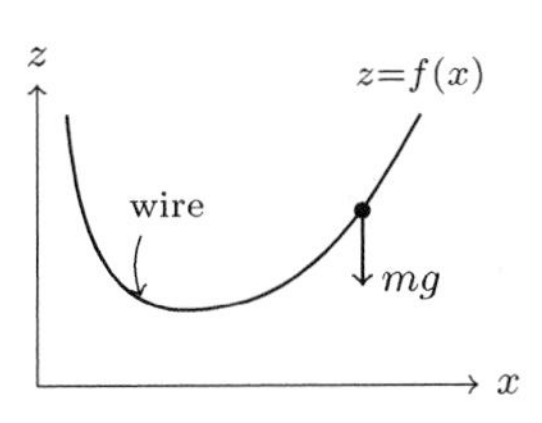

where s denotes the arc length along the wire (from some arbitrary point of reference). Also rewrite this as a second-order differential equation for $x(t)$, using the wire equation $z = f(x)$.

(ii) Obtain the first integral of this equation of motion.

4-11 In the same situation as in Problem 4-10, suppose that the wire is shaped as a cycloid with parametric equations

$$x(\phi) = a(\phi + \sin\phi)\,, \quad z(\phi) = a(1 - \cos\phi).$$

Then, along this cycloidal path, the particle will undergo periodic motion.

(i) In this case, introducing a generalized variable $w = \sin\frac{\phi}{2}$, find the equation of motion satisfied by w.

(ii) Based on the equation derived in (i), discuss what special feature you can observe as regards the period of oscillation when the wire has a cycloidal shape.

4-12 A particle of mass m moves under a harmonic oscillator force with potential energy $\frac{1}{2}k(x^2 + y^2 + z^2)$. Initially it is moving in a circle of radius a. Find the orbital velocity v_0. It is then given a blow of impulse mv_0 in a direction making an angle α with its original velocity. Determine the minimum and maximum distances from the origin during the subsequent motion. Explain your results physically for the two limiting cases $\alpha = 0$ and $\alpha = \pi$.

4-13 As was discussed in the text (Sec. 4.3), orbits of a 2-dimensional anisotropic oscillator can be studied by analyzing the equations

$$x = A_x \sin t, \quad y = A_y \sin(\beta t + \theta_y).$$

Draw Lissajous figures appropriate for values $\beta = \frac{1}{3}$, $\beta = \frac{2}{3}$ and $\beta = \frac{1}{4}$ (for various representative choices of the phase angle θ_y).

4-14 Energy and angular momentum conservations simplify the study of central-force systems a great deal. Now let us contemplate on the following. Our world, at the present time, appears to have only three spatial directions. But, it may not be totally crazy to think about the possibility that the universe, during its long evolution, might once have had more than three, say, n (>3) spatial directions. Then, imagining the case of a central force there, one may write the equation of motion in the form

$$m\frac{d^2 x_i}{dt^2} = f(r)\frac{x_i}{r} \quad \left(i = 1, \cdots, n;\ r = \sqrt{(x_1)^2 + \cdots + (x_n)^2}\right).$$

What can you say about the behavior of this system, especially from the viewpoint of using appropriate conservation laws?

4-15 For a relativistic particle moving in a central force field $\vec{F} = -\vec{\nabla}V(r)$, the equation of motion may be taken to be

$$\frac{d\vec{p}}{dt} = -\vec{\nabla}V(r) \quad \left(\vec{p} \equiv \frac{m_0\vec{v}}{\sqrt{1-\vec{v}^2/c^2}} \text{ with } \vec{v} \equiv \frac{d\vec{r}}{dt}\right)$$

where c is the speed of light.

(i) Show that this system admits the energy conservation law in the form

$$\frac{m_0 c^2}{\sqrt{1-\vec{v}^2/c^2}} + V(r) = E_{\text{rel}}\ (= \text{const.}),$$

where E_{rel}, the relativistic energy, includes the rest energy m_0c^2.

(ii) Using this energy conservation law together with the appropriate angular momentum conservation law ($\vec{L} \equiv \vec{r} \times \vec{p} = m_0\vec{r} \times \frac{\vec{v}}{\sqrt{1-\vec{v}^2/c^2}}$ is a constant of motion), derive the orbit equation satisfied by $u = u(\theta)$ where $u = \frac{1}{r}$ and (r, θ) denote the polar coordinates introduced in the plane of motion.

4-16 Kepler's second law of planetary motion can be viewed as a direct evidence that the force is central. Using the orbit equation valid in the presence of a general central potential $V(r)$ (see (4.84) in the text)

$$\frac{L^2}{2m}\left[\left(\frac{du}{d\theta}\right)^2 + u^2\right] + V\left(\frac{1}{u}\right) = E \quad \left(u \equiv \frac{1}{r}\right)$$

show that Kepler's first law — planetary orbits are ellipses with the Sun at a focus — implies the *inverse-square-force* law.

4-17 Perihelion and aphelion of the dwarf planet Pluto occur at distances of 4.425×10^{12} m and 7.375×10^{12} m from the Sun, respectively. Find the eccentricity of Pluto's orbit and the length of Pluto's planetary year.

4-18 The figure shows the Earth's elliptical orbit around the Sun (point O), with points 'W', 'V', 'S' and 'A' representing the winter solstice (22 December), vernal equinox (21 March), summer solstice (22 June) and autumnal equinox (23 September), respectively. Notice that the interval from autumn to spring via winter (179 days) is less than that from spring to autumn via summer (186 days). This implies, by Kepler's 2nd law, that the Earth passes closest to the Sun in the winter, and so in our figure the winter solstice is taken to occur at the perihelion.

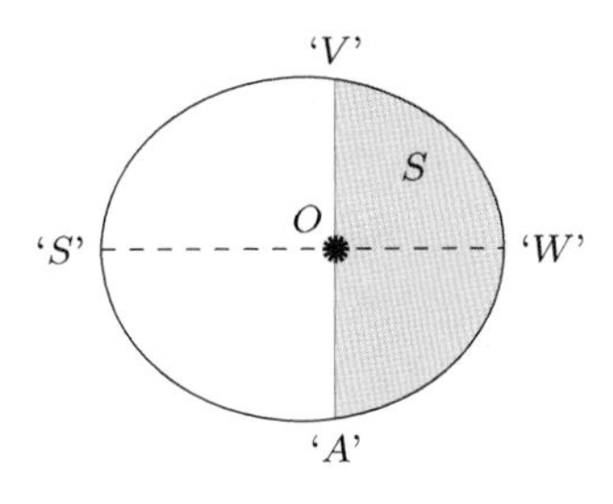

(i) By using the Kepler orbit equation $r = \frac{L^2/m|k|}{e\cos(\theta-\theta_0)+1}$, show that the shaded area of the orbit (see the figure), S, can be expressed as

$$S = a^2(1-e^2)\left(-e + \frac{\cos^{-1} e}{\sqrt{1-e^2}}\right)$$

where a is the semi-major axis of the orbit and e is the orbital eccentricity.

(ii) Then, by invoking Kepler's 2nd law together with the information on the dates of the equinoxes and the solstices, estimate the (present) eccentricity of the Earth's orbit.

4-19 Prove that if a is the semi-major axis of the orbit of a planet, we have $\int \frac{dt}{r(t)} = \frac{T}{a}$ where integration is over a complete planetary year T. Deduce that the time-averaged potential energy of the planet (in an elliptic orbit) is minus twice the time-averaged kinetic energy of the planet. [*Note*: The time average of any periodic function f (with a period T) is defined to be $\bar{f} = \frac{1}{T}\int_0^T f\,dt$.]

4-20 Consider a projectile launched from the surface of the Earth, with an initial speed V at an angle α with the horizontal (see the figure), which follows the Keplerian orbit.

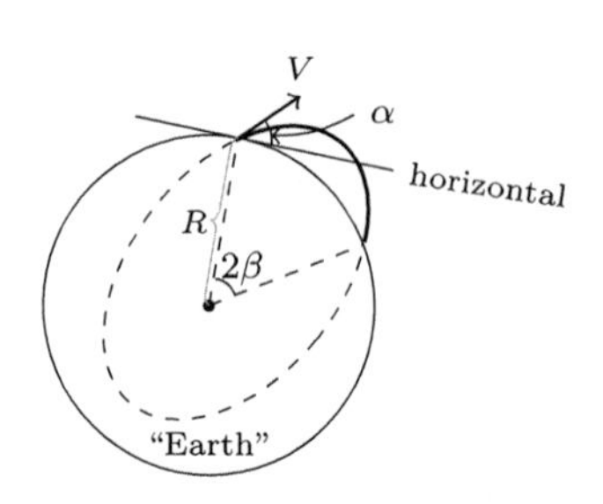

(i) Show that the angular range of projectile, 2β, is determined by

$$\tan\beta = \frac{\frac{1}{2}C^2 \sin 2\alpha}{1 - C^2\cos^2\alpha},$$

where $C^2 = V^2/gR$ with $g = 9.8\,\mathrm{m{\cdot}s^{-2}}$.

(ii) Then verify that, under the appropriate limit, the formula given in (i) reduces to the familiar result obtained for the flat-Earth-with-constant-gravity approximation.

4-21 Given an inverse-square force $\vec{F} = \frac{k}{r^2}\hat{r}$ with $k < 0$, the orbit for bounded motion generally corresponds to an ellipse:

$$\frac{(x+ea)^2}{a^2} + \frac{y^2}{b^2} = 1.$$

Demonstrate that, in the highly elliptical limit of $e \to 1$, this bounded orbit goes over smoothly to a parabolic orbit.

4-22 In an inverse-square central force field

$$\vec{F} = -\frac{k}{r^2}\hat{r} \quad (k > 0),$$

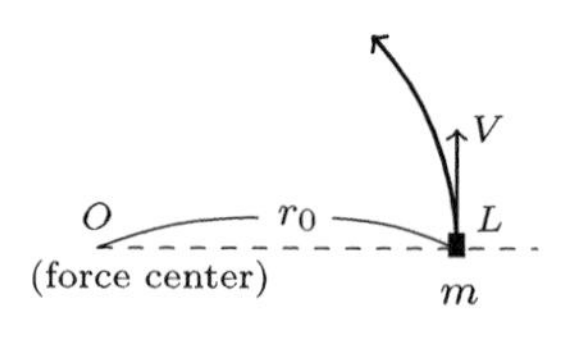

suppose that a particle (of mass m) moves, at radius $r = r_0$, with an initial speed V at right angle to the line directed at the force center. (See the figure). We are here interested in knowing the nature of the resulting orbit as the magnitude of V is changed.

(i) Find the speed value, $V = V_\mathrm{e}$, leading to zero energy (i.e., $E = 0$) and obtain the corresponding orbit. Then discuss how the nature of the orbit changes as the initial speed V is changed from a value above V_e to a value below V_e. (Give also schematic plots of the related orbits.)

(ii) Find the speed value, $V = V_\mathrm{c}$, leading to a circular orbit. Then discuss how the nature of the orbit changes as the initial speed V is changed from a value above V_c to a value below V_c. (Give also schematic plots of the resulting orbits.)

(iii) Discuss the orbit when the initial speed V has a very small, but nonzero value.

4-23 Consider a relativistic particle in an attractive inverse-square force field, governed by the equation of motion

$$\frac{d}{dt}\left[\frac{m_0\vec{v}}{\sqrt{1-\vec{v}^2/c^2}}\right] = -\frac{\kappa}{r^2}\hat{r} \quad (\kappa > 0).$$

(See Problem 4-15.) For this system, discuss the nature of *bounded orbit* in detail. Especially, show that when the relativistic correction is small, the orbit corresponds to a precessing ellipse. What value do you obtain here for the perihelion advance? [If one uses this special relativistic equation of motion for the motion of Mercury orbiting the Sun, the prediction for the perihelion advance is only about one-sixth of the observed value. Hence the equation we have written down is a *wrong* relativistic generalization for the gravity case.]

4-24 Consider a planet of mass m in orbit around the Sun (having mass M). Here pretend that the Sun is surrounded by a uniform dust cloud extending out at least as far as the location of the planet. Then, including the effect of the dust cloud, one can take the total potential (for the planet motion) to be given by the form

$$V(r) = -\frac{GMm}{r} + \frac{1}{2}kr^2 \quad (k > 0).$$

(i) When the planet is in a circular orbit corresponding to angular momentum L, determine the radius of the orbit, r_0, in terms of L, G, M, m and k. (You need not solve the equation.)

(ii) Assuming that the second term in the potential, $\frac{1}{2}kr^2$, is small compared to the first term, now consider an orbit just slightly deviating from the circular orbit of part (i). Show that the orbit is a precessing ellipse and calculate the rate of precession to first order in k.

(iii) Does the axis of the ellipse precess in the same or opposite direction to the direction of revolution?

4-25 For a spaceship to be able to rendezvous with the outer planet after following the Hohmann orbit closely, what should be the angular separation ψ between the destination planet and the Earth at the time of lift off for the spaceship (see the figure)?

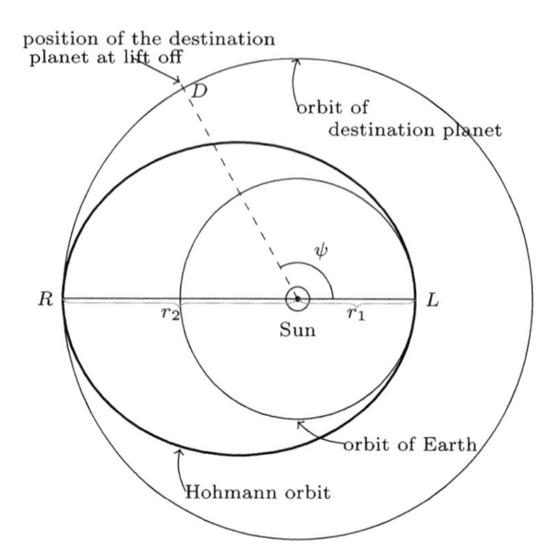

4-26 Consider a potential scattering, in which projectiles experience a spherical-well-type potential of the form

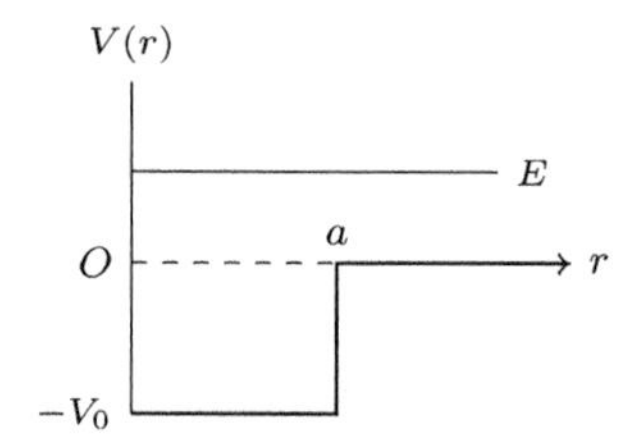

$$V(r) = \begin{cases} 0, & \text{for } r > a \\ -V_0 \ (V_0 > 0), & \text{for } r < a. \end{cases}$$

(i) When $E = \frac{1}{2}mv_0^2$ is the incident beam energy, find $b = b(\theta)$, i.e., determine the impact parameter b as a function of the scattering angle θ. [Note that for given energy E there is a certain maximum value, θ_{max}, for the scattering angle possible.]

(ii) Calculate the differential cross section $\frac{d\sigma}{d\Omega}$ for this case. Also verify that the total cross section, obtained by integrating $\frac{d\sigma}{d\Omega}$ over the solid angle, is equal to the geometrical cross section πa^2.

4-27 The same question as in Problem 4-26, but now assuming a *repulsive* central potential (or a spherical potential "barrier")

$$V(r) = \begin{cases} 0, & \text{for } r > a \\ +V_0 \ (V_0 > 0), & \text{for } r < a \end{cases}$$

for incident energy $E\ (= \frac{1}{2}mv_0^2) > V_0$. [Here note that, for some impact parameter values, the particle gets reflected at the spherical surface at $r = a$ without entering the spherical region.]

4-28 Let us consider scattering of particles (with mass m and incident energy $E = \frac{1}{2}mv_0^2$) in a truncated repulsive Coulomb-type potential

$$V(r) = \begin{cases} \frac{k}{r}, & r > a \\ \frac{k}{a}, & r < a, \end{cases}$$

where k is a positive constant.

(i) For a particle with incident energy $E > \frac{k}{a}$, can you obtain the scattering angle Θ as a function of the impact parameter ratio $b/\bar{b}$, where $\bar{b}$ is the impact parameter value for which the point of closest approach to the origin occurs at $r = a$. (It is sufficient to describe your method to obtain the function.)

(ii) Inferring from the dependence of Θ on $b/\bar{b}$, what can you say about the differential cross section? (Discuss the way this cross section would differ from the Rutherford scattering cross section.)

4-29 Consider a particle of mass m subject to a central inverse-cube force described by the potential energy function $V(r) = \frac{k}{2r^2}$ $(k > 0)$. It enters the region from infinity, with the incident polar angle at $\theta = 0$ when (r, θ) denote polar coordinates introduced in the orbital plane.

(i) Show that the orbit equation can be written in the form

$$r \cos[n(\theta - \theta_0)] = C,$$

where n, θ_0, C are constants.

(ii) Find the scattering angle (as a function of impact parameter b) for particles incident with initial velocity v_0, and then calculate the differential cross section.

4-30 Consider a particle of mass m which is acted on by an attractive central potential

$$V(r) = -\frac{\alpha}{r^4} \quad (\alpha > 0).$$

The particle approaches the force center — the spatial origin $r = 0$ — from a point sufficiently far away.

(i) When the angular momentum of the particle is equal to L, discuss in words the nature of possible motions depending on the particle's total energy value E.

(ii) For the particle(s) coming from infinity with an initial velocity v_0, find the *total cross section for capture*. [*Hint*: Consider the range of impact parameters for which the particle(s) are captured by the potential.]

4-31 The trajectory of a projectile subject to a resistive force proportional to velocity ($\vec{F}_r = -m\gamma\dot{\vec{r}}$) is described by the equation $z = \frac{\gamma v_{0z}+g}{\gamma v_{0x}} x + \frac{g}{\gamma^2}\ln(1 - \frac{\gamma x}{v_{0x}})$ (Sec. 4.6 in the text). By studying this equation for small γ, derive the approximate formula (say, to first order in γ) for the angle of launch α which maximize the range (on level ground) for given launch speed $v_0 = \sqrt{v_{0x}^2 + v_{0y}^2}$.

4-32 A bomb is dropped from an airplane flying horizontally at a height h with speed V_0. Assuming the linear law of resistive force i.e., $\vec{F}_r = -m\gamma\dot{\vec{r}}$, and further assuming that γ is small, show that the time of fall is approximately

$$\sqrt{\frac{2h}{g}}\left(1 + \frac{1}{6}\gamma\sqrt{\frac{2h}{g}}\right).$$

Show also that the horizontal distance travelled by the bomb in its fall is approximately

$$V_0\sqrt{\frac{2h}{g}}\left(1 - \frac{1}{3}\gamma\sqrt{\frac{2h}{g}}\right).$$

4-33 For a projectile subject to a resistive force of the form $\vec{F}_r = -mgCv^2\mathbf{t}$, its motion can be analyzed by numerically integrating $\frac{dx}{d\theta}$, $\frac{dz}{d\theta}$ and $\frac{dt}{d\theta}$ with $v = f(\theta)$ given as

$$\frac{1}{v^2} = \cos^2\theta\{A - C\tanh^{-1}(\sin\theta)\} - C\sin\theta,$$

where A is a constant of integration (see Sec. 4.6). Aside from this, one can also develop analytically the short-time and long-time approximations for the velocity components $\dot{x}(t)$ and $\dot{z}(t)$. When the initial conditions are such that $\dot{x}(t{=}0) = v_{0x}$ $(= v_0\cos\theta_0)$ and $\dot{z}(t{=}0) = v_{0z}$ $(= v_0\sin\theta_0)$, show that

<u>for small t (> 0):</u>

$$\dot{x}(t) = \frac{v_{0x}}{[1 + gCv_0t - \frac{1}{2}g^2C\sin\theta_0 t^2 + \cdots]},$$

$$\dot{z}(t) = \frac{v_{0z} - gt[1 + \frac{1}{2}gCv_0t + \cdots]}{[1 + gCv_0t - \frac{1}{2}g^2C\sin\theta_0 t^2 + \cdots]},$$

for large t:

$$\dot{x}(t) = O\big(e^{-g\sqrt{C}t}\big), \quad \dot{z}(t) = -\frac{1}{\sqrt{C}}\big[1 + O\big(e^{-2g\sqrt{C}t}\big)\big].$$

4-34 In a uniform electric field $\vec{E} = E\mathbf{k}$, the motion of a relativistic charged particle (with rest mass m_0 and electric charge q) is governed by the equation

$$\frac{d}{dt}\left(\frac{m_0\vec{v}}{\sqrt{1-\vec{v}^2/c^2}}\right) = qE\mathbf{k}.$$

With a judicious choice of x- and z-axes, the motion will occur entirely in the xz-plane.

(i) What form does the law of energy conservation take?
(ii) Solve the above equation of motion to find $x(t)$ and $z(t)$, under the initial conditions $x(t{=}0) = x_0$, $z(t{=}0) = z_0$, $\dot{x}(t{=}0) = v_{0x}$ (with $|v_{0x}| < c$), and $\dot{z}(t{=}0) = 0$.
(iii) What do you find for the speed of particle as $t \to \infty$?
(iv) Determine the orbit, and discuss how it differs from that of a corresponding nonrelativistic motion.

4-35 A beam of particles of charge q and given kinetic energy $E = \frac{1}{2}mv_0^2$ is emitted from a point source, roughly parallel to a uniform magnetic field $\vec{B} = B\mathbf{k}$ in the region, but with a small angular dispersion. Show that the effect of the field is to focus the beam to a point at a distance $z = \frac{2\pi m v_0}{|q|B}$ from the source.

4-36 In a region with a uniform magnetic field $\vec{B} = B\mathbf{k}$, consider a charged particle subject to an isotropic harmonic force. The equation of motion is

$$m\frac{d^2\vec{r}}{dt^2} = -k\,\vec{r} + q\,\dot{\vec{r}} \times (B\mathbf{k}) \quad (k > 0).$$

Assuming that the motion lies entirely in the xy-plane, find the general motion. Also discuss how the nature of a typical orbit changes as the magnetic field strength B is varied from zero.

5 The Two-Body Problem, Collision and Many-Particle System

SUMMARY

5.1 The two-body problem

- To discuss the dynamics of a two-body system obeying the equations of the form

$$m_1\ddot{\vec{r}}_1 = m_1\vec{g} + \vec{F}(\vec{r}_1 - \vec{r}_2), \quad m_2\ddot{\vec{r}}_2 = m_2\vec{g} - \vec{F}(\vec{r}_1 - \vec{r}_2),$$

one may utilize the center of mass and relative position variables

$$\vec{R} \equiv \frac{m_1\vec{r}_1 + m_2\vec{r}_2}{m_1 + m_2}, \qquad \vec{r} \equiv \vec{r}_1 - \vec{r}_2$$

to recast the given equations of motion as two one-body-type equations

$$\ddot{\vec{R}} = \vec{g}, \qquad \mu\ddot{\vec{r}} = \vec{F}(\vec{r}),$$

where $\mu \equiv \frac{m_1 m_2}{m_1+m_2}$ is the reduced mass. If $\vec{F}(\vec{r}) = -\vec{\nabla}_{\vec{r}}V(\vec{r})$ (i.e., with a conservative interparticle force), the energy conservation law for the above system becomes (here, $M = m_1 + m_2$)

$$\frac{1}{2}M\dot{\vec{R}}^2 + \frac{1}{2}\mu\dot{\vec{r}}^2 - M\vec{g}\cdot\vec{R} + V(\vec{r}) = E \quad (\text{: const.}).$$

- If two bodies are moving under their mutual gravitational attraction, the relative motion is governed by the equation $\ddot{\vec{r}} = -\frac{GM}{r^2}\hat{r}$. According to this finding, one infers that, in the case of the Earth–Moon system, both the Moon (body 1) and the Earth (body 2) move in ellipses with their center of mass as the common focus, and with semi-major axes $a_1^* = \frac{m_2}{M}a$ and $a_2^* = \frac{m_1}{M}a$ (a is the semi-major axis of the relative orbit), respectively.

5.2 Two-particle collision and the CM and lab cross sections

- If two particles of masses m_1 and m_2 collide elastically along a straight line, the initial and final velocities of the two particles are subject to the kinematical relations

$$\frac{1}{2}m_1 v_{1i}^2 + \frac{1}{2}m_2 v_{2i}^2 = \frac{1}{2}m_1 v_{1f}^2 + \frac{1}{2}m_2 v_{2f}^2$$
(: energy conservation),
$$m_1(v_{1f} - v_{1i}) = -m_2(v_{2f} - v_{2i})$$
(: momentum conservation),

and as a result it follows that $(v_{2f} - v_{1f}) = -(v_{2i} - v_{1i})$. In an inelastic collision, one may use the modified energy equation by including some lost energy Q, i.e.,

$$\frac{1}{2}m_1 v_{1i}^2 + \frac{1}{2}m_2 v_{2i}^2 = \frac{1}{2}m_1 v_{1f}^2 + \frac{1}{2}m_2 v_{2f}^2 + Q$$

together with the (unmodified) momentum equation. [If one defines the coefficient of restitution e by

$$v_{2f} - v_{1f} = -e(v_{2i} - v_{1i}),$$

the energy loss Q is related to the coefficient of restitution e by the equation $Q = \frac{1}{2}\left(\frac{m_1 m_2}{m_1+m_2}\right)(v_{2i} - v_{1i})^2(1 - e^2) > 0$.]
- In the case of elastic two-body collision in three dimensions, kinematics are rather simple in the CM frame (i.e., in the frame of zero total momentum): due to the conservation of total energy and momentum, the magnitudes of each particle's momentum before and after the collision are the same (i.e., $|\vec{p}_i^*| = |\vec{p}_f^*|$) and each particle is scattered through the same angle θ^*. In the lab frame (i.e., with zero initial momentum for particle 2), on the other hand, more complicated kinematical relations follow: for instance, if E_1

(E_1') denote the lab incident (outgoing) energy of particle 1, the scattering angle θ_1 must satisfy the equation

$$\cos\theta_1 = \frac{(m_1 - m_2)E_1 + (m_1 + m_2)E_1'}{2m_1(E_1 E_1')^{1/2}}.$$

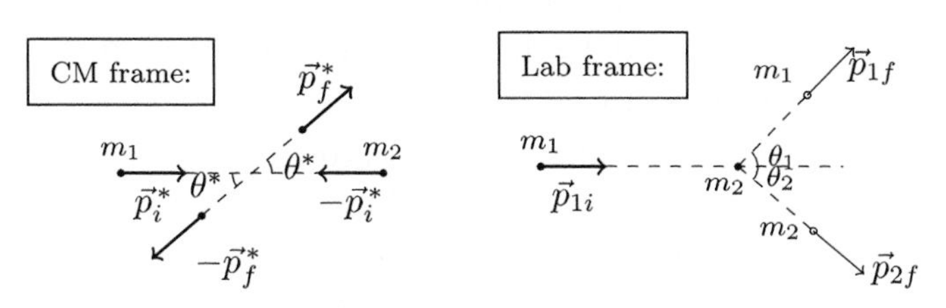

Also note that this lab scattering angle θ_1 is related to the CM scattering angle θ^* by $\tan\theta_1 = \frac{\sin\theta^*}{\cos\theta^* + \frac{m_1}{m_2}}$.

- In relativistic collision problems, one must apply the energy and momentum conservation laws using the relativistic momentum $\vec{p} = \frac{m_0\vec{v}}{\sqrt{1-\vec{v}^2/c^2}}$ and relativistic energy $E = \frac{m_0 c^2}{\sqrt{1-\vec{v}^2/c^2}} = \sqrt{m_0^2 c^4 + c^2\vec{p}^2}$, where m_0 is the rest mass of a particle. With this understanding, the kinematical relations in two-body collisions (involving particles 1 and 2) read

$$\vec{p}_{1i} + \vec{p}_{2i} = \vec{p}_{1f} + \vec{p}_{2f},$$

$$\sqrt{(m_{01})^2c^4 + c^2\vec{p}_{1i}^{\,2}} + \sqrt{(m_{02})^2c^4 + c^2\vec{p}_{2i}^{\,2}} = \sqrt{(m_{01})^2c^4 + c^2\vec{p}_{1f}^{\,2}} + \sqrt{(m_{02})^2c^4 + c^2\vec{p}_{2f}^{\,2}}.$$

In terms of four-momentum P^μ defined by

$$P^\mu = \left(P^0 \equiv \frac{E}{c}, \vec{P} \equiv \vec{p}\right) = \left(\frac{m_0 c}{\sqrt{1-\vec{v}^2/c^2}}, \frac{m_0\vec{v}}{\sqrt{1-\vec{v}^2/c^2}}\right),$$

these conservation laws can be recast as a single four-vector equation

$$P_{1i}^\mu + P_{2i}^\mu = P_{1f}^\mu + P_{2f}^\mu.$$

Notice that the rest mass of a particle is a Lorentz scalar since the four-momentum square given by $(P^0)^2 - \vec{P}^2$, a Lorentz-invariant quantity, equals $m_0^2 c^2$.

- In nonrelativistic two-body collision problems, if one has information on the interparticle force $\vec{F}(\vec{r})$ where $\vec{r} \equiv \vec{r}_1 - \vec{r}_2$ is the relative position, the CM scattering cross section can be obtained as in the fixed-center scattering, taking the equation of motion for the relative coordinate, i.e., $\mu\ddot{\vec{r}} = \vec{F}(\vec{r})$ as the governing equation. Then, from the expression for the CM cross section, that is, $\frac{d\sigma}{d\Omega_{\text{CM}}}(\theta^*, \varphi^*) = \frac{b(\theta^*)}{\sin\theta^*}\left|\frac{db(\theta^*)}{d\theta^*}\right|$ (here, $b = b(\theta^*)$ relates the CM scattering angle θ^* to the impact parameter), the corresponding lab scattering cross section is found as

$$\begin{aligned}\frac{d\sigma}{d\Omega_{\text{Lab}}}(\theta_1, \varphi_1) &= \left|\frac{\sin\theta^*}{\sin\theta_1}\frac{d\theta^*}{d\theta_1}\right| \frac{d\sigma}{d\Omega_{\text{CM}}}(\theta^*, \varphi^*) \\ &= \frac{\{1 + \left(\frac{m_1}{m_2}\right)^2 + 2\left(\frac{m_1}{m_2}\right)\cos\theta^*\}^{3/2}}{\left|1 + \left(\frac{m_1}{m_2}\right)\cos\theta^*\right|} \\ &\quad \times \frac{d\sigma}{d\Omega_{\text{CM}}}(\theta^*, \varphi^*)\end{aligned}$$

based on the connection formula existing between the lab and CM scattering angles, $\tan\theta_1 = \frac{\sin\theta^*}{\cos\theta^* + \frac{m_1}{m_2}}$ (and $\varphi_1 = \varphi^*$).

5.3 Many-body systems and general conservation laws

- Let $\vec{R}$ denote the center of mass (CM) of some material system — a discrete or continuous collection of Newtonian particles — and M be its net mass. Then the total linear momentum of a given object can be identified with $M\dot{\vec{R}}$, and its rate of change is solely governed by the equation

$$\dot{\vec{P}} = M\ddot{\vec{R}} = \vec{F}^{(e)},$$

where $\vec{F}^{(e)}$ is the net external force, i.e., the vector sum of all external forces acting on the object. If there is no net external force, the law of total momentum conservation, $\vec{P} = M\dot{\vec{R}} = \text{const.}$, follows and so the CM of the system must be in a uniform straight motion. These equations concerning the total momentum are especially useful in studying the rocket-like motion in which mass transfer between constituents of a given system occurs.

- To discuss the motion of an object which can rotate, it is useful to consider the angular momentum equation

$$\dot{\vec{L}} = \vec{\Gamma}^{(e)},$$

where $\vec{L} = \sum_i m_i \vec{r}_i \times \dot{\vec{r}}_i$ is the total angular momentum, and $\vec{\Gamma}^{(e)} = \sum_i \vec{r}_i \times \vec{F}^{(e)}$ the net external torque. For an isolated system, this leads to the law of conservation of total angular momentum, i.e., $\frac{d\vec{L}}{dt} = 0$; in a uniform gravitational field, this reduces to $\dot{\vec{L}} = \vec{R} \times (M\vec{g})$. Also, through separating the contributions to $\vec{L}$ from the CM motion and the relative motion by writing $\vec{r}_i = \vec{R} + \vec{r}_i'$, one can have separate angular momentum equations

$$\frac{d}{dt}(M\vec{R} \times \dot{\vec{R}}) = \vec{R} \times \vec{F}^{(e)} \quad \text{(for the CM motion)},$$

$$\frac{d}{dt}\vec{L}' = \sum_i \vec{r}_i' \times \vec{F}^{(e)} \equiv \vec{\Gamma}'^{(e)} \quad \text{(for the relative motion)},$$

where $\vec{L}' = \sum_i \vec{r}_i' \times \dot{\vec{r}}_i'$ represents the angular momentum relative to the CM of the system.

- By applying the angular momentum equation to a rapidly spinning gyroscope (of mass M), one can have an understanding on its steady precessional motion with angular velocity $\Omega = \frac{RMg}{|\vec{L}|}$, where R is the distance from the pivot point to the CM of the body.
- For a system of particles in which interparticle forces $\vec{F}_{ij}$ are derivable from a potential $V(\vec{r}_i - \vec{r}_j)$, we have the energy equation

$$\frac{d}{dt}[T + V_{\text{int}}] = \sum_i \dot{\vec{r}} \cdot \vec{F}_i^{(e)},$$

where $T = \sum_i \frac{1}{2} m_i \dot{\vec{r}}_i^2$ is the total kinetic energy, and $V_{\text{int}} = \sum_{\text{each pair}} V(\vec{r}_i - \vec{r}_j)$ the total internal potential energy. [For a rigid body, V_{int} remains constant and so can be neglected in this equation.] If external forces $\vec{F}_i^{(e)}$ can be derived from a certain potential $V_{\text{ext}}(\vec{r}_i)$ as well, this leads to the law of conservation of total energy, $T + V = E$ (: const.), with $V = V_{\text{int}} + \sum_i V_{\text{ext}}(\vec{r}_i)$. Also, separating the contributions due to the CM motion and the relative motion, the above energy equation can be represented as

$$\frac{d}{dt}\left[\frac{1}{2}M\dot{\vec{R}}^2\right] = \dot{\vec{R}} \cdot \vec{F}^{(e)},$$

$$\frac{d}{dt}(T' + V_{\text{int}}) = \sum_i \dot{\vec{r}}_i' \cdot \vec{F}_i^{(e)},$$

where $T' = \sum \frac{1}{2}m_i\dot{\vec{r}}_i'^2$ is the kinetic energy relative to the CM. $T' + V_{\text{int}}$ corresponds to the total internal energy of the system.

- For an isolated system of relativistic particles, we have the conservation of total four-momentum (from the spacetime translation invariance); but, the additive conservation of the total (rest) mass can be violated.

5.4 The kinetic theory of gases

- In the kinetic theory viewpoint, pressure of a gas is nothing but the (average) perpendicular force per unit area, exerted by randomly moving molecules as they collide with the wall of the gas container. Assuming that intermolecular forces are negligible, it is then possible to relate the pressure P to the average kinetic energy of molecules, $\bar{K} = \frac{1}{2}m\overline{(v^2)}$ (m is the molecular mass), by the formula $P = \frac{2}{3}n\bar{K}$ where $n = \frac{N}{V}$ is the number density. [N denotes the total number of molecules in the box of volume V.] In this basis, the absolute temperature T of an ideal gas — a gas satisfying Boyle's law and the law of Charles and Gay-Lussac — can be identified as a quantity directly proportional to the average kinetic energy $\bar{K}$, i.e., $\bar{K} = \frac{3}{2}k_BT$ where k_B is Boltzmann's constant. In this manner, we arrive at the equation of state for an ideal gas

$$P = nk_BT \quad \text{or } PV = Nk_BT.$$

- The relation $\bar{K} = \frac{3}{2}k_BT$ can be accounted for if at temperature T the velocities of gas molecules have the statistical distribution

$$\Phi(v)dv = \frac{4}{\sqrt{\pi}}\left(\frac{m}{2k_BT}\right)^{3/2} v^2 e^{-mv^2/2k_BT}dv,$$

which is called the Maxwell–Boltzmann speed distribution.

- Under thermal contact a macroscopic system tends to evolve into a state described by an equilibrium energy distribution (at given temperature T); here, not just the center-of-mass kinetic energy (responsible for the value $\bar{K}$) but all other kinds of energy that molecules may have do matter. Hence, if the total internal energy of a gas is written by the expression $U = qNk_BT$, one has $q = \frac{3}{2}$ for a simple atomic gas but a larger value for molecular gases as they can have internal excitations.
- To formulate the law of energy conservation with thermodynamic systems (the first law of thermodynamics), one needs to have a quantitative representation of *heat* — the disordered form of energy flowing from one body to another by means other than through external work. This requires the (nonkinematical) notion like *entropy.*

Problems

5-1 Consider an isolated system of two particles (masses m_1, m_2) subject to a mutual attractive force

$$\vec{F}_{12} = -\vec{F}_{21} = -k(\vec{r}_1 - \vec{r}_2) \quad (k \text{ is a positive constant}).$$

In the center-of-mass frame, describe the general nature of motion — including the respective orbits — for the two particles. Also give the conditions under which the two particles assume circular orbits (with the center at the center of mass of the system).

5-2 Consider a system of three stars with masses m_1, m_2 and m_3. The stars, placed at the corners of an equilateral triangle of side L, rotate under mutual gravitational attractions (while leaving the relative separation of each star unchanged). Determine the rotational period of this system.

5-3 A mass m_1, with initial velocity V_0, strikes a mass–spring system m_2, which is initially at rest but able to recoil. The spring is massless with spring constant k (see the figure) and there is no friction. [In this problem the spring, other than generating appropriate repulsive force if compressed, is not given any dynamical role.]

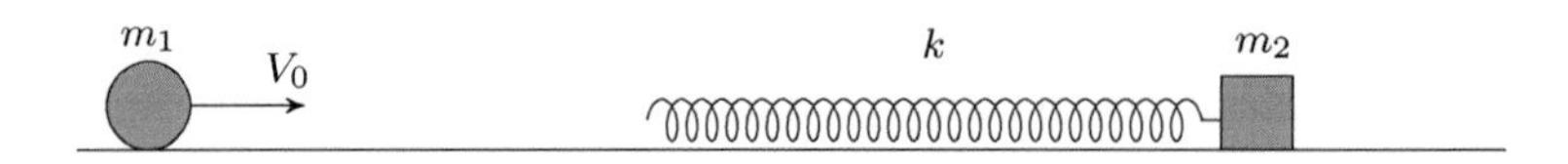

(i) What is the maximum compression of the spring?
(ii) If, long after the collision, both objects travel in the same direction, what are the final velocities V_1 and V_2 of m_1 and m_2, respectively?

5-4 In a one-dimensional collision of two bodies with the coefficient of restitution e, show that the energy loss Q, defined as the deficit found in the total kinetic energy after the collision, is related to the coefficient of restitution by the equation (5.35) in the text, i.e.,

$$Q = \frac{1}{2}\left(\frac{m_1 m_2}{m_1 + m_2}\right)(v_{2i} - v_{1i})^2(1 - e^2),$$

where v_{1i} (v_{2i}) is the initial velocity of particle 1 (particle 2).

5-5 A ball is dropped from height h and bounces vertically. The coefficient of restitution at each bounce is e. Find the velocity immediately after the first bounce, and immediately after the nth bounce. Show that the ball finally comes to rest after a time

$$\frac{1+e}{1-e}\left(\frac{2h}{g}\right)^{1/2}.$$

5-6 In a nonrelativistic elastic scattering of two particles (with masses m_1, m_2), obtain the value of $|\vec{p}_{1f}|/|\vec{p}_{1i}|$ as a function of lab scattering angle θ_1. Here note that, if the mass of the bombarding particle is larger than the target particle mass, there exist *two* possible values of $|\vec{p}_{1f}|/|\vec{p}_{1i}|$ for given θ_1. Give physical explanations for these two possibilities.

5-7 Show that, in an elastic 2-body scattering process, the angle $\theta_1 + \theta_2$ between the emerging particles in the lab system is related to the recoil angle θ_2 by

$$\frac{\tan(\theta_1 + \theta_2)}{\tan \theta_2} = \frac{m_1 + m_2}{m_1 - m_2}.$$

Here, m_1 (m_2) denotes the mass of the bombarding (target) particle and nonrelativistic kinematics may be assumed.

5-8 In a *relativistic* 2-body elastic scattering, show that, if E_1^* and E_2^* denote relativistic energies of the two incident particles in the center-of-mass system, the lab and CM scattering angles (denoted θ_1 and θ^*, respectively) are related by the equation

$$\tan \theta_1 = \frac{m_{02} c^2 \sin \theta^*}{E_1^* + E_2^* \cos \theta^*}.$$

Also, based on this connection, obtain the formula which produces the lab scattering cross section $\frac{d\sigma}{d\Omega_{\text{Lab}}}(\theta_1, \varphi_1)$ from the corresponding CM scattering cross section $\frac{d\sigma}{d\Omega_{\text{CM}}}(\theta^*, \varphi^*)$.

5-9 Consider the Compton scattering in which a photon having energy E collides with a stationary electron (with rest mass m_e). As a result of the collision the direction of the photon's motion is deflected through an angle θ and its energy is reduced to E'. Then, prove that

$$m_e c^2 \left(\frac{1}{E'} - \frac{1}{E} \right) = 1 - \cos \theta.$$

From this, deduce that the wavelength λ of the photon, given by the formula $\lambda = \frac{h}{|\vec{p}|} = \frac{hc}{E}$ (h is Planck's constant), is increased by

$$\Delta \lambda = \frac{2h}{m_e c} \sin^2 \frac{\theta}{2}.$$

5-10 Find the lab differential cross section for the nonrelativistic elastic scattering of two hard spheres (having radii r_1 and r_2, respectively). Also calculate the total cross section.

5-11 Find the location of the center of mass for the following objects:

(i) A disk of uniform density which has a hole cut out of it, as shown in the figure (left).

(ii) A uniform plate of the form shown in the figure (right).

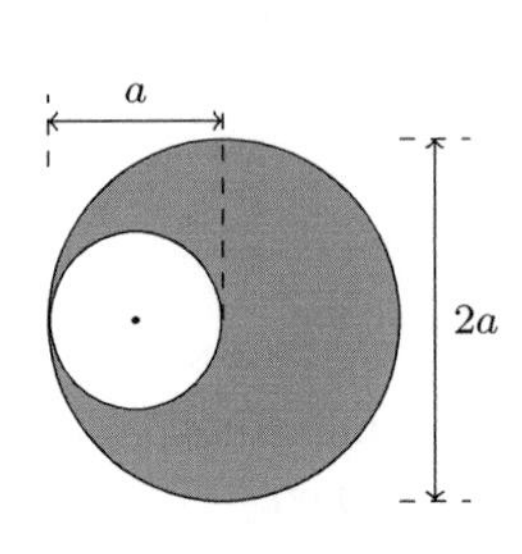

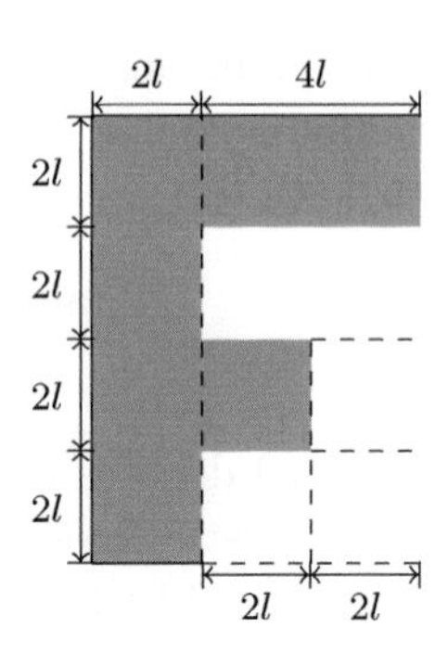

(iii) A solid tetrahedron of uniform mass distribution.

(iv) A homogeneous circular pyramid with base radius a and height h.

5-12 A spherical satellite of radius r is moving with velocity v through a uniform tenuous atmosphere of density ρ_{atm}. Find the retarding force on the satellite if each particle which strikes it (i) adheres to the surface, and (ii) bounces off from it elastically. [Remark: In both situations, the retarding force of $\rho_{\text{atm}}v^2$ times the hard-sphere scattering cross section πr^2 is obtained.]

5-13 A rocket is to be fired vertically upward. The initial mass is M_0, the exhaust velocity (relative to the rocket) u_0 is constant, and the rate of exhaust $-(\frac{dM}{dt}) = A$ is constant. After a total amount of mass ΔM is exhausted, the rocket engine runs out of fuel.

(i) Neglecting air resistance and assuming that the acceleration g of gravity is constant, set up and solve the equation of the rocket motion.

(ii) Show that if M_0, u_0 and ΔM are fixed, then the larger the rate of exhaust A, that is, the faster it uses up its fuel, the greater the maximum altitude reached by the rocket.

5-14 For certain mechanical systems one can study the dynamics rather simply by using the conservation of total energy, while the direct analysis based on Newtonian equations of motion is less obvious. As such an example, consider a uniform flexible chain of mass M and length $2l$. It hangs in equilibrium over a smooth horizontal pin of negligible diameter. Then one end of the chain is given a small vertical displacement so that the chain starts to slip over the pin. See the figure.

(i) Show that, when the longer portion of the chain is of length x, then

$$\dot{x}^2 = \frac{g}{l}(x-l)^2.$$

(ii) Find expressions for the tension T in the chain (at the top position).

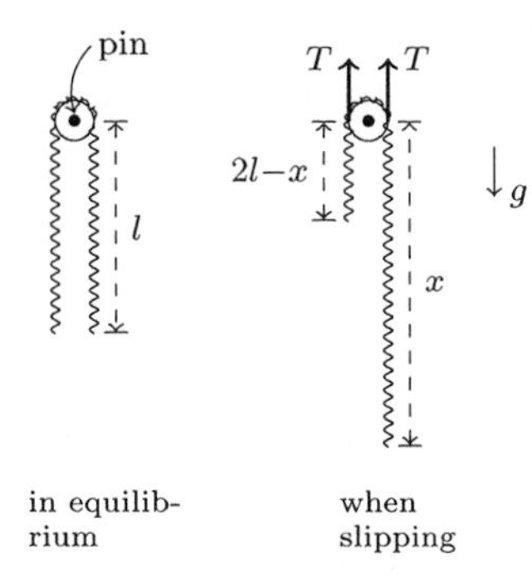

5-15 Threads of lengths h_1, h_2 and h_3 are fastened to the vertices of a homogeneous triangular plate of weight W. The other ends of the threads are fastened to a common point, as shown in the figure. What is the tension in each thread, expressed in terms of the lengths of the threads and the weight of the plate?

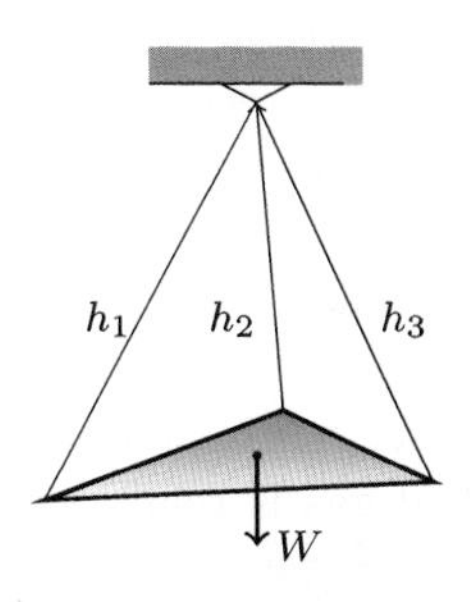

5-16 Show that, in a conservative N-body system, a state of minimal total energy for a *given* total z-component of angular momentum is necessarily one in which the system is rotating as a rigid body about z-axis. [Hint: One may use the method of Lagrange multipliers, and treat the components of the positions $\vec{r}_i$ and velocities $\dot{\vec{r}}_i$ as independent variables.]

5-17 Point masses (all with mass m) are at rest at the corners of a regular n-polygon, as illustrated in the figure for $n = 6$.

(i) How does this system move if only gravitation acts between the bodies?

(ii) How much time elapses before the bodies collide if $n = 2, 3$ and 10? Also examine the limiting case when $n \gg 1$ and $m = M_0/n$, where M_0 is a given total mass.

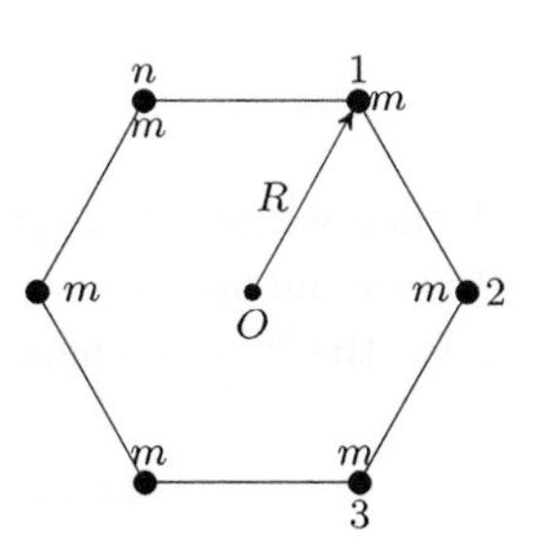

5-18 A proton beam of (relativistic) kinetic energy T enters a hydrogen bubble chamber. Find the threshold energy $T_{\rm th}$ for producing antiprotons in the reaction

$$p + p \to p + p + p + \bar{p},$$

where the target proton is assumed to be at rest. [The rest energy of the proton (p) and of the antiproton ($\bar{p}$) is 938 MeV.]

5-19 An atom at rest in the inertial frame S emits a photon of frequency ν. In so doing, let the initial (rest) energy E of the atom be reduced by an amount ΔE so that $E - \Delta E$ may become the new rest energy of the atom. (Here don't forget that, as the photon is emitted, the atom will recoil (relativistically) and so not remain at rest in S.) Then show that

$$\nu = \frac{\Delta E}{h}\left(1 - \frac{\Delta E}{2E}\right).$$

5-20 Under nonrelativistic kinematics we showed that the pressure of an ideal gas is related to the average kinetic energy $\bar{K}$ by the formula $P = \frac{2}{3}\frac{N}{V}\bar{K}$. Now let us consider an ultrarelativistic gas (such as a photon gas) in which particles in the gas can be assumed to have the kinetic energy given by $\mathcal{E} = |\vec{p}|c$ if the particle has a relativistic momentum $\vec{p}$. In such an ultrarelativistic gas, show that the pressure can be expressed by the formula $P = \frac{1}{3}\frac{N}{V}\bar{\mathcal{E}}$, where $\bar{\mathcal{E}}$ denotes the corresponding average kinetic energy.

5-21 Demonstrate that the Maxwell–Boltzmann speed distribution (see Sec. 5.4) leads to the following average values:

$$\bar{v} \equiv \int_0^\infty v\Phi(v)dv = \sqrt{\frac{8k_BT}{\pi m}},$$

$$\overline{(v^2)} \equiv \int_0^\infty v^2\Phi(v)dv = \frac{3k_BT}{m}.$$

(Hence we get $\overline{K} \equiv \frac{1}{2}m\overline{(v^2)} = \frac{3}{2}k_BT$.) Show also that the Maxwell–Boltzmann speed distribution implies the energy distribution (here E is for the kinetic energy)

$$\Phi_E(E)dE = 2\sqrt{\frac{E}{\pi}}\left(\frac{1}{k_BT}\right)^{3/2} e^{-E/k_BT}dE.$$

6 Gravitational Field Equations

SUMMARY

6.1 Field description for Newtonian gravity

- Consider any small object moving under the influence of gravitational attraction due to some mass distribution in space, described by mass density $\rho_m(\vec{r})$. Then the equation governing the motion of the small object is, regardlessly of its mass,

$$\ddot{\vec{r}} = \vec{g}(\vec{r}) = -\vec{\nabla}\mathcal{G}(\vec{r}),$$

where $\vec{g}(\vec{r})$ and $\mathcal{G}(\vec{r})$, the gravitational field and gravitational potential due to the given mass distribution $\rho_m(\vec{r})$, respectively, are given by

$$\vec{g}(\vec{r}) = -G \iiint \frac{\rho_m(\vec{r}\,')(\vec{r} - \vec{r}\,')}{|\vec{r} - \vec{r}\,'|^3} d^3\vec{r}\,',$$
$$\mathcal{G}(\vec{r}) = -G \iiint \frac{\rho_m(\vec{r}\,')}{|\vec{r} - \vec{r}\,'|} d^3\vec{r}\,'.$$

As a special case, if there is a spherically symmetric mass distribution about the spatial origin, i.e., $\rho_m(\vec{r}\,') = \rho_m(r')$ with $r' = \sqrt{x'^2 + y'^2 + z'^2}$, then the gravitational field is found to be

$$\vec{g}(\vec{r}) = -G\frac{M(r)}{r^3}\vec{r},$$

where $M(r) \equiv \int_0^r \rho_m(r')4\pi r'^2 dr'$ is the total mass existing within a sphere of radius $r = |\vec{r}|$.

- Gravitational field and potential may be specified through local differential equations. For a given mass distribution

$\rho_m(\vec{r})$, the corresponding gravitational field $\vec{g}(\vec{r})$ should satisfy the field equations

$$\vec{\nabla} \times \vec{g} = 0, \quad \vec{\nabla} \cdot \vec{g} = -4\pi \rho_m.$$

The "curl" equation, $\vec{\nabla} \times \vec{g} = 0$, just asserts that $\vec{g}(\vec{r})$ is a conservative vector field. The second "divergence" equation, $\vec{\nabla} \cdot \vec{g} = -4\pi\rho_m$, can be deduced using Gauss's law

$$\iint_{S=\partial V} \vec{g} \cdot \hat{n}\, da = -4\pi G \iiint_V \rho_m(\vec{r}) d^3\vec{r}$$

($S = \partial V$ denotes the boundary surface of an arbitrarily chosen volume V) and the divergence theorem

$$\iint_{S=\partial V} \vec{g} \cdot \hat{n}\, da = \iiint_V \vec{\nabla} \cdot \vec{g}\, d^3\vec{r}.$$

As $\vec{g}(\vec{r})$, a conservative vector field, can always be written in the form $\vec{g}(\vec{r}) = -\vec{\nabla}\mathcal{G}(\vec{r})$, the above field equations for $\vec{g}$ can be turned into Poisson's equation for the gravitational potential

$$\vec{\nabla}^2 \mathcal{G} = 4\pi G \rho_m,$$

where $\vec{\nabla}^2 \equiv \vec{\nabla} \cdot \vec{\nabla}$ is the Laplacian operator. In the region where ρ_m vanishes, the gravitational potential satisfies the Laplace equation, $\vec{\nabla}^2 \mathcal{G} = 0$. This approach is useful not only for purely theoretical reason but also for gravitational boundary value problems.

6.2 Gravitational potential in the large and tidal force

- Suppose we have some localized mass distribution $\rho_m(\vec{r})$ in the vicinity of the spatial origin. Then, with the origin chosen at the CM of the mass distribution, the related gravitational potential at a distance can be described approximately by the form

$$\mathcal{G}(\vec{r}) = -G \left[\frac{M}{r} + \frac{1}{2} \sum_{i,j} Q_{ij} \frac{x_i x_j}{r^5} + O\left(\frac{1}{r^4}\right) \right],$$

where M $(= \int d^3\vec{r}'\rho_m(\vec{r}'))$ is the total mass, and Q_{ij} (gravitational quadrupole moments) are given by

$$Q_{ij} = Q_{ji} = \int d^3\vec{r}'(3x_i'x_j' - \vec{r}'^2\delta_{ij})\rho_m(\vec{r}').$$

If the mass distribution happens to be axially symmetric (as is the case for the Earth which is flattened at the poles), one may choose the z-axis in the direction of the symmetry axis and use spherical coordinates to write $\rho_m(\vec{r}') = \rho_m(r', \theta')$. Then the above quadrupole-dependent potential is simplified as

$$\mathcal{G}_{\text{quad}}(\vec{r}) = -\frac{GQ}{4r^3}(3\cos^2\theta - 1)$$

where $Q \equiv Q_{33} = \int d^3\vec{r}'(2z'^2 - x'^2 - y'^2)\rho_m(r', \theta')$. Note that, for the Earth which is approximately an oblate spheroid, we have $Q < 0$.

- Given a certain localized mass distribution in the presence of a smooth, but not strictly uniform, external gravitational potential $\mathcal{G}_{\text{ex}}(\vec{r})$, it is possible to express its potential energy approximately by

$$V_{\text{ex}}(\vec{R}) = M\mathcal{G}_{\text{ex}}(\vec{R}) - \frac{1}{6}Q_{ij}\frac{\partial g_{\text{ex},j}(\vec{R})}{\partial R_i} + \cdots.$$

The quadrupole moment thus couples with the external field gradient.

- On an extended body like the Earth, there are tidal effects caused by not-strictly-uniform gravitational fields due to massive objects not too far away, such as the Moon or the Sun. If we denote the position of the Moon (or Sun) from the Earth's center by $\vec{R}$ and its mass by m, the gravitational potential at a point $\vec{r}$ on the Earth is approximately given as

$$\begin{aligned}\mathcal{G}(\vec{r}) &= -\frac{Gm}{|\vec{R} - \vec{r}|} \\ &= -Gm\left[\frac{1}{R} + \frac{r}{R^2}\cos\theta + \frac{r^2}{2R^3}(3\cos^2\theta - 1) + O\left(\frac{1}{R^4}\right)\right],\end{aligned}$$

where θ is the angle between $\vec{r}$ and $\vec{R}$. From the term proportional to $r^2(3\cos^2\theta - 1)$ which describes the leading tidal effects, we are led to the tidal gravitational field with component expressions

$$g_r = \frac{Gmr}{R^3}(3\cos^2\theta - 1), \quad g_\theta = -\frac{3Gmr}{R^3}\cos\theta\sin\theta.$$

This expression, together with Earth's rotation, can be used to account for some basic features of ocean tides, including the twice-daily periodicity of the tides.

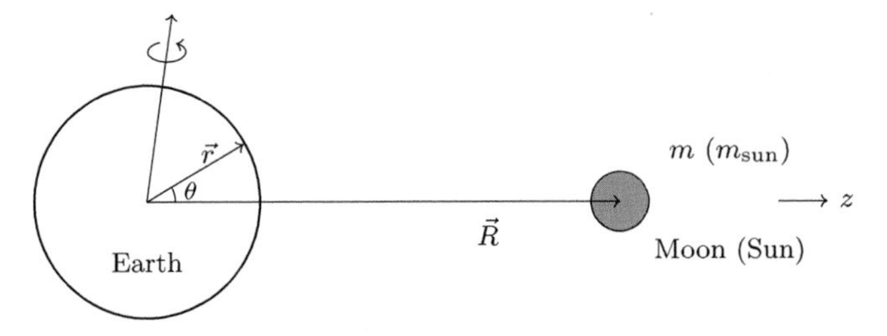

6.3 The equivalence principle and general relativity

- The Newtonian theory of gravitation, with the instantaneous action at a distance, is not consistent with the relativity principle of Einstein. Also, from the equality between the inertial mass and the gravitational mass, one is led to the equivalence principle, that is, to the physical equivalence (or interchangeability) of gravitational and inertial accelerations.
- Demanding that physical phenomena inside a small laboratory freely falling in a gravitational field (as seen by an observer falling with the laboratory) are described by the same physical laws that apply to corresponding phenomena as observed in an inertial frame in the absence of a gravitational field, one is led to the metric theory of gravity: in the presence of a gravitational field, the invariant line element for the spacetime is altered from the Minkowski line element $ds^2 = -c^2dt^2 + dx^2 + dy^2 + dz^2$ to a more general

form involving a nontrivial metric tensor $g_{\mu\nu}(x)$

$$ds^2 = \sum_{\mu=0}^{3}\sum_{\nu=0}^{3} g_{\mu\nu}dx^\mu dx^\nu,$$

where $x^\mu = (x^0 \equiv ct, x^1 \equiv x, x^2 \equiv y, x^3 \equiv z)$. The full information on the gravitational field is contained in the metric tensor. [Note that, in a sufficiently small neighborhood of a given spacetime point, it is always possible to find a coordinate system with respect to which the line element at least in that small neighborhood assumes the Minkowskian form.]

- Einstein obtained (generally covariant) dynamical laws for the metric tensor by demanding that the so-called geodesic equation for the spacetime geometry yield an equation comparable to the Newtonian equation of motion for a test particle in a weak gravitational field. This leads to Einstein's field equations (in general relativity).

Problems

6-1 Assume that the density of a star is a function only of the radius r measured from the center of the star, and is given by

$$\rho_m(r) = \frac{Ma^2}{2\pi r(r^2 + a^2)^2}, \quad 0 \le r < \infty,$$

where M is the mass of the star, and a is a constant which determines the size of the star. For this mass distribution, find the gravitational field $\vec{g}(\vec{r})$ and the gravitational potential $\mathcal{G}(\vec{r})$ in space.

6-2 A uniform solid sphere of mass M and radius R is fixed a distance h above a thin infinite sheet of surface mass density σ. With what force does the sheet attract the sphere?

6-3 A sphere of uniform density ρ has within it a spherical cavity whose center is at a distance a from the center of the sphere. Show that the gravitational field within the cavity is uniform, and determine its magnitude and direction.

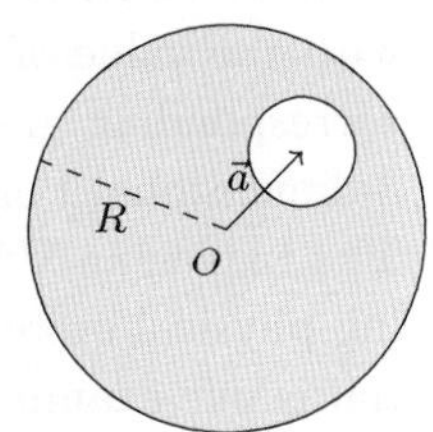

6-4 A system of point masses m_i $(i = 1, \ldots, N)$ are located initially far apart from one another. Show that the work done in bringing them to some given positions $\vec{r}_1, \ldots, \vec{r}_N$ can be identified with $\frac{1}{2}\sum_{i=1}^{N} m_i \mathcal{G}(\vec{r}_i)$, where $\mathcal{G}(\vec{r}_i)$ is the gravitational potential at the position of the ith particle due to all the other masses. [Why a factor $\frac{1}{2}$ in front?]. Also generalize the expression to the case of a continuous mass distribution with mass density $\rho_m(\vec{r})$.

6-5 A hole is bored in a straight line through the Earth from Seoul to Paris, and a ball bearing is dropped in at the Seoul end. Assuming that frictional and air resistance forces are negligible, and that the Earth may be taken as a uniform-density sphere of radius 6400 km, how long does it take the ball bearing to arrive at Paris? [Neglect any effects due to the rotation of the Earth, and assume the acceleration due to gravity at the Earth's surface to be $9.8\,\mathrm{m\,s^{-2}}$.]

6-6 In 3 dimensions the gravitational field due to any spherically symmetric mass distribution is exactly equal to that of total (interior) mass accumulated at its center position — a feature that originates from the inverse-square-law nature of the gravitational force. Now imagine an n-dimensional Euclidean space. Then, assuming that the force is proportional to the inverse pth power of the distance between the masses and directed along a straight line joining them, one may represent the gravitational field — an n-dimensional vector field — at field point $\vec{r} = (x_1, x_2, \ldots, x_n)$ due to some mass distribution $\rho(\vec{r}\,')$ by an expression like

$$\vec{g}(\vec{r}) = -\kappa \int \frac{\rho(\vec{r}\,')(\vec{r} - \vec{r}\,')}{|\vec{r} - \vec{r}\,'|^{p+1}} d^n\vec{r}\,',$$

where κ is a constant and $|\vec{r} - \vec{r}\,'| \equiv \sqrt{(x_1 - x_1')^2 + \cdots + (x_n - x_n')^2}$. In this n-dimensional world, determine the value of power p by requiring that the gravitational field due to a hyperspherically symmetric mass distribution (i.e., when $\rho(\vec{r})$ can be written, say, as $\rho(r)$ with $r = \sqrt{(x_1)^2 + \cdots + (x_n)^2}$), in the region outside the mass distribution, be equal to a point mass located at its center position. Also, with the value of p fixed by this requirement, can you write for the corresponding gravitational potential an equation analogous to the 3-dimensional Poisson's equation?

6-7 The Laplacian $\vec{\nabla}^2 \equiv \frac{\partial^2}{\partial x^2} + \frac{\partial^2}{\partial y^2} + \frac{\partial^2}{\partial z^2}$ often enters equations describing physical systems. Poisson's equation for the gravitational potential is one example. Give three other examples of equations with the

Laplacian, and suggest a kind of general reason(s) why the Laplacian appears in physical equations so often.

6-8 Let the mass density be nonzero only in the region $z < 0$. We then wish to find the gravitational potential $\mathcal{G}(x,z)$ in the region $z > 0$, which reduces to some given function $\mathcal{G}(x, z = 0)$ on the boundary.

(i) Using a Fourier integral expansion of the potential in the variable x

$$\mathcal{G}(x,z) = \int_{-\infty}^{\infty} \widetilde{\mathcal{G}}(k,z) e^{ikx} dk$$

with the Laplace equation, show that the Fourier component $\widetilde{\mathcal{G}}(k,z)$ must satisfy for $z > 0$ the following differential equation

$$\frac{\partial^2 \widetilde{\mathcal{G}}(k,z)}{\partial z^2} - k^2 \widetilde{\mathcal{G}}(k,z) = 0.$$

(ii) Based on the information provided in (i), show that we can write the solution to the given boundary value problem in the form

$$\mathcal{G}(x,z) = \int_{-\infty}^{\infty} \tilde{\mathcal{G}}(k, z{=}0)\, e^{ikx} e^{-k|z|} dk,$$

where $\tilde{\mathcal{G}}(k, z{=}0) = \frac{1}{2\pi} \int \mathcal{G}(x, z{=}0)\, e^{-ikx} dx$. (This is (6.43) in the text.)

6-9 The Earth has approximately the shape of an oblate spheroid whose polar diameter $2a(1-\eta)$ is slightly shorter than its equatorial diameter $2a$. (Roughly, $\eta \approx 0.0034$.) Assume that the mass density, taken to be uniform over the spheroid, is equal to ρ_0. Then the gravitational potential $\mathcal{G}(r,\theta)$ (here the Earth's center corresponds to $r = 0$, and θ is the colatitude) far from the earth, i.e., for $r \gg a$ can approximately be described by the form

$$\mathcal{G}(r,\theta) = -G\frac{M}{r} - \frac{GQ}{4r^3}(3\cos^2\theta - 1).$$

Determine M and Q in terms of a, η and ρ_0. (It is sufficient to consider these quantities to first order in η.)

6-10 Let us consider an Earth satellite (of mass m) moving in a nearly circular orbit (say, of radius a_0) in the equatorial plane. Based on the approximate gravitational potential found in Problem 6-9, calculate the rate of precession of the perigee (i.e., the point of closest approach) for this satellite.

6-11 A satellite moves in a circular orbit of radius a_0 whose plane is inclined so that its normal makes an angle α with the polar axis of the Earth.

(i) Using the approximate gravitational potential found in Problem 6-9, calculate the average torque on the satellite during a revolution. (Assume that the orbit of the satellite is very little affected in one revolution.)

(ii) Show that the effect of such a torque is to make the normal to the orbit precess in a cone of half angle α about the polar axis, and find a formula for the rate of precession in degrees per revolution.

6-12 Suppose we have a small axially symmetric object of a prolate spheroidal shape, with the quadrupole moment $Q > 0$, in an external gravitational field due to a (distant) fixed point mass M. (Here, Q refers to the value of Q_{33} in the gravitational quadrupole moment when the symmetry axis is taken as the 3rd axis.) When the symmetry axis is free to change its direction, show (using the formula (6.60) in the text for instance) that the gravitational potential is lowest when the symmetry axis points in the direction of the fixed point mass.

6-13 The tidal bulge in the ocean may be regarded as adding a surface mass distribution of the form $\sigma = \sigma_0 \left(\frac{3}{2}\cos^2\theta - \frac{1}{2}\right)$ (θ denotes the angle between the lines from the Earth's center to the given point on the surface and to the Moon or Sun, see Sec. 6.2), positive near $\theta = 0$ and π and negative near $\theta = \frac{\pi}{2}$. Find the quadrupole moment of this distribution. Then, taking into account the resulting gravitational potential, determine how much the effective height of the tides is increased over the expression $h(\theta) \sim \frac{mr^4}{MR^3}\left(\frac{3}{2}\cos^2\theta - \frac{1}{2}\right)$ (which was found in the text). Give approximate numerical values for your expression.

6-14 Two small identical uniform spheres of density ρ and radius r are orbiting the Earth in a circular orbit of radius a ($\gg r$). Given that the spheres are just touching, with their centers in line with the Earth's center, and that the only force between them is gravitational, show that they will be pulled apart by the Earth's tidal force if a is less than $a_c = 2\left(\rho_E/\rho\right)^{\frac{1}{3}} R_E$, where ρ_E is the mean density of the Earth, and R_E its radius.

6-15 To understand the gravitational shift of spectral lines (as first predicted by Einstein in 1907), suppose that (at $t = 0$) a laboratory of height h is released from rest into free fall and let a photon of frequency ν_e $(= \frac{\omega_e}{2\pi})$ be emitted from an atom at the floor at $t = 0$ and travel upwards (see the figure). Then, according to the equivalence principle, it hits the ceiling at time $t = \frac{h}{c}$ and an observer fixed to the ceiling (and hence falling with the laboratory) should see the photon with frequency ν_e again. As it reaches the position of the ceiling, now let ν_o be the frequency of the photon as measured by an observer outside the falling laboratory (who is moving upwards relative to the laboratory, with speed $v = gt = gh/c$). Here assume that $v \ll c$.

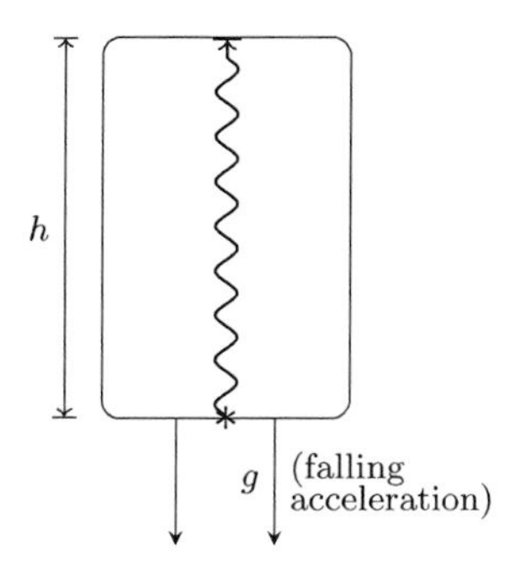

(i) Taking into account the fact that the outside observer sees a Doppler-shifted photon (see footnote 16, Chapter 5 in the text), show that there should exist a frequency difference $\Delta\nu \equiv \nu_o - \nu_e$ as given by

$$\frac{\Delta\nu}{\nu_e} \sim -\frac{v}{c} = -\frac{gh}{c^2}.$$

(Note that gh is the change in the Newtonian gravitational potential experienced by the photon.)

(ii) What do you expect for a photon falling from the ceiling rather than moving upwards?

7 Rigid Body Dynamics I

SUMMARY

7.1 Basic principles and rotational equilibrium

- For the motion of a rigid body, we have two basic equations, viz.,

$$\dot{\vec{P}} = M\ddot{\vec{R}} = \sum_i \vec{F}_i^{(e)} \quad (\equiv \vec{F}^{(e)})$$

for its CM motion, and

$$\dot{\vec{L}} = \sum_i \vec{r}_i \times \vec{F}_i^{(e)} \quad (\equiv \vec{\Gamma}^{(e)})$$

for rotational motion about some fixed origin or about the CM of the body. In a conservative external force $\vec{F}_i^{(e)} = -[\vec{\nabla}V_{\text{ext}}(\vec{r})]\Big|_{\vec{r}=\vec{r}_i}$, a consequence of these equations is the energy conservation law

$$T + V = \left(\frac{1}{2}\sum_i m_i \dot{\vec{r}}_i^{\,2}\right) + \left(\sum_i V_{\text{ext}}(\vec{r}_i)\right) = E \quad (\text{: const.}).$$

Notice that there is no potential energy change associated with the internal forces in a rigid body as it moves.

- For a rigid body in static equilibrium, the necessary and sufficient conditions are the fulfillment of

$$\sum_i \vec{F}_i^{(e)} = 0 \quad \text{(: force balance equation)},$$

$$\sum_i \vec{r}_i \times \vec{F}_i^{(e)} = 0 \quad \text{(: torque balance equation)}.$$

7.2 Fixed-axis rotation

- The motion of a rigid body rotating about a fixed axis in space (the z-axis) is governed by a simple one-dimensional equation

$$\frac{dL_z}{dt} = I_z\dot{\omega} = \Gamma_z^{(e)}$$

where $\omega \equiv \dot{\varphi}$ denotes the angular velocity of the body and I_z, the moment of inertia about the z-axis, is a constant given by

$$I_z = \sum_i m_i(x_i^2 + y_i^2)$$

or, in terms of mass density $\rho_m(x, y, z)$, by the integral

$$I_z = \iiint_{\text{(body)}} \rho_m(x, y, z)(x^2 + y^2)\, d^3\vec{r}.$$

- As a special case of fixed-axis rotation, we have a compound (or physical) pendulum, the problem involving an arbitrary rigid body pivoted about a horizontal axis and moving under gravity. The related equation of motion becomes

$$I_z\ddot{\varphi} = -MgR\sin\varphi,$$

where R is the distance from the pivot point O to the CM of the body. This is of the same form as the equation satisfied by a simple pendulum of length $l = I_z/MR$. The point O' at a distance $l = I_z/MR$ from the pivot O along the line through the CM is called the *center of oscillation* and the period of small oscillation is given by $T = 2\pi\sqrt{\frac{l}{g}} = 2\pi\sqrt{\frac{I_z}{MgR}}$.

- In a compound pendulum, the force $\vec{F}'$ given to the body from the axis (and the corresponding reaction force, $-\vec{F}'$, given to the axis) is determined by the CM equation for the body

$$M\frac{d^2\vec{R}}{dt^2} = M\vec{g} + \vec{F}'.$$

From this one then finds that the center of oscillation (O') is also the *center of percussion*; no impulsive reaction is felt at the pivot (O) if a sudden blow is given at the point O', i.e., at the center of percussion.

- For the calculation of the moment of inertia, following theorems may prove useful:
  (i) (Parallel axis theorem) The moment of inertia of a body about any given axis is the moment of inertia about a parallel axis through the CM plus the moment of inertia about the given axis if all the mass of the body were located at the CM.
  (ii) (Perpendicular axis theorem) The sum of the moments of inertia of a plane lamina about any two perpendicular axes in the plane is equal to the moment of inertia about an axis through their point of intersection perpendicular to the lamina.

- If the external force, not including the force given at the axis, is conservative, i.e., derivable from a potential V, one can also discuss the motion of a rigid body undergoing fixed-axis rotation on the basic of the energy conservation law; for instance, for a cylinder of radius a rolling down an inclined plane, we have the equation

$$\frac{1}{2}M\dot{x}^2 + \frac{1}{2}I_z\dot{\varphi}^2 - Mgx\sin\theta_0 = E \quad (\text{: const.}),$$

where θ_0 is the inclination angle of the plane, and x ($= a\varphi$, from the condition of rolling) represents the displacement at time t of the center of the cylinder along the inclined plane.

7.3 Rotation of a rigid body about an arbitrary axis

- To discuss general rotation of a rigid body about some fixed point or about the CM, consider a body-fixed triad $(\mathbf{e}'_1, \mathbf{e}'_2, \mathbf{e}'_3)$ — a frame moving with the rigid body and related to the space-fixed frame $(\mathbf{e}_1, \mathbf{e}_2, \mathbf{e}_3)$ by a certain time-dependent orthogonal transformation. The position of any particle of the rigid body is represented relative to this body-fixed frame as

$$\vec{r}(t) = x'_1\mathbf{e}'_1(t) + x'_2\mathbf{e}'_2(t) + x'_3\mathbf{e}'_3(t)$$
$$(x'_1, x'_2, x'_3\text{: constant scalars}).$$

For the velocity, i.e., for $\dot{\vec{r}}(t) = x'_1\dot{\mathbf{e}}'_1(t) + x'_2\dot{\mathbf{e}}'_2(t) + x'_3\dot{\mathbf{e}}'_3(t)$, one then has the equation

$$\dot{\vec{r}}(t) = \vec{\omega}(t) \times \vec{r}(t),$$

where $\vec{\omega}(t)$, representing the angular velocity of the rigid body, is a vector that can be chosen to satisfy the equations

$$\frac{d}{dt}\mathbf{e}'_i(t) = \vec{\omega}(t) \times \mathbf{e}'_i(t) \quad (i = 1, 2, 3).$$

- Thanks to the formula $\dot{\vec{r}} = \vec{\omega} \times \vec{r}$, the angular momentum of a rigid body can be represented as

$$\vec{L} = \sum_i m_i[r_i^2\vec{\omega} - (\vec{r}_i \cdot \vec{\omega})\vec{r}_i].$$

This shows that, for the components of the vector $\vec{L}$ and $\vec{\omega}$ *relative to a body-fixed frame*, we have the relationship

$$L_m = \sum_{n=1}^{3} I_{mn}\omega_n \quad (m = 1, 2, 3)$$

where we defined $I_{mn} = \sum_i m_i[r_i^2\delta_{mn} - (x_m)_i(x_n)_i]$ $(m, n = 1, 2, 3)$. (Here, x_m, x_n refer to coordinates relative to the body-fixed axes.) Regarding the nine elements I_{mn} as components of a (rank-2) tensor $\mathbf{I}$ $(= I_{mn}\mathbf{e}_m\mathbf{e}_n)$, this relationship can also be presented in the form $\mathbf{L} = \mathbf{I}\cdot\boldsymbol{\omega}$, an equation without reference to any specific body-fixed basis. Similarly, the kinetic energy of a rigid body can be expressed in terms of the angular velocity and the inertia tensor:

$$T = \frac{1}{2}\vec{\omega} \cdot \vec{L} = \frac{1}{2}\sum_{m,n} I_{mn}\omega_m\omega_n \quad \left(\equiv \frac{1}{2}\boldsymbol{\omega} \cdot \mathbf{I} \cdot \boldsymbol{\omega}\right).$$

- A rigid body can be in the state of free rotation (about some given axis in space), i.e., execute a fixed-$\vec{\omega}$ rotation in the absence of any external torque, if the angular momentum of the body happens to point in the same direction as $\vec{\omega}$ (i.e., when $\vec{\omega} \times \vec{L} = 0$). In a generic case, such state is possible only when $\vec{\omega}$ points in the direction of certain particular body axes, called principal axes, of the given body. For an arbitrary rigid body, three, mutually perpendicular, principal axes can always be found. If one chooses those three principal axes for one's body-fixed basis $(\mathbf{e}_1, \mathbf{e}_2, \mathbf{e}_3)$, the inertia tensor can be represented by a diagonal form $\mathbf{I} = I_1\mathbf{e}_1\mathbf{e}_1 + I_2\mathbf{e}_2\mathbf{e}_2 + I_3\mathbf{e}_3\mathbf{e}_3$ or by a diagonal matrix

$$I \equiv (I_{mn}) = \begin{pmatrix} I_1 & 0 & 0 \\ 0 & I_2 & 0 \\ 0 & 0 & I_3 \end{pmatrix}$$

(I_1, I_2 and I_3 are called principal moments of inertia). In this basis, the relation $\mathbf{L} = \mathbf{I} \cdot \boldsymbol{\omega}$ simplifies to

$$L_1 = I_1\omega_1, \quad L_2 = I_2\omega_2, \quad L_3 = I_3\omega_3,$$

and the kinetic energy also acquires the simple expression

$$T = \frac{1}{2}I_1\omega_1^2 + \frac{1}{2}I_2\omega_2^2 + \frac{1}{2}I_3\omega_3^2.$$

7.4 General methods of finding principal axes of inertia

- Given any real symmetric tensor $\mathbf{T} = T_{ij}\mathbf{e}_i\mathbf{e}_j$ such as our inertia tensor, there always exist a set of mutually perpendicular axes with respect to which $\mathbf{T}$ is diagonal, i.e., of the form $\mathbf{T} = \lambda_i\mathbf{e}_i\mathbf{e}_i$. This is the case since, if $\mathbf{T}$ is such a tensor in n-dimensional Euclidean space, one can find a set of n linearly independent eigenvectors $(\mathbf{A}_1, \mathbf{A}_2, \cdots, \mathbf{A}_n)$; here

$\mathbf{A}_l$ is a nonzero solution of the equation $\mathbf{T} \cdot \mathbf{A} = \lambda \mathbf{A}$ with eigenvalue $\lambda = \lambda_l$.

- The equation for an eigenvector, $\mathbf{T} \cdot \mathbf{A} = \lambda \mathbf{A}$, is equivalent to

$$(\mathbf{T} - \lambda \mathbf{1}) \cdot \mathbf{A} = 0 \quad (\text{with } \mathbf{1} = \delta_{ij} \mathbf{e}_i \mathbf{e}_j \equiv \mathbf{e}_i \mathbf{e}_i)$$

and this has a nontrivial solution if λ, one of the eigenvalues, is a root of the related secular or characteristic equation

$$\det(\mathbf{T} - \lambda \mathbf{1}) = \begin{vmatrix} T_{11} - \lambda & T_{12} & \cdots & T_{1n} \\ T_{21} & T_{22} - \lambda & \cdots & T_{2n} \\ \vdots & \vdots & \ddots & \vdots \\ T_{n1} & T_{n2} & \cdots & T_{nn} - \lambda \end{vmatrix} = 0.$$

- For a real symmetric tensor, all of its eigenvalues are real and eigenvectors corresponding to different eigenvalues are necessarily orthogonal to one another. In fact, even when some of the eigenvalues correspond to a multiple root of the characteristic equation, one can show that a complete basis consisting of mutually orthogonal normalized eigenvectors of the tensor exist. For an inertia tensor (defined in 3-dimensional space), we have three principal axes $(\mathbf{e}_1, \mathbf{e}_2, \mathbf{e}_3)$ in direct correspondence with 3 normalized, mutually orthogonal eigenvectors of the tensor; also the eigenvalues of the inertia tensor are identified with principal moments of inertia.
- A rigid body for which two of the principal moments are equal (say, $I_1 = I_2$) is called a symmetric body. Note that, when $\mathbf{e}_3$ is the symmetry axis, the other two principal axes $\mathbf{e}_1$ and $\mathbf{e}_2$ can be any pair of perpendicular axes in the plane normal to $\mathbf{e}_3$.

7.5 Motions of a rigid body without a fixed rotation axis: Discussions

- To specify the configuration of a rigid body at any given time, one should be able to tell as a function of time the orientation of the chosen principal axes $(\mathbf{e}_1, \mathbf{e}_2, \mathbf{e}_3)$ in space. For that purpose, one will have to solve the angular momentum equation $\frac{d\vec{L}}{dt} = \vec{\Gamma}^{(e)}$ (with $\vec{L}(t) = I_1\omega_1(t)\mathbf{e}_1(t) + I_2\omega_2(t)\mathbf{e}_2(t) + I_3\omega_3(t)\mathbf{e}_3(t)$) for given external torque $\vec{\Gamma}^{(e)}$. But this is highly nontrivial, the main difficulty originating from the fact that our principal axes $(\mathbf{e}_1, \mathbf{e}_2, \mathbf{e}_3)$ themselves are rotating. For a systematic study, one may thus rely on more advanced tools to be considered in Chapter 12.

- By applying the formalism of this chapter more carefully, one can verify that the steady precessional motion of a rapidly spinning top in a uniform gravitational field is indeed a true consequence of the angular momentum equation $\frac{d\vec{L}}{dt} = \vec{\Gamma}^{(e)}$.

Problems

7-1 A uniform brick of length L is laid on a smooth horizontal surface. Other equal bricks are piled on (as shown in the figure), so that the sides form continuous planes, but the ends are offset at each brick from the previous brick by a distance L/n, where n is an integer. How many bricks can be used in this manner before the pile topples over?

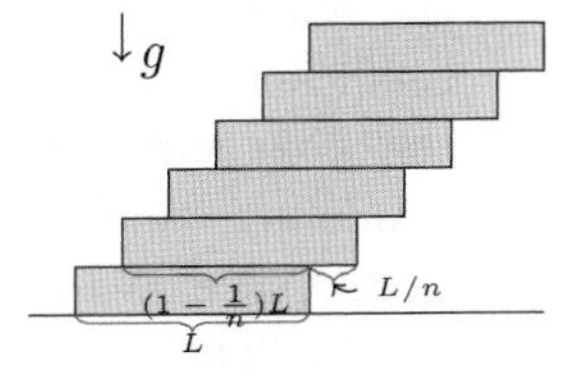

7-2 A light ladder of length L is supported on a rough floor against a smooth, i.e., frictionless wall. The coefficient of static friction at the floor is μ and the inclination of the ladder to the horizontal is α. How far up the ladder can a man of mass M climb without making the ladder slip? Also, how does the answer get modified if both wall and floor are rough (with the same coefficient of friction μ)?

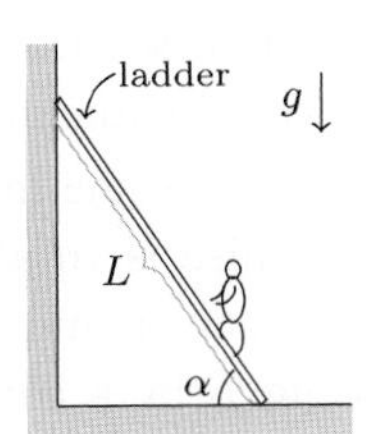

7-3 A uniform cylinder rests on two other identical cylinders, these cylinders resting on a rough horizontal floor. The axes of the three cylinders are parallel and the angle between the lines, drawn in a vertical plane, joining the axes of the upper cylinder to the axes of the lower cylinders is 2α. (See the figure for cross section of the three cylinders). If the coefficient of static friction between the cylinders is μ and between the cylinder and floor is μ', show that the following two inequalities involving the angle α must hold:

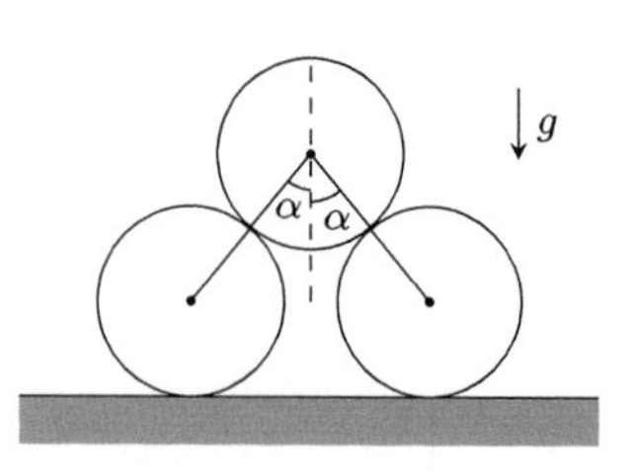

$$\frac{3\sin\alpha}{1+\cos\alpha} \leq \mu, \qquad \frac{\sin\alpha}{1+\cos\alpha} \leq \mu'.$$

7-4 In the figure we have a pole (or uniform ladder) AB, of length $2a$ and total mass M. The end A rests on the ground, and the end B rests against a rough vertical wall. The perpendicular from A to the wall meets it at P, and the angle $\measuredangle BAP = \alpha$. The coefficient of friction between the pole and the wall is μ. Assuming that no slipping takes place at A, determine the inclination θ of PB to the vertical when B is on the point of sliding along the wall.

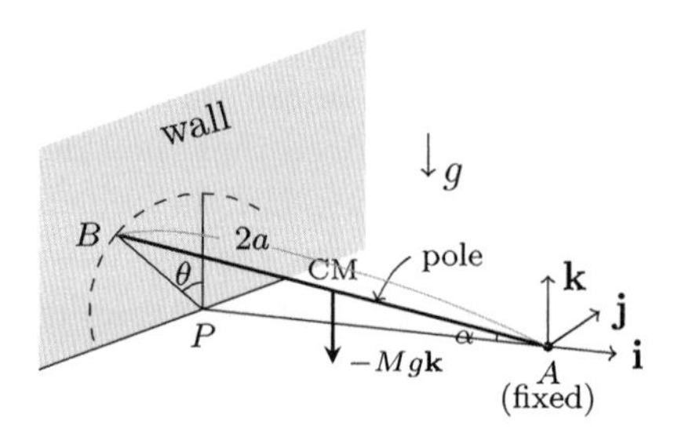

7-5 A uniform circular disc of mass M and radius a rotates freely in a vertical plane about its center O, the vertical plane and center being fixed in some inertial frame. Initially the disc is rotating with an angular velocity ω_0 and is reduced to rest in a time T by a constant frictional force F applied tangentially to the rim of the disc. Show that

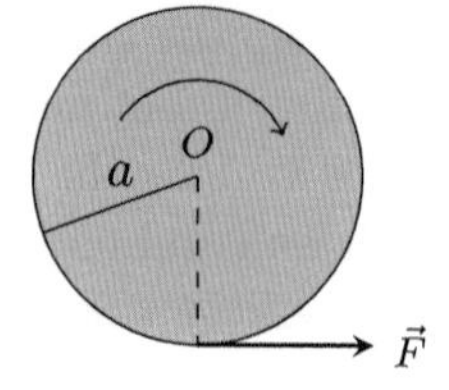

$$F = \frac{Ma\omega_0}{2T}.$$

7-6 Consider a uniform solid cylinder of net mass M (see the figure) which can rotate about the x-axis under the influence of gravity. (Assume that the x-axis meets the cylinder axis at the right angle.) Calculate the moment of inertia I_x and then discuss the allowed motion.

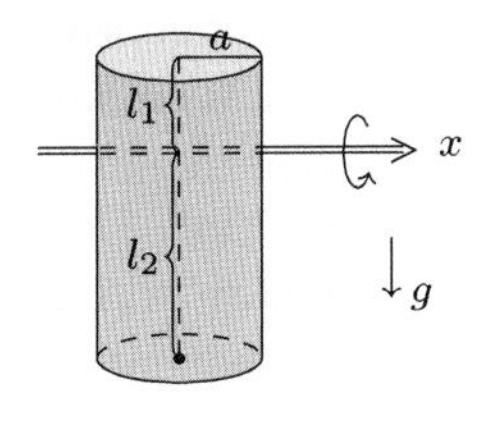

7-7 A uniform rod of mass M and length $2a$ hangs from a smooth hinge at one end. Find the length l of the equivalent simple pendulum. It is struck sharply, with impulse I, at a point a distance b below the hinge. Use the angular momentum equation to find the initial value of the angular velocity. Find also the initial momentum. Determine the point at which the rod may be struck without producing any impulsive reaction at the hinge. Show that, if the rod is struck elsewhere, the direction of the impulsive reaction depends on whether the point of impact is above or below this point.

7-8 On a frictionless horizontal table, a uniform rigid rod of mass M and length $2l$, at rest, is hit by a projectile of mass m and speed v_0. Let the rod be struck perpendicularly at some distance a from its center. (See the figure.)

(i) Assuming that the collision is elastic, find out how the projectile and rod move after the collision. Also determine the distance a that will make the rod acquire maximum angular velocity of rotation.

(ii) How does the answer get modified if the collision is totally inelastic (i.e., the projectile sticks to the rod as it collides)?

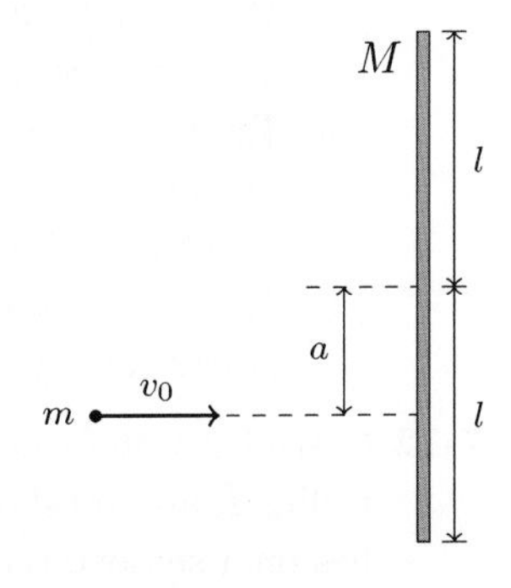

7-9 Equation of motion for a simple pendulum is readily obtained by using the equation for the rate of the total angular momentum, $\dot{\vec{L}} = \vec{\Gamma}^{(e)}$, with $\vec{L}$ and $\vec{\Gamma}^{(e)}$ defined relative to the fixed suspension point. Use the appropriate generalization of the latter formula to find the equation of motion for a simple pendulum whose suspension point O is now undergoing a slow vertical oscillation according to $z_O = a \sin \beta t$.

7-10 A uniform ladder of length $2a$ lies in a vertical plane with one end against a smooth (i.e., frictionless) wall, the other end being supported on a smooth horizontal floor. The ladder is released from the rest when inclined at an angle α to the horizontal (see the figure).

(i) Set up the equations of motion which determine the subsequent motion of the ladder.

(ii) Show that the ladder will cease to touch the wall when its upper end has fallen a distance equal to one third of its original height.

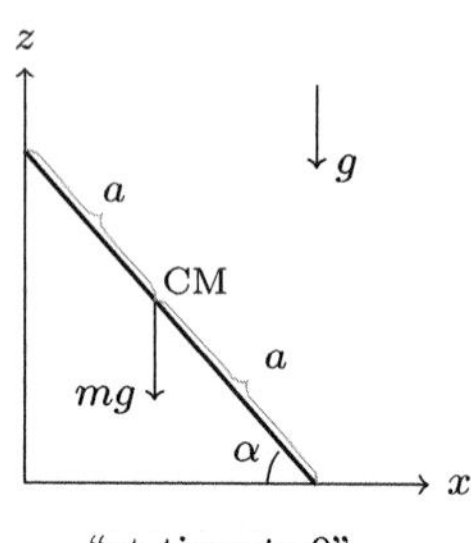

7-11 Consider a solid billiard ball of radius a which is in motion on a horizontal table. The table is sufficiently rough, the coefficient of kinetic friction being equal to μ. At time $t = 0$, we suppose that the center of the ball has a horizontal velocity $\vec{v}_0$ and the ball has an angular velocity $\vec{\omega}_0$. [Assume that, at $t = 0$, the velocity of the particle at the point of contact P is not equal to zero.]

(i) Write the equations of motion which determine the horizontal velocity of the center, $\vec{v}(t)$, and the angular velocity $\vec{\omega}(t)$ for time $t > 0$.

(ii) Find $\vec{v}(t)$ explicitly and also the velocity of the particle at the contact point P, $\vec{v}\,'(t)$.

(iii) How long does it take before the ball starts rolling?

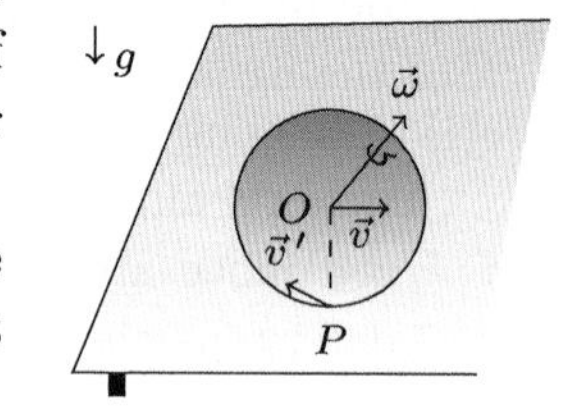

7-12 Consider a uniform hemisphere of radius R and total mass M, which lies on a smooth (i.e., frictionless) horizontal surface in the field of uniform gravity (see the figure).

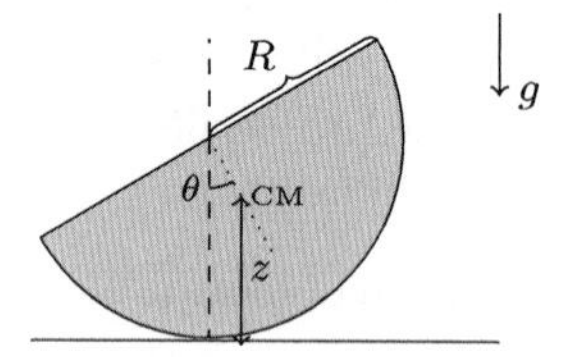

(i) Express the height z of the CM above the plane in terms of the angle θ over which the hemisphere is turned.

(ii) Find the moment of inertia about the axis passing through the CM.

(iii) Obtain the equation of motion for θ and determine from it the frequency of small oscillations.

7-13 Show that, for a uniform ellipsoid $\frac{x^2}{a^2} + \frac{y^2}{b^2} + \frac{z^2}{c^2} \leq 1$ of net mass M, the moments of inertia tensor about the center is given by

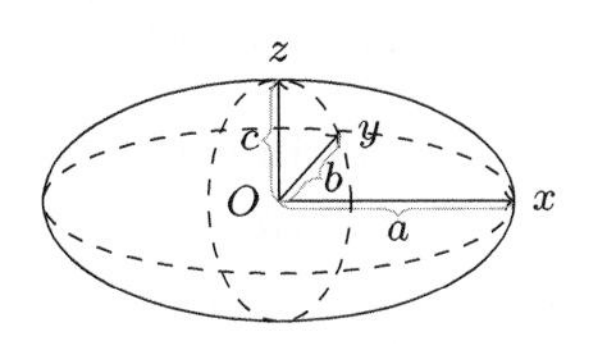

$$I = \begin{pmatrix} \frac{1}{5}M(b^2+c^2) & 0 & 0 \\ 0 & \frac{1}{5}M(c^2+a^2) & 0 \\ 0 & 0 & \frac{1}{5}M(a^2+b^2) \end{pmatrix}.$$

7-14 Consider a solid *homogeneous* cube of mass M.

(i) Find the moment of inertia about an arbitrary line (having the direction $\hat{u} = \cos\alpha\mathbf{i} + \cos\beta\mathbf{j} + \cos\gamma\mathbf{k}$) through its center O.

(ii) Determine the principal moments of inertia when a corner point A is taken as the rotation origin.

(iii) What are the principal axes of inertia at the corner point A?

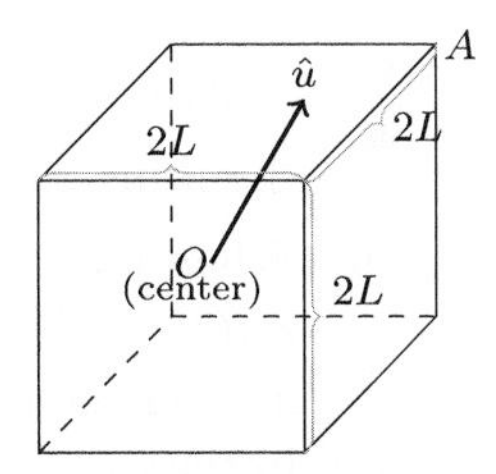

7-15 For a sphere of radius R inside of which there is a spherical cavity of radius r (see the figure), determine the principal moments of inertia with respect to the CM of the body. [Assume a uniform mass density.]

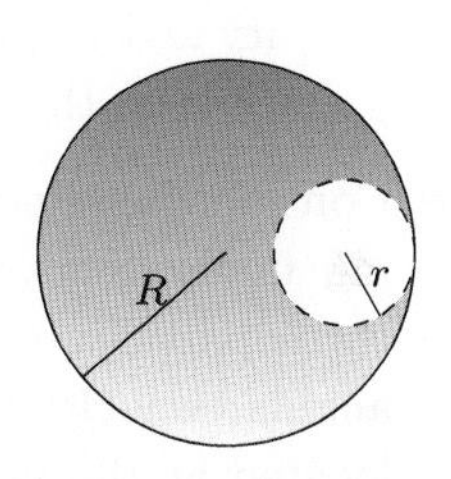

7-16 Consider a rigid spherical object of radius a in the presence of some external gravitational potential $\mathcal{G}_{ext}(\vec{r})$ produced by certain distant source. With the spatial origin chosen at the center of the given rigid object, assume that the object has a spherically symmetric mass distribution $\rho(r)$ (with the total mass equal to $\int_0^a 4\pi r^2 \rho(r) dr = M$). It can have both translational and rotational motions.

(i) Show that the net force $\vec{F}_{\rm tot}$ felt by the rigid sphere is equal to

$$\vec{F}_{\rm tot} = -M\left[\vec{\nabla}\mathcal{G}_{\rm ext}(\vec{r})\right]\Big|_{\vec{r}=0}.$$

(ii) Show that the net torque $\vec{\Gamma}_{ext}$ on the rigid sphere (about the origin) vanishes.

(iii) Discuss the implication of the results (i) and (ii) on the translational and rotational motions of the sphere.

7-17 Consider a uniform rectangular slab of mass m with sides a and b ($b > a$), which is rotating about the diagonal. (The diagonal points in the fixed direction $\hat{z}$.) Ignore gravity.

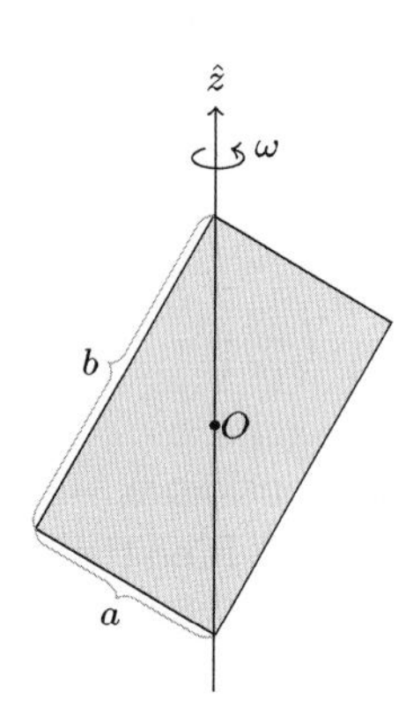

(i) Indicate three principal axes through the center O. Also find the corresponding principal moments.

(ii) When the slab rotates about the diagonal with angular velocity ω (see the figure), represent the angular momentum vector $\vec{L}$ for the slab (about the point O) using the body-fixed basis chosen along the principal axes.

(iii) When the $\hat{z}$-component of the external torque (with the origin at O), $\Gamma_z^{(e)}(t)$, is given as a function of time, what equation of motion should the angular velocity $\omega(t)$ satisfy?

(iv) In the case that the slab is rotating with constant angular velocity $\omega = \omega_0$, find the external torque $\vec{\Gamma}^{(e)}$ which must be applied to keep the axis of rotation fixed.

7-18 Consider three noncollinear particles of masses m_1, m_2 and m_3, moving under their mutual gravitational attractions. Show that these three particles can *move as a rigid body* (rotating with a constant angular velocity vector about the center of mass) only when they are located at the vertices of an equilateral triangle. [We note that the angular velocity of the rotation should equal $\sqrt{G\frac{(m_1+m_2+m_3)}{s^3}}$ when s denotes the length of the triangle's side: this was Problem 5-2.]

8 Elements of Fluid Mechanics

SUMMARY

8.1 Fluids in equilibrium

- For a fluid in static equilibrium there should be no shearing or tangential stress — only the normal stress in the form of the pressure (p) is allowed. In equilibrium the (surface) force due to pressure on any small fluid volume V, $\vec{F}_p = -\iint_{S=\partial V} p(\vec{r})\hat{n}da$, must be balanced by the externally given force on the volume, $\vec{F}_b = \iiint_V \vec{f}(\vec{r})d^3\vec{r}$ ($\vec{f}(\vec{r})$ represents the related body force density). For this to be the case, the pressure gradient in a fluid must be equal to the body force density, i.e.,

$$\vec{\nabla}p(\vec{r}) = \vec{f}(\vec{r}).$$

- In a uniform gravitational field, we have $\vec{f}(\vec{r}) = -\rho(\vec{r})g\mathbf{k}$ ($\rho(\vec{r})$ is the mass density) for the body force density. If we have $\rho(\vec{r}) = \rho$ (: const.) as in a liquid, we then find, from $\vec{\nabla}p(\vec{r}) = -\rho g\mathbf{k}$, the equilibrium pressure distribution $p(\vec{r}) = p_0 - \rho g z$. In the case of the isothermal atmosphere for which Boyle's law tells us that $\frac{\rho(\vec{r})}{\rho_0} = \frac{p(\vec{r})}{p_0}$, the hydrostatic balance equation gives the pressure in the form

$$p(z) = p_0 e^{-\frac{\rho_0}{p_0}gz}.$$

(This is equivalent to the barometric formula $p(z) = p_0 e^{-\frac{mgz}{k_B T}}$, where m denotes the average mass of an air molecule.)

- In the presence of a conservative body force described by the form $\vec{f}(\vec{r}) = -\rho\vec{\nabla}\mathcal{G}(\vec{r})$ ($\mathcal{G}(\vec{r})$ is a suitable potential, and

ρ a constant), the hydrostatic equation is integrated to yield

$$\mathcal{G}(\vec{r}) + \frac{1}{\rho} p(\vec{r}) = \text{const.}$$

Based on this equation one infers that the free surface of a liquid in static equilibrium corresponds to an equipotential surface.

- (Archimedes' principle) A body of volume V immersed in a fluid is acted on by a buoyant force — the force due to pressure from the surrounding fluid (i.e., $\vec{F}_p = -\iint_{\partial V} p(\vec{r})\,\hat{n} da$) which should be of equal magnitude as the weight of the volume of the fluid displaced.

8.2 Fluids in motion and dynamical equations for an ideal fluid

- In a moving fluid, each fluid element undergoes certain deformation due to stresses (that is, *surface* forces per unit area) from the surrounding fluid. In terms of the velocity field $\vec{v}(\vec{r}, t)$ for the fluid, velocity gradients $\frac{\partial v_j(\vec{r},t)}{\partial x_i}$ contain information on short-time deformation of fluid elements (located at position $\vec{r}$ at time t); for instance, $\vec{\nabla} \cdot \vec{v} = 0$ for a fluid which is incompressible, and the shearing motion of fluid elements is related to the symmetric and traceless part of the matrix $\left(\frac{\partial v_j}{\partial x_i}\right)$. In an ideal or perfect fluid it is assumed that (as in a fluid in equilibrium) there is no shearing or tangential stress — normal stress involving pressure $p(\vec{r}, t)$ is the only kind allowed.
- The law of conservation of mass applied to a fluid element in a small volume δV (and hence mass $\rho\,\delta V$) implies that

$$0 = \frac{D}{Dt}(\rho\,\delta V),$$

where $\frac{D}{Dt} \equiv \frac{\partial}{\partial t} + \vec{v} \cdot \vec{\nabla}$ is the material or hydrodynamic derivative. Then, using $\frac{d\,\delta V}{dt} = \vec{\nabla} \cdot \vec{v}\,\delta V$, one is led to the equation of continuity

$$\frac{\partial \rho}{\partial t} + \vec{\nabla} \cdot (\rho \vec{v}) = 0.$$

If the fluid is incompressible, we also have $\vec{\nabla} \cdot \vec{v} = 0$ and accordingly this continuity equation reduces to $\frac{D\rho}{Dt} = \frac{\partial \rho}{\partial t} + \vec{v} \cdot \vec{\nabla} \rho = 0$.

- Applying Newton's second law to a fluid element of an ideal fluid occupying a small region δV yields

$$(\rho\, \delta V) \frac{D\vec{v}}{Dt} = (-\vec{\nabla} p + \vec{f})\, \delta V,$$

where $\vec{f}\, \delta V$ denotes the external body force. This leads to Euler's equation of motion for an ideal fluid

$$\frac{\partial \vec{v}}{\partial t} + (\vec{v} \cdot \vec{\nabla})\vec{v} = -\frac{1}{\rho} \vec{\nabla} p + \frac{1}{\rho} \vec{f}.$$

- Aside from the equation of continuity and Euler's equation of motion, we need a separate equation of the form $f(\rho, p) = 0$, describing the thermal state of the fluid elements. The most commonly adopted for this purpose are the followings: (i) (uniform incompressible flow) $\rho =$const., as should be appropriate with many liquids, and (ii) (isentropic flow) $s = s(\rho, p) =$const. (this may be put into the form $p = p(\rho)$) where s denotes the thermodynamic entropy per unit mass. For an isentropic flow one can equate $\frac{1}{\rho} \frac{\partial p}{\partial t}$ with $\frac{\partial h}{\partial t}$ and $\frac{1}{\rho} \vec{\nabla} p$ with $\vec{\nabla} h$ (h is the enthalpy per unit mass). In fact, the case with $\rho =$ const. can also be regarded as a special case of an isentropic flow — that with $h = \frac{p}{\rho} +$ const.
- Consider a steady flow, i.e., a flow for which $\frac{\partial \vec{v}}{\partial t} = 0$. If the body force per unit mass, $\frac{1}{\rho} \vec{f}$, is conservative, i.e., $\frac{1}{\rho} \vec{f} = -\vec{\nabla} \mathcal{G}$ with a suitable potential $\mathcal{G}$, one may then use Euler's equation (for an isentropic flow) with the identity

$$(\vec{v} \cdot \vec{\nabla})\vec{v} = -\vec{\nabla} \left(\frac{1}{2} \vec{v}^2 \right) - \vec{v} \times (\vec{\nabla} \times \vec{v})$$

to deduce an equation of the form

$$\vec{v}\cdot\vec{\nabla}\left(\frac{1}{2}\vec{v}^2+h+\mathcal{G}\right)=0.$$

(If ρ = const., replace h by $\frac{p}{\rho}$.) This leads to Bernoulli's theorem for an ideal fluid: the quantity $\frac{1}{2}\vec{v}^2+h+\mathcal{G}$ (or $\frac{1}{2}\vec{v}^2+\frac{p}{\rho}+\mathcal{G}$, for a fluid with ρ = const.) is invariant along the streamline. If the fluid is irrotational, i.e., the vorticity $\vec{\Omega}\equiv\vec{\nabla}\times\vec{v}$ vanishes everywhere, Bernoulli's theorem takes a more restrictive form: the quantity $\frac{1}{2}\vec{v}^2+h+\mathcal{G}$ assumes the same constant value everywhere in the fluid. Bernoulli's theorem, which expresses the conservation of the total energy per unit mass in an ideal fluid, provides a simple understanding for many interesting behaviors concerning steady fluids.

- The mathematical equations obtained in this section were applied to a few cases of ideal fluid motions — the boundary shape of a rotating ideal fluid in a cylindrical vessel, the steady flow of an ideal fluid through a straight horizontal pipe with a symmetrical constriction, and the propagation of sound waves in a fluid medium like air were studied in particular.

8.3 Kelvin's theorem and potential flows

- A fluid flow is irrotational if the vorticity $\vec{\Omega}\equiv\vec{\nabla}\times\vec{v}$ vanishes everywhere. An ideal fluid with zero vorticity defines a potential flow since the velocity field can then be expressed as $\vec{v}=\vec{\nabla}\phi$, $\phi(\vec{r},t)$ being the velocity potential.
- For a general isentropic ideal fluid, it is shown that the velocity circulation over any closed *fluid contour* C_t, $\Gamma(t)=\oint_{C_t}\vec{v}\cdot d\vec{l}$ $(=\int_{S_t}\vec{\Omega}\cdot d\vec{a}$, if $\partial S_t=C_t)$, remains unchanged, i.e., $\frac{d\Gamma(t)}{dt}=0$. (This is Kelvin's theorem.) According to this theorem, vorticity creation or destruction is prohibited in this fluid; the elements of an ideal fluid, which are free of vorticity at some instant, remain free of vorticity indefinitely. This provides a strong case for studying potential flows, and at the same time shows that viscosity effects are not to be neglected in *real* fluids.

- For an incompressible potential flow (and hence $\vec{\nabla} \cdot \vec{v} = 0$), the corresponding velocity potential should satisfy the Laplace equation, $\vec{\nabla}^2 \phi = 0$. This equation may be studied with appropriate boundary conditions at the surfaces where the fluid meets solid bodies: at *fixed* solid surfaces, the fluid velocity component normal to the surface should vanish, i.e., $v_n = \hat{n} \cdot \vec{\nabla}\phi = 0$. For an incompressible potential flow with $\rho =$ const., one can then utilize Euler's equation of motion to express the pressure by the form

$$p = -\rho \left(\frac{1}{2}\vec{v}^2 + \mathcal{G} + \frac{\partial \phi}{\partial t} \right)$$

(with a judiciously chosen velocity potential $\phi(\vec{r}, t)$).

- For a two-dimensional incompressible flow, one may introduce the stream function $\psi(x, y)$ in such a way that $v_x = \frac{\partial \psi}{\partial y}$ and $v_y = -\frac{\partial \psi}{\partial x}$; then, streamlines are simply the family of curves described by the equation $\psi(x, y) =$ const. For a two-dimensional, incompressible, potential flow, we thus have two scalar functions, the velocity potential ϕ and the stream function ψ, related to each other through

$$\frac{\partial \phi}{\partial x} = \frac{\partial \psi}{\partial y}, \qquad \frac{\partial \phi}{\partial y} = -\frac{\partial \psi}{\partial x}.$$

These are the Cauchy–Riemann conditions for the expression $F = \phi + i\psi$ to be an analytic function of the complex argument $z = x + iy$; $F(z) = \phi + i\psi$ is called the complex potential, and $\frac{dF}{dz}$ $(= v_x - iv_y)$ the complex velocity. One can give a compact description for many two-dimensional potential flows by finding appropriate complex potential $F = F(z)$.

- The complex potential of the form $F(z) = Az^n$ $(n > \frac{1}{2})$ can be used to describe a potential flow near an angle (or wedge) formed by two intersecting plane boundaries. Also the complex potential

$$F(z) = -v_0 \left(z + \frac{R^2}{z} \right) \qquad (|z| \geq R)$$

represents the potential flow around an infinite circular cylinder (of radius R) placed in an initially uniform flow towards the negative-x direction. In polar coordinates the related velocity potential can be written as

$$\phi = -v_0 \left(r + \frac{R^2}{r} \right) \cos\theta \quad (r \geq R),$$

and hence the velocity field is given by

$$\vec{v}(r,\theta) = -v_0 \left(1 - \frac{R^2}{r^2} \right) \cos\theta \, \hat{r} + v_0 \left(1 + \frac{R^2}{r^2} \right) \sin\theta \, \hat{\theta} \quad (r \geq R).$$

But, rather paradoxically, no force is given on the cylinder from the incident fluid — this is due to our neglecting viscosity effects, especially near the solid body.

- As an interesting three-dimensional potential flow problem, we have studied the flow around a moving sphere (of radius a and with instantaneous velocity $\vec{u}(t)$) in a fluid which is at rest at large distance from the sphere. The velocity potential is found to have the form

$$\phi = -\frac{1}{2} a^3 \frac{\vec{u} \cdot \vec{r}'}{r'^3}$$

and hence the velocity field becomes

$$\vec{v} = -\frac{1}{2} \frac{a^3}{r'^3} \vec{u} + \frac{3}{2} \frac{a^3}{r'^5} (\vec{u} \cdot \vec{r}') \vec{r}',$$

where $\vec{r}'$ denotes the spatial coordinates measured from the instantaneous position of the center of the sphere. The force that the fluid exerts on the sphere can be calculated, with the result $\vec{F}_p = -\frac{2\pi}{3} \rho a^3 \frac{d\vec{u}}{dt}$; this force effectively renormalizes the mass of the sphere by increasing its value by $\Delta M = \frac{1}{2}(\frac{4}{3}\pi\rho a^3)$ (ΔM is the *induced or added mass*).

8.4 Viscous flows and the Navier–Stokes equation

- In real fluid flow, viscous friction between adjacent fluid layers should not be neglected: it causes a shearing stress which tends to reduce their relative velocity. To take into account the effects due to such shearing stress, Euler's equation (for an ideal fluid) should be replaced by a more general equation that contains an additional force term related to the viscous stress tensor σ_{ij}:

$$\frac{\partial v_i}{\partial t} + v_j \frac{\partial v_i}{\partial x_j} = \frac{1}{\rho}\left[-\frac{\partial p}{\partial x_i} + \frac{\partial}{\partial x_j}\sigma_{ij} + f_i\right].$$

Assuming for the viscous tensor a phenomenologically viable form

$$\sigma_{ij} = \eta\left(\frac{\partial v_j}{\partial x_i} + \frac{\partial v_i}{\partial x_j}\right) + \mu\delta_{ij}\frac{\partial v_k}{\partial x_k}$$

(η, μ are constants), one is led to the Navier–Stokes equation

$$\rho\left[\frac{\partial \vec{v}}{\partial t} + (\vec{v}\cdot\vec{\nabla})\vec{v}\right] = -\vec{\nabla}p + \eta\vec{\nabla}^2\vec{v} + (\eta+\mu)\vec{\nabla}(\vec{\nabla}\cdot\vec{v}) + \vec{f}.$$

The constant η is identified with the viscosity of the fluid. Ignoring the term $(\eta+\mu)\vec{\nabla}(\vec{\nabla}\cdot\vec{v})$ which is absent for an incompressible fluid, this reduces to

$$\frac{\partial \vec{v}}{\partial t} + (\vec{v}\cdot\vec{\nabla})\vec{v} = -\frac{1}{\rho}\vec{\nabla}p + \nu\vec{\nabla}^2\vec{v} + \frac{1}{\rho}\vec{f},$$

where $\nu \equiv \eta/\rho$ is called the kinematic viscosity.

- At the boundary of fluid, the Navier–Stokes equation should be supplemented by appropriate boundary conditions. With the Navier–Stokes equation which has second-order spatial derivative terms of $\vec{v}$ (in contrast to the case with Euler's equation of motion which involves only first-order derivatives), we demand that both the normal and tangential components of the velocity vanish on the fixed solid surface (the "no-slip condition"), i.e., $\vec{v} = 0$ on the boundary. Implication of this no-slip boundary condition is: even a very small viscosity assumes a non-negligible role near a solid surface.

8.5 Some examples of viscous fluid flow

- Exact, steady or non-steady, solutions to the Navier–Stokes equation satisfying appropriate geometric boundary conditions are difficult to find. Even when one has obtained certain steady solutions, one must worry about the stability of the solutions. Generally, the stable steady solutions are the ones in which the effect of the viscous friction term $\nu\vec{\nabla}^2\vec{v}$ (in the Navier–Stokes equation) is dominant over that due to the *inertial* term $(\vec{v}\cdot\vec{\nabla})\vec{v}$. Here an important dimensionless parameter is the Reynolds number $\mathcal{R} = uL/\nu$, where u is some characteristic flow velocity and L is a characteristic length scale for the velocity change. For relatively small values of $\mathcal{R}$ (a lamina flow regime), the flow is smooth and gives rise to a steady flow; on the other hand, if $\mathcal{R}$ becomes larger than some critical value, the flow becomes unstable and enters a chaotic or turbulent regime. We discussed this issue for some exact solutions of the Navier–Stokes equation, one describing a flow along an inclined plane under the influence of gravity and the other concerning a steady flow in a pipe of appropriate cross section (the Poiseuille flow included).
- The steady solution describing an incompressible flow past a fixed sphere of radius R (with the fluid having a constant velocity $\vec{u}$ at infinity) has been studied, neglecting the inertial term $(\vec{v}\cdot\vec{\nabla})\vec{v}$ (i.e., for $\mathcal{R} \ll 1$). For this case, one has to solve a linear equation

$$\vec{\nabla}p - \eta\vec{\nabla}^2\vec{v} = 0$$

(together with $\vec{\nabla}\cdot\vec{v} = 0$) under the boundary conditions $\vec{v}(\vec{r})\big|_{r=R} = 0$ and $\vec{v}(\vec{r})\big|_{r\to\infty} = \vec{u}$. Exploiting spherical symmetry in the problem, the corresponding velocity distribution is shown to have the form

$$\vec{v}(\vec{r}) = \vec{u} - \frac{3}{4}R\frac{\vec{u} + \hat{r}(\vec{u}\cdot\hat{r})}{r} + \frac{1}{4}R^3\frac{3\hat{r}(\vec{u}\cdot\hat{r}) - \vec{u}}{r^3}.$$

The pressure distribution is given by $p(\vec{r}) = p_0 - \frac{3}{2}\eta R\frac{\vec{u}\cdot\hat{r}}{r^2}$. If this expression is used to calculate the force exerted on the sphere by the moving fluid (or, equivalently, the drag on the sphere as it moves through the fluid with uniform velocity), a resistive force of the form $F = 6\pi\eta R u$ (Stokes' law) follows, which is linear in the relative velocity.

Problems

8-1 A fluid of volume V and arbitrary shape is acted on by some uniform external hydrostatic pressure.

(i) Suppose that the volume of this system undergoes a small change $V \to V + dV$ against a given pressure p. (See the figure, where the case with $dV > 0$, i.e., expansion is shown). Then show that the net work done on the system is given by the formula

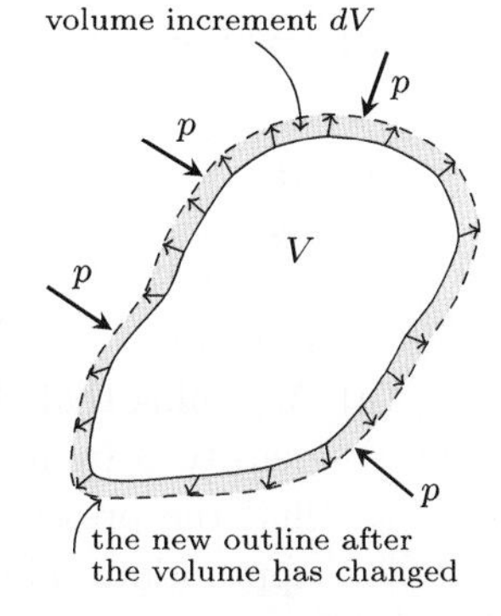

$$\delta W = -p\,dV.$$

(ii) Now consider a system consisting of an ideal gas for which the total internal energy is given by the formula $U = q\,pV$ where q is a constant. The system is allowed to undergo an *adiabatic change* (that is, without heat transfer to or from the outside). Then, by using the energy conservation (or the first law of thermodynamics), show that the pressure and the volume in adiabatic processes may vary in accordance with

$$pV^{\gamma} = \text{const.}$$

(or $p = (\text{const.})\,\rho^{\gamma}$, if ρ is the mass density), where $\gamma \equiv 1 + \frac{1}{q}$ is the adiabatic index. [*Remark*: For air we have $q \approx \frac{5}{2}$ and so $\gamma \approx 1.4$].

8-2 The formula $p(z) = p_0 e^{-\frac{\rho_0}{p_0} gz}$ for the pressure variation in the Earth's atmosphere does not apply when the temperature variation with the altitude cannot be neglected. (In fact, this formula is not a particularly good representation, say, at the height not far from the surface.) For a better representation, one can consider the air medium for which the temperature at the altitude z is described by some nonconstant function $T = T(z)$. Find the pressure dependence on the altitude when

(i) $T = T_0\left(1 - \frac{z}{H}\right), \quad z < H$;

(ii) $T = T_0\left(1 - \frac{z^2}{H^2}\right), \quad z < H$.

8-3 Consider a spherical object that is held together by its self gravity, and is balanced against collapse by pressure gradients. Write $p(r)$ and $\rho(r)$ for the pressure and density at radius r from the center of the spherical object.

(i) Use the hydrostatic equilibrium condition to obtain the relation

$$\frac{dp(r)}{dr} = -\frac{GM(r)\rho(r)}{r^2},$$

where $M(r) = 4\pi \int_0^r \rho(r')r'^2 dr'$ denotes the mass interior to the radius r.

(ii) Assuming that the Earth is an incompressible liquid of constant density ρ (and neglecting the effect due to Earth's rotation), find the pressure in the Earth as a function of r. Also, using appropriate values for the Earth's mass and radius, estimate the pressure at the Earth's center.

8-4 For a spherical star that is held together by its self gravity, $p(r)$ and $\rho(r)$ — the pressure and mass density at radius r from the center of the star — should satisfy the hydrostatic equilibrium condition obtained in Problem 8-3(i). Define the surface of the star as the radius $r = r_*$ at which the pressure goes to zero.

(i) For this system, show that the volume-averaged pressure, $\bar{p}$, is related to the gravitational self-energy E_{gr} by $\bar{p} = -\frac{1}{3}\frac{1}{V}E_{\mathrm{gr}}$, where $V = \frac{4\pi}{3}r_*^3$ and E_{gr} is given by (see Problem 6-4)

$$E_{\mathrm{gr}} = -\frac{G}{2} \int\limits_{r' \equiv |\vec{r}\,'| < r_*} d^3\vec{r}\,' \int\limits_{r \equiv |\vec{r}| < r_*} d^3\vec{r}\, \frac{\rho(r')\,\rho(r)}{|\vec{r}\,' - \vec{r}|}.$$

(ii) Assume that this star is composed of a monoatomic ideal gas. Then, according to the result of (i), what would happen to the total thermal energy of the star as it contracts (and its gravitational self-energy becomes more negative)? Explain.

8-5 A circular hole of radius r at the bottom of an initially full water container is sealed by a ball of mass m and radius R ($> r$). The depth of the water is now slowly reduced, and when it reaches a certain value, h_0, the ball rises out of the hole. Find h_0. (Denote the mass density of water by ρ.)

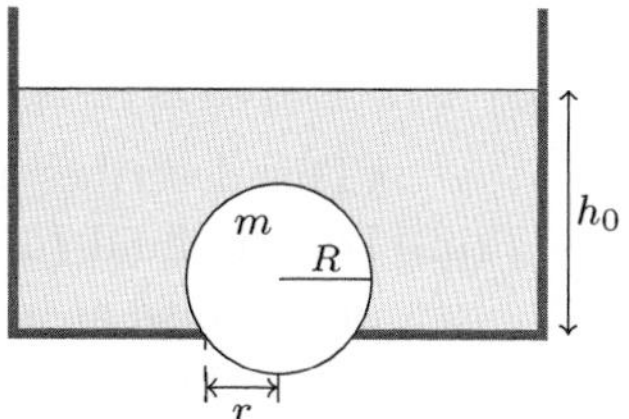

8-6 When external forces are absent, basic dynamical equations for ideal fluids consist of

(i) (the continuity equation) $\frac{\partial \rho}{\partial t} + \vec{\nabla} \cdot (\rho \vec{v}) = 0$, and
(ii) (Euler's equation) $\frac{\partial \vec{v}}{\partial t} + (\vec{v} \cdot \vec{\nabla})\vec{v} = -\frac{1}{\rho}\vec{\nabla} p$.

Show that these equations are form-invariant under Galilean transformations (and so can be assumed in any Newtonian inertial frame). [Note that, under Galilean boost $\vec{x} \to \vec{x}' = \vec{x} - \vec{u}t$, the mass density, pressure, and velocity fields may get transformed as

$$\rho(\vec{x}, t) \to \rho'(\vec{x}', t) = \rho(\vec{x}, t)\Big|_{\vec{x}=\vec{x}'+\vec{u}t},$$
$$p(\vec{x}, t) \to p'(\vec{x}', t) = p(\vec{x}, t)\Big|_{\vec{x}=\vec{x}'+\vec{u}t},$$
$$\vec{v}(\vec{x}, t) \to \vec{v}'(\vec{x}', t) = \vec{v}(\vec{x}, t)\Big|_{\vec{x}=\vec{x}'+\vec{u}t} - \vec{u}.]$$

8-7 Consider an incompressible flow of an ideal fluid with the vorticity in space given by the form

$$\vec{\Omega}(\vec{r}) = w(\rho)\,\mathbf{k},$$

where $w(\rho)$, a function of $\rho \equiv \sqrt{x^2 + y^2}$ only, vanishes sufficiently fast as $\rho \to \infty$. Assume that there is no external force acting on the fluid.

(i) Show that, for the fluid velocity field, one can assume the form $\vec{v}(\vec{r}) = f(\rho)\hat{\varphi}$ (with an appropriate function $f(\rho)$), where $\hat{\varphi} = \frac{1}{\rho}(-y\mathbf{i} + x\mathbf{j})$.

(ii) Determine the velocity field explicitly when $w(\rho)$ (in $\vec{\Omega}(\vec{r})$) is given as

$$w(\rho) = \begin{cases} 2\omega_0, & \text{if } \rho \leq a \\ 0, & \text{if } \rho \geq a. \end{cases}$$

(This describes a vortex tube.)

(iii) For the fluid (of constant density ρ_0) having the velocity field determined as in (ii), determine the pressure at each point of the fluid. (The value of the pressure on the z-axis, i.e., at $x = y = 0$, may be denoted as p_0.)

8-8 For an isentropic flow of an ideal fluid in the presence of a conservative body force per unit mass (i.e., $\frac{1}{\rho}\vec{f} = -\vec{\nabla}\mathcal{G}$ with a suitable time-independent potential $\mathcal{G}(\vec{r})$), one can define the energy density, i.e., the energy per unit volume $\mathcal{E}$ by the equation

$$\mathcal{E} = \frac{1}{2}\rho\vec{v}^2 + \rho u + \rho\mathcal{G},$$

where u is the internal energy per unit mass.

(i) Show, using the continuity equation $\frac{\partial\rho}{\partial t} + \vec{\nabla}\cdot(\rho\vec{v}) = 0$ and Euler's equation of motion for an ideal fluid, that the (local) rate of change of the energy density satisfies the equation

$$\frac{\partial\mathcal{E}}{\partial t} = -\vec{\nabla}\cdot\left[\rho\vec{v}\left(\frac{1}{2}\vec{v}^2 + h + \mathcal{G}\right)\right],$$

where h ($\equiv u + p/A$) is the enthalpy per unit mass. [*Hint*: Use the relation $d(\rho u) = h\,d\rho$, valid for an isentropic flow, if necessary.]

(ii) Integrating the equation found in (i) over a fixed volume V, one is led to the integral form of the general energy conservation law for an ideal fluid:

$$\frac{d}{dt}\int_V \left(\frac{1}{2}\rho\vec{v}^2 + \rho u + \rho\mathcal{G}\right) dxdydz$$
$$= -\int_{S=\partial V} \rho\left(\frac{1}{2}\vec{v}^2 + h + \mathcal{G}\right)\vec{v}\cdot\hat{n}\,da.$$

Based on the structure of this equation, provide appropriate physical interpretation.

(iii) Then, restricting one's attention to a *steady flow*, use the thus-obtained energy conservation law to prove Bernoulli's theorem:

along each streamline, the value of the quantity $\frac{1}{2}\vec{v}^2 + h + \mathcal{G}$ remains fixed.

8-9 The free surface of a liquid in a vessel is at a level z_0 above its bottom, and the vessel has a small hole at level z (see the figure). Determine the distance at which the outflowing liquid jet reaches the plane of the bottom ($z = 0$). Also determine z (with z_0 fixed) when this distance attains the maximum value.

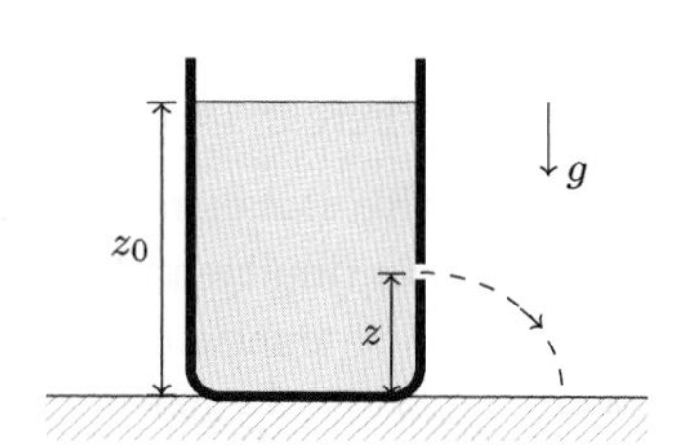

8-10 A water jar in the form of a surface of revolution is fixed with its axis vertical so that water runs out of a small orifice on the axis of symmetry (see the figure). What shape must the jar be for the level of water to descend with a uniform speed? (The ancient water clock, known as the clepsydra, operated on this basis.)

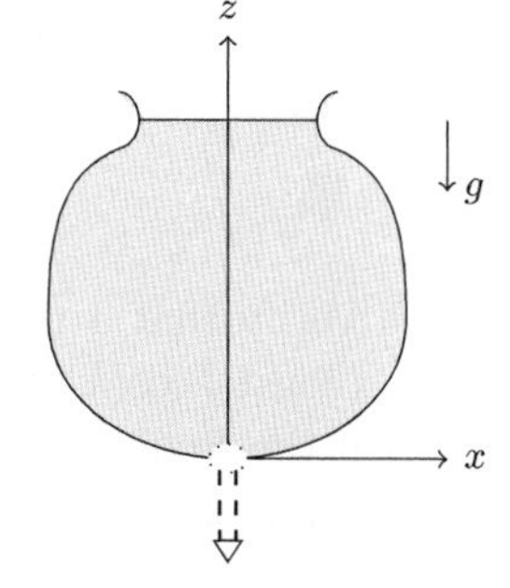

8-11 Show that, for an isentropic flow of an ideal gas in a constant gravitational field (i.e., $\vec{f}/\rho = -g\mathbf{k}$), Bernoulli's theorem implies that the quantity $\frac{1}{2}\vec{v}^2 + \frac{\gamma}{\gamma-1}\frac{p}{\rho} + gz$, where γ is the adiabatic index of the gas, remains constant along each streamline.

8-12 Consider a uniform stationary fluid with constant density ρ_0 and constant pressure p_0, in a region where the external body force can be neglected. If this fluid is subject to a small disturbance, the density, pressure, and velocity will acquire small fluctuation terms ρ', p' and $\vec{v}'$, i.e., $\rho = \rho_0 + \rho'(\vec{r}, t)$, $p = p_0 + p'(\vec{r}, t)$, $\vec{v} = 0 + \vec{v}'(\vec{r}, t)$. The fluid motion, including the small fluctuations, satisfies the equations appropriate to an isentropic flow of an ideal fluid. We are here interested in the time-dependent dynamics of the fluctuation fields (or the phenomenon of *sound waves*), as determined by the fluid dynamic equations considered to *first orders in the fluctuations* ρ', p' and $\vec{v}'$. (Note that, for an isentropic flow, the pressure p can be assumed to be some definite function of density ρ.)

(i) For the fluctuations in pressure and density, i.e., p' and ρ' of an isentropic flow, explain under what ground the two are related by

$$p'(\vec{r},t) = c^2\rho'(\vec{r},t),$$

where $c^2 = \left(\frac{\partial p}{\partial \rho}\right)_s\Big|_{\rho=\rho_0}$ (the subscript s signifies that the derivative is taken at constant entropy). Then, from the equation of continuity and Euler's equation of motion, show that the pressure fluctuation $p'(\vec{r},t)$ is required to satisfy the *wave equation* of the form

$$\vec{\nabla}^2 p' - \frac{1}{c^2}\frac{\partial^2 p'}{\partial t^2} = 0.$$

(ii) If the fluid is also irrotational, show that the velocity $\vec{v}'$ also satisfies the wave equation

$$\vec{\nabla}^2 \vec{v}' - \frac{1}{c^2}\frac{\partial^2 \vec{v}'}{\partial t^2} = 0.$$

(iii) Verify that a monochromatic plane wave of the form $p'(\vec{r},t) =$ (const.) $\cos(\vec{k}\cdot\vec{r}-\omega t+\alpha)$ becomes a solution to the wave equation if the frequency ω is related to the wave vector $\vec{k}$ by $\omega = c|\vec{k}|$. Find also the related velocity $\vec{v}'(\vec{r},t)$.

8-13 In Problem 8-12, a sound wave propagating through a *stationary* uniform medium of an ideal fluid was studied. Now suppose that the given medium is moving with a constant velocity $\vec{v}_0$. To discuss sound waves in this moving medium, one will have to consider the fluctuation fields ρ', p' and $\vec{v}'$ defined by

$$\rho = \rho_0 + \rho'(\vec{r},t), \quad p = p_0 + p'(\vec{r},t), \quad \vec{v} = \vec{v}_0 + \vec{v}'(\vec{r},t).$$

Assume an isentropic flow and $|\vec{v}_0| < c$ (c is the speed defined in Problem 8-12).

(i) In this case, derive the differential equation satisfied by the pressure fluctuation $p'(\vec{r},t)$ and find the corresponding monochromatic wave solution.

(ii) Interpret the findings of (i) in the light of the Galilean invariance considered in Problem 8-6. (This is relevant for the *Doppler shift* of a sound wave.)

8-14 The velocity field of a two-dimensional incompressible flow can be represented in terms of the stream function $\psi = \psi(x,y)$: $v_x = \frac{\partial \psi}{\partial y}$ and $v_y = -\frac{\partial \psi}{\partial x}$. The flux of the flow (or the flow rate) $\mathcal{N}$ across the plane curve connecting two given points $P_1 = (x_1, y_1)$ and $P_2 = (x_2, y_2)$ is defined by

$$\mathcal{N} = \int_1^2 v_n \, dl,$$

where v_n is the velocity projection on the normal to the curve, and dl denotes the length element. Show that this flux is equal to the difference of the values of the stream function at the endpoints (and is independent of the curve chosen), i.e.,

$$\mathcal{N} = \psi(x_2, y_2) - \psi(x_1, y_1).$$

8-15 The velocity potential $\phi(x, y) = -v_0 x\left(1 + \frac{R^2}{x^2+y^2}\right)$, describing the potential flow past an infinite right circular cylinder placed in an initially uniform flow, was obtained with the help of the complex potential method in the text (see Sec. 8.3). We will here consider an alternative approach to the problem.

(i) In terms of polar coordinate (r, θ) introduced in the xy-plane of flow, show that the Laplace equation for the velocity potential $\phi(r, \theta)$ assumes the form

$$\frac{1}{r}\frac{\partial}{\partial r}\left(r\frac{\partial \phi}{\partial r}\right) + \frac{1}{r^2}\frac{\partial^2 \phi}{\partial \theta^2} = 0.$$

(ii) Show that any function of the form

$$\phi(r, \theta) = a_0 + b_0 \ln r + (c_0 + d_0 \ln r)\theta + \sum_{n=1}^{\infty}\left[\left(a_n r^n + \frac{b_n}{r^n}\right)\cos n\theta + \left(c_n r^n + \frac{d_n}{r^n}\right)\sin n\theta\right]$$
$$(r \neq 0)$$

$(a_0, b_0, c_0, d_0, a_1, b_1, \cdots$ are arbitrary constants) is a solution to the Laplace equation.

(iii) Then, by imposing the boundary conditions $\left.\frac{\partial \phi(r,\theta)}{\partial r}\right|_{r=R} = 0$ and $\phi(r, \theta) \to -v_0 r \cos\theta$ as $r \to \infty$, obtain the velocity potential (describing the flow past an infinite cylinder at $r = R$) in the form

$$\phi(r, \theta) = a_0 + k\theta - v_0\left(r + \frac{R^2}{r}\right)\cos\theta \quad (r \geq R),$$

where a_0, k are arbitrary constants. [If one restricts oneself to the case $k = 0$ (as the expression $k\theta$ does not correspond to a single-valued function of position), the potential form $\phi = -v_0\left(r + \frac{R^2}{r}\right)\cos\theta$ (see (8.97) in the text) is recovered; but, as will be discussed in the next problem, the case with $k \neq 0$ also produces an interesting flow.]

8-16 A two-dimensional potential flow is described by the velocity potential (obtained in Problem 8-15 (ii))

$$\phi(r,\theta) = a_0 + k\theta - v_0\left(r + \frac{R^2}{r}\right)\cos\theta \quad (r \geq R).$$

(i) Find the associated velocity field $\vec{v}(r,\theta)$ (which is single-valued for any k, everywhere in the region $r \geq R$), and show that the circulation $\Gamma = \oint_C \vec{v}\cdot d\vec{r}$ around any circle $r =$ const. is equal to $2\pi k$. [Notice that, although the flow is irrotational, the circulation may not vanish, since the point $r = 0$ about which the fluid tends to rotate is not a point in the fluid itself.]

(ii) Show that the streamlines are given by the equation

$$-v_0 y\left(1 - \frac{R^2}{x^2+y^2}\right) - \frac{k}{2}\ln(x^2+y^2) = \text{constant},$$

in rectangular coordinates. Exhibit these streamlines schematically for $k = 0$ and for $k \neq 0$.

(iii) Where are the stagnation points (i.e., the points at which the flow velocity vanishes) in the flow?

8-17 Suppose that an infinite circular cylinder which is *spinning about its axis* is placed in an initially uniform flow. Assuming a potential flow, one may then describe the fluid motion in terms of the velocity potential obtained in Problem 8-16, that is, by

$$\phi(r,\theta) = a_0 + \frac{\Gamma}{2\pi}\theta - v_0\left(r + \frac{R^2}{r}\right)\cos\theta \quad (r \geq R)$$

with a suitable nonzero value for the circulation Γ (evaluated around any circle $r =$ const.).

(i) Find the pressure distribution over the cylindrical surface.

(ii) In contrast to the case with zero circulation, the fluid now exerts on the cylinder a nonvanishing force in the y-direction, i.e., in the direction perpendicular to the incident fluid velocity $\vec{v}_0 = -v_0\mathbf{i}$.

Find the magnitude of this force. [This force, often called the *Magnus force*, is present for any cylindrical body (not only of circular cross section).]

8-18 Consider an infinite circular cylinder of radius R which is moving with velocity $\vec{u}(t)$ (in the direction perpendicular to its axis) in an incompressible ideal fluid of constant density ρ.

(i) Determine the potential flow of the fluid past the cylinder.
(ii) Show that, if the cylinder moves with a constant velocity (i.e., $\vec{u} = \text{const.}$), the fluid exerts no force on the cylinder.
(iii) Find the induced mass per unit length of the cylinder when it is accelerating, i.e., $\frac{d\vec{u}}{dt} \neq 0$.

8-19 If a system involving a spring (or pendulum) is placed in an ideal fluid of constant density ρ, its frequency of oscillation gets altered because of the effect from the surrounding fluid. Assuming that the potential flow description is valid, derive such frequency change when the system corresponds to (a) a small heavy ball of mass m (with uniform mass density ρ_0) connected to a horizontal spring of spring constant k, and (b) a simple pendulum with the small ball attached to a light string of length l.

8-20 The presence of viscosity necessarily leads to the dissipation of energy. It is especially simple to demonstrate this for an incompressible viscous flow (with constant density ρ and no external force) satisfying the Navier–Stokes equation $\frac{\partial \vec{v}}{\partial t} + (\vec{v}\cdot\vec{\nabla})\vec{v} = -\frac{1}{\rho}\vec{\nabla}p + \nu\vec{\nabla}^2\vec{v}$ or, equivalently, the equation

$$\frac{\partial v_i}{\partial t} = -v_k \frac{\partial v_i}{x_k} - \frac{1}{\rho}\frac{\partial p}{\partial x_i} + \frac{1}{\rho}\frac{\partial \sigma_{ij}}{\partial x_j},$$

where $\sigma_{ij} = \eta\left(\frac{\partial v_j}{\partial x_i} + \frac{\partial v_i}{\partial x_j}\right)$.

(i) Show that the time derivative of the kinetic energy density $\frac{1}{2}\rho\vec{v}^2$ can be expressed as

$$\frac{\partial}{\partial t}\left(\frac{1}{2}\rho\vec{v}^2\right) = -\vec{\nabla}\cdot\left[\rho\vec{v}\left(\frac{1}{2}\vec{v}^2 + \frac{p}{\rho}\right) - \vec{v}\cdot\vec{\sigma}\right] - \sigma_{ij}\frac{\partial v_i}{\partial x_j},$$

where $\vec{v}\cdot\vec{\sigma}$ denotes the vector whose ith component is $v_j\sigma_{ji}$.

(ii) By integrating the equation obtained in (i) over the whole volume of the fluid, show that, when the fluid velocity vanishes at infinity, the rate of dissipation for the total kinetic energy

$E_{\text{kin}} = \int \frac{1}{2}\rho\vec{v}^2\, d^3\vec{r}$ is given by

$$\frac{dE_{\text{kin}}}{dt} = -\frac{\eta}{2}\int\left(\frac{\partial v_j}{\partial x_i} + \frac{\partial v_i}{\partial x_j}\right)\left(\frac{\partial v_j}{\partial x_i} + \frac{\partial v_i}{\partial x_j}\right) d^3\vec{r}$$

$$= -\eta\int(\vec{\nabla}\times\vec{v})^2\, d^3\vec{r}\,.$$

(Notice that, since we must have $\frac{dE_{\text{kin}}}{dt} < 0$, the viscosity η must always be positive.)

8-21 Consider a viscous flow of constant density ρ in a pipe of annular cross section, the internal and external radii being R_1, R_2. When Δp is the pressure difference between the ends of the pipe and l is its length, find the velocity distribution in the pipe. Also determine the (mass) discharge.

8-22 Let us consider the motion of a viscous fluid (having constant density ρ) between two infinite coaxial cylinders of radii R_1 and R_2 ($R_2 > R_1$), rotating about their axis with angular velocities ω_1 and ω_2.

(i) Find the velocity and pressure distributions in the fluid.

(ii) Also determine the frictional force acting on unit length of each cylinder.

8-23 For a steady flow of a viscous fluid past a fixed sphere of radius R (Stokes flow), it was found that the velocity field has the form (see Sec. 8.5)

$$\vec{v}(\vec{r}) = \vec{u} - \frac{3}{4}R\frac{\vec{u} + \hat{r}(\vec{u}\cdot\hat{r})}{r} + \frac{1}{4}R^3\frac{3\hat{r}(\vec{u}\cdot\hat{r}) - \vec{u}}{r^3},$$

where $\vec{u}$ denotes the fluid velocity at infinity. Then show that, for the fluid having viscosity η, the pressure field corresponding to this flow should be given as $p(\vec{r}) = p_0 - \frac{3}{2}\eta R\frac{\vec{u}\cdot\hat{r}}{r^2}$.

9 Motion in a Non-Inertial Reference Frame

SUMMARY

9.1 Moving coordinate systems and inertial forces

- There is often a practical need to have dynamical laws written down with respect to certain non-inertial reference frames — an observer might very well be stuck with a specific reference frame the spatial origin of which is in an accelerated motion and/or the spatial axes of which are rotating relative to those of an inertial frame. Mechanical laws given in a non-inertial frame are marked by the appearance of inertial or fictitious/apparent forces in the equations of motion.
- Let $\vec{r}$, $\vec{r}'$ denote the position of a moving particle with respect to different spatial origins, O and O', where O is fixed in some inertial frame and O' is moving relative to O (so that, if one writes $\overrightarrow{OO'} = \vec{h}$, one can associate O' with a point having $\vec{v}_h \equiv \dot{\vec{h}}$ and $\vec{a}_h \equiv \ddot{\vec{h}}$). For the velocity and acceleration of the particle, one will then have $\vec{v} = \vec{v}' + \vec{v}_h$ and $\vec{a} = \vec{a}' + \vec{a}_h$ where $\vec{v} \equiv \frac{d\vec{r}}{dt}$, $\vec{v}' \equiv \frac{d\vec{r}'}{dt}$, $\vec{a} \equiv \frac{d^2\vec{r}}{dt^2}$, and $\vec{a}' \equiv \frac{d^2\vec{r}'}{dt^2}$. Newton's second law, $m\frac{d^2\vec{r}}{dt^2} = \vec{F}$ in an inertial frame, assumes in the primed system the form

$$m\frac{d^2\vec{r}'}{dt^2} = \vec{F} - m\vec{a}_h,$$

where $-m\vec{a}_h$ is the related inertial (or fictitious) force. Therefore, a uniformly accelerated observer finds the uniform gravitational force, i.e., $\vec{F} = m\vec{g}$ in an inertial frame, to be represented effectively by the form $\vec{F}' = m\vec{g}'$ where $\vec{g}' = \vec{g} - \vec{a}_h$.

- Suppose there is an observer stuck with the rotating axes $(\mathbf{e}_i^*) \equiv (\mathbf{i}^*, \mathbf{j}^*, \mathbf{k}^*)$, obeying the equations $\frac{d}{dt}\mathbf{e}_i^* = \vec{\omega} \times \mathbf{e}_i^*$ $(i = 1, 2, 3)$ according to an inertial observer. Then, for a general vector quantity $\vec{A}(t)$ with the representation $\vec{A}(t) = A_i^*(t)\mathbf{e}_i^*$, the rotating observer may use the formula $\frac{d^*\vec{A}}{dt} = \dot{A}_i^*\mathbf{e}_i^*$ for its rate of change, while for an inertial observer the rate of change for this vector should be given by

$$\frac{d\vec{A}}{dt} = \dot{A}_i^*\mathbf{e}_i^* + A_i^*\dot{\mathbf{e}}_i^* = \frac{d^*\vec{A}}{dt} + \vec{\omega} \times \vec{A}.$$

Applying this formula with $\vec{A} = \vec{r}$ and $\vec{v}$ ($\vec{r}$, $\vec{v}$ denote the position and velocity of a particle), one finds that

$$\vec{v} = \vec{v}^* + \vec{\omega} \times \vec{r} \qquad \left(\vec{v}^* \equiv \frac{d^*\vec{r}}{dt}\right)$$

and, for particle's acceleration,

$$\vec{a} = \vec{a}^* + 2\vec{\omega} \times \vec{v}^* + \vec{\omega} \times (\vec{\omega} \times \vec{r}) + \frac{d\vec{\omega}}{dt} \times \vec{r} \qquad \left(\vec{a}^* \equiv \frac{d^{*2}\vec{r}}{dt^2}\right).$$

Newton's second law in an inertial frame, $m\frac{d^2\vec{r}}{dt^2} = \vec{F}$, thus assumes the following form in a rotating frame:

$$m\frac{d^{*2}\vec{r}}{dt^2} = \vec{F} - 2m\vec{\omega} \times \frac{d^*\vec{r}}{dt} - m\vec{\omega} \times (\vec{\omega} \times \vec{r}) - m\frac{d\vec{\omega}}{dt} \times \vec{r}.$$

The term $-2m\vec{\omega} \times \frac{d^*\vec{r}}{dt}$ is called the Coriolis force, and $-m\vec{\omega} \times (\vec{\omega} \times \vec{r})$ the centrifugal force: they are inertial forces.

- We discussed various mechanical problems where the centrifugal force plays an important role; e.g., the conical pendulum rotating at fixed angular velocity, and the shape of spinning liquid in a bucket under uniform gravity. Also, in a frame rigidly fixed at a specific location on the Earth's surface (say, at position $\vec{r}$ from the center of the Earth), the gravitational force $\vec{F} = m\vec{g}$ and the centrifugal force may be combined into the effective gravitational force $\vec{F}_{\text{eff}} = m\vec{g}^*$, with $\vec{g}^* = \vec{g} - \vec{\omega} \times (\vec{\omega} \times \vec{r})$, where $\vec{\omega}$ is the Earth's rotational angular velocity.

9.2 Coriolis force effects

- The Coriolis force $-2m\vec{\omega} \times \frac{d^*\vec{r}}{dt}$, a velocity-dependent fictitious force, is significant only when the velocity (as observed in the given rotating system) is not small. As such, it is responsible for some large-scale meteorological effects on the Earth such as cyclones and trade winds.
- The effect of the Coriolis force on a freely falling object can also be found. Near the surface of the Earth in colatitude θ, the relevant equation of motion can be written

$$m\frac{d^*\vec{r}}{dt^2} = m\vec{g}^* - 2m\vec{\omega} \times \frac{d^*\vec{r}}{dt},$$

  where $\vec{g}^* = g^*\mathbf{k}^*$ is the effective gravitational acceleration, and $\vec{\omega}$ $(= \omega \sin\theta \mathbf{j}^* + \omega \cos\theta \mathbf{k}^*)$ denotes the Earth's rotational angular velocity. Writing $\vec{r} = x\mathbf{i}^* + y\mathbf{j}^* + z\mathbf{k}^*$ for the particle's position, this equation of motion in the small-ω approximation yields

$$m\ddot{x} = 2m\omega g^* t \sin\theta, \quad m\ddot{y} = 0, \quad m\ddot{z} = -mg^*.$$

  This predicts that a particle, dropped from rest at a height h from the ground, will hit the ground at a point east of the vertical below its point of release, at distance $x = \frac{1}{3}\omega \sin\theta\left(\frac{8h^3}{g^*}\right)^{1/2}$.
- In colatitude θ on the Earth's surface, suppose there is a heavy bob hanging from a long string which can swing freely in any vertical plane. If it started swinging in a definite vertical plane, the plane of swinging is seen to precess slowly about the vertical axis with angular velocity equal to $\omega \cos\theta$, where ω is the Earth's rotational velocity. This is Foucault's pendulum, and we show explicitly that the precession is due to the Coriolis force.

9.3 Other uses of rotating frames

- For certain mechanical problems, a description of the system in a suitably chosen rotating frame can serve a useful purpose. Suppose we want to know the effect of a weak constant magnetic field on a particle of mass m and charge q, moving in a bounded orbit about a fixed charge $-q'$. The related equation of motion

$$m\frac{d^2\vec{r}}{dt^2} = -\frac{k}{r^2}\hat{r} + q\frac{d\vec{r}}{dt} \times \vec{B} \quad \left(k \equiv \frac{qq'}{4\pi\epsilon_0}\right)$$

can be transformed, in terms of quantities of a rotating frame (with relative angular velocity $\vec{\omega} = -\frac{q}{2m}\vec{B}$), into that for a particle just in an inverse-square force field

$$m\frac{d^{*2}\vec{r}}{dt^2} = -\frac{k}{r^2}\hat{r},$$

dropping a term of $O(\vec{B}^2)$. Hence the orbit in the rotating frame is an ellipse; in the original frame, a slowly precessing ellipse (the Larmor effect) with the precession rate given by the Larmor frequency $\omega_L = \frac{q|\vec{B}|}{2m}$.
- A rotating frame description is also convenient for studying the restricted three-body problem, i.e., the motion of a body with very small mass m in the time-dependent gravitational field produced by two massive bodies (of masses M_1, M_2) which are revolving in circles under their mutual gravitational attraction. (Assume that m remains in the same plane, the xy-plane, in which M_1 and M_2 revolve.) One may choose the coordinate origin at the CM of M_1 and M_2, and the coordinate axes rotating with angular velocity $\omega = \sqrt{\frac{G(M_1+M_2)}{a^3}}$ (a denotes the separation distance between M_1 and M_2); then, both M_1 and M_2 appear to be "at rest", M_1 at $\vec{r}_1 = x_1\mathbf{i}^*$ with $x_1 = \frac{M_2}{M_1+M_2}a$ and M_2 at $\vec{r}_2 = x_2\mathbf{i}^*$ with $x_2 = -\frac{M_1}{M_1+M_2}a$. With respect to this rotating frame, the equation of motion for m can be written

$$m\ddot{x} = -\frac{\partial}{\partial x}V_{\text{eff}}(x,y) + 2m\omega\dot{y},$$

$$m\ddot{y} = -\frac{\partial}{\partial y}V_{\text{eff}}(x,y) - 2m\omega\dot{x},$$

with

$$V_{\text{eff}}(x,y) = -\frac{GmM_1}{[(x-x_1)^2+y^2]^{1/2}} - \frac{GmM_2}{[(x-x_2)^2+y^2]^{1/2}} - \frac{Gm(M_1+M_2)}{2a^3}(x^2+y^2),$$

where the velocity-dependent force is due to the Coriolis force. Since no work is done by the Coriolis force, we have the energy conservation law in this system,

$$\frac{1}{2}m(\dot{x}^2+\dot{y}^2) + V_{\text{eff}}(x,y) = \text{“}E\text{”} \quad (:\text{ const.}).$$

For the given effective potential $V_{\text{eff}}(x,y)$, the equations $\frac{\partial V_{\text{eff}}}{\partial x} = \frac{\partial V_{\text{eff}}}{\partial y} = 0$ have five distinct solution points: these points, called the Lagrange points, represent equilibrium positions in the rotating frame. Three of them (collinear solutions) are on the x-axis, while the other two (noncollinear solutions) are the ones for which the three masses M_1, M_2 and m form equilateral triangles. Studying the stability of these points is not quite trivial as the effect of the Coriolis force term must be taken into account. According to some careful analysis, only the equilateral triangle solutions can be stable (if certain conditions on the mass ratio M_1/M_2 are met).

Problems

9-1 Consider, in the xz-plane, the motion of the bob of a simple pendulum (of length l) which is attached to a support moving with some given acceleration $\vec{a}(t) = a_x(t)\mathbf{i} + a_z(t)\mathbf{k}$. (See the figure.)

(i) Show that the equation of motion for the angular motion of the pendulum is given by the form

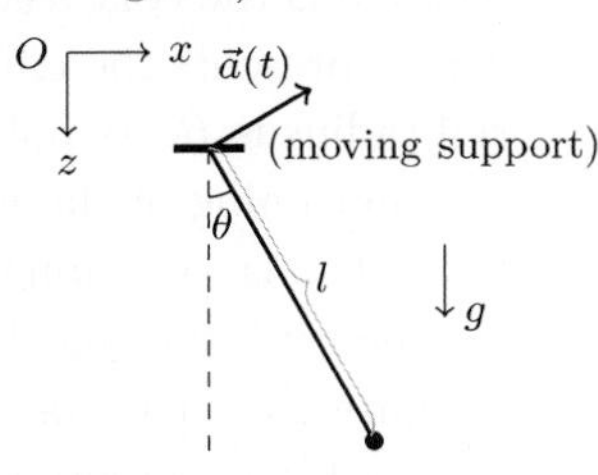

$$\ddot{\theta} + \frac{1}{l}[g - a_z(t)]\sin\theta = -\frac{1}{l}a_x(t)\cos\theta.$$

(ii) Now, concentrating on the case that the acceleration of the support is equal to a constant vector (i.e., $\vec{a} = \vec{a}_0 = \text{const.}$), interpret the result of (i) in view of the principle of equivalence.

9-2 A car of mass m travelling at speed v moves on a horizontal track so that the center of mass describes a circle of radius a. (Assume that the track is sufficiently rough to prevent the car from skidding.) Show that the limiting speed beyond which the car will overturn is given by

$$v^2 = gad/h,$$

where $2d$ ($\ll a$) is the separation of the inner and outer wheels and h is the height of the center of mass above the ground. [It will be instructive to answer this question first by working in the inertial frame, and then by considering in the frame of reference that rotates at the same rate as the car.]

9-3 In a closed system of many particles interacting through mutual central forces, the total angular momentum (defined in an inertial frame), $\vec{L} = \sum_i m_i \vec{r}_i \times \vec{v}_i$, is conserved, i.e., $\frac{d\vec{L}}{dt} = 0$. Suppose that such system is seen by an observer whose spatial axes are rotating with angular velocity $\vec{\omega}$ relative to an inertial frame. When the observer takes for his total angular momentum vector the expression $\vec{L}^* = \sum_i m_i \vec{r}_i \times \vec{v}_i^*$ where $\vec{v}_i^* = \frac{d^* \vec{r}_i}{dt}$, find the equation giving the rate $\frac{d^* \vec{L}^*}{dt}$ (that has the same content as the conservation equation $\frac{d\vec{L}}{dt} = 0$ in the internal frame). Also write the (appropriately generalized) equation valid for $\frac{d^* \vec{L}^*}{dt}$ when there are some external torques acting on the system.

9-4 On the Earth's surface the gravitational potential is not precisely of the form $-\frac{GM}{r}$ (M is the mass of the Earth), but can be better approximated by including a small quadrupole term, i.e., by the form

$$\mathcal{G}(r, \theta) = -\frac{GM}{r} - \frac{GQ}{4r^3}(3\cos^2\theta - 1),$$

where θ is the colatitude and $Q < 0$. The quadrupole correction term here represents the effect of Earth's flattening; if the Earth's equatorial radius is R, its polar radius is equal to $R(1 - \eta)$, with $\eta \approx 0.0034$. The flattening of the Earth is a natural consequence of Earth's rotation; that is, over long periods of time, the Earth is capable of plastic deformations and behaves more like a viscous liquid than a solid (despite the fact that its crust does have some rigidity). Then it is reasonable to suppose that one can determine the Earth's shape by demanding it be in equilibrium under the combined gravitational and centrifugal forces, just as in the case of liquid's shape inside a rotating vessel.

(i) Adopting the above picture, write the equation giving the shape of the Earth and also near the surface the expression for $\vec{g}^*$, the apparent gravitational acceleration (including the effect of Earth's flattening) at a point at colatitude θ.
(ii) Based on the shape equation for the Earth, obtain the approximate value for the quadrupole moment Q using the known values for M, R, ω and η.
(iii) Determine the angle α between the direction of the apparent gravitational acceleration $\vec{g}^*$ and the radial (i.e., towards the Earth's center) direction, in colatitude θ. [*Remark*: It is to be noted that the resulting formula for α, which generalizes the formula (9.31) in the text, provides a good fit to the actual plumb-line deviation at any latitude.]

9-5 The wind speed at colatitude θ is v. By considering the forces on a small volume of air, show that the pressure gradient required to balance the horizontal component of the Coriolis force, and thus to maintain a constant wind direction, is $\frac{dP}{dx} = 2\rho\omega v \cos\theta$, where ρ is the density of the air and ω the rotational angular velocity of the Earth.

9-6 A heavy object is thrown *up* into the air. Calculate the deflection of the object (to first order in ω) when it hits the ground due to the Coriolis force. Compare the result to that of an object dropped at rest from its maximum height.

9-7 A projectile is launched due north (the direction of $\mathbf{j}^*$) from a point at colatitude θ at an angle $\frac{\pi}{4}$ to the horizontal, and aimed at a target whose distance is y_0. Assume that y_0 is small compared to the Earth's radius.

(i) Show that, if one does not take the effects of the Coriolis force into account, the projectile will miss the target by a distance

$$x = \omega \left(\frac{2y_0^3}{g^*} \right)^{1/2} \left(\cos\theta - \frac{1}{3}\sin\theta \right),$$

where ω is the rotational angular velocity of the Earth and the direction of $\mathbf{i}^*$ points to the east.

(ii) Give simple explanations for the fact that the deviation is to the east near the north pole, but to the west both at the equator and near the south pole.

9-8 Solve the equation of motion of a particle falling freely from height h to *second order* in ω, and show that there is a deviation to the south (in the northern hemisphere), given by

$$y = -\left(\frac{2\omega^2 h^2}{3g^*}\right) \sin\theta \cos\theta$$

as well as that to the east. (Assume the same notation as used in our discussion in Sec. 9.2.) Calculate both components of the deviation for $h = 400$ m and latitude 50° N. How would an *inertial* observer interpret this deviation to the south?

9-9 Let us consider the motion of a projectile, assuming that the projectile is subject to air resistance proportional to its velocity. The equation of motion in the reference frame fixed on the surface of the Earth can then be written

$$m\frac{d^{*2}\vec{r}}{dt^2} = -b\frac{d^*\vec{r}}{dt} + m\vec{g}^* - 2m\vec{\omega} \times \frac{d^*\vec{r}}{dt} \quad (b > 0),$$

using the same notational convention as used in Sec. 9.2. Solve this equation of motion to first order in ω under the initial conditions $\vec{r}(t{=}0) = 0$ and $\frac{d^*\vec{r}}{dt}\Big|_{t=0} = v_{0y}\mathbf{j}^* + v_{0z}\mathbf{k}^*$. From the resulting motion, indicate notable features that stem from the existence of the Coriolis force.

9-10 A small object of mass m is ejected from an orbital station that stays in a circular orbit of radius r_0 around the Earth (in such a way that the object, after the ejection, transfers to another Kepler orbit around the Earth). We are here interested in knowing how *the motion of the small object would appear to the astronauts on the station.* Assume that the astronauts use the reference frame whose origin lies in the station and the axes $(\mathbf{i}^*, \mathbf{j}^*, \mathbf{k}^*)$ given as in the figure. That is, $\mathbf{k}^*$ points in the direction perpendicular to the plane of the orbit (i.e., parallel to the angular velocity vector $\vec{\Omega}$ of revolution of the station); $\mathbf{i}^*$ lies in the plane of the orbit and extends radially outward from the center of the Earth; $\mathbf{j}^*$ is taken to be parallel to the orbital velocity $\vec{v}_s$ of the station.

(i) When $\vec{r}(t) = x(t)\mathbf{i}^* + y(t)\mathbf{j}^* + z(t)\mathbf{k}^*$ denotes the position of the ejected object relative to the astronaut's frame of reference, write

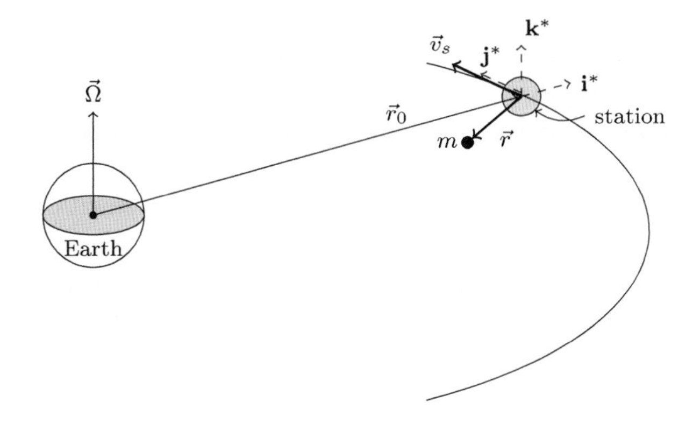

the equation giving the acceleration (in the astronaut's frame) $\vec{a}^* \equiv \frac{d^{*2}\vec{r}}{dt^2}$.

(ii) Show that, when the ejected object is at distance not too far away from the station so that $|\vec{r}|$ can be taken to be sufficiently small compared to r_0, the equation of (i) implies, approximately,

$$\ddot{x} = 3\Omega^2 x + 2\Omega y, \quad \ddot{y} = -2\Omega\dot{x}, \quad \ddot{z} = -\Omega^2 z$$

when $\vec{\Omega} = \Omega\mathbf{k}^*$.

(iii) Based on the equations obtained in (ii), discuss the motion with following initial conditions:

(a) $x(0) = y(0) = z(0) = \dot{x}(0) = \dot{y}(0) = 0$, but $\dot{z}(0) = v_0\,(\neq 0)$ (i.e., ejected from the station in a direction perpendicular to the plane of orbit),

(b) $x(0) = y(0) = z(0) = \dot{y}(0) = \dot{z}(0) = 0$, $\dot{x}(0) = -v_0$ with $v_0 > 0$ (i.e., ejected from the station in the radial direction toward the Earth), and

(c) $x(0) = y(0) = z(0) = \dot{x}(0) = \dot{z}(0) = 0$, $\dot{y}(0) = v_0\,(\neq 0)$ (i.e., ejected from the station parallel or antiparallel to the orbital velocity $\vec{v}_s$).

9-11 A charged particle (e.g., an electron) with a nonzero internal angular momentum, i.e., spin $\vec{S}$ can possess an intrinsic magnetic dipole moment $\vec{m} = \gamma\vec{S}$, the constant γ being the gyromagnetic ratio. In the presence of an external magnetic field $\vec{B}$ (which may be time-dependent), the equation of motion for the spin $\vec{S}$ is

$$\frac{d\vec{S}}{dt} = \vec{m} \times \vec{B} = \gamma\vec{S} \times \vec{B}.$$

(i) Show that in a frame rotating at a certain angular velocity, the effect of the magnetic field can be made to vanish. Find this angular velocity $\vec{\omega}_0$.

(ii) Suppose that the magnetic field has two (mutually perpendicular) components, viz., $\vec{B} = B_0\mathbf{k} + \vec{B}_1(t)$, where B_0 is a constant and $\vec{B}_1(t)$ is a rotating magnetic field of small magnitude: $\vec{B}_1(t) = B_1(\cos\omega t\,\mathbf{i} + \sin\omega t\,\mathbf{j})$, with a certain constant angular frequency ω and $|B_1| \ll |B_0|$. Find the equation of motion valid in a rotating frame $(\mathbf{i}^*, \mathbf{j}^*, \mathbf{k}^*)$ where $\mathbf{k}^* = \mathbf{k}$ and $\mathbf{i}^*$ coincides with the direction of $\vec{B}_1(t)$ (i.e., $\mathbf{i}^* = \cos\omega t\mathbf{i} + \sin\omega t\mathbf{j}$) and then solve the equation. Describe qualitatively what happens to the spin vector $\vec{S}$ if the angular frequency ω of the rotating field approaches the angular velocity value ω_0 (considered in (i)). [*Comment*: This kind of consideration is important in the study of nuclear magnetic resonance (NMR).]

9-12 Let us consider a system in two dimensions, defined by the equations of motion

$$\ddot{x} = -x\cos 2\omega t - y\sin 2\omega t,$$
$$\ddot{y} = -x\sin 2\omega t + y\cos 2\omega t$$

or, writing $\mathbf{x} = \begin{pmatrix} x \\ y \end{pmatrix}$,

$$\ddot{\mathbf{x}} = -\mathbf{S}(t)\mathbf{x}, \quad \text{with } \mathbf{S}(t) = \begin{pmatrix} \cos 2\omega t & \sin 2\omega t \\ \sin 2\omega t & -\cos 2\omega t \end{pmatrix}.$$

(i) Show that the above equations of motion describe the motion of a unit mass in the plane under the influence of a *rotating saddle potential* — the potential whose graph is obtained by rotating the graph of $z = \frac{1}{2}(x^2 - y^2) \equiv U_0(\mathbf{x})$ around the z-axis with angular velocity ω; that is, they are realized as

$$\ddot{\mathbf{r}} = -\nabla U(\mathbf{x}, t)$$

where $\vec{r} = x\mathbf{i} + y\mathbf{j}$, and

$$U(x, y, t) = U_0(x\cos\omega t + y\sin\omega t, -x\sin\omega t + y\cos\omega t).$$

(ii) Rewrite the equations of motion taking as dynamical variables the new coordinates (X, Y) which are related to the initial coordinates (x, y) by

$$X = x\cos\omega t + y\sin\omega t, \quad Y = -x\sin\omega t + y\cos\omega t.$$

Can you solve the resulting equations of motion?

(iii) Regarding ω as an adjustable parameter, discuss the stability of this system at the critical point $x = y = 0$. [Notice that an unstable point of an unrotated potential can turn into a stable point of the rotated potential.]

9-13 To determine explicitly the collinear Lagrange points L_1, L_2 and L_3 (as shown in the figure in Sec. 9.3), it is necessary to solve appropriate quintic equations resulting from

$$-\frac{\xi_2(\xi - \xi_1)}{|\xi - \xi_1|^3} + \frac{\xi_1(\xi - \xi_2)}{|\xi - \xi_2|^3} - \xi = 0$$

(see (9.57a) in the text, with η set to zero), where $\xi_1 = \frac{M_2}{M_1+M_2}$ and $\xi_2 = -\frac{M_1}{M_1+M_2}$. Solve this equation numerically for the Earth–Moon system, and represent schematically the corresponding orbits of mass m in the Earth–Moon orbital plane in the inertial reference frame.

9-14 The general three-body problem (involving masses m_1, m_2 and m_3) in the center-of-mass frame is defined by the equations

$$\ddot{\vec{r}}_1 = -Gm_2\frac{\vec{r}_1 - \vec{r}_2}{|\vec{r}_1 - \vec{r}_2|^3} - Gm_3\frac{\vec{r}_1 - \vec{r}_3}{|\vec{r}_1 - \vec{r}_3|^3},$$

$$\ddot{\vec{r}}_2 = -Gm_3\frac{\vec{r}_2 - \vec{r}_3}{|\vec{r}_2 - \vec{r}_3|^3} - Gm_1\frac{\vec{r}_2 - \vec{r}_1}{|\vec{r}_2 - \vec{r}_1|^3},$$

$$\ddot{\vec{r}}_3 = -Gm_1\frac{\vec{r}_3 - \vec{r}_1}{|\vec{r}_3 - \vec{r}_1|^3} - Gm_2\frac{\vec{r}_3 - \vec{r}_2}{|\vec{r}_3 - \vec{r}_2|^3}$$

with the position vectors related by

$$m_1\vec{r}_1 + m_2\vec{r}_2 + m_3\vec{r}_3 = 0.$$

(i) Show that, in terms of the relative position vectors $\vec{s}_1$, $\vec{s}_2$, $\vec{s}_3$ defined by

$$\vec{s}_1 = \vec{r}_3 - \vec{r}_2, \quad \vec{s}_2 = \vec{r}_1 - \vec{r}_3, \quad \vec{s}_3 = \vec{r}_2 - \vec{r}_1$$

(so that we have $\vec{s}_1+\vec{s}_2+\vec{s}_3=0$), the above equations of motion can be written in the form

$$\ddot{\vec{s}}_1 = -GM\frac{\vec{s}_1}{|\vec{s}_1|^3} + Gm_1\vec{S},$$

$$\ddot{\vec{s}}_2 = -GM\frac{\vec{s}_2}{|\vec{s}_2|^3} + Gm_2\vec{S},$$

$$\ddot{\vec{s}}_3 = -GM\frac{\vec{s}_3}{|\vec{s}_3|^3} + Gm_3\vec{S},$$

where $M \equiv m_1+m_2+m_3$, and

$$\vec{S} \equiv \frac{\vec{s}_1}{|\vec{s}_1|^3} + \frac{\vec{s}_2}{|\vec{s}_2|^3} + \frac{\vec{s}_3}{|\vec{s}_3|^3}.$$

(ii) The system of equations found in (i) decouples into a set of three similar two-body equations if $\vec{S}=0$, and the condition $\vec{S}=0$ is clearly realized if $|\vec{s}_1|^2=|\vec{s}_2|^2=|\vec{s}_3|^2$. Based on this observation, obtain the most general equilateral-triangle solutions allowed for this system. [*Remark*: Note that the masses can assume more general Kepler orbits than circles.]

(iii) The general collinear solutions can be obtained if the collinearity conditions

$$\vec{s}_1 = \lambda\vec{s}_3, \quad \vec{s}_2 = -(1+\lambda)\vec{s}_3$$

(λ is a scalar to be determined) are assumed with the solutions to the equations of (i). Show that such collinear solutions are possible only when λ corresponds to a root of the quintic equation

$$\begin{aligned}(m_1+m_2)\lambda^5 &+ (3m_1+2m_2)\lambda^4 + (3m_1+m_2)\lambda^3 \\ &- (m_2+3m_3)\lambda^2 - (2m_2+3m_3)\lambda \\ &- (m_2+m_3) = 0.\end{aligned}$$

Then, by studying roots allowed for this quintic equation, demonstrate that there are three distinct families of collinear three-body solutions (which may be identified as L_1-, L_2- and L_3-points). [Again, we remark that the three masses, while being collinear, can assume more general Kepler orbits than circles.]

10 Lagrangian Mechanics

SUMMARY

10.1 The calculus of variations

- Newtonian equations of motion can be reformulated as Lagrange equations, an approach to mechanics that proved to be very useful in making further theoretical and practical considerations. These equations can be derived from a variational principle — a principle which states that, for desired solutions, some quantity should have a *stationary* (or *extremal*) value.
- When $y = y(x)$ denotes a yet unspecified function of x, consider an integral of the form

$$J = \int_{x_0}^{x_1} f(y, y'; x) dx,$$

where $f(y, y'; x)$ is a specified function of y, $y'(\equiv \frac{dy}{dx})$ and x. Let $y = \bar{y}(x)$ be a specific function, with fixed end-point values $\bar{y}(x_0) = y_0$ and $\bar{y}(x_1) = y_1$, that makes this integral stationary. Then, under a small variation made on the function, i.e., for $y(x) = \bar{y}(x) + \delta y(x)$, this integral should be unchanged to first order in $\delta y(x)$ (i.e., $\delta J = 0$, when δJ represents the first-order variation of J). From this requirement one sees that the function $y = \bar{y}(x)$, which makes the integral J stationary, should be the one satisfying the Euler–(Lagrange) equation

$$\frac{\delta J}{\delta y(x)} \equiv \frac{\partial f}{\partial y} - \frac{d}{dx}\left(\frac{\partial f}{\partial y'}\right) = 0.$$

If $\frac{\partial f}{\partial x} = 0$, i.e., with an integrand of the form $f = f(y, y')$, the stationary condition $\delta J = 0$ is actually equivalent to

$$f - y' \frac{\partial f}{\partial y'} = \text{const.}$$

(This is called the second form of the Euler equation.)

- The calculus of variations finds applications, say, in determining a geodesic on a surface, i.e., the shortest path along the surface between two given points. As an example, for a geodesic on the sphere of radius R, one may represent any path along the sphere by positing a function $\theta = \theta(\varphi)$ (θ, φ denote usual spherical angles) and then consider an integral

$$J = R \int_{\varphi_1}^{\varphi_2} \sqrt{\left(\frac{d\theta}{d\varphi}\right)^2 + \sin^2\theta}\; d\varphi,$$

expressing the total length of the path. Then, by solving the related Euler–Lagrange equation (or from the second form of the Euler equation), one deduces that the geodesic on the sphere corresponds to a great circle connecting two given points.

- For mechanical applications, one needs to consider an integral J whose integrand involves multiple coordinate variables $q_1(t), \ldots, q_n(t)$ (representing a general path in the configuration space) and their time derivatives, viz.,

$$J = \int_{t_0}^{t_1} f(q_1, \ldots, q_n, \dot{q}_1, \ldots, \dot{q}_n; t) dt.$$

The condition for the stationary path, $\delta J = 0$, subject to the fixed-end conditions $\delta q_i(t_0) = \delta q_i(t_1) = 0$ for all i are now expressed by n Euler–Lagrange equations

$$\frac{\delta J}{\delta q_i(t)} \equiv \frac{\partial f}{\partial q_i} - \frac{d}{dt}\left(\frac{\partial f}{\partial \dot{q}_i}\right) = 0, \quad \text{for each } i = 1, 2, \ldots, n.$$

These equations have the remarkable property of being invariant under reparametrization for the paths (i.e., one obtains an equivalent set of equations irrespectively of the specific coordinate variables used to represent the paths).

10.2 Hamilton's principle and Lagrange equations

- For a particle satisfying the Newtonian equation of motion $m\ddot{\vec{r}} = -\vec{\nabla}V$, one may define the related Lagrange function (or Lagrangian) as

$$L = T - V = \frac{1}{2}m(\dot{x}^2 + \dot{y}^2 + \dot{z}^2) - V(x, y, z, t).$$

Then the equation of motion can be presented as Euler–Lagrange equations

$$\frac{d}{dt}\left(\frac{\partial L}{\partial \dot{x}_i}\right) - \frac{\partial L}{\partial x_i} = 0 \quad (i = 1, 2, 3)$$

for the action integral $S = \int_{t_0}^{t_1} L(x, y, z, \dot{x}, \dot{y}, \dot{z}; t)\, dt$. This is Hamilton's principle (or the principle of stationary action): the classical Newtonian path of a particle is that for which the action integral $S = \int_{t_0}^{t_1} L\, dt$ is stationary, i.e., $\delta S = 0$ under arbitrary variations δx_i $(i = 1, 2, 3)$ which vanish at the limits of integration t_0 and t_1. Then, since within the variational approach arbitrary reparametrization for the path is allowed, one may write the equation of motion in arbitrary curvilinear coordinates q_1, q_2 and q_3 (e.g., spherical coordinates with $q_1 = r$, $q_2 = \theta$ and $q_3 = \varphi$) by simliar Lagrange equations, viz.,

$$\frac{d}{dt}\left(\frac{\partial L}{\partial \dot{q}_i}\right) - \frac{\partial L}{\partial q_i} = 0 \quad (i = 1, 2, 3)$$

where $L = L(\mathbf{q}, \dot{\mathbf{q}}; t)$ — the Lagrangian in generalized coordinates q_1, q_2 and q_3 — is obtained from $L(x, y, z, \dot{x}, \dot{y}, \dot{z}; t)$ simply by rewriting it using variables q_1, q_2 and q_3. Accordingly, Newtonian equations of motion in spherical coordinates for a particle moving in an external potential $V(r, \theta, \varphi)$ follow immediately, from the related Lagrangian

$$L = \frac{1}{2}m(\dot{r}^2 + r^2\dot{\theta}^2 + r^2 \sin^2\theta\dot{\varphi}^2) - V(r, \theta, \varphi).$$

- The Lagrangian approach can also be utilized for many-particle systems, by taking the Lagrangian to be given by the total kinetic energy minus the total potential energy. For an N-particle system in three dimensions, one then has a Lagrangian $L(q_1, \ldots, q_{3N}, \dot{q}_1, \ldots, \dot{q}_{3N}; t)$ (the q's denote arbitrarily chosen generalized coordinates in the $3N$-dimensional configuration space) in such a way that $3N$ Lagrange equations

$$\frac{d}{dt}\left(\frac{\partial L}{\partial \dot{q}_\alpha}\right) - \frac{\partial L}{\partial q_\alpha} = 0 \quad (\alpha = 1, 2, \ldots, 3N)$$

contain the same information as full Newtonian equations of motion for the system. [Note that, using generalized momenta $p_\alpha \equiv \frac{\partial L}{\partial \dot{q}_\alpha}$, the Lagrange equations are also presented as $\dot{p}_\alpha = \frac{\partial L}{\partial q_\alpha}$.] In the case of two particles for instance, consider a system defined by the Lagrangian

$$L = \frac{1}{2}m_1\dot{\vec{r}}_1^2 + \frac{1}{2}m_2\dot{\vec{r}}_2^2 + m_1\vec{g}\cdot\vec{r}_1 + m_2\vec{g}\cdot\vec{r}_2 - V_{\text{int}}(\vec{r}_1 - \vec{r}_2).$$

Choosing the center-of-mass position $\vec{R} = \frac{1}{M}(m_1\vec{r}_1 + m_2\vec{r}_2)$ (here, $M \equiv m_1 + m_2$) and relative position $\vec{r} = \vec{r}_1 - \vec{r}_2$ as new generalized coordinates, this Lagrangian can be rewritten as

$$L = \frac{1}{2}M\dot{\vec{R}}^2 + M\vec{g}\cdot\vec{R} + \frac{1}{2}\mu\dot{\vec{r}}^2 - V_{\text{int}}(\vec{r}) \quad \left(\mu \equiv \frac{m_1 m_2}{m_1 + m_2}\right)$$

thus resulting in the Lagrange equations

$$M\ddot{\vec{R}} = M\vec{g}, \qquad \mu\ddot{\vec{r}} = -\vec{\nabla}_{\vec{r}}V_{\text{int}}(\vec{r}),$$

which are familiar forms from the Newtonian approach.

- The Lagrangian approach can be generalized into the form not requiring the existence of a suitable potential, by considering fixed-end variations (or virtual displacements) about actual classical paths for the integral involving the total kinetic energy, $J = \int_{t_0}^{t_1}(\sum_i \frac{1}{2}m_i\dot{x}_i^2)dt$. If $F_i(\mathbf{x}, t)$ are Newtonian forces and $x_i = x_i^*(t)$ describe the actual classical path of the problem, then

$$\delta J = \int_{t_0}^{t_1} \sum_i m_i\dot{x}_i^*(t)\,\delta\dot{x}_i(t)\,dt$$

$$= -\int_{t_0}^{t_1} \sum_i m_i \ddot{x}_i^*(t)\, \delta x_i(t)\, dt$$

$$= -\int_{t_0}^{t_1} \sum_i F_i(\mathbf{x}^*(t), t)\, \delta x_i(t)\, dt \quad \left(\equiv -\int_{t_0}^{t_1} \delta W\, dt\right),$$

where $\delta W = \sum_i F_i(\mathbf{x}^*(t), t)\, \delta x_i$ represents the *virtual work* done by the forces F_i. Now, if we introduce generalized coordinates $(q_1, \ldots, q_n)$ so that the total kinetic energy is represented as $T(q_1, \ldots, q_n, \dot{q}_1, \ldots, \dot{q}_n; t)$, the above finding can be translated into

$$\int_{t_0}^{t_1} \sum_i \left(\frac{\partial T}{\partial q_i} - \frac{d}{dt}\frac{\partial T}{\partial \dot{q}_i} \right)\bigg|_{\mathbf{q}=\mathbf{q}^*(t)} \delta q_i(t)\, dt$$

$$= -\int_{t_0}^{t_1} \sum_i Q_i(\mathbf{q}^*(t), t) \delta q_i(t)\, dt$$

where $Q_i \equiv \sum_j F_j \frac{\partial x_j}{\partial q_i}$ are appropriate generalized forces. We thus obtain the general equations

$$\frac{d}{dt}\left(\frac{\partial T}{\partial \dot{q}_i}\right) - \frac{\partial T}{\partial q_i} = Q_i \quad (i = 1, \ldots, n).$$

For generalized forces of the form $Q_i = -\frac{\partial V}{\partial q_i}$ (i.e., derived from a potential $V(q_1, \ldots, q_n, t)$), these equations reduce to the standard Lagrange equations for the Lagrange function $L = T - V$.

- If generalized forces Q_i in a system can be expressed, using a suitable velocity-dependent potential $V(q_1, \ldots, q_n, \dot{q}_1, \ldots, \dot{q}_n; t)$, in the form

$$Q_i = -\frac{\partial V}{\partial q_i} + \frac{d}{dt}\left(\frac{\partial V}{\partial \dot{q}_i}\right),$$

such system can also be described in terms of standard Lagrange equations with the Lagrangian $L = T - V(q_1, \ldots, q_n, \dot{q}_1, \ldots, \dot{q}_n; t)$. This case arises especially if one wishes to study mechanics of a charged particle in the presence of electromagnetic fields. This is the case since the Lorentz force in component forms can be written as

$$q\,[\vec{E}(\vec{r},t) + \dot{\vec{r}} \times \vec{B}(\vec{r},t)]_i = -\frac{\partial V}{\partial x_i} + \frac{d}{dt}\left(\frac{\partial V}{\partial \dot{x}_i}\right)$$

with a velocity-dependent potential $V = q\phi(\vec{r},t) - q\dot{\vec{r}} \cdot \vec{A}(\vec{r},t)$, where $\phi(\vec{r},t)$ and $\vec{A}(\vec{r},t)$ are scalar and vector potentials in terms of which electromagnetic fields are given by $\vec{E} = -\vec{\nabla}\phi - \frac{\partial \vec{A}}{\partial t}$ and $\vec{B} = \vec{\nabla} \times \vec{A}$.

- Lagrange equations can be used to discuss relativistic mechanics of a particle (of rest mass m_0) in an external potential $V(\vec{r},t)$, if a Lagrangian which is not of the $(T-V)$-form is chosen, i.e.,

$$L = -m_0 c^2 \sqrt{1 - \dot{\vec{r}}^2/c^2} - V(\vec{r},t).$$

Note that, in the absence of any external force, the classical path should satisfy the stationary condition $\delta S = 0$, with the action S in this case given by $S = -m_0 c^2 \int_{t_0}^{t_1} d\tau$ where $d\tau = \sqrt{1 - \dot{\vec{r}}^2/c^2}\,dt$ represents the proper-time interval taken along the trajectory. This implies that the path of a free particle corresponds to a geodesic of the 4-dimensional (i.e., Minkowski) world.

10.3 Lagrange equations for systems with constraints

- The Lagrangian approach has a big advantage over the direct Newtonian approach in dealing with constrained systems, i.e., when the variables $x_1, \ldots, x_n$ of a given system happen to be bound (by *a priori* unspecified constraint forces) during the motion not to violate certain holonomic constraint conditions

$$h_l(x_1, \ldots, x_n; t) = 0 \quad (l = 1, \ldots, k).$$

One may then choose a new set of generalized coordinates $q_1, \ldots, q_n$ such that the given k constraint conditions become equivalent to the conditions $q_{n-k+1} = q_{n-k+2} = \cdots = q_n = 0$, thereby leaving $q_1, q_2, \ldots, q_{n-k}$

as unconstrained variables. Here it is important to note that, for virtual displacements observing the constraint conditions (i.e., for variations $\delta q_1, \dots, \delta q_{n-k}$), no virtual work is done by any ideal constraint forces. Because of this, dynamics in the original n-dimensional configuration space with constraint forces can be reduced to that in the $(n-k)$-dimensional configuration space (with coordinates $q_1, \dots, q_{n-k}$) on which constraint forces can be neglected. Let $\bar{T}(q_1, \dots, q_{n-k}, \dot{q}_1, \dots, \dot{q}_{n-k}; t)$ be the total kinetic energy in the reduced space (i.e., the expression obtained from the original total kinetic energy by setting $q_{n-k+1} = q_{n-k+2} = \cdots = q_n = 0$), and let $\bar{Q}_i$ $(i = 1, \dots, n-k)$ denote the explicitly given generalized force associated with the unconstrained virtual displacement δq_i (with all $q_{n-k+1}, \dots, q_n$ set to zero). Then the given constrained system can be described in terms of the equations of motion involving only unconstrained variables (and no constraint forces), viz.,

$$\frac{d}{dt}\left(\frac{\partial \bar{T}}{\partial \dot{q}_i}\right) - \frac{\partial \bar{T}}{\partial q_i} = \bar{Q}_i \quad (i = 1, \dots, n-k).$$

If $\bar{Q}_i$ are derivable from a (possibly velocity-dependent) potential $\bar{V}(q_1, \dots, q_{n-k}, \dot{q}_1, \dots, \dot{q}_{n-k}; t)$, these equations can be changed into the standard Lagrange equations with the Lagrangian $\bar{L} = \bar{T} - \bar{V}$. [For practical problems, one usually also write T, V and L for $\bar{T}$, $\bar{V}$ and $\bar{L}$.]

- The above observation has an implication on the equilibrium state of a constrained system also. In this regard, we have the principle of virtual work: a system under holonomic constraints is in the state of static equilibrium if and only if the (net) virtual work done by all explicitly applied forces vanishes for an arbitrary infinitesimal displacement satisfying the constraints. This has many useful applications — for example, the equilibrium conditions for a rigid body follow immediately from this principle.
- To explain how the reduced configuration-space description works, we have studied some concrete examples

of constrained systems (the simple pendulum, Atwood's machine, and the problem of a uniform cylinder rolling on another cylinder, etc.). Also considered is the way to determine the constraint forces within this scheme, when there is a need for them.

- There is another way of dealing with constrained systems, that using the method of Lagrange multipliers in the context of constrained variational problems. According to this method, given k constraint conditions $h_l(q_1, \ldots, q_n; t) = 0$ $(l = 1, \ldots, k)$ with the Lagrangian $L(q_1, \ldots q_n, \dot{q}_1, \ldots, \dot{q}_n; t)$, one may introduce the modified Lagrangian

$$\widetilde{L} = L + \sum_{l=1}^{k} \lambda^l h_l(q_1, \ldots, q_n; t)$$

(the k λ's are Lagrange multiplier variables), and consider equations of motion following from the stationary condition $\delta[\int_{t_0}^{t_1} \widetilde{L}\, dt] = 0$ for unrestricted variations $\delta q_i(t)$ $(i = 1, \ldots, n)$ satisfying just the fixed-end conditions. As a result, one gets $(n + k)$ equations

$$\frac{d}{dt}\left(\frac{\partial L}{\partial \dot{q}_i}\right) - \frac{\partial L}{\partial q_i} = \sum_{l=1}^{k} \lambda^l \frac{\partial h_l}{\partial q_i} \quad (i = 1, \ldots, n)$$

$$h_l(q_1, \ldots, q_n; t) = 0 \quad (l = 1, \ldots, k)$$

for $(n+k)$ unknowns $(q_1, \ldots, q_n, \lambda^1, \ldots, \lambda^k)$. This method can be effective for dealing with rather complicated constraints. Some examples are treated using this method.

10.4 Symmetries and conservation laws

- Various symmetries in the Lagrangian lead to conservation laws in the system. If the Lagrangian has no explicit time dependence (i.e., invariant under the time translation $t \to t + \alpha$), the energy given by the general expression

$$E = \sum_i \dot{q}_i \frac{\partial L}{\partial \dot{q}_i} - L$$

is a constant of motion, i.e., $\frac{dE}{dt} = 0$. If a specific generalized coordinate q_k does not appear in the Lagrangian L, i.e., L is invariant under the shift $q_k \to q_k + \alpha$ (the coordinate q_k in this case is said to be ignorable or cyclic), the related generalized momentum $p_k \equiv \frac{\partial L}{\partial \dot{q}_k}$ is a constant of motion. For example, for a particle with the associated Lagrangian (in cylindrical coordinates) $L = \frac{1}{2}m(\dot{\rho}^2 + \rho^2\dot{\varphi}^2 + \dot{z}^2) - V(\rho, z)$, the angle φ is cyclic and hence the related generalized momentum $p_\varphi \equiv \frac{\partial L}{\partial \dot{\varphi}} = m\rho^2\dot{\varphi}$ is a conserved quantity.

- More generally, if the Lagrangian $L(q_1, \ldots q_n, \dot{q}_1, \ldots, \dot{q}_n; t)$ of a system is invariant under the infinitesimal transformation of the form

$$q_i(t) \to q_i'(t) = q_i(t) + \epsilon f_i(\mathbf{q}, t) \qquad (i = 1, \ldots, n;\ \epsilon \text{ is infinitesimal})$$

one can show that the quantity given as $I = \sum_{i=1}^{n} \frac{\partial L}{\partial \dot{q}_i} f_i$ becomes a constant of motion. This is Noether's theorem.

Problems

10-1 Show that the geodesic on the surface of a right circular cylinder is a helix.

10-2 On the ellipsoid of revolution which is described by the equation

$$\frac{x^2}{a^2} + \frac{y^2}{a^2} + \frac{z^2}{c^2} = 1$$

in 3-dimensional Euclidean space, determine the geodesic curves connecting two given points on it. (Note that the given surface is generated by revolving the curve $x = g(z) = a\sqrt{1 - z^2/c^2}$ about the z-axis.)

10-3 Use Fermat's principle to show that, in an inhomogeneous medium in which the index of refraction n varies continuously in the y-direction (see the figure), the paths of light rays are such that Snell's law may hold:

$$n \cos \psi = \text{const.}$$

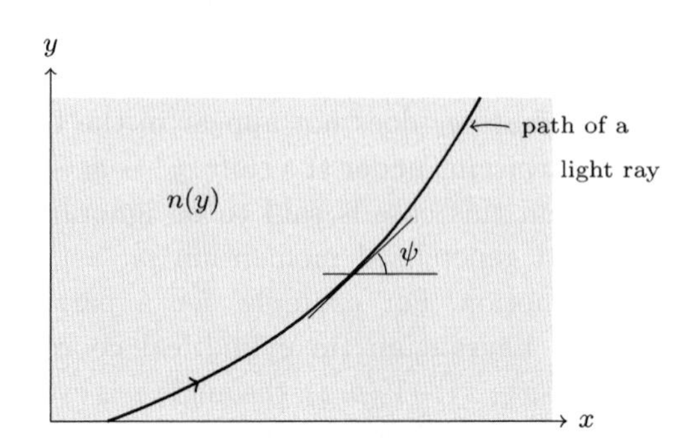

10-4 Suppose that, in a 3-dimensional space, a curve $y = y(x)$ in the xy-plane connecting the points (x_1, y_1) and (x_2, y_2) is revolved around the y-axis (see the figure) to generate a surface of revolution. Can you find the curve $y(x)$ which will generate the surface of revolution with the minimum surface area? [This problem is relevant in determining the shape of a soap film supported by a pair of coaxial rings.]

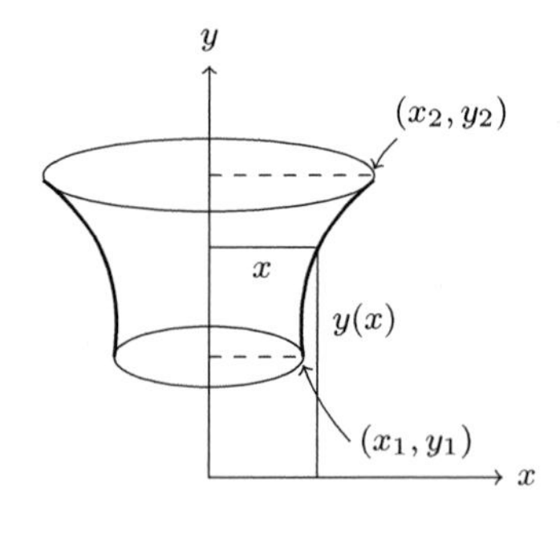

10-5 Consider a uniform chain or cable (of constant line mass density λ and of fixed length L), hanging under its own weight from two fixed points A and B (see the figure). In static equilibrium it will assume the shape which minimizes the gravitational potential energy (for given fixed length L). Set up the differential equation or "shape equation" whose solution gives the desired shape of the chain. Also find the shape (which is known as the *catenary*).

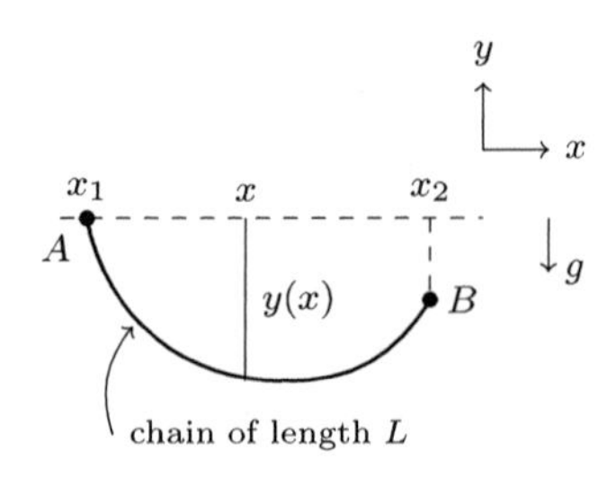

10-6 Imagine forming a closed loop (on a plane) with a string of given length l. Determine the shape of the loop for which the largest enclosed area results.

10-7 Using spherical coordinates (r, θ, φ), the Lagrangian for a particle in an external potential assumes the form

$$L = \frac{1}{2}m(\dot{r}^2 + r^2\dot{\theta}^2 + r^2\sin^2\theta\,\dot{\varphi}^2) - V(r, \theta, \varphi).$$

Write the corresponding Lagrange equations, and verify that they are fully consistent with the equations of motion in spherical coordinates obtained by rewriting $m\frac{d^2\vec{r}}{dt^2} = \vec{F}$ directly.

10-8 Consider a particle described by the Lagrangian

$$L = \frac{1}{2}m\dot{\vec{r}}^2 - V(\vec{r}, t),$$

where $\vec{r} = (x, y, z)$ denotes position coordinates with respect to an inertial frame. Suppose that this system is seen by a rotating observer, whose axes rotate with angular velocity $\vec{\omega}$. Find by the Lagrangian method the appropriate equations of motion, using the position coordinates defined relative to the axes of the rotating observer as generalized coordinates.

10-9 Let $L(q_1, \ldots, q_n, \dot{q}_1, \ldots, \dot{q}_n; t) \equiv L(\mathbf{q}, \dot{\mathbf{q}}; t)$ denote a Lagrange function given in terms of generalized coordinates $\mathbf{q} = (q_1, \ldots, q_n)$, and $\widetilde{L}(q'_1, \ldots, q'_n, \dot{q}'_1, \ldots, \dot{q}'_n; t) \equiv \widetilde{L}(\mathbf{q}', \dot{\mathbf{q}}'; t)$ the very function rewritten in terms of a new set of generalized coordinates $\mathbf{q}' = (q'_1, \ldots, q'_n)$ which are related to the old coordinates through certain invertible relations $q_i = f_i(\mathbf{q}'; t)$ $(i = 1, \ldots, n)$, that is

$$\widetilde{L}(\mathbf{q}', \dot{\mathbf{q}}'; t) = L(\mathbf{q}, \dot{\mathbf{q}}; t)\Big|_{q_i = f_i(\mathbf{q}'; t),\ \dot{q}_i = \frac{\partial f_i}{\partial q'_j}\dot{q}'_j + \frac{\partial f_i}{\partial t}}.$$

Then verify directly that we have the relations

$$\frac{\partial \widetilde{L}}{\partial q'_i} - \frac{d}{dt}\left(\frac{\partial \widetilde{L}}{\partial \dot{q}'_i}\right) = \sum_{j=1}^{n}\left[\frac{\partial L}{\partial q_j} - \frac{d}{dt}\left(\frac{\partial L}{\partial \dot{q}_j}\right)\right]\frac{\partial f_j}{\partial q'_i} \quad (i = 1, \ldots, n).$$

[This shows that if $\mathbf{q} = \bar{\mathbf{q}}(t)$ solves Lagrange equations $\frac{\partial L}{\partial q_i} - \frac{d}{dt}\left(\frac{\partial L}{\partial \dot{q}_i}\right) = 0 \quad (i = 1, \ldots, n)$, then $\mathbf{q}' = \bar{\mathbf{q}}'(t)$, with $\bar{\mathbf{q}}'(t)$ related to $\bar{\mathbf{q}}(t)$ by $\bar{q}_i(t) = f_i(\bar{\mathbf{q}}'(t); t)$, solves corresponding Lagrange equations $\frac{\partial \widetilde{L}}{\partial q'_i} - \frac{d}{dt}\left(\frac{\partial \widetilde{L}}{\partial \dot{q}'_i}\right) = 0$ $(i = 1, \ldots, n)$.]

10-10 To describe dynamics of a charged particle which obeys the Lorentz force law, one needs to first express the electromagnetic fields in terms of the scalar and vector potentials $(\phi(\vec{r}, t), \vec{A}(\vec{r}, t))$ according to $\vec{E} = -\vec{\nabla}\phi - \frac{\partial \vec{A}}{\partial t}$, $\vec{B} = \vec{\nabla} \times \vec{A}$.

(i) Verify the relation $qE_i + q\epsilon_{ijk}\dot{x}_j B_k = -\frac{\partial V}{\partial x_i} + \frac{d}{dt}\left(\frac{\partial V}{\partial \dot{x}_i}\right)$, when V denotes the velocity-dependent potential $V = q\phi(\vec{r},t) - q\dot{\vec{r}}\vec{A}(\vec{r},t)$.

(ii) Show that, given a uniform magnetic field $\vec{B} = B_0\mathbf{k}$ in space (but no electric field), one may choose the related vector potential to have the "symmetrical" form $\vec{A}(\vec{r}) = -\frac{1}{2}\vec{r}\times(B_0\mathbf{k})$ *or* an asymmetric form $\vec{A}(\vec{r}) = -B_0 y\mathbf{i}$. Also, given a uniform electric field $\vec{E} = E_0\mathbf{i}$ (but no magnetic field), show that it may be described in terms of a scalar potential $\phi(\vec{r}) = -E_0 x$ *or* by a time-dependent vector potential $\vec{A}(\vec{r}) = -E_0 t\mathbf{i}$.

(iii) The ambiguities present in the choice of the potentials for given electromagnetic fields are called *gauge transformation* ambiguities, which take the general form

$$\vec{A}(\vec{r},t) \to \vec{A}'(\vec{r},t) = \vec{A}(\vec{r},t) - \vec{\nabla}\Lambda(\vec{r},t),$$

$$\phi(\vec{r},t) \to \phi'(\vec{r},t) = \phi(\vec{r},t) + \frac{\partial\Lambda(\vec{r},t)}{\partial t},$$

where $\Lambda(\vec{r},t)$ can be an arbitrary given function of $\vec{r}$ and t. What are the effects of these transformations on the Lagrangian form (see Sec. 10.2) and to related Lagrange equations?

(iv) What gauge function $\Lambda(\vec{r},t)$ connects the two potential forms used to represent a uniform magnetic field in (ii)? Answer the analogous question for the uniform electric field case considered in (ii).

10-11 As mentioned in Sec. 10.2, a straight-line trajectory in free space corresponds to a geodesic in the 4-dimensional flat spacetime (i.e., Minkowski space). This sort of view can be extended to a nonrelativistic trajectory of a particle in the presence of a Newtonian gravitational potential, as governed by the equation of motion

$$\ddot{\vec{r}} = -\vec{\nabla}\mathcal{G}(\vec{r}) \quad (\text{with } \mathcal{G}(\vec{r}) \to 0 \text{ as } |\vec{r}| \to \infty).$$

To make such connection, consider a 4-dimensional "curved" spacetime with the line element of the form

$$\begin{aligned} ds^2 &\equiv -c^2 d\tau^2 \\ &= -\left\{1 + \frac{2}{c^2}\mathcal{G}(\vec{r})\right\}c^2 dt^2 + \left\{1 - \frac{2}{c^2}\mathcal{G}(\vec{r})\right\}(dx^2 + dy^2 + dz^2). \end{aligned}$$

Show that, to first order in $1/c^2$, the above nonrelativistic trajectory can be viewed as a geodesic of this 4-dimensional space in the sense that it follows from the condition for an extremal path connecting two given spacetime points '1' and '2',

$$\delta\left[\int_1^2 d\tau\right] = 0$$

to first nontrivial order in $\frac{1}{c^2}$. [What is discussed here constitutes the nonrelativistic limit of the formalism taken in Einstein's general relativity.]

10-12 A uniform ladder of length L and weight W is in equilibrium at an angle θ with the floor. If the wall and floor are smooth, the ladder must be held in place by a rope fastened at the bottom of the ladder and to a point of the wall at a height h above the floor. (See the figure.) Find the tension in the rope, with the help of the principle of the virtual work.

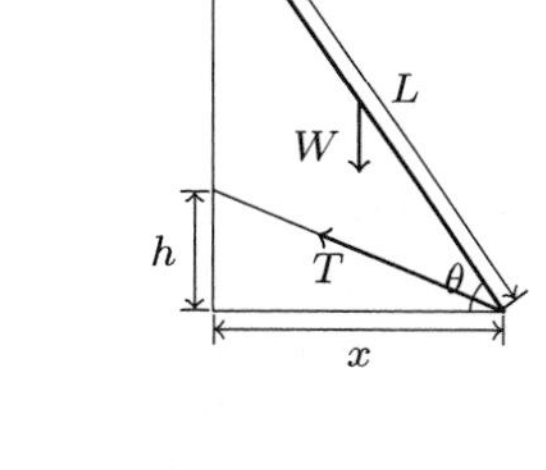

10-13 We have three logs of equal cross section (a circle of radius R) and weight (equal to Mg), placed between two vertical walls as shown in the figure. By using the principle of the virtual work (or other method), find the horizontal forces which the logs exert on the vertical walls.

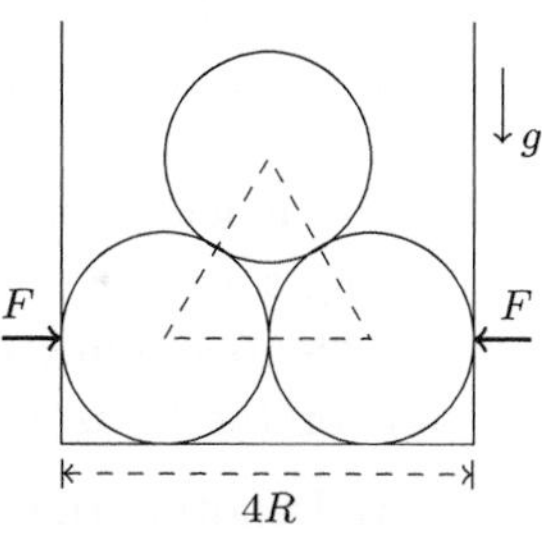

10-14 Consider a chain of n equal uniform rods (each rod has mass m and length a), smoothly jointed together and suspended from one end A_1. (See the figure.) A horizontal force $\vec{F}$ is applied to the other end A_{n+1} of this chain. Using the principle of the virtual work, find the equilibrium configuration. (Determine the angles $\theta_1, \theta_2, \ldots, \theta_n$.)

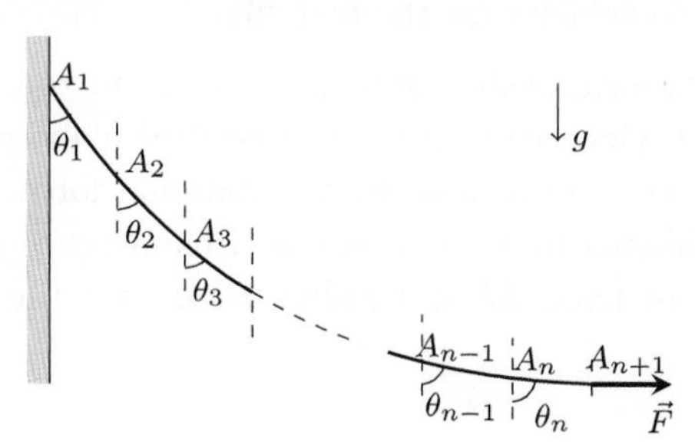

10-15 A bead of mass m slides freely on a frictionless circular wire (with negligible mass) of radius b that rotates in a horizontal plane about a point P on the circular wire with a constant angular velocity ω. (See the figure.) Obtain the Lagrange equation governing dynamics of the bead and discuss the character of the motion.

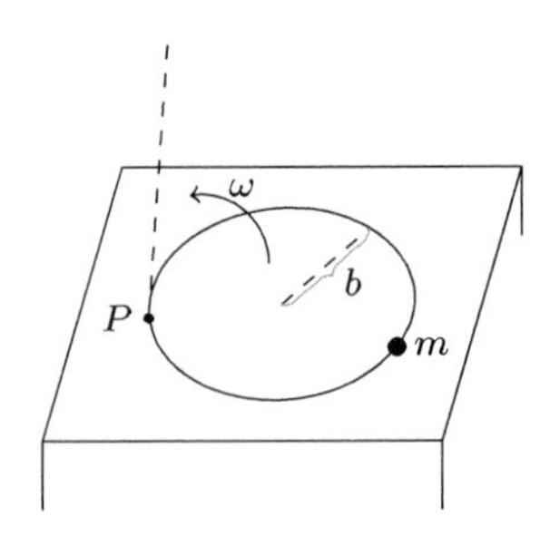

10-16 A light inextensible string is passed over a light smooth pulley, and carries a mass $4m$ on one end. The other end supports a second pulley with a string over it carrying masses $3m$ and m on the two ends. (See the figure.) Using a suitable pair of generalized coordinates, write down the Lagrange function for the system and the Euler–Lagrange equations. Then determine the downward accelerations of the three masses. Also find the tensions in the strings.

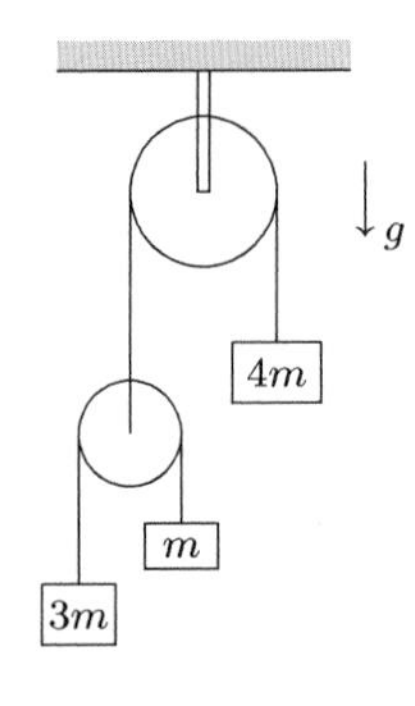

10-17 Two particles of masses m_1 and m_2, connected by a light spring with spring constant k (and the natural length l_0), can move on a frictionless one-dimensional track. Assume that the track is a straight line on the horizontal plane.

(i) Denoting the positions of the two masses as x_1 and x_2, write the Lagrangian for the system, $L(x_1, x_2, \dot{x}_1, \dot{x}_2)$. Then derive the equations of motion that the two masses satisfy.

(ii) For the system originally at rest, suppose that an impulse I for very short duration is delivered on m_1 (so that m_1 may start moving). Then, how far will m_2 move before it comes to have zero velocity for the first time?

10-18 A mechanical system involves a uniform rod AB of mass m and length l which can rotate in a vertical plane about the point A in the presence of a uniform gravitational force. The point B of the rod is hinged to a point on the circumference of a uniform circular lamina of mass M and radius R so that the circular lamina can rotate freely in the same vertical plane about the point B. (See the figure.) Obtain the Lagrangian of the system using, as your generalized coordinates, the angles of inclination θ and ϕ of AB and BO to the downward vertical. (Here, O is the center of the circular lamina.)

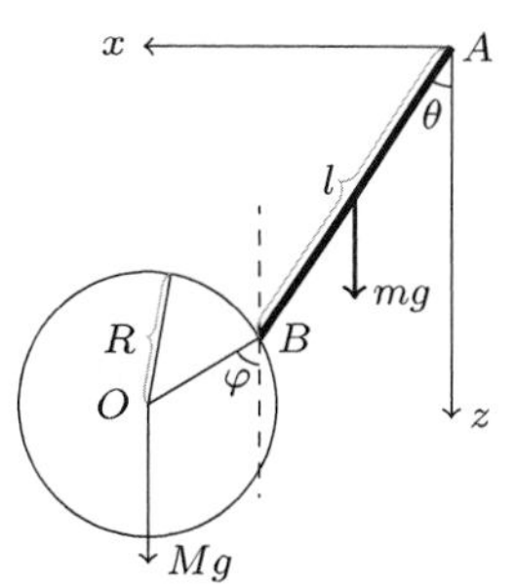

10-19 In the presence of a uniform gravitational force, consider a homogeneous cylinder of radius a and mass M, which is rolling inside a fixed cylinder of radius R. (See the figure.) Find the Lagrangian of the system using the angle ϕ between the vertical and the line joining the centers of the cylinders as generalized coordinates. Also discuss the nature of the allowed motion.

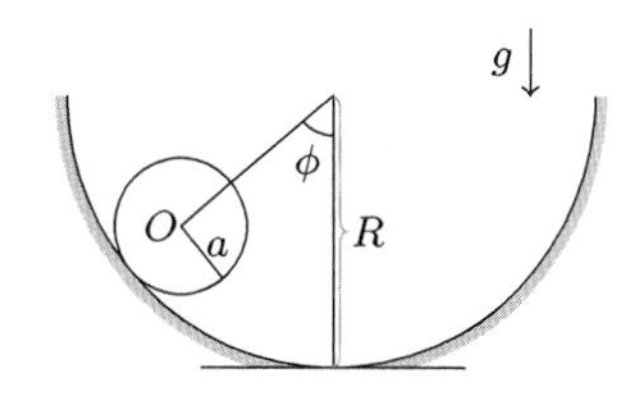

10-20 A uniform sphere of mass m and radius R rolls without slipping down a triangular block of mass M that is free to move on a frictionless horizontal surface, as shown in the figure.

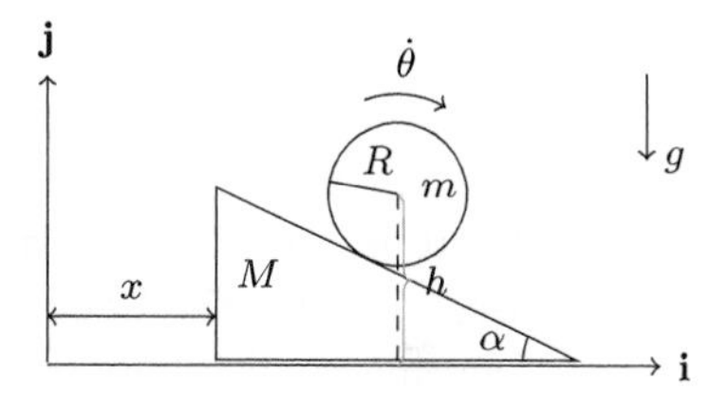

(i) Find the Lagrangian and obtain the Lagrange equations for this system, choosing x and θ (the angle of rotation of the sphere) as generalized coordinates.

(ii) Determine the motion of the system, given that all objects are initially at rest and the sphere's center is at a distance h above the horizontal surface.

10-21 Consider two particles (of masses m_1 and m_2, respectively) on a vertical plane, which are connected by a very light rod of length l. Assume further that particle 1 is constrained to move on a smooth rail on the x-axis.

(i) Taking x_1 (the position of mass 1 on the rail) and the angle θ (see the figure) as generalized coordinates, write the Lagrangian describing the system.

(ii) Derive the equation of motion that the angle $\theta(t)$ satisfies. Then discuss the behavior of this system when the angle θ undergoes small-amplitude oscillations about the vertical.

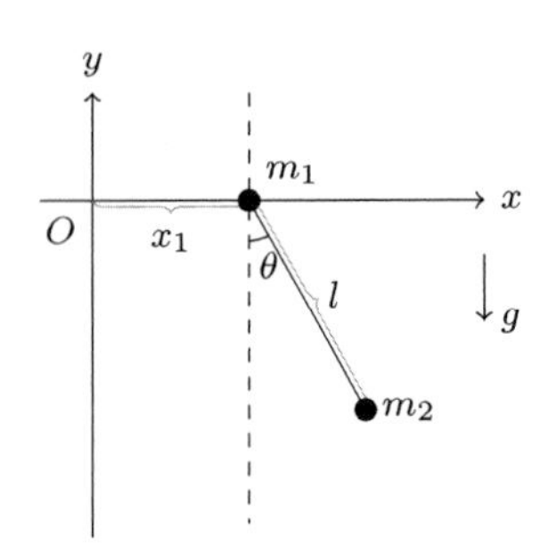

10-22 A particle of mass m rests on a smooth plane. The plane is raised to an inclination angle θ at a constant rate α (assume $\theta = 0$ at $t = 0$), causing the particle to move down the plane. Determine the motion of the particle with the help of the Lagrangian method.

10-23 Analyze the dynamics of the combined pulley system in Problem 10-16 with the help of the method of Lagrange multipliers. What are the constraint forces represented by the Lagrange multipliers here?

10-24 A particle in three dimensions is constrained to move on the surface of a sphere

$$x^2 + y^2 + z^2 = a^2.$$

There is no other external force. To describe this system one may use, with the free Lagrangian $L = \frac{1}{2}m(\dot{x}^2 + \dot{y}^2 + \dot{z}^2)$, the method of Lagrange multipliers to incorporate the above constraint condition. Discuss the allowed motion by this method.

10-25 The Lagrangian approach, together with the Lagrange multiplier method, may be employed to analyze dynamics of a mechanical system which is subject to *nonholonomic* constraints of the form

$$\sum_{j=1}^{n} a_{lj}(\mathbf{q}, t)\dot{q}_j + b_l(\mathbf{q}, t) = 0 \quad (l = 1, \ldots, k)$$

or, in differential notations,

$$\sum_{j=1}^{n} a_{lj}(\mathbf{q}, t)dq_j + b_l(\mathbf{q}, t)dt = 0 \quad (l = 1, \ldots, k).$$

(Assume that the starting Lagrangian, i.e., before considering the constraints, is $L(q_1, \ldots, q_n, \dot{q}_1, \ldots, \dot{q}_n; t) \equiv L(\mathbf{q}, \dot{\mathbf{q}}; t)$.) These constraints will genuinely be nonholonomic if these conditions cannot be integrated into the forms $h_l(\mathbf{q}, t) = 0$ $(l = 1, \ldots, k)$ with suitable h_l.

(i) For the above system, physical classical trajectories $\mathbf{q} = \mathbf{q}^*(t)$ may be chosen to be those satisfying the stationary condition

$$0 = \delta \left[\int_{t_0}^{t_1} L(\mathbf{q}, \dot{\mathbf{q}}; t) dt \right]$$

with respect to arbitrary infinitesimal (virtual) displacements $\delta q_i(t)$ fulfilling the conditions

$$\sum_{j=1}^{n} a_{ij}(\mathbf{q}, t) \delta q_j = 0 \quad (l = 1, \ldots, k).$$

Then, by repeating our reasoning detailed in Sec. 10.3 in the summary, show that related equations of motion are given by $n + k$ equations

$$\frac{d}{dt}\left(\frac{\partial L}{\partial \dot{q}_i}\right) - \frac{\partial L}{\partial q_i} = \sum_{l=1}^{k} \lambda^l a_{li} \quad (i = 1, \ldots, n)$$

$$\sum_{j=1}^{n} a_{lj} \dot{q}_j + b_l = 0 \quad (l = 1, \ldots, k).$$

There are $(n+k)$ unknowns $q_1, \ldots, q_n, \lambda^1, \ldots, \lambda^k$, where the λ's are Lagrange multipliers. [This is sometimes called the *D'Alembert–Lagrange method.*]

(ii) Use the method of (i) to study the motion of a uniform circular disk of mass m and radius a that rolls without slipping on the horizontal plane — the xy-plane — under the simplifying assumption that the disk stands perpendicular to the plane. Is the finding consistent with that from a direct application of Newton's equation of motion? [*Hint*: To describe rotational motion of the disk about its center, two angle coordinates, ψ and φ, may be used: ψ denotes the rotation angle of the disk around its symmetry axis $\mathbf{e}_3$, which points in the direction $\mathbf{e}_3 = \cos\varphi \mathbf{i} + \sin\varphi \mathbf{j}$ in space. Then one has the angular velocity vector for the disk represented by $\vec{\omega} = \dot{\psi}\mathbf{e}_3 + \dot{\varphi}\mathbf{k}$. (Can you see why?).]

10-26 The Lagrangian for a charged particle moving in the presence of a static electromagnetic field (with $\vec{E}(\vec{r}) = -\vec{\nabla}\phi(\vec{r})$, $\vec{B}(\vec{r}) = \vec{\nabla} \times \vec{A}(\vec{r})$)

is given by the form (see Sec. 10.2)

$$L = \frac{1}{2}m\dot{\vec{r}}^2 - q\phi(\vec{r}) + q\dot{\vec{r}} \cdot \vec{A}(\vec{r}).$$

What is the conserved energy function for this system?

10-27 With a mechanical Lagrangian it is possible to generalize the notion of its invariance slightly.

(i) Suppose that under infinitesimal transformations $q_i(t) \to q_i'(t) = q_i(t) + \epsilon f_i(\mathbf{q}, t)$ the Lagrangian $L(\mathbf{q}, \dot{\mathbf{q}}; t)$ changes by

$$\delta L = \epsilon \frac{d}{dt}\Lambda(\mathbf{q}, t) + O(\epsilon^2),$$

where $\Lambda(\mathbf{q}, t)$ denotes a certain function of $q_1, q_2, \ldots, q_n, t$. Then show that

$$I = \frac{\partial L}{\partial \dot{q}_i} f_i - \Lambda$$

is a constant of motion.

(ii) For a free Lagrangian $L = \frac{1}{2}\sum_{i=1}^{n} m_i \dot{\vec{r}}_i^2$, use the above observation to derive the constants of motion related to the infinitesimal Galilean transformations $\vec{r}_i \to \vec{r}_i{}' = \vec{r}_i - \epsilon\vec{u}\,t$ $(i = 1, \ldots, n)$. What do these conserved quantities represent physically?

11 Application of the Lagrangian Method: Small Oscillations

SUMMARY

11.1 Problem-solving by the Lagrangian method

- In the Lagrangian approach to mechanics, everything follows from a single scalar quantity, i.e., from the Lagrangian of a system and equations of motion can be written down using arbitrary generalized coordinates. Any holonomic constraint can be incorporated into the formalism in a very natural way, and in integrating the equations of motion one can take full advantage of conservation laws originating from symmetries of the given Lagrangian. These are illustrated with some sample problems.
- For a bead on a rotating hoop of radius a (with the axis of rotation along the vertical diameter), one can choose as the generalized coordinate the angle θ that the bead makes with the downward vertical, to have the Lagrangian

$$L = \frac{1}{2}ma^2\dot{\theta}^2 + \frac{1}{2}ma^2\omega^2 \sin^2\theta + mga\cos\theta,$$

where ω is the angular velocity of the hoop. As L has no explicit time dependence, there is the energy conservation law

$$\frac{1}{2}ma^2\dot{\theta}^2 + V_{\text{eff}}(\theta) = E \quad (\text{: const.})$$

with the effective potential $V_{\text{eff}}(\theta) = -\frac{1}{2}ma^2\omega^2\sin^2\theta - mga\cos\theta$. The nature of allowed motions may be inferred by studying the behavior of $V_{\text{eff}}(\theta)$. Especially, every equilibrium point $\theta = \theta_0$ is found by solving the equation $\frac{dV_{\text{eff}}}{d\theta}\Big|_{\theta=\theta_0} = 0$ and its stability by studying the expansion of $V_{\text{eff}}(\theta)$ about $\theta = \theta_0$. For $|\omega| < \sqrt{\frac{g}{a}}$ the stable equilibrium is found at $\theta = 0$, but for $|\omega| > \sqrt{\frac{g}{a}}$ the stable equilibrium is found at $\theta = \cos^{-1}\left(\frac{g}{a\omega^2}\right)$.

- For a spherical pendulum of length l which has two unconstrained degrees of freedom, one can choose the spherical angles θ, φ as the generalized coordinates. With the Lagrangian given by

$$L = \frac{1}{2}ml^2(\dot{\theta}^2 + \sin^2\theta\,\dot{\varphi}^2) + mgl\cos\theta,$$

the angle φ is cyclic and hence the related conjugate momentum is conserved, i.e., $p_\varphi \equiv ml^2\sin^2\theta\,\dot{\varphi} = L_z$ (: const.). Then, using this information with the conservation of total energy, one obtains a first integral involving the angle θ only; explicitly,

$$\frac{1}{2}ml^2\dot{\theta}^2 + V_{\text{eff}}(\theta) = E \quad (\text{: const.})$$

where $V_{\text{eff}}(\theta) = \frac{L_z^2}{2ml^2\sin^2\theta} - mgl\cos\theta$. The problem is thus like that of a one-dimensional motion. For the angle $\theta = \theta_0$ where $\frac{dV_{\text{eff}}(\theta)}{d\theta}$ vanishes (i.e., at the minimum of $V_{\text{eff}}(\theta)$), one finds the value $|\dot{\varphi}| = \left|\frac{L_z}{ml^2\sin^2\theta_0}\right| = \sqrt{\frac{g}{l\cos\theta_0}}$; viz., at energy $E = V_{\text{eff}}(\theta_0)$ $(\equiv E_0)$, a uniform circular rotation at a fixed angle θ_0 results. If the energy E is slightly larger than the value E_0, the angle θ will undergo a harmonic oscillation about the value θ_0, with frequency $\omega = \sqrt{\frac{g}{l}\frac{1+3\cos^2\theta_0}{\cos\theta_0}}$ (as found by studying the expansion of $V_{\text{eff}}(\theta)$ about $\theta = \theta_0$).

11.2 Small oscillations and orthogonal coordinates

- As degrees of freedom increase with a mechanical system, it becomes quickly impossible to integrate the motion.

Even for the case of two coupled pendulums restricted to move in a vertical plane, a complete integration is not possible, due to nonlinearity and the lack of enough conservation laws. In such circumstance, it will be useful to identify small oscillation patterns about the equilibrium point(s) which are characteristic of a given system. A systematic treatment on these can be given.

- Consider a mechanical system described in terms of generalized coordinates $q_1, \ldots, q_n$. If its equilibrium position is at $q_1 = q_2 = \cdots = q_n = 0$, one can represent the Lagrangian $L = T - V$ governing allowed near-equilibrium dynamics by taking the quadratic forms

$$T = \frac{1}{2}\sum_{i,j} a_{ij}\dot{q}_i\dot{q}_j, \qquad V = \frac{1}{2}\sum_{i,j} k_{ij}q_iq_j$$

with some constants a_{ij} and k_{ij}. In studying related equations of motion, it is desirable to have the kinetic energy term expressed by a sum of squares with no cross product terms, i.e., in a form $T = \frac{1}{2}\sum_i a_i'(\dot{q}_i')^2$, by introducing new generalized coordinates $q_1', \ldots, q_n'$. The new coordinates, called orthogonal coordinates, always exist: to eliminate all cross product terms involving $\dot{q}_1$ from the expression $\frac{1}{2}\sum_{i,j} a_{ij}\dot{q}_i\dot{q}_j$, consider a new coordinate q_1' (instead of q_1) given by $q_1' = q_1 + \frac{a_{12}}{a_{11}}q_2 + \cdots + \frac{a_{1n}}{a_{11}}q_n$. Note that, through rescaling of the coordinates q_i', the form $T = \frac{1}{2}\sum_i a_i'(\dot{q}_i')^2$ may then be put to the standardized form $T = \frac{1}{2}\sum_i (\dot{q}_i)^2$. With the Lagrangian $\frac{1}{2}\sum_i \dot{q}_i^2 - \frac{1}{2}\sum_{i,j} k_{ij}q_iq_j$, the equations of motion become

$$\ddot{q}_i = -\sum_{j=1}^{n} k_{ij}q_j \qquad (\text{for each } i = 1, 2, \ldots, n).$$

- For example, in the case of planar double pendulums (of length l_1 and l_2), one may choose as generalized coordinates the inclinations of the two pendulums to the vertical,

θ and φ; then the full Lagrangian is

$$L = \frac{1}{2}m_1 l_1^2\dot{\theta}^2 + \frac{1}{2}\{l_1^2\dot{\theta}^2 + l_2^2\dot{\varphi}^2 + 2l_1 l_2\dot{\theta}\dot{\varphi}\cos(\varphi - \theta)\}$$
$$- (m_1 + m_2)gl_1(1 - \cos\theta) - m_2 g l_2(1 - \cos\varphi).$$

The equilibrium position is at $\theta = \varphi = 0$. To study small oscillations, one may utilize orthogonal coordinates $x_1 = l_1\theta$ and $x_2 = l_1\theta + l_2\varphi$ to obtain the small-oscillation Lagrangian in the form

$$L = \frac{1}{2}m_1\dot{x}_1^2 + \frac{1}{2}m_2\dot{x}_2^2 - \frac{(m_1 + m_2)g}{2l_1}x_1^2 - \frac{m_2 g}{2l_2}(x_2 - x_1)^2.$$

The related equations of motion are given as

$$\begin{pmatrix}\ddot{x}_1\\ \ddot{x}_2\end{pmatrix} = \begin{pmatrix} -\frac{m_1+m_2}{m_1}\frac{g}{l_1} - \frac{m_2}{m_1}\frac{g}{l_2} & \frac{m_2}{m_1}\frac{g}{l_2}\\ \frac{g}{l_2} & -\frac{g}{l_2}\end{pmatrix}\begin{pmatrix}x_1\\ x_2\end{pmatrix}.$$

11.3 Normal modes

- The general solution for the linear homogeneous system, $\ddot{q}_i = -\sum_{j=1}^{n} k_{ij}q_j$ $(i = 1, \ldots, n)$, can be given once one has a set of n linearly independent solutions of the equations. To obtain such a set, one may try a solution of the form $q_i(t) = A_i e^{i\omega t}$, $i = 1, \ldots, n$ (with a tacit understanding that the real part of the expression is to be taken at the end), i.e., a normal mode in which all generalized coordinates q_i oscillate with a certain common frequency ω. Then, for the (in general complex) amplitudes A_i of a normal mode, one gets the equations

$$\sum_j (k_{ij} - \omega^2\delta_{ij})A_j = 0 \quad (i = 1, \ldots, n)$$

and these admit a nontrivial solution only when the frequency ω of a normal mode satisfies the secular (or characteristic equation)

$$\det(\mathbf{K} - \omega^2\mathbf{1}) = 0,$$

$\mathbf{K} = (k_{ij})$ and $\mathbf{1} = (\delta_{ij})$ being $n \times n$ matrices. For the equilibrium point to be stable, all roots ω^2 of this equation should be positive. With ω^2 chosen equal to one of the roots, the *relative* magnitudes for all A_i's now get fixed to some definite real values, leaving two arbitrary real constants with each normal mode (that is, a real solution in the form $q_i(t) = a_i C \cos(\omega t + \alpha)$, $i = 1, \ldots, n$ with certain definite values a_i, that contains two arbitrary real constants C and α). The general solution is then given by a superposition of all the normal modes (n modes) one finds in association with different roots ω^2 of the secular equation. Also, if the normal mode with frequency ω_k is denoted by $q_i(t) = a_i^{(k)} C_k \cos(\omega_k t + \alpha_k)$, $i = 1, \ldots, n$, one may contemplate on making the coordinate transformation from $(q_1, \ldots, q_n)$ to $(\Theta_1, \ldots, \Theta_n)$ according to $q_i(t) = \sum_{k=1}^{n} a_i^{(k)} \Theta_k(t)$, $i = 1, \ldots, n$; then the new coordinate Θ_k undergoes independent harmonic motion with frequency ω_k. These new coordinates are often referred to as normal coordinates of oscillation.

- For small oscillations of the double pendulum, the secular equation is

$$0 = \begin{vmatrix} \frac{m_1+m_2}{m_1}\frac{g}{l_1} + \frac{m_2}{m_1}\frac{g}{l_2} - \omega^2 & -\frac{m_2}{m_1}\frac{g}{l_2} \\ -\frac{g}{l_2} & \frac{g}{l_2} - \omega^2 \end{vmatrix}$$

$$= \omega^4 - \frac{m_1+m_2}{m_1}\left(\frac{g}{l_1} + \frac{g}{l_2}\right)\omega^2 + \frac{m_1+m_2}{m_1}\frac{g^2}{l_1 l_2}.$$

For $m_1 \gg m_2$ and l_2 not too close to l_1, two roots of this equation are approximately given by $\omega^2 \approx \frac{g}{l_2}$ (with the corresponding amplitude ratio $\frac{A_1}{A_2} \approx \frac{m_2}{m_1}\frac{l_1}{l_2-l_1}$) and $\omega^2 \approx \frac{g}{l_1}$ (with the amplitude ratio $\frac{A_1}{A_2} \approx \frac{l_1-l_2}{l_1}$). Using these normal modes, one has the general solution to the small oscillation problem given as

$$\begin{pmatrix} x_1(t) \\ x_2(t) \end{pmatrix} = C_2 \begin{pmatrix} \frac{m_2}{m_1}\frac{l_1}{l_2-l_1} \\ 1 \end{pmatrix} \cos\left(\sqrt{\frac{g}{l_2}}\, t + \alpha_2\right)$$
$$+ C_1 \begin{pmatrix} \frac{l_1-l_2}{l_1} \\ 1 \end{pmatrix} \cos\left(\sqrt{\frac{g}{l_1}}\, t + \alpha_1\right).$$

With $m_1 \ll m_2$, on the other hand, one gets by an analogous method the general solution of the form

$$\begin{pmatrix} x_1(t) \\ x_2(t) \end{pmatrix} = C_1 \begin{pmatrix} \frac{l_1}{l_1+l_2} \\ 1 \end{pmatrix} \cos\left(\sqrt{\frac{g}{l_1+l_2}}\, t + \alpha_1\right) + C_2 \begin{pmatrix} 1 \\ -\frac{m_1}{m_2}\frac{l_1}{l_1+l_2} \end{pmatrix} \cos\left(\sqrt{\frac{m_2}{m_1}\left(\frac{g}{l_1}+\frac{g}{l_2}\right)}\, t + \alpha_2\right).$$

Based on this information, one can also find appropriate normal coordinates of oscillation.

- Transverse or vertical oscillations of a loaded string is another interesting problem. This concerns a light string of length $(N+1)l$, stretched to a tension T, with equal masses m placed at regular intervals l and the string ends fastened at the horizontal axis. The corresponding small-oscillation Lagrangian can be expressed in the form

$$L = \frac{1}{2} m \sum_{j=1}^{N} \dot{y}_j^2 - \frac{1}{2} m\omega_0^2 \left[y_1^2 + (y_2 - y_1)^2 + \cdots + (y_N - y_{N-1})^2 + y_N^2 \right],$$

where $\omega_0^2 \equiv \frac{T}{ml}$. This leads to the equations of motion

$$\ddot{y}_j = \omega_0^2 (y_{j-1} - 2y_j + y_{j+1}) \quad (j = 1, 2, \ldots, N)$$

assuming the boundary conditions at the ends, $y_0 = y_{N+1} = 0$. Now, for small enough N, normal modes may be found by applying directly the method described above (i.e., for the frequencies of normal modes, solve the related secular equation, etc.). But this becomes impractical if N is not small. One can here utilize the symmetry in the system — specifically, the invariance of the equations of motion under the variable shifts $y_j \to y_{j+1}$ (for all j simultaneously) if the boundary conditions are neglected. Based on this symmetry argument, normal modes for arbitrary N have been obtained: there are N normal modes, with associated frequencies ω_n ($n = 1, \ldots, N$) and amplitudes

$a_j^{(n)}$ given by

$$\omega_n = 2\omega_0 \sin\left[\frac{n\pi}{2(N+1)}\right], \qquad a_j^{(n)} = 2\sin\left(\frac{jn\pi}{N+1}\right).$$

These results can be presented in an alternative way. If one uses the horizontal coordinate $x = jl$ and the *wave number* $k = \frac{n\pi}{L}$ (with $L \equiv (N+1)l$, the total length of the string) instead of the mode index n, the amplitudes $a_j^{(n)}$ can be represented by a function proportional to $\sin kx$, and the related normal-mode frequency ω_n by $\omega(k) = 2\omega_0 \sin\left(\frac{kl}{2}\right)$. [Notice that, for a sufficiently small wave number k, $\omega(k) \approx \omega_0 kl \ (= \sqrt{\frac{Tl}{m}}k)$, i.e., ω is linear in k.]

11.4 Dynamics of a continuous string and classical fields

- In the loaded string problem, consider the continuum limit, i.e., let $N \to \infty$, $l \to 0$ and $m \to 0$ in such a way that $L = (N+1)l$ (the total length of the string) and $\mu = \frac{m}{l}$ (mass per unit length) remain finite. Then, in the Lagrangian, it would make sense to replace $y_{j+1}(t) - y_j(t)$ by $ly'(x = jl, t)$, and $m\sum_{j=1}^{N}$ by $\mu\int_0^L dx$. This gives rise to the continuum Lagrangian involving the (classical) field $y(x,t)$

$$L = \int_0^L \left[\frac{1}{2}\mu\dot{y}^2 - \frac{1}{2}Ty'^2\right] dx \quad \left(\equiv \int \mathcal{L}\, dx\right),$$

where $\mathcal{L} = \frac{1}{2}\mu\dot{y}^2 - \frac{1}{2}Ty'^2$ is the Lagrangian density of this system. If one applies the continuum limit to the equations of motion of the loaded string, one gets the one-dimensional wave equation

$$\ddot{y} = v^2 y'' \qquad (v^2 = T/\mu).$$

- For a dynamical equation involving the field variable $y(x,t)$, one can also bring in the variational principle, i.e., look for stationary field configurations with the action integral $I = \int_{t_0}^{t_1}\int_0^L \mathcal{L}(y, \dot{y}, y')\, dx dt$ (under appropriate boundary conditions). This results in an equation of the general form

$$\frac{\partial \mathcal{L}}{\partial y} - \frac{\partial}{\partial t}\left(\frac{\partial \mathcal{L}}{\partial \dot{y}}\right) - \frac{\partial}{\partial x}\left(\frac{\partial \mathcal{L}}{\partial y'}\right) = 0,$$

which is a (classical) field equation for $y(x,t)$. With $\mathcal{L} = \frac{1}{2}\mu\dot{y}^2 - \frac{1}{2}Ty'^2$, the wave equation $\ddot{y} = v^2 y''$ follows immediately from this.

- A direct solution of the wave equation $\ddot{y} = v^2 y''$, subject to the boundary conditions $y(x=0,t) = y(x=L,t) = 0$, is known to have the general form

$$y = f(x+vt) - f(vt-x),$$

where $f(w)$ should satisfy the periodicity requirement $f(w+2L) = f(w)$. An equivalent solution can be obtained by the normal mode method. Here, normal mode solutions may be represented in the form

$$y(x,t) = A(x)e^{i\omega t}.$$

Then, from the wave equation and the boundary conditions, the amplitude function $A(x)$ should satisfy the equations

$$A''(x) + k^2 A(x) = 0 \quad \left(k = \frac{\omega}{v}\right)$$
$$A(0) = A(L) = 0.$$

These lead to the solutions of the form $A(x) = (\text{const.})\sin kx$, with k restricted to values $k = \frac{n\pi}{L}$ $(n = 1, 2, 3, \ldots)$. Hence the solution corresponding to the nth normal mode — the mode with the wave number $k = \frac{n\pi}{L}$ and angular frequency $\omega = \frac{n\pi v}{L}$ — can be written as the real part of

$$y(x,t) = A_n \sin\left(\frac{n\pi x}{L}\right)e^{i\frac{n\pi v}{L}t}.$$

The above direct solution is fully accounted for if one considers an arbitrary superposition of all these normal modes; they are related by the Fourier series expansion.

Problems

11-1 A rigid body having a rounded (e.g., spherical) shape undergoes simple rolling motion on a rough horizontal surface. (Assume that the point of contact on the horizontal surface lies on a straight line.) Let R be the radius of curvature of the base, and let the CM of the body (the point G) be a distance d downward from the center of curvature (the point O). In the figure, the body is shown in a position where the line $\overline{OG}$ makes an angle θ with the vertical. The rigid body has mass M, and the moment of inertia about the axis (through the CM) perpendicular to the plane of motion is I_{CM}.

(i) Find the Lagrangian of the system, choosing the angle θ as the relevant generalized coordinate.

(ii) Give a gross description on the motion of this body, based on the statement of energy conservation.

(iii) Determine the angular frequency as the body undergoes small oscillations about the position $\theta = 0$.

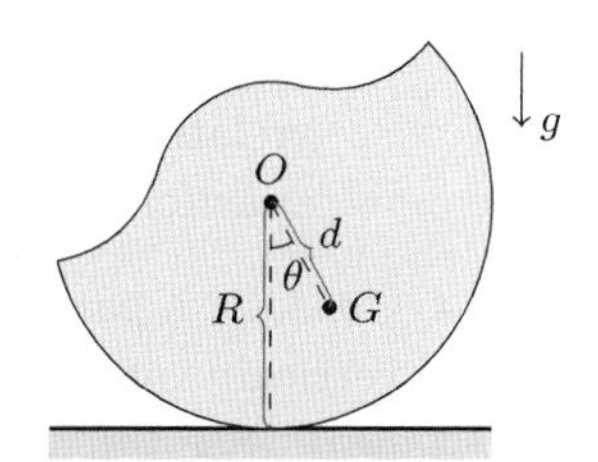

11-2 Two masses m and M ($m \neq M$) are connected by an inextensible string of length l which passes through a small hole O in a horizontal table (see the figure). Assume that the string and mass m move without friction on the table.

(i) Write the Lagrangian of the system, choosing polar coordinates (r, θ) of the mass m as generalized coordinates.

(ii) When the mass m is observed to be in a circular motion of radius R, determine the corresponding period of revolution T_0.

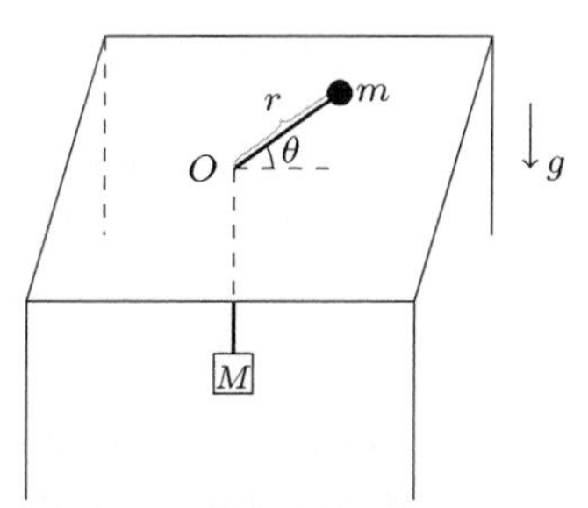

(iii) When the radius r of the mass m is not strictly constant but oscillates slightly about the value R (of the circular orbit case), determine the period of this small radial oscillation in terms of m, M and T_0 (found in (ii)).

11-3 Consider a particle of mass m which is constrained to move on the surface of a sphere, $r \equiv \sqrt{x^2 + y^2 + z^2} = l$, with no other explicit force acting. Using spherical angels (θ, φ), we then have the

Lagrangian

$$L = \frac{1}{2} m l^2 (\dot{\theta}^2 + \sin^2 \theta \, \dot{\varphi}^2).$$

(i) Show that the above Lagrangian is invariant under the infinitesimal transformations

$$\delta\theta = -\epsilon_1 \sin\varphi + \epsilon_2 \cos\varphi,$$
$$\delta\varphi = -\epsilon_1 \cot\theta \cos\varphi - \epsilon_2 \cot\theta \sin\varphi + \epsilon_3,$$

where ϵ_1, ϵ_2 and ϵ_3 denote three independent infinitesimal parameters. Demonstrate that these invariances are directly related to the 3-dimensional rotation symmetry.

(ii) Find the associated conserved quantities, and then use the related conservation laws to determine the motion of the particle.

11-4 Consider a particle on the surface of a cone (of half angle α) placed vertically and with vertex downwards in a uniform gravitational field. (See the figure.)

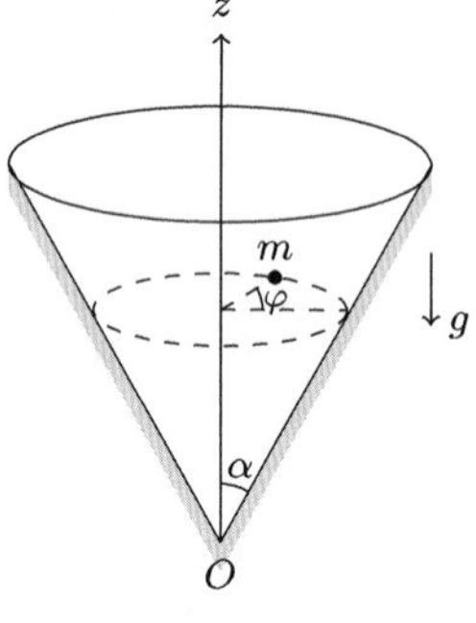

(i) Find the Lagrangian, choosing spherical coordinates $r, \theta = \alpha$ (: fixed) and φ (with the origin at the vertex) as generalized coordinates.

(ii) Identify all constants of motion, and then integrate the resulting equation of motion.

(iii) Under what initial conditions does the orbit of the particle become a circle? Discuss also the motion of the particle when its orbit deviates from a circular orbit slightly.

11-5 Consider a point mass m that is constrained to move on the surface of a torus (i.e., a donut shape) of major radius a and minor radius b whose axis is vertical (see the figure). Using a cylindrical coordinate system (ρ, φ, z) (with the origin at the center of the torus and the z axis vertically upwards),

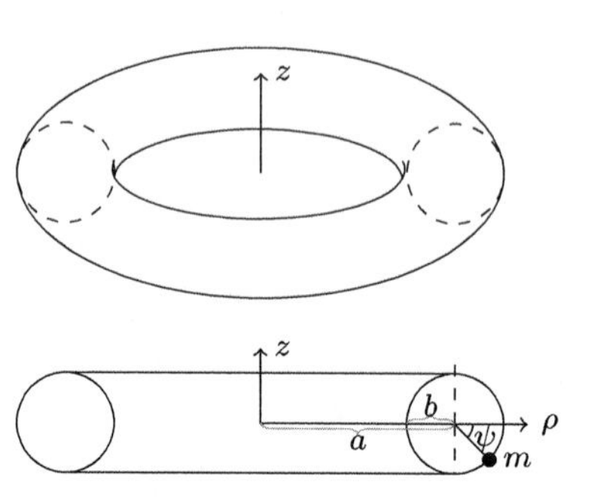

a point on the surface of the torus can be described by two angular coordinates, one of which is the azimuthal angle φ in the cylindrical coordinate system. The other, which we denote by ψ, is the angle measured with respect to the plane $z = 0$ in a vertical plane that contains the point as well as the axis, as shown in the bottom figure.

(i) Using φ and ψ as two independent coordinates, write the Lagrangian for the system.

(ii) Show that, if the point mass is in the state of uniform circular motion at $\psi = \psi_0$, its angular velocity Ω ($\equiv \dot{\varphi}$) is given by

$$\Omega^2 = \frac{g}{a \tan \psi_0 + b \sin \psi_0}.$$

(iii) Find the angular frequency of small oscillations about uniform circular motion of the point mass.

11-6 Consider a simple pendulum (in the xz-plane) of mass m and length l, whose point of support (x_0, z_0) is not fixed but oscillates over a small distance d (a) along the vertical by the formula $(x_0(t) = 0, z_0(t) = d \sin \beta t)$, or (b) along the horizontal by the formula $(x_0(t) = d \sin \beta t, z_0(t) = 0)$.

(i) Taking as the generalized coordinate the angle θ which the pendulum string makes with the downward vertical, use the Lagrangian method to deduce the appropriate equation of motion for each case mentioned above. Are the resulting equations of motion consistent with the result of Problem 9-1?

(ii) When the above vertical or horizontal oscillation of the support occurs with a high frequency β ($\gg \sqrt{\frac{g}{l}}$), discuss what effects such rapid oscillation would produce on the motion of the pendulum involving the angle θ. [Can you understand the fact that, with the support subject to a rapid vertical oscillation, even the position $\theta = \pi$ of the pendulum can turn into a point of stable equilibrium or, with a rapid horizontal oscillation, the position $\theta = 0$ of the pendulum can become an unstable point?]

11-7 A uniform disk of mass M and radius R slides without friction on a horizontal surface. Another uniform disk of mass m and radius r ($< R$) is *pinned* through its center to a point off the center of the first disk by a distance b, so that it can rotate smoothly on the first disk. This system is shown schematically in the figure.

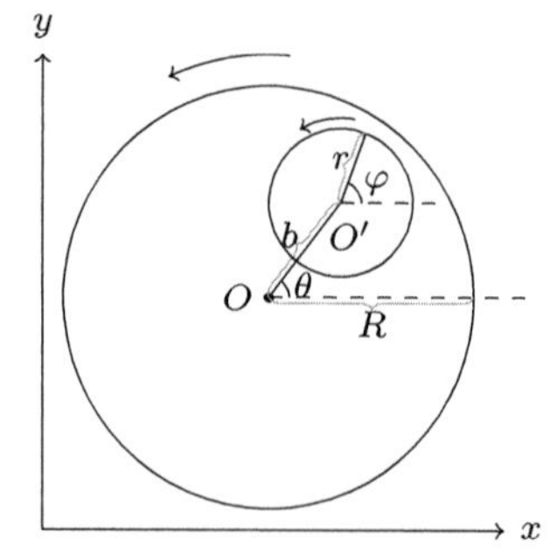

(i) Write the Lagrangian for this two-disk system, choosing as generalized coordinates the coordinates (x, y) of the CM of the larger disk, the angles of rotation of the larger disk (angle θ), and the angle φ describing the rotation of the smaller disk.

(ii) Describe the motion, identifying the constants of motion as needed.

11-8 Consider a three-body system (of masses m, m and M) obeying the following equations of motion

$$m\frac{d^2\vec{r}_1}{dt^2} = k(\vec{r}_2 - \vec{r}_1) + K(\vec{r}_3 - \vec{r}_1),$$
$$m\frac{d^2\vec{r}_2}{dt^2} = k(\vec{r}_1 - \vec{r}_2) + K(\vec{r}_3 - \vec{r}_2),$$
$$M\frac{d^2\vec{r}_3}{dt^2} = K(\vec{r}_1 - \vec{r}_3) + K(\vec{r}_2 - \vec{r}_3)$$

where k, K are certain positive constants. Write the corresponding Lagrangian in terms of $\vec{r}_1$, $\vec{r}_2$, $\vec{r}_3$, $\dot{\vec{r}}_1$, $\dot{\vec{r}}_2$ and $\dot{\vec{r}}_3$. Then, by introducing suitable generalized coordinate for 3 *independent* systems, discuss the motion of the system.

11-9 A particle of charge q and mass m is free to slide on a smooth horizontal plane. Assume that two *fixed* charges q are placed at $\pm a\mathbf{j}$, and two fixed charges $6q$ at $\pm 2a\mathbf{i}$. For the particle which can freely slide, find the electrostatic potential near the origin. Is the origin a position of stable equilibrium? Find the frequencies of possible normal modes of oscillation near it.

11-10 In two dimensions, consider a particle of mass m in a potential given by

$$V(x, y) = -\frac{1}{2}kx^2 - kxy + \frac{1}{2}ky^2 + \frac{1}{4}\lambda x^4$$

(k, λ are positive constants)

where x, y are Cartesian coordinates.

(i) Find the position(s) (x_0, y_0) for the particle to be in *stable* equilibrium.
(ii) Determine the Lagrangian appropriate for describing small oscillations about such stable equilibrium point(s).
(iii) Find the normal mode frequencies associated with the small oscillations.

11-11 A bead of mass m slides on a smooth circular hoop of mass M and radius a, which is pivoted on its rim, so that it can swing freely in its plane in the presence of uniform gravity.

(i) Write the Lagrangian of the system in terms of the angle of inclination θ of the diameter through the pivot and the angular position φ of the bead measured from the vertical (see the figure).
(ii) Find the frequencies of the normal modes, and discuss the nature of motion in various possible configurations of hoop and bead.

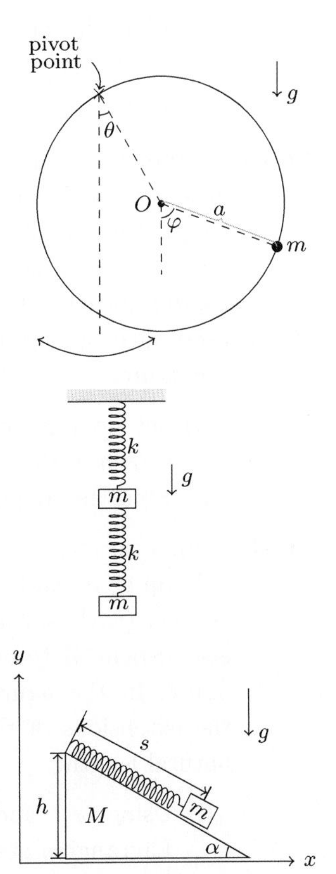

11-12 Consider two particles (of mass m) which are connected to the ceiling by two identical light springs of spring constant k, as shown in the figure. The springs can move only vertically. Find the normal modes and corresponding frequencies.

11-13 A block of mass m is attached to a wedge of mass M by a spring with spring constant k. The inclined frictionless surface of the wedge makes an angle α to the horizontal. The wedge is free to slide on a horizontal frictionless surface, as shown in the figure.

(i) Given the relaxed length of the spring alone is d, find the value s_0 when both the block and the wedge are at rest.

(ii) Find the Lagrangian for the system as a function of the x-coordinate of the wedge and the length of spring s. Write the equations of motion.

(iii) What is the natural frequency of oscillation?

11-14 A thin uniform rod having mass m and length $\frac{3}{2}l$ is suspended by a string of length l and negligible mass, in the presence of uniform gravity. Assume that the motion occurs in a vertical plane.

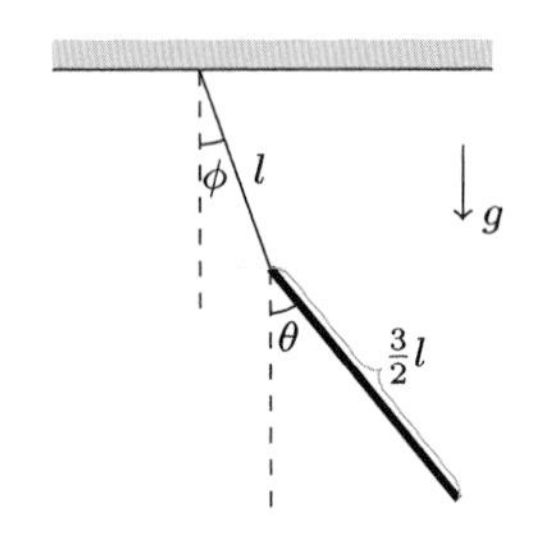

(i) Choosing two angles ϕ and θ (shown in the figure) as generalized coordinates, write the Lagrangian for the system.

(ii) Find the normal modes and corresponding frequencies for small oscillations.

11-15 A simple pendulum consisting of a mass m and weightless string of length l is mounted on a support of mass M which is attached to a horizontal spring with force constant k (and natural length x_0), as shown in the figure.

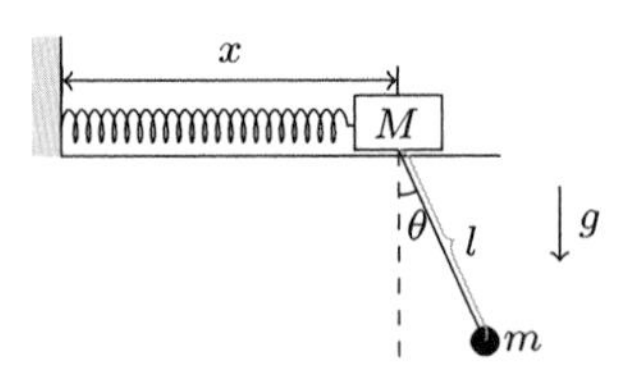

(i) Set up Lagrange's equations, choosing the variables x and θ shown in the figure as generalized coordinates.

(ii) Find the frequencies of small oscillations.

11-16 A uniform rod of mass m and length l is held up at its ends by two identical light springs (with spring constant k) which are suspended from the ceiling. See the figure. In the figure, x_1 and x_2 denote the extensions of the springs from their natural length.

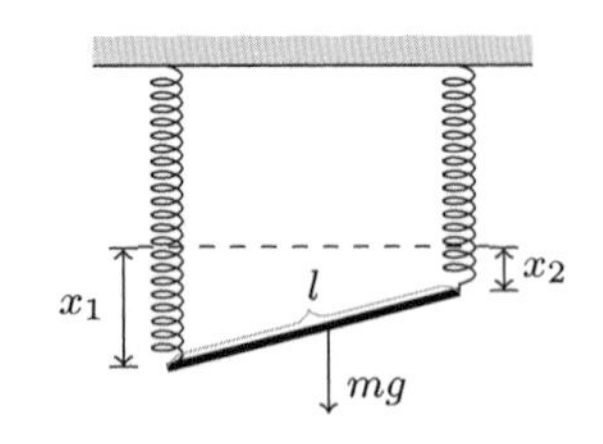

(i) Using x_1 and x_2 as generalized coordinates, write the Lagrangian describing small oscillations of the system (about the equilibrium positions $x_{10} = x_{20} = mg/(2k)$.

(ii) Find normal modes and normal coordinates.

(iii) Give physical interpretations for the normal modes obtained.

11-17 Consider three objects of equal mass m, which are connected by springs of spring constant k. (See the figure.) The motion is confined to one dimension. Mass A is further subjected to an external driving force $F(t)$.

(i) When x_1, x_2 and x_3 represent the displacements from the respective equilibrium positions for the three objects A, B and C, write the Lagrangian for the system.

(ii) Suppose that at $t = 0$ the three masses are at rest at their equilibrium positions and the driving force on mass A for $t > 0$ is of the form

$$F(t) = f \cos \omega t \quad (f, \omega\text{: given constants}).$$

In this case, determine the motion of mass C for $t > 0$. [Hint: use the normal mode method.]

11-18 On a horizontal table, consider three point particles, two of mass m and one of mass M, that are constrained to lie on a fixed circular hoop of radius r. They are mutually connected by identical light springs of spring constant k, as shown in the figure. (Assume that each spring has relaxed length equal to $\frac{2\pi}{3}r$, and the spring follow the arc of the hoop.) Find the frequencies of the normal modes and also obtain a set of three normal coordinates for the system.

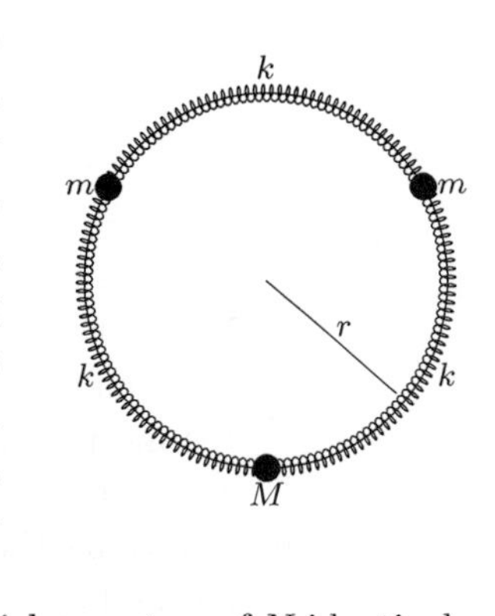

11-19 Consider the following three setups by which a system of N identical particles with masses m are connected by identical springs with spring constants k to make an array.

(i) (Setup 1) A straight-line array between two fixed walls:

(ii) (Setup 2) A straight-line array with one *free* end:

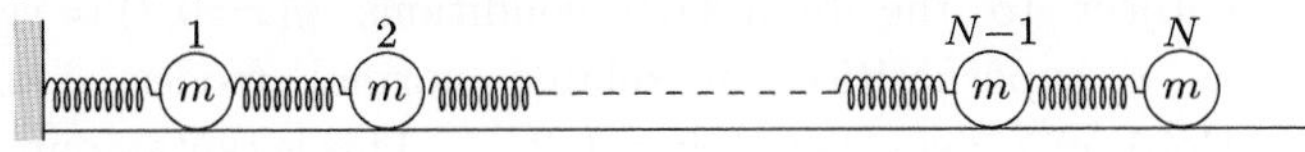

(iii) (Setup 3) A circular array:

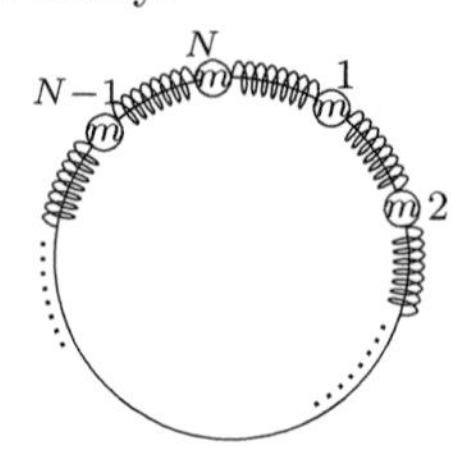

Find the normal modes and corresponding frequencies for the respective cases, and interpret the differences existing between these cases.

11-20 Suppose that there are N rows of pendulums, each of mass m and length l, joined by light springs of spring constant k as shown in the figure. The distances between the pivots are chosen so that the springs are unstretched when the pendulums are vertical, i.e., situated at their equilibrium positions in the absence of the springs. Gravity acts in the vertical direction.

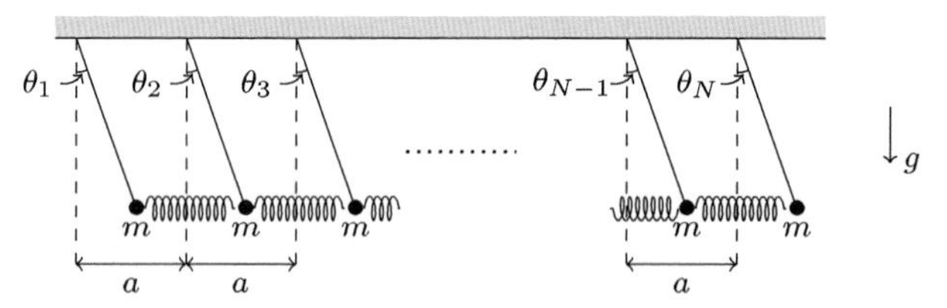

(i) When the system undergoes small-amplitude oscillations about the equilibrium configuration, find the Lagrangian describing the motion (with the angles $\theta_1, \theta_2, \ldots, \theta_N$ shown in the figure as generalized coordinates).

(ii) Determine the frequencies of normal-mode oscillations and corresponding normal modes for general N.

(iii) For sufficiently large N, discuss the characteristic features that the dispersion relation for this system shows.

11-21 In Sec. 11.3 we have found the general solution to the wave equation

$$\ddot{y} - v^2 y'' = 0 \quad (0 < x < L)$$

subject to the boundary conditions $y(x{=}0, t) = y(x{=}L, t) = 0$, as a superposition of related normal modes, i.e., $y(x,t) = \mathrm{Re}[A_n e^{i\frac{n\pi v}{L}t} \sin\left(\frac{n\pi x}{L}\right)]$, $n = 1, 2, \ldots$. This is tantamount to the series expression

$$y(x,t) = \sum_{n=1}^{\infty} \left(a_n \cos \frac{n\pi v}{L} t + b_n \sin \frac{n\pi v}{L} t \right) \sin \left(\frac{n\pi x}{L} \right),$$

where a_n, b_n are real constants.

(i) Let the initial conditions for the string displacement field $y(x,t)$ be $y(x,t=0) = f(x)$ and $\frac{\partial y(x,t)}{\partial t}|_{t=0} = 0$, where

$$f(x) = \begin{cases} \frac{2h}{L} x, & 0 < x < \frac{L}{2} \\ \frac{2h}{L}(L-x), & \frac{L}{2} < x < L. \end{cases}$$

Obtain the corresponding solution $y(x,t)$ valid for all $t > 0$, using the above series expression.

(ii) Show that the series solution of (i) can be summed into the form

$$y(x,t) = \frac{1}{2}\left[f^*(x - vt) + f^*(x + vt) \right],$$

where $f^*(w)$ represents the odd-periodic extension of the function $f(w)$, $0 < w < L$, with the period $2L$, i.e.,

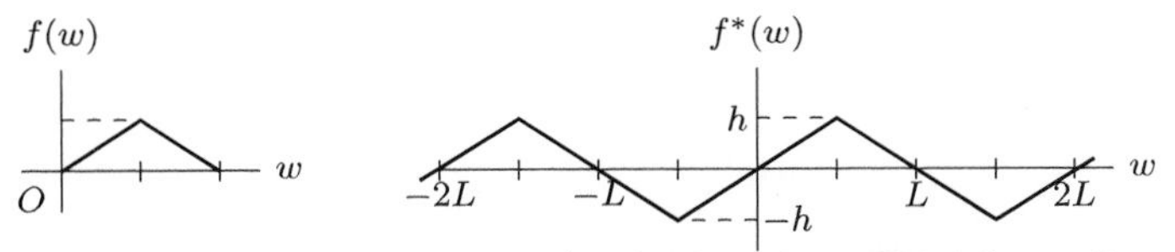

(iii) Draw the actual string shapes (with the above initial conditions) at times $t = \frac{L}{4v}, \frac{L}{2v}, \frac{3L}{4v}$ and $\frac{L}{v}$.

11-22 Show that, when $y(x,t)$ $(0 \leq x \leq L)$ is the string displacement field satisfying the equation of motion

$$\ddot{y} - v^2 y'' = 0 \quad \left(v \equiv \sqrt{T/\mu} \right)$$

the energy defined by

$$E(t) = \int_0^L \left\{ \frac{\mu}{2} [\dot{y}(x,t)]^2 + \frac{T}{2} [y'(x,t)]^2 \right\} dx$$

is a constant of motion. Then evaluate this energy when $y(x,t)$ is given by the series

$$y(x,t) = \sum_{n=1}^{\infty} \left(a_n \cos \frac{n\pi v}{L} t + b_n \sin \frac{n\pi v}{L} t \right) \sin \left(\frac{n\pi x}{L} \right).$$

What can you conclude from the result obtained?

expansion

where

(iii) Draw the actual string shapes (with the above initial condi-

What can you conclude from the result obtained?

12 Rigid Body Dynamics II

SUMMARY

12.1 Introducing Euler's equations and Euler's angles

- For a rigid body rotating about a fixed pivot or about its CM, one may represent its angular momentum by the expression $\vec{L} = I_1\omega_1\mathbf{e}_1 + I_2\omega_2\mathbf{e}_2 + I_3\omega_3\mathbf{e}_3$ where $(\mathbf{e}_1, \mathbf{e}_2, \mathbf{e}_3)$ denote body-fixed principal axes (which rotate with the body). Then the basic dynamical equation $\frac{d\vec{L}}{dt} = \vec{\Gamma}^{(e)}$ can be rewritten as

$$\frac{d^*\vec{L}}{dt} + \vec{\omega} \times \vec{L} = \vec{\Gamma}^{(e)}$$

with $\frac{d^*\vec{L}}{dt} = I_1\dot{\omega}_1\mathbf{e}_1 + I_2\dot{\omega}_2\mathbf{e}_2 + I_3\dot{\omega}_3\mathbf{e}_3$. ($\frac{d^*\vec{L}}{dt}$ is the rate of change of $\vec{L}$ relative to the frame rotating with the body.) The components of the above equation

$$I_1\dot{\omega}_1 + (I_3 - I_2)\omega_2\omega_3 = \Gamma_1^{(e)},$$
$$I_2\dot{\omega}_2 + (I_1 - I_3)\omega_3\omega_1 = \Gamma_2^{(e)},$$
$$I_3\dot{\omega}_3 + (I_2 - I_1)\omega_1\omega_2 = \Gamma_3^{(e)}$$

are called Euler's equations for a rigid body motion.
- Euler's equations may be used to discuss the stability of a rigid body, executing free (i.e., with $\vec{\Gamma}^{(e)} = 0$) stationary rotation about one of the principal axes. Specifically, if I_1, I_2 and I_3 are all different, it is not difficult to see from linearized Euler's equations that a body rotating about, say, the axis $\mathbf{e}_3$ is stable if I_3 is either the largest or the

smallest of the three principal moments, but not if it is the middle one.

- To discuss dynamics of a rigid body, a certain scheme to specify the orientation of the three principal axes in space is required. For this, a particularly convenient way is to employ three Euler's angles (φ, θ, ψ). Two of these angles, θ and φ, are used to specify the direction of one of the principal axes, say, $\mathbf{e}_3$ while the third, ψ, is used to specify the angle through which the body has been rotated from a standard position about the singled-out axis $\mathbf{e}_3$. (For a symmetric body, the symmetry axis is usually taken to be the singled-out axis $\mathbf{e}_3$.) Precise specification of these angles is made according to the following formula connecting the principal axes $(\mathbf{e}_1, \mathbf{e}_2, \mathbf{e}_3)$ (at any given instant), via the intermediate axes $(\mathbf{e}'_1, \mathbf{e}'_2, \mathbf{e}_3)$, to the fixed axes in space $(\mathbf{i}, \mathbf{j}, \mathbf{k})$:

$$\begin{pmatrix} \mathbf{e}_1 \\ \mathbf{e}_2 \\ \mathbf{e}_3 \end{pmatrix} = \begin{pmatrix} \cos\psi & \sin\psi & 0 \\ -\sin\psi & \cos\psi & 0 \\ 0 & 0 & 1 \end{pmatrix} \begin{pmatrix} \mathbf{e}'_1 \\ \mathbf{e}'_2 \\ \mathbf{e}_3 \end{pmatrix},$$

$$\begin{pmatrix} \mathbf{e}'_1 \\ \mathbf{e}'_2 \\ \mathbf{e}_3 \end{pmatrix} = \begin{pmatrix} \cos\theta & 0 & -\sin\theta \\ 0 & 1 & 0 \\ \sin\theta & 0 & \cos\theta \end{pmatrix} \begin{pmatrix} \cos\varphi & \sin\varphi & 0 \\ -\sin\varphi & \cos\varphi & 0 \\ 0 & 0 & 1 \end{pmatrix} \begin{pmatrix} \mathbf{i} \\ \mathbf{j} \\ \mathbf{k} \end{pmatrix}.$$

The angular velocity $\vec{\omega}$ of the rigid body is determined by the rates of changes of these three angles, the exact formula being $\vec{\omega} = \dot{\varphi}\mathbf{k} + \dot{\theta}\mathbf{e}'_2 + \dot{\psi}\mathbf{e}_3$. For the components of $\vec{\omega}$ relative to the principal axes $(\mathbf{e}_1, \mathbf{e}_2, \mathbf{e}_3)$ (i.e., for ω_1, ω_2 and ω_3 entering Euler's equations above), one has following results:

$$\omega_1 = -\dot{\varphi}\sin\theta\cos\psi + \dot{\theta}\sin\psi,$$
$$\omega_2 = \dot{\varphi}\sin\theta\sin\psi + \dot{\theta}\cos\psi,$$
$$\omega_3 = \dot{\varphi}\cos\theta + \dot{\psi}.$$

- For a symmetric body with the symmetry axis taken as $\mathbf{e}_3$ (and so $I_1 = I_2$), the representations of $\vec{\omega}$ and $\vec{L}$ relative to the axes $(\mathbf{e}_1', \mathbf{e}_2', \mathbf{e}_3)$ may also prove useful, viz.,

$$\vec{\omega} = -\dot{\varphi}\sin\theta\mathbf{e}_1' + \dot{\theta}\mathbf{e}_2' + (\dot{\psi} + \dot{\varphi}\cos\theta)\mathbf{e}_3,$$
$$\vec{L} = I_1(-\dot{\varphi}\sin\theta\mathbf{e}_1' + \dot{\theta}\mathbf{e}_2') + I_3(\dot{\psi} + \dot{\varphi}\cos\theta)\mathbf{e}_3.$$

The rotational kinetic energy of a symmetric body is given as

$$T = \frac{1}{2}\vec{\omega}\cdot\vec{L} = \frac{1}{2}I_1(\dot{\varphi}^2\sin^2\theta + \dot{\theta}^2) + \frac{1}{2}I_3(\dot{\psi} + \dot{\varphi}\cos\theta)^2.$$

As an application of these representations, consider the case of a symmetric body rotating freely (i.e., with no external torque); then, $\frac{d\vec{L}}{dt} = 0$ and accordingly one may take

$$\vec{L} = L\mathbf{k} = -L\sin\theta\,\mathbf{e}_1' + L\cos\theta\,\mathbf{e}_3.$$

Then, equating this with the above representation of $\vec{L}$, one finds its general motion (for $\theta \neq 0$ or π) to be characterized by

$$I_1\dot{\varphi} = L, \quad \dot{\theta} = 0, \quad I_3(\dot{\psi} + \dot{\varphi}\cos\theta) = L\cos\theta.$$

Hence the symmetry axis, $\mathbf{e}_3$, rotates around the direction of $\vec{L}$ $(= L\mathbf{k})$ at a constant rate $\dot{\varphi} = \frac{L}{I_1}$, maintaining a constant angle θ to it, and in addition the body spins about its axis with constant angular velocity $\dot{\psi} = L(\frac{1}{I_3} - \frac{1}{I_1})\cos\theta$. This finding can be used to explain the free precession exhibited by the rotation axis of the flattened Earth, the so-called Chandler wobble.

12.2 Lagrangian treatment of a rigid body motion

- In treating rigid body dynamics, the Lagrangian approach using three Euler's angles as members of generalized coordinates is usually superior to that based on Euler's equations (with angular velocity expressed in terms of Euler's angles). Given a symmetric top of mass M which is pivoted at a point on the axis of symmetry ($\mathbf{e}_3$) and moving under uniform gravity, the Lagrangian is given by the form

$$L = \frac{1}{2}I_1(\dot{\theta}^2 + \dot{\varphi}^2 \sin^2\theta) + \frac{1}{2}I_3(\dot{\psi} + \dot{\varphi}\cos\theta)^2 - MgR\cos\theta,$$

where R is the distance from the pivot point to the CM of the body. For this system we have three conservation laws: generalized momenta conjugate to two cyclic variables φ and ψ should be conserved, i.e.,

$$I_1\dot{\varphi}\sin^2\theta + I_3(\dot{\psi} + \dot{\varphi}\cos\theta)\cos\theta = p_\varphi \quad (\text{: const.}),$$
$$I_3(\dot{\psi} + \dot{\varphi}\cos\theta) \equiv I_3\omega_3 = p_\psi \quad (\text{: const.}),$$

and the conservation of the total energy

$$\frac{1}{2}I_1(\dot{\theta}^2 + \dot{\varphi}^2 \sin^2\theta) + \frac{1}{2}I_3(\dot{\psi} + \dot{\varphi}\cos\theta)^2 + MgR\cos\theta = E \quad (\text{: const.}).$$

From these conservation laws, we are led to a θ-equation of the one-dimensional energy-conservation-law type:

$$\frac{1}{2}I_1\dot{\theta}^2 + V_{\text{eff}}(\theta) = E - \frac{p_\psi^2}{2I_3} \quad (\equiv E'),$$

where $V_{\text{eff}}(\theta) = \frac{(p_\varphi - p_\psi\cos\theta)^2}{2I_1\sin^2\theta} + MgR\cos\theta$. The last equation provides especially useful information as regards what kind of motion is generally allowed for a top with some given value for the effective energy E'. Suppose that the effective potential $V_{\text{eff}}(\theta)$ has minimum at $\theta = \theta_0$ ($\neq 0, \pi$). Then the body with the effective energy $E' = V_{\text{eff}}(\theta_0)$ will precess uniformly at angle $\theta = \theta_0$ with the vertical, and with constant angular velocity $\dot{\varphi} = \frac{p_\varphi - p_\psi\cos\theta_0}{I_1\sin^2\theta_0}$; if E' is larger than this value, the body will undergo a nutation or oscillation of the axis of the top in the θ-direction as it precesses. Also, based on the above equations, one can discuss in some detail what motion would follow if a top, spinning about its axis with angular velocity ω_3, is held with its axis initially at rest at certain angle θ_1 and then released.

- It is possible to give an analogous treatment on the problem called the precession of the equinoxes, i.e., to understand the slow precession of the Earth's rotational axis about the normal to the ecliptic (over the period of ~26 000 years). In this case the (spheroidal) Earth — a kind of a symmetric top (having principal moments $I_1, I_2 = I_1$ and I_3) with the Earth's rotation axis as its symmetric axis $\mathbf{e}_3$ — is subject to the external torque due to gravitational forces from the Sun and Moon. As it turns out, it is possible to describe the net effects of this external torque approximately by including in the Lagrangian the potential energy of the form

$$V = \frac{1}{2}\Omega^2(I_3 - I_1)\frac{1}{2}(1 - 3\cos^2\theta),$$

where θ denotes the angle that the Earth's rotation axis, $\mathbf{e}_3$, makes with the normal to the ecliptic, and Ω^2 a certain numerical factor. In this way, the problem is reduced to the study of a system described by the Lagrangian

$$L = \frac{1}{2}I_1(\dot{\theta}^2 + \dot{\varphi}^2\sin^2\theta) + \frac{1}{2}I_3(\dot{\psi} + \dot{\varphi}\cos\theta)^2 - \frac{1}{4}\Omega^2(I_3 - I_1)(1 - 3\cos^2\theta).$$

For uniform precession without nutation (i.e., $\dot{\theta} = \ddot{\theta} = 0$), one then finds from the Lagrange equation for the variable θ

$$I_3\dot{\psi}\dot{\varphi}\sin\theta + \frac{3}{2}\Omega^2(I_3 - I_1)\cos\theta\sin\theta = 0,$$

if the quadratic terms in $\dot{\varphi}$ are neglected (as they are very small). This gives the approximate formula for the precession rate: $\dot{\varphi} \approx -\frac{3}{2}\left(\frac{I_3 - I_1}{I_3}\right)\Omega^2\frac{\cos\theta}{\dot{\psi}}$. If appropriate numerical values $\frac{I_3 - I_1}{I_3} \approx \frac{1}{304}$, $\theta \approx 23.45°$, $\dot{\psi} = \frac{2\pi}{1\,\text{day}} \approx 7.27 \times 10^{-5}\,\text{s}^{-1}$ and $\Omega^2 \approx 12.44 \times 10^{-14}\,\text{s}^{-2}$ are substituted into this formula, one obtains the value $\dot{\varphi} \approx -7.76 \times 10^{-12}\,\text{rad}\,\text{s}^{-1}$ or the precession period $T = \frac{2\pi}{|\dot{\varphi}|} \approx$ 25 700 years, a result quite close to the observed value.

Problems

12-1 Let us consider the torque-free motion of a symmetric body about some fixed pivot point. The axis of cylindrical symmetry is the principal axis $\mathbf{e}_3$ with moment of inertia I_3. (The other two principal axes are perpendicular to $\mathbf{e}_3$ and $I_1 = I_2$.)

(i) Solve explicitly the corresponding Euler's equations to find the expressions for $\omega_1(t)$, $\omega_2(t)$ and $\omega_3(t)$ (i.e., the components of the angular velocity vector with respect to three principal axes).
(ii) Based on the finding in (i), what can you say about the actual motion of the body in space?

12-2 A uniform rectangular board of mass m with sides a and b ($b > a$) and negligible thickness rotates with constant angular velocity ω about a diagonal (see the figure). Ignore gravity. Show that the external torque $\left|\vec{\Gamma}^{(e)}\right| = \frac{m(b^2-a^2)ab\omega^2}{12(a^2+b^2)}$ must be applied to keep the axis of rotation fixed.

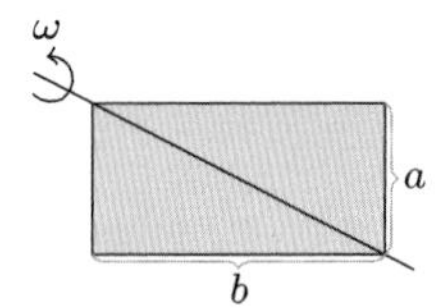

12-3 A symmetric top with one point fixed has mass M and principal moments of inertia $I_1 = I_2 > I_3$. The top, which is set to spin around its symmetry axis with the spin velocity s $(= \dot{\psi})$, is precessing steadily at a constant angular velocity Ω $(= \dot{\varphi})$ with the symmetry axis inclined at an angle $\theta = \theta_0$ $(\neq 0)$ from the vertical. (Here, φ, θ and ψ represent Euler's angles). The center of mass for the top is at a distance R from the pivot point.

(i) Obtain the relation between the spin velocity s, the precession rate Ω, and the inclination angle θ_0 with the help of Euler's equations.
(ii) What is the minimum spin velocity for the top to precess steadily at an inclination θ_0?

12-4 A thin uniform disk of radius a and mass m rolls without slipping on a rough horizontal plane. Consider the disk in the state of steady motion, with the center of mass of the disk moving in a horizontal circle of radius b and the normal to the disk, $\mathbf{e}_3$, forming a fixed angle θ_0 with the vertical direction (see the figure). Determine the angular velocity Ω of the center of mass of the disk.

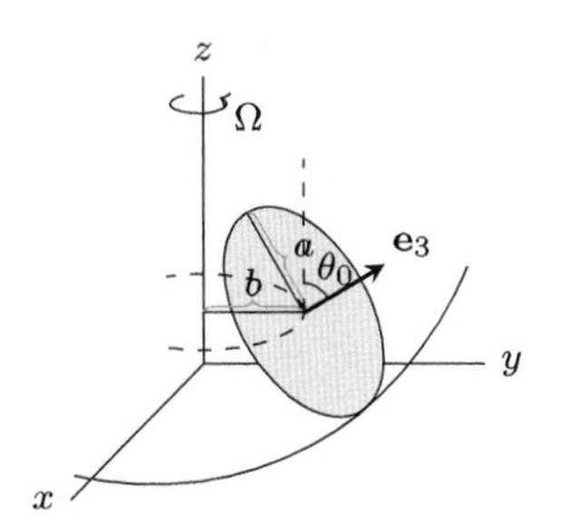

12-5 A uniform sphere of mass M and radius a rolls without slipping on the inside of a circular cone (with the half angle α), under the influence of a uniform gravitational field acting in the direction of the symmetry axis of the cone. See the figure.

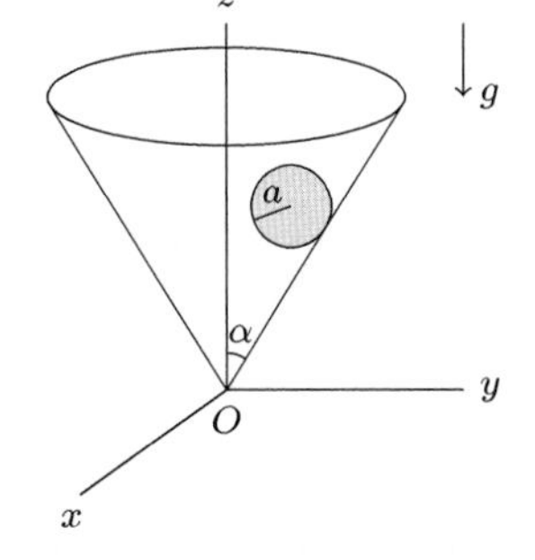

(i) Obtain the equations of motion for the sphere (by applying a direct Newtonian reasoning), and look for all constants of motion. [*Hint*: Cylindrical coordinates may be employed to simplify the analysis; also note that, for a totally symmetric body, there is no distinctive advantage in expressing the angular velocity vector in terms of components relative to body-fixed axes.]

(ii) Give a detailed study on the nature of circular orbits possible, including the aspects concerning small oscillations about them.

12-6 For a symmetric body moving under uniform gravity, the Lagrangian method was used to deduce the following equations of motion satisfied by the Euler's angles φ, θ and ψ (see Sec. 12.2):

$$I_1\ddot{\theta} = I_1\dot{\varphi}^2 \sin\theta\cos\theta - I_3(\dot{\psi} + \dot{\varphi}\cos\theta)\dot{\varphi}\sin\theta + MgR\sin\theta,$$

$$\frac{d}{dt}\Big[I_1\dot{\varphi}\sin^2\theta + I_3(\dot{\psi} + \dot{\varphi}\cos\theta)\cos\theta\Big] = 0,$$

$$\frac{d}{dt}\Big[I_3(\dot{\psi} + \dot{\varphi}\cos\theta)\Big] = 0.$$

Verify that these equations also follow from Euler's equation for a rigid body (Sec. 12.1) if one has the angular velocity components ω_1, ω_2, and ω_3 written in terms of Euler's angles (see Sec. 12.1).

12-7 Consider a rapidly spinning symmetric top under uniform gravity whose axis is initially at rest at an angle θ_1 and then released. Its subsequent nutation can be studied on the basis of the energy conservation equation

$$\frac{1}{2}I_1\dot{\theta}^2 + V_{\text{eff}}(\theta) = MgR\cos\theta_1$$

with the effective potential

$$V_{\text{eff}}(\theta) = \frac{I_3^2\omega_3^2}{2I_1}\left[\frac{(\cos\theta_1 - \cos\theta)^2}{\sin^2\theta} + \alpha\cos\theta\right],$$

where $\alpha \equiv \frac{2I_1MgR}{I_3^2\omega_3^2}$. Show that when ω_3 is sufficiently large so that $\alpha \ll 1$, the above effective potential can be approximated by the harmonic potential

$$V_{\text{eff}}(\theta) = V_{\text{eff}}(\theta_0) + \frac{1}{2}\frac{I_3^2}{I_1}\omega_3^2(\theta - \theta_0)^2$$

where $\theta_0 = \theta_1 + \frac{1}{2}\alpha\sin\theta_1$.

12-8 Consider a spinning symmetric top of mass M, the apex of which is constrained to remain in contact with a *smooth* horizontal plane. For this "skating" top, the horizontal velocity of the center of mass will be fixed in time and we may therefore without loss of generality suppose that this horizontal component is zero. The center of mass of the top is at a distance R from the apex.

(i) With Euler's angles (θ, φ, ψ) chosen as generalized coordinates, write the Lagrangian for the system and indicate all the first integrals.

(ii) If the top is released from $\theta = \theta_1$ with $\dot{\theta} = \dot{\varphi} = 0$ and $\dot{\psi} = \omega_3$, discuss the motion.

12-9 The *gyrocompass* is a very important navigational instrument on ships, for it finds true north as determined by Earth's rotation (without being affected, unlike a magnetic compass, by nearby ferromagnetic materials). A gyrocompass consists of a symmetric rigid body which is mounted on a gimbal support so that its center of mass remains fixed and *its symmetry axis is constrained to move in a horizontal plane* at the Earth's surface.

(i) Choose a pair of angle coordinates and set up the Lagrange function when the gyrocompass is at a fixed location on the

Earth's surface of colatitude θ_0. Effects due to Earth's rotation should be included, but ignore friction.

(ii) Show that the angular velocity component ω_3 along the symmetry axis remains constant, and that if $\omega_3 > \left(\frac{I_1}{I_3}\right)\omega_0 \sin\theta_0$, where ω_0 is the rotational angular velocity of the Earth, then the symmetry axis oscillates in the horizontal plane about a north–south axis. Find the frequency of small oscillations. [In an actual gyrocompass, the rotor must be driven to make up for frictional torques about the symmetry axis, while frictional torques in the horizontal plane damp the oscillations of the symmetry axis, which comes to rest in a north–south line.]

12-10 In Problem 12-4 we considered the steady motion of a rolling tilted disk, where the point of contact with the horizontal plane was moving in a circle. For the same system, obtain full equations of motion by the Lagrangian method. [Note that (nonholonomic) rolling constraints in this problem may be incorporated by the method described in Problem 10-25.] Then, in this framework, identify possible steady motions and also study the stability of these solutions and small oscillations about them.

12-11 If an extended symmetric top is placed in the gravitational potential produced by a point mass $\bar{M}$ at large distance R, show that the quadrupole-dependent part of the related gravitational potential energy can be expressed by the form

$$V_2(\Theta) = \frac{G\bar{M}(I_3 - I_1)}{2R^3}(3\cos^2\Theta - 1),$$

where Θ denotes the angle between the symmetry axis $\mathbf{e}_3$ and the direction to the distant point mass. See the figure.

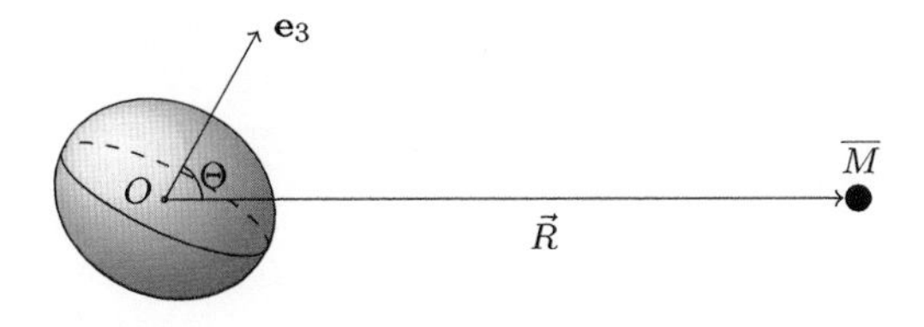

13 Hamiltonian Mechanics

SUMMARY

13.1 Motion in phase space: Hamilton's equations

- Hamiltonian mechanics is another approach to mechanics, whose pervasive role in modern physics cannot be overemphasized. In this approach, dynamical variables are assumed by generalized coordinates $\mathbf{q} = (q_1, \ldots, q_n)$ *and* corresponding generalized momenta $\mathbf{p} = (p_1, \ldots, p_n)$, and each mechanical system has its own Hamiltonian function $H(\mathbf{q}, \mathbf{p})$. Given a Lagrangian $L(\mathbf{q}, \dot{\mathbf{q}}, t)$ for a system, two sets of equations can be written down: the first are $p_i = \frac{\partial L}{\partial \dot{q}_i}$ (defining generalized momentum conjugate to q_i), and the second — those specifying dynamical development of the system — are Lagrange equations that may be written as $\dot{p}_i = \frac{\partial L}{\partial q_i}$. To form the Hamiltonian function from a Lagrangian L, solve the equations $p_i = \frac{\partial L}{\partial \dot{q}_i}$ $(i = 1, \ldots, n)$ for the $\dot{q}$'s (i.e., have $\dot{q}_i$ expressed as a function of the q's and p's, at each given t) and define the Hamiltonian — a function of the phase-space coordinates q's and p's (and possibly t) — by

$$H(\mathbf{q}, \mathbf{p}) = \sum_{i=1}^{n} p_i \dot{q}_i - L.$$

(This procedure, involving appropriate change of variables, corresponds to a Legendre transform.) It is to be noted that the Hamiltonian is nothing but the energy function $E(\mathbf{q}, \dot{\mathbf{q}})$ written in terms of the q's and p's (and possibly t). With the thus-constructed Hamiltonian, one can then write $2n$ first-order differential equations, specifying the time development of phase-space coordinates variables, as

$$\dot{q}_i = \frac{\partial H}{\partial p_i} \quad \dot{p}_i = -\frac{\partial H}{\partial q_i} \qquad (i = 1, \ldots, n).$$

These are Hamilton's (canonical) equations of motion.

- If the generalized momentum variables are eliminated from Hamilton's equations, one gets back Lagrange equations. Also, if a particular q_k is a cyclic coordinate of the Lagrangian (i.e., $\frac{\partial L}{\partial q_k} = 0$), the very q_k is also a cyclic coordinate of the Hamiltonian (i.e., $\frac{\partial H}{\partial q_k} = 0$); this is a simple consequence of the general relation $\frac{\partial H}{\partial q_k}|_{p\text{'s}} = -\frac{\partial L}{\partial q_k}|_{\dot{q}\text{'s}}$. If q_k is cyclic, the related conservation law, $p_k = \text{const.}$, is now a part of Hamilton's equations of motion. Also, if the Hamiltonian does not involve time t explicitly, the value of the Hamiltonian H, i.e., the total energy of the system is a constant of motion. Actually one can prove the more general relation, $\frac{dH}{dt} = \frac{\partial H}{\partial t}$, which indicates that H changes with time only in virtue of the explicit time dependence.
- Hamiltonians for some representative systems are as follows.

(i) *A single particle in an external potential:*

$$H = \frac{1}{2m}\vec{p}^{\,2} + V(\vec{r})$$

(in Cartesian coordinates $\vec{r} = (x, y, z)$),

$$H = \frac{p_r^2}{2m} + \frac{p_\theta^2}{2mr^2} + \frac{p_\varphi^2}{2mr^2 \sin^2\theta} + V(r, \theta, \varphi)$$

(in spherical coordinates).

(ii) *Two particles under mutual interaction:*

$$H = \frac{1}{2m_1}\vec{p}_1^{\,2} + \frac{1}{2m_2}\vec{p}_2^{\,2} + V(\vec{r}_1 - \vec{r}_2)$$

(using individual coordinates),

$$H = \frac{\vec{P}^2}{2M} + \frac{\vec{p}^2}{2\mu} + V(\vec{r}), \text{ with } M = m_1 + m_2 \text{ and}$$
$$\mu = \frac{m_1 m_2}{m_1 + m_2}$$

(using the CM and relative position variables

$$\vec{R} = \frac{m_1\vec{r}_1 + m_2\vec{r}_2}{M}, \ \vec{r} = \vec{r}_1 - \vec{r}_2 \text{ with}$$
$$\vec{P} = M\dot{\vec{R}}, \ \vec{p} = \mu\dot{\vec{r}}).$$

Here, $\vec{R}$ being cyclic, the CM momentum $\vec{P}$ becomes a constant of motion.

(iii) *A symmetric top under uniform gravity*:

$$H = \frac{(p_\varphi - p_\psi \cos\theta)^2}{2I_1 \sin^2\theta} + \frac{p_\theta^2}{2I_1} + \frac{p_\psi^2}{2I_3} + MgR\cos\theta$$

(using Euler's angles $\varphi, \theta,$ and ψ).

In this case, φ and ψ are clearly cyclic and so the corresponding conjugate momenta, p_φ and p_ψ, become constants of motion.

(iv) *A charged particle in an external electromagnetic field* $(\vec{E}, \vec{B})$:

$$H = \frac{1}{2m}[\vec{p} - q\vec{A}(\vec{r}, t)]^2 + q\phi(\vec{r}, t),$$
$$\text{with } \vec{E} = -\vec{\nabla}\phi - \frac{\partial\vec{A}}{\partial t} \text{ and } \vec{B} = \vec{\nabla} \times \vec{A}.$$

Hamiltonian formulation for a (continuous) field system can also be given.

13.2 Phase flows of Hamiltonian systems: Liouville's theorem

- In a system with n degrees of freedom, a point in the $(\mathbf{q}, \mathbf{p})$-phase space, representing the instantaneous state of the given system, can be represented by a $2n$-dimensional vector $\boldsymbol{\xi} \equiv (q_1, \ldots, q_n, p_1, \ldots p_n)$. Then, as time progresses, any specific point in the phase space will move along its phase trajectory $\boldsymbol{\xi}(t)$, and in a Hamiltonian system with the Hamiltonian $H(\mathbf{q}, \mathbf{p})$ this movement is governed by Hamilton's equations which prescribe the phase velocity vector $\mathbf{u}$ (at each given instant) as

$$\vec{u} \equiv \dot{\boldsymbol{\xi}} = \left(\frac{\partial H}{\partial p_1}, \ldots, \frac{\partial H}{\partial p_n}, -\frac{\partial H}{\partial q_1}, \ldots, -\frac{\partial H}{\partial q_n}\right).$$

Phase flows of Hamiltonian systems are special in that they are always volume-preserving (Liouville's theorem). That is, any given volume of "phase fluid" moves with time as if the fluid is incompressible, as the related phase velocity field $\mathbf{u}(\boldsymbol{\xi})$ has a vanishing ($2n$-dimensional) divergence

$$\begin{aligned}\vec{\nabla} \cdot \mathbf{u} &\equiv \sum_{\alpha=1}^{2n} \frac{\partial \dot{\xi}_\alpha}{\partial \xi_\alpha} \\ &= \frac{\partial}{\partial q_1}\left(\frac{\partial H}{\partial p_1}\right) + \cdots + \frac{\partial}{\partial q_n}\left(\frac{\partial H}{\partial p_n}\right) \\ &\quad + \frac{\partial}{\partial p_1}\left(-\frac{\partial H}{\partial q_1}\right) + \cdots + \frac{\partial}{\partial p_n}\left(-\frac{\partial H}{\partial q_n}\right) = 0.\end{aligned}$$

- For a system with too numerous degrees of freedom, it becomes impossible to specify precisely the instantaneous state of the system (i.e., the values of the q's and p's at given time). In such circumstances, it will be useful to introduce the probability density function $\rho(\mathbf{q}, \mathbf{p}, t)$, representing the probability for the system to be found within a unit phase volume about the phase point $(\mathbf{q}, \mathbf{p})$ at time t. Now, assuming a Hamiltonian system, there is a direct implication on this function from Liouville's theorem: it must satisfy Liouville's equation.

$$\frac{\partial \rho}{\partial t} = \frac{\partial \rho}{\partial p_i}\frac{\partial H}{\partial q_i} - \frac{\partial \rho}{\partial q_i}\frac{\partial H}{\partial p_i}.$$

This equation is of much importance in statistical mechanics. Especially, for an equilibrium distribution (i.e., with $\frac{\partial \rho}{\partial t} = 0$), this equation selects the form $\rho = \rho\big(H(\mathbf{q}, \mathbf{p})\big)$ as a particularly natural choice. Note that the famous Boltzmann distribution $\rho \propto e^{-H(\mathbf{q},\mathbf{p})/k_B T}$ is of this type. As another implication of Liouville's theorem, one can prove Poincaré's recurrence theorem: in a Hamiltonian system, the time evolution of a state is such that (if its phase trajectory lies in a bounded region of phase space) it passes arbitrarily close to its initial point in phase space at some later time. But, for a system with a sufficiently large number of degrees of freedom, this theorem has absolutely no observable consequences (since the Poincaré recurrence time can easily be longer than the lifetime of the universe).

13.3 Poisson brackets and canonical transformations

- In phase space with coordinates $\mathbf{q} = (q_1, \ldots, q_n)$ and $\mathbf{p} = (p_1, \ldots, p_n)$, one can define the Poisson bracket of two functions $u(\mathbf{q}, \mathbf{p})$ and $v(\mathbf{q}, \mathbf{p})$ by

$$[u, v]_{\mathbf{q},\mathbf{p}} = \sum_{i=1}^{n} \left(\frac{\partial u}{\partial q_i} \frac{\partial v}{\partial p_i} - \frac{\partial u}{\partial p_i} \frac{\partial v}{\partial q_i} \right).$$

Then, if H is the Hamiltonian of the system, the time evolution of any observable $\Phi(\mathbf{q}, \mathbf{p}, t)$ is governed by $\frac{d\Phi}{dt} = [\Phi, H]_{\mathbf{q},\mathbf{p}} + \frac{\partial \Phi}{\partial t}$. Especially, Hamilton's equations of motion can be written as

$$\dot{q}_i = [q_i, H]_{\mathbf{q},\mathbf{p}}, \quad \dot{p}_i = [p_i, H]_{\mathbf{q},\mathbf{p}} .$$

Using the compact notation $\boldsymbol{\xi} = (\xi_1 \equiv q_1, \ldots, \xi_n \equiv q_n, \xi_{n+1} \equiv p_1, \ldots, \xi_{2n} \equiv p_n)$ and omitting the subscript attached to the Poisson bracket, one may express these relations by

$$[u, v] = \sum_{\alpha,\beta=1}^{2n} \frac{\partial u}{\partial \xi_\alpha} J_{\alpha\beta} \frac{\partial v}{\partial \xi_\beta}, \quad \dot{\xi}_\alpha = [\xi_\alpha, H] = J_{\alpha\beta} \frac{\partial H}{\partial \xi_\beta},$$

where $J_{\alpha\beta}$ are certain numerical numbers, equal to 1 (-1) if $\alpha = i$ and $\beta = n + i$ ($\alpha = n + i$ and $\beta = i$) and 0 otherwise. The so-called fundamental Poisson brackets for canonical coordinates $\boldsymbol{\xi} \equiv (\mathbf{q}, \mathbf{p})$ consist of the relations

$$[q_i, q_j] = [p_i, p_j] = 0, \quad [q_i, p_j] = \delta_{ij} \quad (\text{or } [\xi_\alpha, \xi_\beta] = J_{\alpha\beta}).$$

- A change of coordinates in phase space, from a given set $\boldsymbol{\xi} \equiv (\mathbf{q}, \mathbf{p})$ to another set of coordinate variables $\boldsymbol{\xi}' \equiv (\mathbf{Q} = \mathbf{Q}(\mathbf{q}, \mathbf{p}, t), \mathbf{P} = \mathbf{P}(\mathbf{q}, \mathbf{p}, t))$ is called a canonical transformation if the Poisson brackets for new variables $\mathbf{Q}(\mathbf{q}, \mathbf{p}, t)$ and $\mathbf{P}(\mathbf{q}, \mathbf{p}, t)$ with respect to $(\mathbf{q}, \mathbf{p})$ yield the results consistent with the fundamental Poisson brackets appropriate for canonical coordinates $(\mathbf{Q}, \mathbf{P})$, i.e.,

$$[Q_i, Q_j]_{\mathbf{q},\mathbf{p}} = [P_i, P_j]_{\mathbf{q},\mathbf{p}} = 0, \qquad [Q_i, P_j]_{\mathbf{q},\mathbf{p}} = \delta_{ij}$$

(or, equivalently, $[\xi'_\alpha, \xi'_\beta]_{\boldsymbol{\xi}} = J_{\alpha\beta}$ for $\xi'_\alpha = \xi'_\alpha(\boldsymbol{\xi}, t)$). If the coordinates $(\mathbf{Q}, \mathbf{P})$ are related to $(\mathbf{q}, \mathbf{p})$ by a canonical transformation, one finds that (i) (for general observables u and v) $[u, v]_{\mathbf{Q},\mathbf{P}} = [u, v]_{\mathbf{q},\mathbf{p}}$, and (ii) the equations of motion for the new variables $\mathbf{Q}, \mathbf{P}$ are again given by standard Hamilton's equations

$$\dot{Q}_i = \frac{\partial H'}{\partial P_i}, \quad \dot{P}_i = -\frac{\partial H'}{\partial Q_i}.$$

[If the transformation does not have explicit t-dependence, the new Hamiltonian $H'(\mathbf{Q}, \mathbf{P}, t)$ can be taken to be equal to the old-variable Hamiltonian, i.e., $H' = H\big(\mathbf{q}(\mathbf{Q}, \mathbf{P}), \mathbf{p}(\mathbf{Q}, \mathbf{P}), t\big)$.]

- A simple way to produce canonical transformations is to use generating functions. If the new and old phase-space coordinates, i.e., $(\mathbf{Q}, \mathbf{P})$ and $(\mathbf{q}, \mathbf{p})$, are related using an arbitrary generating function $F_2(\mathbf{q}, \mathbf{P}, t)$ through the equations

$$Q_i = \frac{\partial F_2(\mathbf{q}, \mathbf{P}, t)}{\partial P_i}, \quad p_i = \frac{\partial F_2(\mathbf{q}, \mathbf{P}, t)}{\partial q_i},$$

the given two sets are related by a canonical transformation and the Hamiltonian H' with the new variables $(\mathbf{Q}, \mathbf{P})$ can be taken as $H' = H(\mathbf{q}, \mathbf{p}, t) + \frac{\partial F_2(\mathbf{q}, \mathbf{P}, t)}{\partial t}$.

Other types of generating functions can also be considered: for instance, if the two sets of coordinates $(\mathbf{Q}, \mathbf{P})$ and $(\mathbf{q}, \mathbf{p})$ are related using another type of generating function $F_1(\mathbf{q}, \mathbf{Q}, t)$ through

$$p_i = \frac{\partial F_1}{\partial q_i}, \quad P_i = -\frac{\partial F_1}{\partial Q_i}, \quad H' = H + \frac{\partial F_1}{\partial t},$$

they are related by a canonical transformation. Note that, with the generating function $F_2 = \sum_i q_1 P_i$, the identity transformation, i.e., $(Q_1 = q_1, P_i = p_i)$ results; with the choice $F_1 = \sum_i q_i Q_i$, on the other hand, the resulting canonical transformation becomes $(Q_i = p_i,\ P_i = -q_i)$, i.e., the usual coordinates and momenta are interchanged (up to the minus sign).

13.4 Symmetries and conservation laws again

- In phase space, one can represent an infinitesimal canonical transformation generated by an (arbitrary) function $G(\mathbf{q}, \mathbf{p}, t)$ by the form

$$\delta q_i = \frac{\partial G}{\partial p_i}\delta\lambda, \qquad \delta p_i = -\frac{\partial G}{\partial q_i}\delta\lambda,$$

where $\delta\lambda$ is an infinitesimal parameter. The effect of this infinitesimal canonical transformation on any given function $F(\mathbf{q}, \mathbf{p}, t)$ becomes $\delta F = [F, G]\,\delta\lambda$; hence, any quantity F satisfying the condition $[F, G] = 0$ is left invariant under this transformation. This has the following implication: if the Hamiltonian H of the system is left invariant under the (time-independent) transformation generated by $G(\mathbf{q}, \mathbf{p})$, i.e., $[H, G] = 0$, the very symmetry generator $G(\mathbf{q}, \mathbf{p})$ is a conserved quantity, i.e., $\frac{dG(\mathbf{q},\mathbf{p})}{dt} = 0$. If the Hamiltonian possesses several independent symmetries, the set of all symmetry generators is closed under Poisson brackets. (This is Poisson's theorem.)
- The Hamiltonian H of a rotationally invariant one-particle system (in three dimensions) is invariant under infinitesimal transformations generated by angular momenta

$L_i \equiv \epsilon_{ijk} x_j p_k$ ($i = 1, 2, 3$), i.e., $[H, L_i] = 0$. The three rotation generators L_i define a closed system under Poisson brackets, with

$$[L_i, L_j] = \epsilon_{ijk} L_k.$$

In quantum theory, there are direct analogues of these Poisson bracket relations (as relations involving operator commutators instead).

- The Hamiltonian of the three-dimensional Kepler problem, $H = \frac{1}{2m}\vec{p}^2 - \frac{\kappa}{r}$ ($\kappa > 0$), is special in that (in addition to the angular momentum $\vec{L} = \vec{r} \times \vec{p}$) the system admits another conserved quantity, the (Laplace)–Runge–Lenz vector $\vec{M} = \frac{1}{m}\vec{p} \times \vec{L} - \frac{\kappa\vec{r}}{r}$. It is not difficult to verify that $[M_i, H] = 0$ ($i = 1, 2, 3$), and the conserved nature of this vector is responsible for the fact that only *closed* orbits are possible for bounded motion in this system. But it is not easy to describe the nature of symmetry transformations generated by the Runge–Lenz vector in simple physical terms. The three kinds of constants of motion — $\vec{L}$, $\vec{M}$ and the Kepler Hamiltonian H — form a closed system under Poisson brackets, satisfying

$$[L_i, L_j] = \epsilon_{ijk} L_k, \quad [L_i, M_j] = \epsilon_{ijk} M_k,$$
$$[M_i, M_j] = -\frac{2}{m} H \epsilon_{ijk} L_k$$

in addition to $[H, L_i] = [H, M_i] = 0$.

Problems

13-1 Given a Lagrange function $L(\mathbf{q}, \dot{\mathbf{q}}, t)$ where $\mathbf{q} = (q_1, q_2, \ldots, q_n)$, one introduces the momentum variables $\mathbf{p} = (p_1, \ldots, p_n)$ by setting $p_i = \frac{\partial L}{\partial \dot{q}_i}$ ($i = 1, \ldots, n$) and obtains the Hamiltonian — a function of $\mathbf{q}, \mathbf{p}$ and t — by the Legendre transform, i.e, $H = \sum_{i=1}^{n} p_i \dot{q}_i - L$ with $\dot{q}_i = \dot{q}_i(\mathbf{q}, \mathbf{p}, t)$ derived using the (supposedly invertible) equations $p_i = \frac{\partial L}{\partial \dot{q}_i}$ ($i = 1, \ldots, n$). Now, with the thus-obtained Hamiltonian $H(\mathbf{q}, \mathbf{p})$, suppose one introduces the velocity variables $\dot{\mathbf{q}} = (\dot{q}_1, \ldots, \dot{q}_n)$ according to $\dot{q}_i = \frac{\partial H}{\partial p_i}$ ($i = 1, \ldots, n$) and defines a function $\widetilde{L}(\mathbf{q}, \dot{\mathbf{q}}, t)$ by the Legendre transform, i.e.,

$\widetilde{L} = \sum_{i=1}^{n} \dot{q}_i p_i - H$ with $p_i = p_i(\mathbf{q}, \dot{\mathbf{q}}, t)$ derived using the equations $\dot{q}_i = \frac{\partial H}{\partial p_i}$ $(i = 1, \ldots, n)$. Show that $\widetilde{L} = L$, i.e., one gets back the original Lagrangian.

13-2 Given a mechanical system with the Lagrangian $L(q_1, \ldots, q_n, \dot{q}_1, \ldots, \dot{q}_n, t)$, the corresponding Hamiltonian is $H(\mathbf{q}, \mathbf{p}) = \sum_{i=1}^{n} p_i \dot{q}_i - L$ with the $\dot{q}$'s expressed in terms of the q's and p's by using the relations $p_i = \frac{\partial L}{\partial \dot{q}_i}$ $(i = 1, \ldots, n)$. Let us study the form of Hamiltonian as a different set of generalized coordinates $q'_i = q_i(\mathbf{q}, t)$ $(i = 1, \ldots, n)$ are used to describe the same system.

(i) Show that the canonical momentum conjugate to q'_i, i.e., $p'_i = \frac{\partial L}{\partial \dot{q}'_i}$ is given by

$$p'_i = \sum_{j=1}^{n} \frac{\partial q_j}{\partial q'_i} p_j$$

(which implies also $p_i = \sum_{j=1}^{n} \frac{\partial q'_j}{\partial q_i} p'_j$).

(ii) Show that the Hamiltonian function in the primed variables, $H'(\mathbf{q}', \mathbf{p}') = \sum_{i=1}^{n} p'_i \dot{q}'_i - L$, satisfies the relation

$$H' = H + \sum_{i=1}^{n} \frac{\partial q'_i}{\partial t} p'_i.$$

[Hence the Hamiltonian is invariant under arbitrary *time-independent* coordinate transformations.]

(iii) Verify explicitly that Hamilton's equations in the old variables

$$\frac{\partial H}{\partial p_i} = \dot{q}_i, \quad \frac{\partial H}{\partial q_i} = -\dot{p}_i \quad (i = 1, \ldots, n)$$

imply Hamilton's equations in the new variables

$$\frac{\partial H'}{\partial p'_i} = \dot{q}'_i, \quad \frac{\partial H'}{\partial q'_i} = -\dot{p}'_i \quad (i = 1, \ldots, n).$$

13-3 Consider a system of two nonrelativistic particles in a region where a uniform electric field $\vec{E} = E_0 \mathbf{k}$ is present. The two particles, with masses and electric charges given by (m_1, q) and $(m_2, -q)$ respectively, feel not only the external field $\vec{E}$ but also mutual Coulomb attraction. (Ignore gravitational force between the particles.)

(i) Using the particle's positions $\vec{r}_1$ and $\vec{r}_2$ (in an inertial frame) and corresponding conjugate momenta $\vec{p}_1$ and $\vec{p}_2$, write the Hamiltonian for the system.

(ii) Also obtain the Hamiltonian when one takes the CM position $\vec{R}$ $(\equiv \frac{m_1\vec{r}_1+m_2\vec{r}_2}{m_1+m_2})$, the relative position $\vec{r}$ $(\equiv \vec{r}_1 - \vec{r}_2)$, and corresponding conjugate momenta as one's phase-space variables.

(iii) Rewrite the Hamiltonian of (ii), now replacing the relative position $\vec{r}$ (and related conjugate momentum) by their spherical coordinate counterparts.

(iv) Based on the Hamiltonian form found in (iii), deduce what sort of conservation laws hold for this system.

13-4 For a relativistic particle in the presence of an external potential $V(\vec{r})$, the Lagrangian is

$$L = -m_0c^2\sqrt{1 - \dot{\vec{r}}^2/c^2} - V(\vec{r}).$$

Obtain the corresponding Hamiltonian, and also write the related Hamilton's equations of motion.

13-5 Suppose we are given a Lagrangian of the general form

$$\begin{aligned} L &= \frac{1}{2}\sum_{i,j=1}^{n} a_{ij}(\mathbf{q})\dot{q}_i\dot{q}_j - U(\mathbf{q}) \quad (a_{ij}(\mathbf{q}) = a_{ji}(\mathbf{q})) \\ &\equiv \frac{1}{2}\dot{\mathbf{q}}^{\mathrm{T}}\mathbf{A}(\mathbf{q})\dot{\mathbf{q}} - U(\mathbf{q}), \end{aligned}$$

where, in the second line, $\dot{\mathbf{q}}$ denotes a related n-column vector and $\mathbf{A}(\mathbf{q}) = (a_{ij}(\mathbf{q}))$ an $n \times n$ matrix.

(i) Show that, using the conjugate momentum vector $\mathbf{p} = \mathbf{A}(\mathbf{q})\dot{\mathbf{q}}$ (this implies $\dot{\mathbf{q}} = \mathbf{A}^{-1}(\mathbf{q})\mathbf{p}$), the Hamiltonian can be expressed by the general form

$$H(\mathbf{q}, \mathbf{p}) = \frac{1}{2}\mathbf{p}^{\mathrm{T}}\mathbf{A}^{-1}(\mathbf{q})\mathbf{p} + U(\mathbf{q}).$$

(ii) Let us apply this to the case of a general, nonsymmetric, rigid body in the presence of a uniform gravity. In this case, the Lagrangian using Euler's angles is (cf. Sec. 12.2)

$$\begin{aligned} L = &\frac{1}{2}I_1(-\dot{\varphi}\sin\theta\cos\psi + \dot{\theta}\sin\psi)^2 + \frac{1}{2}I_2(\dot{\varphi}\sin\theta\sin\psi + \dot{\theta}\cos\psi)^2 \\ &+ \frac{1}{2}I_3(\dot{\varphi}\cos\theta + \dot{\psi})^2 - MgR\cos\theta, \end{aligned}$$

where we used the general kinetic energy expression given in (12.22) in the text. Using the result of (i) (or by a direct attack), show that the related Hamiltonian is given by

$$H = \frac{1}{2\sin^2\theta}(p_\varphi - p_\psi \cos\theta)^2 \left(\frac{\cos^2\psi}{I_1} + \frac{\sin^2\psi}{I_2}\right) + \frac{1}{2}p_\theta^2 \left(\frac{\sin^2\psi}{I_1} + \frac{\cos^2\psi}{I_2}\right) + \frac{1}{2I_3}p_\psi^2 - \frac{\sin\psi\cos\psi}{2\sin\theta} p_\theta (p_\varphi - p_\psi\cos\theta)\left(\frac{1}{I_1} - \frac{1}{I_2}\right) + MgR\cos\theta.$$

13-6 Consider a particle of mass m moving under the influence of some external potential $V(\vec{r}, t)$. An observer, who is attached to a uniformly rotating frame of reference (with angular velocity $\vec{\omega}$), will describe the motion of this particle using the Lagrangian of the form (this was Problem 10-8)

$$L = \frac{1}{2}m\vec{v}^2 + m\vec{v}\cdot\vec{\omega}\times\vec{r} + \frac{1}{2}(\vec{\omega}\times\vec{r})^2 - V,$$

where $\vec{r}(t) = (x(t), y(t), z(t))$ denotes the radius vector of the particle at time t as seen in the rotating frame, and $\vec{v}(t) \equiv (\dot{x}(t), \dot{y}(t), \dot{z}(t))$. Find the corresponding Hamiltonian. Then, after writing out the related Hamilton's equations of motion, identify the centrifugal and Coriolis forces from them.

13-7 In a $2n$-dimensional phase space where points may be represented by $\boldsymbol{\xi} \equiv (q_1, \ldots, q_n, p_1, \ldots, p_n)$, we can take the volume of a specific region R in this space to be given by

$$V = \int_R \prod_{\alpha=1}^{2n} d\xi_\alpha \equiv \int_R \prod_{i=1}^{n} dq_i\, dp_i.$$

Then, when $q_i' = q_i'(\mathbf{q}, t)$ $(i = 1, \ldots, n)$ denote different generalized coordinates and $p_i' = \sum_{j=1}^n \frac{\partial q_j}{\partial q_i'} p_j$ the related conjugate moment (see Problem 13-2), prove that the volume of the given region R can be expressed using the variables $\boldsymbol{\xi}' \equiv (q_1', \ldots, q_n', p_1', \ldots, p_n')$ by the identical formula as above, i.e.,

$$V = \int_R \prod_{\alpha=1}^{2n} d\xi_\alpha' \equiv \int_R \prod_{i=1}^{n} dq_i'\, dp_i'.$$

13-8 The Hamiltonian for a particle of mass m in a uniform gravitational field is $H = \frac{1}{2m}p^2 + mgz$ (p is the momentum conjugate to z), considering a vertical motion only. We are here interested in the flow of corresponding phase points, i.e., $z(t)$ and $p(t)$ for some initial values of z and p.

(i) Calculate the initial phase-space area A, covered by the phase points having the momentum value within the range $p_1 \leq p \leq p_2$ and with energy $E_1 \leq E \leq E_2$.

(ii) Show explicitly that the corresponding phase-space area A', covered by the phase points $z' = z(t)$ and $p' = p(t)$ when their initial values z and p cover the area A of (i), has the same size as A for any t (> 0).

13-9 If we have an ideal gas of N atoms under the influence of a certain external potential $V(\vec{r})$, we can take the Hamiltonian of the system to be given by the sum of one-particle Hamiltonians, viz.,

$$H(\vec{r}_1, \ldots, \vec{r}_N, \vec{p}_1, \ldots, \vec{p}_N) = \sum_{n=1}^{N} \left\{ \frac{1}{2m} \vec{p}_n^{\,2} + V(\vec{r}_n) \right\}$$

$$\equiv \sum_{n=1}^{N} H^{(1)}(\vec{r}_n, \vec{p}_n).$$

Here, let $\mathcal{n}(\vec{r}, \vec{p}) d^3\vec{r}\, d^3\vec{p}$ denote the average number of atoms with position and momentum in the ranges $(x, x + dx)$, $(y, y + dy)$, $(z, z + dz)$ and $(p_x, p_x + dp_x)$, $(p_y, p_y + dp_y)$, $(p_z, p_z + dp_z)$.

(i) Show that Boltzmann equilibrium distribution $\rho \propto \exp(-H(\vec{r}_1, \ldots, \vec{r}_N, \vec{p}_1, \ldots, \vec{p}_N)/k_B T)$ with the above Hamiltonian gives rise to the following expression for $\mathcal{n}(\vec{r}, \vec{p})$:

$$\mathcal{n}(\vec{r}, \vec{p}) = N \frac{e^{-H^{(1)}(\vec{r}, \vec{p})/k_B T}}{[\int e^{-H^{(1)}(\vec{r}, \vec{p})/k_B T} d^3\vec{r} d^3\vec{p}]}.$$

(ii) When the potential of the one-particle Hamiltonian is given as

$$V(\vec{r}) = \begin{cases} 0, & \text{for any } \vec{r} \text{ in the range } (0 < x < a,\ 0 < y < b, \\ & \quad 0 < z < c) \\ \infty, & \text{otherwise} \end{cases}$$

(i.e., the gas is confined to a rectangular box of volume $V = abc$, but subject to no other external potential), use the above understanding to derive the Maxwell–Boltzmann velocity distribution:

$$\bar{\Phi}(\vec{v})d^3\vec{v} = \left(\frac{M}{2\pi k_B T}\right)^{3/2} e^{-(\frac{1}{2}m\vec{v}^2)/k_B T} d^3\vec{v}$$

(iii) Next, with the potential of the one-particle Hamiltonian given as

$$V(\vec{r}) = \begin{cases} mgz, & \text{for } \vec{r} \text{ in the range } (0 < x < a,\ 0 < y < b, \\ & 0 < z < c) \\ \infty, & \text{otherwise} \end{cases}$$

(i.e., the gas is subject to an additional uniform gravitational potential), find the explicit form of the corresponding distribution $\mathcal{n}(\vec{r}, \vec{p})$ and interpret the result.

13-10 When [,] denote Poisson brackets, show that the following relations hold generally:

$$\text{(i) } [u, vw] = [u, v]w + v[u, w],$$
$$\text{(ii) } [u, [v, w]] + [v, [w, u]] + [w, [u, v]] = 0.$$

13-11 It is convenient to employ generating functions to discuss canonical transformations from one set of canonical coordinates, say $(\mathbf{q}, \mathbf{p})$, to another set $(\mathbf{Q}, \mathbf{P})$. With a generating function $F_2(\mathbf{q}, \mathbf{P}, t)$ which may have essentially arbitrary dependences on $\mathbf{q} = (q_1, \ldots, q_n)$, $\mathbf{P} = (P_1, \ldots, P_n)$ and t, suppose that the new coordinates $(\mathbf{Q}, \mathbf{P})$ are given in terms of $\mathbf{q}, \mathbf{p}$ and t through the equations

$$Q_i = \frac{\partial F_2(\mathbf{q}, \mathbf{P}, t)}{\partial P_i}, \quad p_i = \frac{\partial F_2(\mathbf{q}, \mathbf{P}, t)}{\partial q_i} \quad (i = 1, \ldots, n).$$

Show that this is a canonical transformation, that is, the Poisson bracket relations

$$[Q_i, Q_j]_{\mathbf{q},\mathbf{p}} = [P_i, P_j]_{\mathbf{q},\mathbf{p}} = 0, \quad [Q_i, P_j]_{\mathbf{q},\mathbf{p}} = \delta_{ij}$$

hold.

[Hint: Using the $n \times n$ matrices $\mathbf{A}$, $\mathbf{B}$ and $\mathbf{C}$ formed by

$$\frac{\partial^2 F_2}{\partial q_i \partial q_j} = A_{ij}\ (= A_{ji}), \quad \frac{\partial^2 F_2}{\partial P_i \partial q_j} = B_{ij}, \quad \frac{\partial^2 F_2}{\partial P_i \partial P_j} = C_{ij}\ (= C_{ji}),$$

the following relations are true:

$$\left.\frac{\partial P_i}{\partial q_j}\right|_{\mathbf{p}} = -(\mathbf{AB}^{-1})_{ji}, \quad \left.\frac{\partial P_i}{\partial p_j}\right|_{\mathbf{q}} = (\mathbf{B}^{-1})_{ji},$$
$$\left.\frac{\partial Q_i}{\partial q_j}\right|_{\mathbf{p}} = B_{ij} - (\mathbf{AB}^{-1}\mathbf{C})_{ji}, \quad \left.\frac{\partial Q_i}{\partial p_j}\right|_{\mathbf{q}} = (\mathbf{B}^{-1}\mathbf{C})_{ji}.]$$

13-12 Given a system with a Hamiltonian $H(\mathbf{q}, \mathbf{p}, t)$, Hamilton's equations of motion assume the form

$$\dot{q}_i = \frac{\partial H}{\partial p_i}, \qquad \dot{p}_i = -\frac{\partial H}{\partial q_i} \quad (i = 1, \ldots, n).$$

Let $(\mathbf{Q}, \mathbf{P})$ denote a new set of canonical coordinates obtained with the help of a generating function $F_2(\mathbf{q}, \mathbf{P}, t)$ which may have explicit time dependence (see Problem 13-11). Show that, for the equations of motion using the new coordinates to be given also by the standard form, i.e.,

$$\dot{Q}_i = \frac{\partial H'}{\partial P_i}, \qquad \dot{P}_i = -\frac{\partial H'}{\partial Q_i} \quad (i = 1, \ldots, n),$$

with a suitably chosen Hamiltonian $H'(\mathbf{Q}, \mathbf{P}, t)$, one needs to choose the new Hamiltonian in accordance with the equation

$$H' = H + \frac{\partial}{\partial t} F_2(\mathbf{q}, \mathbf{P}, t).$$

13-13 Prove that, in a system with one degree of freedom, a rotation in the (q, p)-space

$$Q = q\cos\alpha - p\sin\alpha, \qquad P = q\sin\alpha + p\cos\alpha$$

is a canonical transformation. Also verify that this phase-space rotation is effected with the help of the generating function

$$F_1 = \frac{1}{2\sin\alpha}(q^2\cos\alpha - 2qQ + Q^2\cos\alpha).$$

13-14 For a one-dimensional harmonic oscillator with the Hamiltonian $H = \frac{1}{2m}p^2 + \frac{1}{2}m\omega^2 q^2$, consider a canonical transformation based on the generating function

$$F_2(q, P, t) = -\frac{1}{2}m\omega q^2 \tan[\omega(t - P)].$$

Determine P, Q and H' as functions of q, p and t. What do you find for the solutions to Hamilton's equations in new coordinates?

13-15 Given the Hamiltonian H, the time evolution of a function $\Phi(\mathbf{q}, \mathbf{p}, t)$ in phase space is governed by

$$\frac{d\Phi}{dt} = [\Phi, H] + \frac{\partial \Phi}{\partial t}$$

where [,] denotes the Poisson bracket with respect to the canonical variables $\mathbf{q}, \mathbf{p}$. Now, when two functions $G_1(\mathbf{q}, \mathbf{p}, t)$ and $G_2(\mathbf{q}, \mathbf{p}, t)$ are given, consider the third function $G_3(\mathbf{q}, \mathbf{p}, t)$ defined by $G_3 = [G_1, G_2]$. Show that

$$\frac{dG_3}{dt} = \left[\frac{dG_1}{dt}, G_2\right] + \left[G_1, \frac{dG_2}{dt}\right].$$

(This is useful in proving Poisson's theorem.).

13-16 For a Hamiltonian system (with its Hamiltonian denoted $H(\mathbf{q}, \mathbf{p}, t)$), let us consider an infinitesimal canonical transformation represented by

$$\delta q_i = \frac{\partial G}{\partial p_i}\delta\lambda, \qquad \delta p_i = -\frac{\partial G}{\partial q_i}\delta\lambda$$

where the generator $G \equiv G(\mathbf{q}, \mathbf{p}, t)$ may contain explicit time dependence.

(i) Show that the above infinitesimal transformation corresponds to a *symmetry of equations of motion* (i.e., a symmetry of Hamilton's equations of motion, given in specific forms of differential equations involving $\mathbf{q}(t), \mathbf{p}(t)$) only when the Hamiltonian of the system possesses the property

$$\begin{aligned} \delta H(\mathbf{q}, \mathbf{p}, t) &\equiv H(\mathbf{q} + \delta\mathbf{q}, \mathbf{p} + \delta\mathbf{p}, t) - H(\mathbf{q}, \mathbf{p}, t) \\ &= \frac{\partial G}{\partial t}\delta\lambda + O((\delta\lambda)^2). \end{aligned}$$

(ii) Show that if we have a time-dependent symmetry in the sense of (i), the corresponding generator $G(\mathbf{q}, \mathbf{p}, t)$ is conserved, i.e., satisfies $\frac{d}{dt}G(\mathbf{q}(t), \mathbf{p}(t), t) = 0$.

(iii) For an isolated system of N nonrelativistic particles, the Galilean relativity requires that the equations of motion be unchanged under the infinitesimal transformations

$$\delta\vec{r}_i = t\hat{u}\,\delta\lambda, \qquad \delta\vec{p}_i = m_i\hat{u}\,\delta\lambda \quad (i = 1, \ldots, N)$$

where m_i denotes the mass of the ith particle, and $\hat{u}$ a unit vector along any chosen direction. Find the corresponding generator $G(\vec{\mathbf{r}}, \vec{\mathbf{p}}, t)$ and provide an interpretation of the related conservation law. Also, based on the result of (i) above, discuss the restriction imposed on the allowed form of the Hamiltonian for a Galilean-invariant N-particle system.

13-17 For the Kepler Hamiltonian $H = \frac{1}{2m}\vec{p}^2 - \frac{\kappa}{r}$ $(\kappa > 0)$, we have two conserved vector quantities, i.e., the angular momentum $\vec{L} = \vec{r} \times \vec{p}$ and the (Laplace)–Runge–Lenz vector $\vec{M} = \frac{1}{m}\vec{p} \times \vec{L} - \frac{\kappa\vec{r}}{r}$. By using the associated conservation laws, obtain explicit forms of allowed Kepler orbits.

13-18 Like the case of the Kepler Hamiltonian, the two-dimensional isotropic harmonic oscillator, with the Hamiltonian $H = \frac{1}{2m}(p_x^2 + p_y^2) + \frac{m}{2}\omega^2(x^2+y^2)$, also admits some additional constants of motion (beyond the angular momentum $L = xp_y - yp_x$). Let us study the related aspects here.

(i) Prove that (by calculating the Poisson brackets with the Hamiltonian) the components of a symmetric tensor A_{ij} defined as

$$A_{ij} = \frac{1}{2m}(p_i p_j + m^2\omega^2 x_i x_j) \quad (i, j = 1, 2)$$

are conserved quantities.

(ii) Show, as a consequence of these additional conservation laws, that only *closed orbits* are possible for this system.

(iii) Actually, this system has a hidden three-dimensional-rotation-like symmetry. To get some idea on this, verify that if we consider three conserved quantities S_1, S_2, S_3 given by

$$S_1 = \frac{1}{2\omega}(A_{12} + A_{21}) = \frac{1}{2m\omega}(p_x p_y + m^2\omega^2 xy),$$
$$S_2 = \frac{1}{2\omega}(A_{22} - A_{11}) = \frac{1}{4m\omega}\left[p_y^2 - p_x^2 + m^2\omega^2(y^2 - x^2)\right],$$
$$S_3 = \frac{L}{2} = \frac{1}{2}(xp_y - yp_x),$$

they satisfy the Poisson-bracket relations (akin to those satisfied by the three components of the three-dimensional angular momentum vector)

$$[S_i, S_j] = \sum_{k=1}^{3} \epsilon_{ijk} S_k.$$

13-19 For the three-dimensional Kepler system, verify the Poisson-bracket relations

$$[L_i, M_j] = \sum_{k=1}^{3} \epsilon_{ijk} M_k,$$

$$[M_i, M_j] = -\frac{2}{m}\left(\frac{1}{2m}\vec{p}^2 - \frac{\kappa}{r}\right)\sum_{k=1}^{3} \epsilon_{ijk} L_k$$

(with $L_i = \epsilon_{ijk} x_j p_k$ and $M_i = \frac{1}{m}\epsilon_{ijk} p_j L_k - \frac{\kappa}{r} x_i$), by calculating the Poisson brackets involved explicitly.

13-20 For a general rigid body with a fixed pivot, we can have the Hamiltonian expressed as a function of three Euler's angles (θ, φ, ψ) and related canonical momenta $(p_\theta, p_\varphi, p_\psi)$. (See Problem 13-5 (ii).) For this system we have as a useful physical quantity the total angular momentum $\vec{L}$, which can be represented as $\vec{L} = \sum_{i=1}^{3} L_i \mathbf{e}_i$ where $(\mathbf{e}_1, \mathbf{e}_2, \mathbf{e}_3)$ denote the body-fixed basis directed along the principal axes of the body, or as $\vec{L} = \sum_{i=1}^{3} L_i^{(0)} \mathbf{e}_i^{(0)}$ with a space-fixed basis $(\mathbf{e}_1^{(0)} \equiv \mathbf{i},\ \mathbf{e}_2^{(0)} \equiv \mathbf{j},\ \mathbf{e}_3^{(0)} \equiv \mathbf{k})$. An interesting problem is to find out what algebraic relations hold between different components of this vector $\vec{L}$ under Poisson brackets.

(i) Show that, in terms of canonical variables $(\theta,\ \varphi,\ \psi,\ p_\theta,\ p_\varphi,\ p_\psi)$, components of the vector $\vec{L}$ relative to the body-fixed basis $\{\mathbf{e}_i\}$ can be expressed by

$$L_1 = -\frac{\cos\psi}{\sin\theta}(p_\varphi - p_\psi \cos\theta) + p_\theta \sin\psi,$$

$$L_2 = \frac{\sin\psi}{\sin\theta}(p_\varphi - p_\psi \cos\theta) + p_\theta \cos\psi,$$

$$L_3 = p_\psi,$$

and components of $\vec{L}$ relative to the space-fixed basis $\{\mathbf{e}_i^{(0)}\}$ by

$$L_1^{(0)} = \frac{\cos\varphi}{\sin\theta}(p_\psi - p_\varphi \cos\theta) - p_\theta \sin\varphi,$$

$$L_2^{(0)} = \frac{\sin\varphi}{\sin\theta}(p_\psi - p_\varphi \cos\theta) + p_\theta \cos\varphi,$$
$$L_3^{(0)} = p_\varphi.$$

[Hint: Note that $L_1 = I_1\omega_1$, $L_2 = I_2\omega_2$ and $L_3 = I_3\omega_3$ with $(\omega_1, \omega_2, \omega_3)$ expressed in terms of Euler's angles, while we can set $p_\theta = \frac{\partial T}{\partial \dot{\theta}}$, $p_\varphi = \frac{\partial T}{\partial \dot{\varphi}}$, and $p_\psi = \frac{\partial T}{\partial \dot{\psi}}$ in terms of the kinetic energy function $T = T(\theta, \varphi, \psi, \dot{\theta}, \dot{\varphi}, \dot{\psi})$ (see Sec. 12.1).]

(ii) Based on the expressions obtained in (i), show that we have the following Poisson-bracket relations between different components of the angular momentum vector:

$$\left[L_i^{(0)}, L_j^{(0)}\right] = \sum_{k=1}^{3} \epsilon_{ijk} L_k^{(0)},$$
$$\left[L_i^{(0)}, L_j\right] = 0,$$
$$[L_i, L_j] = -\sum_{k=1}^{3} \epsilon_{ijk} L_k.$$

Can you provide some simple interpretations of these relations?

(iii) Consider a *freely-rotating* rigid body with a fixed pivot. In this case, show that one can obtain Euler's equations for a rigid body (cf. Sec. 12.1 with $\vec{\Gamma}^{(e)} = 0$) by considering our general time evolution equation $\frac{d\Phi}{dt} = [\Phi, H] + \frac{\partial \Phi}{\partial t}$ for $\Phi = L_i$ $(i = 1, 2, 3)$.

Part II

Solutions

1 In Three-Dimensional Space: Vector Description

1-1 Supposing that the length of each side of the cube is a, we have the vector $\overrightarrow{OA} = (a, a, a)$ along a long diagonal and the vector $\overrightarrow{OB} = (0, a, a)$ along an adjacent face diagonal. Hence the desired cosine of the angle is determined as

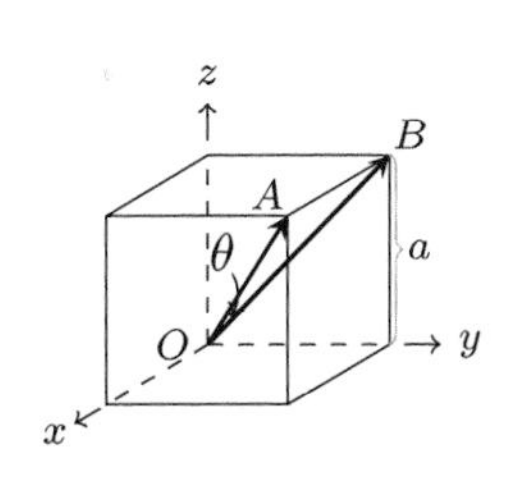

$$\cos\theta = \frac{\overrightarrow{OA}\cdot\overrightarrow{OB}}{|OA|\,|OB|} = \frac{2a^2}{(\sqrt{3}a)(\sqrt{2}a)} = \sqrt{\frac{2}{3}}.$$

1-2 (i) The unit vector which is normal to the plane can be taken as

$$\hat{n} = \left(\frac{A}{\sqrt{A^2+B^2+C^2}}, \frac{B}{\sqrt{A^2+B^2+C^2}}, \frac{C}{\sqrt{A^2+B^2+C^2}}\right).$$

(Note that the negative of this vector is also normal to the plane.)

(ii) Let us consider a vector $\vec{r}$ denoting any point on the plane. When $\vec{r}_P \equiv (x_0, y_0, z_0)$ denotes the position of the point P, the projection of the vector $(\vec{r}-\vec{r}_P)$ on the normal vector $\hat{n}$ will give the shortest distance. Hence the desired distance is given by

$$d = |(\vec{r}-\vec{r}_P)\cdot\hat{n}| = \frac{|A(x-x_0)+B(y-y_0)+C(z-z_0)|}{\sqrt{A^2+B^2+C^2}}$$
$$= \frac{|Ax_0+By_0+Cz_0+D|}{\sqrt{A^2+B^2+C^2}},$$

since $\vec{r} = (x, y, z)$ is a point on the plane $Ax+By+Cz+D=0$.

(iii) The angle between two planes is equal to the angle between two vectors which are normal to the two planes respectively. The inner product of two unit normal vectors determines the cosine of the angle: hence, using the result of (i),

$$|\cos\theta| = |\hat{n}_1 \cdot \hat{n}_2| = \frac{|Aa + Bb + Cc|}{\sqrt{A^2+B^2+C^2}\sqrt{a^2+b^2+c^2}}.$$

1-3 (i) We may first consider a right-handed orthonormal triplet including the unit vector $\mathbf{u} = \frac{1}{\sqrt{2}}(\mathbf{e}_1 + \mathbf{e}_2)$, say $(\mathbf{u}, \mathbf{v}, \mathbf{w})$ with

$$\mathbf{v} = \frac{1}{\sqrt{2}}(-\mathbf{e}_1 + \mathbf{e}_2), \quad \mathbf{w} = \mathbf{e}_3. \tag{1}$$

Then, using $(\mathbf{u}, \mathbf{v}, \mathbf{w})$, we may describe the rotation indicated in the problem by

$$\begin{aligned} \mathbf{u}'\left(= \frac{1}{\sqrt{2}}(\mathbf{e}'_1 + \mathbf{e}'_2)\right) &= \mathbf{u} \\ \mathbf{v}'\left(= \frac{1}{\sqrt{2}}(-\mathbf{e}'_1 + \mathbf{e}'_2)\right) &= \cos\theta\,\mathbf{v} + \sin\theta\,\mathbf{w} \\ \mathbf{w}'\,(= \mathbf{e}'_3) &= -\sin\theta\,\mathbf{v} + \cos\theta\,\mathbf{w}. \end{aligned} \tag{2}$$

Using (1) with (2), we now see that $\{\mathbf{e}'_i\}$ are related to $\{\mathbf{e}_i\}$ by

$$\begin{pmatrix} \frac{1}{\sqrt{2}} & \frac{1}{\sqrt{2}} & 0 \\ -\frac{1}{\sqrt{2}} & \frac{1}{\sqrt{2}} & 0 \\ 0 & 0 & 1 \end{pmatrix} \begin{pmatrix} \mathbf{e}'_1 \\ \mathbf{e}'_2 \\ \mathbf{e}'_3 \end{pmatrix} = \begin{pmatrix} \frac{1}{\sqrt{2}} & \frac{1}{\sqrt{2}} & 0 \\ -\frac{1}{\sqrt{2}}\cos\theta & \frac{1}{\sqrt{2}}\cos\theta & \sin\theta \\ \frac{1}{\sqrt{2}}\sin\theta & -\frac{1}{\sqrt{2}}\sin\theta & \cos\theta \end{pmatrix} \times \begin{pmatrix} \mathbf{e}_1 \\ \mathbf{e}_2 \\ \mathbf{e}_3 \end{pmatrix}.$$

This in turn leads to

$$\begin{aligned} \begin{pmatrix} \mathbf{e}'_1 \\ \mathbf{e}'_2 \\ \mathbf{e}'_3 \end{pmatrix} &= \begin{pmatrix} \frac{1}{\sqrt{2}} & -\frac{1}{\sqrt{2}} & 0 \\ \frac{1}{\sqrt{2}} & \frac{1}{\sqrt{2}} & 0 \\ 0 & 0 & 1 \end{pmatrix} \begin{pmatrix} \frac{1}{\sqrt{2}} & \frac{1}{\sqrt{2}} & 0 \\ -\frac{1}{\sqrt{2}}\cos\theta & \frac{1}{\sqrt{2}}\cos\theta & \sin\theta \\ \frac{1}{\sqrt{2}}\sin\theta & -\frac{1}{\sqrt{2}}\sin\theta & \cos\theta \end{pmatrix} \\ &\quad \times \begin{pmatrix} \mathbf{e}_1 \\ \mathbf{e}_2 \\ \mathbf{e}_3 \end{pmatrix} \\ &= \begin{pmatrix} \frac{1}{2}(1+\cos\theta) & \frac{1}{2}(1-\cos\theta) & -\frac{1}{\sqrt{2}}\sin\theta \\ \frac{1}{2}(1-\cos\theta) & \frac{1}{2}(1+\cos\theta) & \frac{1}{\sqrt{2}}\sin\theta \\ \frac{1}{\sqrt{2}}\sin\theta & -\frac{1}{\sqrt{2}}\sin\theta & \cos\theta \end{pmatrix} \begin{pmatrix} \mathbf{e}_1 \\ \mathbf{e}_2 \\ \mathbf{e}_3 \end{pmatrix}. \end{aligned} \tag{3}$$

As the components of a vector transform (under basis rotation) in the same way as the basis vectors do, the desired matrix O should be

$$O = \begin{pmatrix} \frac{1}{2}(1+\cos\theta) & \frac{1}{2}(1-\cos\theta) & -\frac{1}{\sqrt{2}}\sin\theta \\ \frac{1}{2}(1-\cos\theta) & \frac{1}{2}(1+\cos\theta) & \frac{1}{\sqrt{2}}\sin\theta \\ \frac{1}{\sqrt{2}}\sin\theta & -\frac{1}{\sqrt{2}}\sin\theta & \cos\theta \end{pmatrix}.$$

(ii) These are straightforward calculations.

1-4 (i) For every possible choice of j and k ($\neq j$), the quantity $\epsilon_{ijk}\epsilon_{ljk}$ ($\equiv \sum_{j,k}\epsilon_{ijk}\epsilon_{ljk}$) will be nonvanishing only when both i and l differ from the indices j and k. This implies that the given quantity is nonzero only when $i = l$ (since only three choices are available for each index). Hence we can set $\epsilon_{ijk}\epsilon_{ljk}$ to be $C_i\delta_{il}$ (i not summed). But, for any given $i\,(= l)$, say $i = 1$, we have

$$\epsilon_{1jk}\epsilon_{1jk} = \epsilon_{123}\epsilon_{123} + \epsilon_{132}\epsilon_{132} = 2,$$

i.e., $C_1 = 2$. Similarly, we find $C_2 = C_3 = 2$ so that $\epsilon_{ijk}\epsilon_{ljk} = 2\delta_{il}$. Now, $\epsilon_{ijk}\epsilon_{ijk} = \sum_i 2\delta_{ii} = 6 = 3!$.

(ii) For every choice of the index k, $\epsilon_{ijk}\epsilon_{lmk}$ ($\equiv \sum_k \epsilon_{ijk}\epsilon_{lmk}$) will be nonvanishing only when (i, j) and (l, m) are pairwise equal, i.e., $(i = l,\ j = m)$ or $(i = m,\ j = l)$, while i and j ($\neq i$) differ from k. Hence we may write

$$\epsilon_{ijk}\epsilon_{lmk} = C_{ij}\delta_{il}\delta_{jm} + D_{ij}\delta_{im}\delta_{jl}\,.$$

But, from $\epsilon_{ijk}\epsilon_{lmk} = -\epsilon_{jik}\epsilon_{lmk}$, we should have $D_{ij} = -C_{ij}$, i.e.,

$$\epsilon_{ijk}\epsilon_{lmk} = C_{ij}(\delta_{il}\delta_{jm} - \delta_{im}\delta_{jl}) \text{ (with } C_{ij} = 0 \text{ if } i = j).$$

Then, with $i = l$ and $j = m$ ($\neq i$), $\epsilon_{ijk}\epsilon_{ijk}$ (i, j not summed) should be equal to 1 as k may take the remaining index different from i, j. On the other hand, with $i = m$ and $j = l$ ($\neq i$), $\epsilon_{ijk}\epsilon_{jik}$ (i, j not summed) equal to -1, since $\epsilon_{jik} = -\epsilon_{ijk}$. All these are accounted for by writing

$$\epsilon_{ijk}\epsilon_{lmk} = \delta_{il}\delta jm - \delta_{im}\delta_{jl}\,.$$

(iii) According to the usual definition of the determinant for a matrix $A = (A_{ij})$, we may write

$$\det A = \begin{vmatrix} A_{11} & A_{12} & A_{13} \\ A_{21} & A_{22} & A_{23} \\ A_{31} & A_{32} & A_{133} \end{vmatrix} = \epsilon_{lmn} A_{1l} A_{2m} A_{3n} \,.$$

Then, due to the antisymmtry of the ϵ-symbol, we have

$$\begin{aligned} \epsilon_{lmn} A_{2l} A_{1m} A_{3n} &= -\epsilon_{mln} A_{1m} A_{2l} A_{3n} = -\det A, \\ \epsilon_{lmn} A_{3l} A_{1m} A_{2n} &= \epsilon_{mnl} A_{1m} A_{2n} A_{3l} = \det A, \end{aligned}$$

etc.; these allow us to write $\epsilon_{lmn} A_{il} A_{jm} A_{kn} = \epsilon_{ijk} \det A$. The relation $\epsilon_{lmn} A_{li} A_{mj} A_{nk} = \epsilon_{ijk} \det A$ should also be true since the definition for the determinant given above can also be written as (verify this)

$$\det A = \begin{vmatrix} A_{11} & A_{21} & A_{31} \\ A_{12} & A_{22} & A_{32} \\ A_{13} & A_{23} & A_{33} \end{vmatrix} = \epsilon_{lmn} A_{l1} A_{m2} A_{n3} = \det A^{\mathrm{T}},$$

where A^{T} denotes the matrix transpose of A with $(A^{\mathrm{T}})_{ij} = A_{ji}$.

(iv) If C denotes the matrix product AB (i.e., $C_{ij} = A_{ik} B_{kj}$), then

$$\begin{aligned} \det(AB) &= \epsilon_{lmn} C_{1l} C_{2m} C_{3n} \\ &= \epsilon_{lmn} A_{1p} B_{pl} A_{2q} B_{qm} A_{3r} B_{rn} \\ &= A_{1p} A_{2q} A_{3r} \epsilon_{lmn} B_{pl} B_{qm} B_{rn} \\ &= \epsilon_{pqr} A_{1p} A_{2q} A_{3r} \det B \\ &= \det A \cdot \det B, \end{aligned}$$

where we made use of the relations obtained in (iii).

1-5 Using the identity $(\vec{A} \times \vec{B}) \cdot \vec{A} = (\vec{A} \times \vec{A}) \cdot \vec{B} = 0$, one may easily find the following relations

$$\begin{aligned} (\vec{A} \times \vec{B}) \cdot \vec{D} &= \gamma (\vec{A} \times \vec{B}) \cdot \vec{C} \\ (\vec{B} \times \vec{C}) \cdot \vec{D} &= \alpha (\vec{B} \times \vec{C}) \cdot \vec{A} \\ (\vec{C} \times \vec{A}) \cdot \vec{D} &= \beta (\vec{C} \times \vec{A}) \cdot \vec{B}. \end{aligned}$$

From these relations and the well-known properties of the scalar triple product, we get the coefficients

$$\alpha = \frac{\vec{D}\cdot(\vec{B}\times\vec{C})}{\vec{A}\cdot(\vec{B}\times\vec{C})}, \quad \beta = \frac{\vec{A}\cdot(\vec{D}\times\vec{C})}{\vec{A}\cdot(\vec{B}\times\vec{C})}, \quad \gamma = \frac{\vec{A}\cdot(\vec{B}\times\vec{D})}{\vec{A}\cdot(\vec{B}\times\vec{C})}.$$

[Comment: As we use the column notations $\vec{A} = \begin{pmatrix} A_1 \\ A_2 \\ A_3 \end{pmatrix}$, $\vec{B} = \begin{pmatrix} B_1 \\ B_2 \\ B_3 \end{pmatrix}$, $\vec{C} = \begin{pmatrix} C_1 \\ C_2 \\ C_3 \end{pmatrix}$ and $\vec{D} = \begin{pmatrix} D_1 \\ D_2 \\ D_3 \end{pmatrix}$, the given equation for these vectors implies

$$\begin{aligned} A_1\alpha + B_1\beta + C_1\gamma &= D_1, \\ A_2\alpha + B_2\beta + C_2\gamma &= D_2, \\ A_3\alpha + B_3\beta + C_3\gamma &= D_3. \end{aligned}$$

Then, clearly, our expressions obtained for α, β and γ above are equivalent to those given by the well-known Cramer's rule.]

1-6 When $\vec{a}$, $\vec{b}$ and $\vec{c}$ denote the vectors as shown in the figure, we have $\vec{c} = \vec{b} - \vec{a}$. Taking a vector product with $\vec{c}$ then gives $\vec{c}\times\vec{b} = \vec{c}\times\vec{a}$ and hence $|\vec{c}\times\vec{b}| = |\vec{c}\times\vec{a}|$, or

$$c\,b\sin\alpha = c\,a\sin\beta.$$

This implies

$$\frac{a}{\sin\alpha} = \frac{b}{\sin\beta}.$$

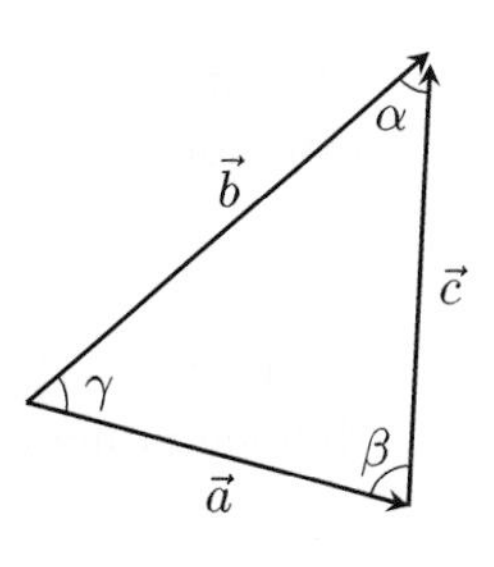

Similarly, from the relation $\vec{a}\times\vec{c} = \vec{a}\times(\vec{b} - \vec{a}) = \vec{a}\times\vec{b}$, we find

$$\frac{c}{\sin\gamma} = \frac{b}{\sin\beta}.$$

Hence the sine law of trigonometry is established.

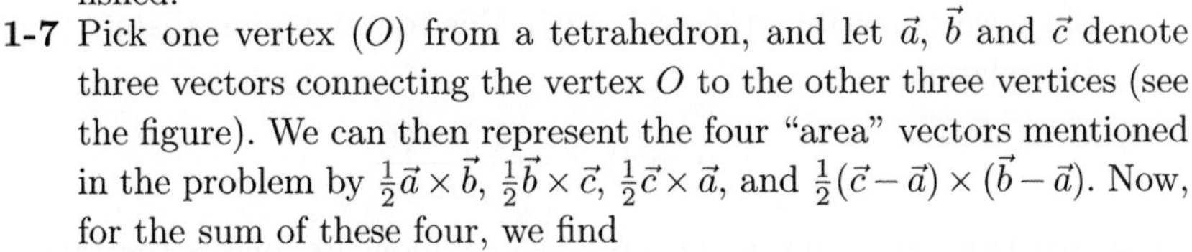

1-7 Pick one vertex (O) from a tetrahedron, and let $\vec{a}$, $\vec{b}$ and $\vec{c}$ denote three vectors connecting the vertex O to the other three vertices (see the figure). We can then represent the four "area" vectors mentioned in the problem by $\frac{1}{2}\vec{a}\times\vec{b}$, $\frac{1}{2}\vec{b}\times\vec{c}$, $\frac{1}{2}\vec{c}\times\vec{a}$, and $\frac{1}{2}(\vec{c}-\vec{a})\times(\vec{b}-\vec{a})$. Now, for the sum of these four, we find

$$\frac{1}{2}[\vec{a}\times\vec{b} + \vec{b}\times\vec{c} + \vec{c}\times\vec{a} + (\vec{c}-\vec{a})\times(\vec{b}-\vec{a})] = 0.$$

Also any polyhedron can be viewed as a collection of tetrahedrons put together. Since the sum of four outward area vectors from any tetrahedron vanishes, the sum of such area vectors from all tetrahedrons (making the given polyhedron) should also be a zero vector. Here, not all area vectors of tetrahedrons will be the ones that can be associated with outer faces of the polyhedron — there enter also area vectors related to inner faces in the sum. But note that, for each inner face, we can always think of two tetrahedrons having it as one of their faces. This means that, in association with each inner face, there appear (in our sum) two area vectors, differing only by sign, and they thus cancel. We are thus left with the area vectors associated with outer faces, and hence a theorem: *the sum of all area vectors associated with the faces of any polyhedron is zero.* (This can also be shown with the help of Stoke's theorem, discussed in the appendix to Chapter 8.)

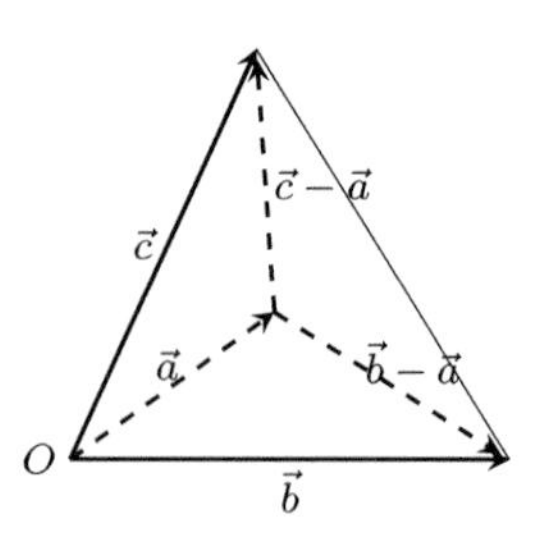

1-8 Denoting the center of the sphere as $\vec{r}_0$, the position vector $\vec{r}$ of the particle is subject to the equation

$$(\vec{r}-\vec{r}_0)\cdot(\vec{r}-\vec{r}_0)=R^2,$$

since it lies on the surface of the sphere. Taking a derivative of this equation with respect to t yields

$$\vec{v}(t)\cdot(\vec{r}(t)-\vec{r}_0)=0$$

where $\vec{v}(t)=\frac{d}{dt}\vec{r}(t)$, i.e., represents the velocity of the particle. After differentiating once more, we have

$$\vec{a}(t)\cdot(\vec{r}(t)-\vec{r}_0)+\vec{v}(t)\cdot\vec{v}(t)=0.$$

Since we can write $(\vec{r}(t)-\vec{r}_0)=\mathbf{n}R$, this equation immediately yields the relation

$$\mathbf{n}\cdot\vec{a}(t)=-\frac{\vec{v}(t)\cdot\vec{v}(t)}{R}.$$

1-9 Acceleration at any point have two components: one is the tangential component and the other is the normal component to the path. The tangential component is zero since uniform speed is assumed.

The normal component is $a_n = -\frac{v^2}{R}$, R denoting the radius of curvature; this is related to the rate of change in the tangent *directions*. Therefore, at the point A where the path has a maximal curvature, the magnitude of acceleration acquires a maximal value.

1-10 Let $v(t)$ denote the instantaneous forward speed of the wheel at time t (so that $\dot{v} = a_0$). If P denotes a mass point on the rim at angle θ from the highest point, its position from the center of the wheel can be represented by

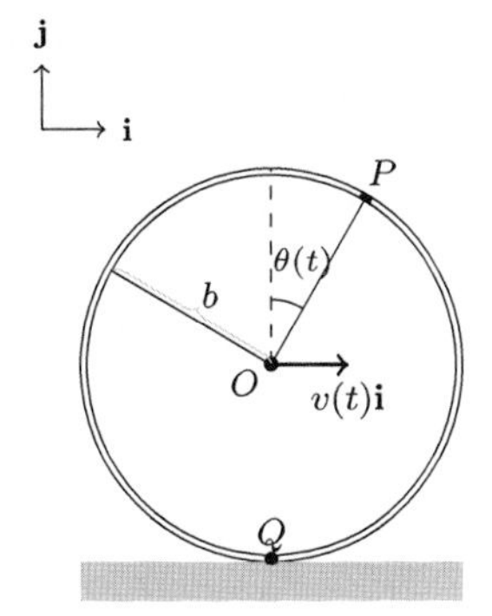

$$\overrightarrow{OP}(t) = b(\sin\theta(t)\mathbf{i} + \cos\theta(t)\mathbf{j}).$$

So the velocity of this mass point relative to the ground is given by

$$\begin{aligned}\vec{v}_P(t) &= v\mathbf{i} + \frac{d}{dt}\overrightarrow{OP}(t)\\ &= (v + b\dot{\theta}\cos\theta)\mathbf{i} - b\dot{\theta}\sin\theta\mathbf{j}.\end{aligned}$$

But, as the wheel is rolling, we must have $\vec{v}_Q = 0$ at the contact point (Q) with the ground (i.e., at $\theta = \pi$) and this demands that $v - b\dot{\theta} = 0$, i.e., $\dot{\theta} = \frac{v}{b}$. Based on this, the acceleration of the mass point P relative to the center of the wheel is

$$\begin{aligned}\frac{d^2}{dt^2}\overrightarrow{OP} &= \frac{d}{dt}\left[b\left(\frac{v}{b}\right)(\cos\theta\mathbf{i} - \sin\theta\mathbf{j})\right]\\ &= a_0(\cos\theta\mathbf{i} - \sin\theta\mathbf{j}) - \frac{v^2}{b}(\sin\theta\mathbf{i} + \cos\theta\mathbf{j})\\ &= \left(a_0\cos\theta - \frac{v^2}{b}\sin\theta\right)\mathbf{i} - \left(a_0\sin\theta + \frac{v^2}{b}\cos\theta\right)\mathbf{j},\end{aligned}$$

where we have used $\dot{v} = a_0$, and hence the magnitude $|\frac{d^2}{dt^2}\overrightarrow{OP}| = [(a_0\cos\theta - \frac{v^2}{b}\sin\theta)^2 + (a_0\sin\theta + \frac{v^2}{b}\cos\theta)^2]^{1/2} = \sqrt{a_0^2 + \frac{v^4}{b^2}}$. The acceleration relative to the ground, i.e., $\frac{d}{dt}\vec{v}_P(t)$, is $a_0\mathbf{i} + \frac{d^2}{dt^2}\overrightarrow{OP}$, and this vector has the magnitude

$$\begin{aligned}&\sqrt{\left(a_0 + a_0\cos\theta - \frac{v^2}{b}\sin\theta\right)^2 + \left(a_0\sin\theta + \frac{v^2}{b}\cos\theta\right)^2}\\ &\quad = a_0\sqrt{2 + 2\cos\theta + \frac{v^4}{a_0^2 b^2} - \left(\frac{2v^2}{a_0 b}\right)\sin\theta}.\end{aligned}$$

From this expression follows that the greatest acceleration relative to the ground occurs when the point P is at angle $2\pi - \tan^{-1}\left(\frac{v^2}{a_0 b}\right)$.

1-11 If the center of the wheel moves with a uniform velocity $\vec{v} = v\mathbf{i}$, the wheel in the state of rolling should rotate with a constant angular speed $\omega = \frac{v}{b}$ (b is the radius of the wheel). The vector representing the center position of the wheel is, clearly, $\vec{r}_0(t) = (x_0 + vt)\mathbf{i} + b\mathbf{j}$. Then a particular point P on the rim, satisfying the given initial condition, will have the position given by

$$\begin{aligned}\vec{r}_P(t) &= \vec{r}_0(t) + b(\sin\omega t\mathbf{i} + \cos\omega t\mathbf{j}) \\ &= (x_0 + b\omega t + b\sin\omega t)\mathbf{i} + (b + b\cos\omega t)\mathbf{j}.\end{aligned}$$

Hence its trajectory is described by the two equations

$$x = x_0 + b(\omega t + \sin\omega t), \quad y = b(1 + \cos\omega t),$$

which corresponds to a *cycloid*. [Note that, at $t = \frac{\pi}{\omega}, \frac{3\pi}{\omega}, \frac{5\pi}{\omega}, \ldots$, the point is at the lowest position (i.e., in contact with the ground) and the instantaneous velocity is equal to zero there.] A rough sketch of this trajectory is given below.

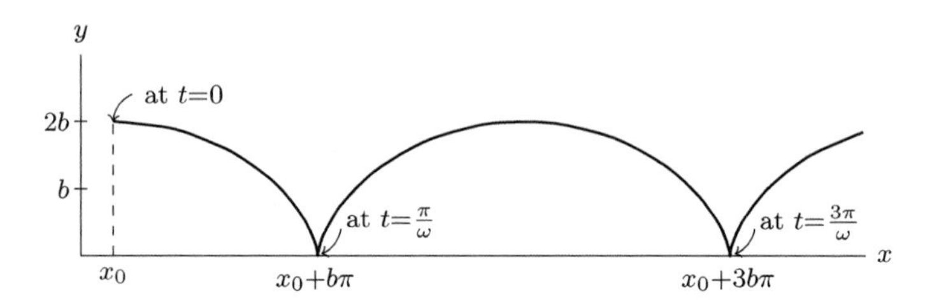

1-12 The trajectory of the bullet is described by

$$x(t) = (v_0\cos\theta)t, \quad z(t) = h + (v_0\sin\theta)t - \frac{1}{2}gt^2.$$

The time $\bar{t}$ for the bullet to reach the level plain, $z = 0$, can be determined by solving the equation

$$h + (v_0\sin\theta)\bar{t} - \frac{1}{2}g\bar{t}^2 = 0.$$

Instead of solving this quadratic equation, we use the relation $\bar{t} = \frac{R}{v_0\cos\theta}$ where R denotes the horizontal range, and insert it into above the equation to get

$$R^2 - \frac{v_0^2}{g}(2\sin\theta\cos\theta)R - \frac{2hv_0^2}{g}\cos^2\theta = 0. \tag{1}$$

This equation determines R as a function of θ. To obtain a maximum value of $R(\theta)$, we impose $\frac{dR(\theta)}{d\theta} = 0$:

$$-\frac{2v_0^2}{g}\cos(2\theta)R + \frac{2hv_0^2}{g}\sin(2\theta) = 0.$$

Hence,

$$R = h\tan(2\theta).$$

Plugging this into (1) and eliminating R, we find that

$$h^2\tan^2 2\theta - \frac{v_0^2 h}{g}\frac{\sin^2 2\theta}{\cos 2\theta} - \frac{2v_0^2 h}{g}\cos^2\theta = 0.$$

This can be simplified to

$$h\frac{\sin^2 2\theta}{\cos 2\theta} - \frac{v_0^2}{g}(1+\cos 2\theta) = 0$$

or

$$4h\sin^2\theta - \frac{2v_0^2}{g}(1 - 2\sin^2\theta) = 0.$$

Hence the desired angle of elevation must be chosen such that

$$\csc^2\theta = \frac{1}{\sin^2\theta} = 2\left(\frac{gh}{v_0^2} + 1\right).$$

1-13 Let the inclined angle and the initial speed be θ and V_0, respectively. We will then have the trajectory given by

$$x = V_0(\cos\theta)t, \quad y = V_0(\sin\theta)t - \frac{1}{2}gt^2 \quad (t \geq 0).$$

This means that, at $x = L$ (i.e., at the time $t = \frac{L}{V_0\cos\theta}$), the ball is at the height

$$\begin{aligned} y|_{x=L} &= \frac{L\sin\theta}{\cos\theta} - \frac{1}{2}g\frac{L^2}{V_0^2\cos^2\theta} = L\tan\theta - \frac{1}{2}\frac{gL^2}{V_0^2}(1+\tan^2\theta) \\ &= -\frac{1}{2}\frac{gL^2}{V_0^2}\left(\tan\theta - \frac{V_0^2}{gL}\right)^2 + \frac{1}{2}\frac{V_0^2}{g} - \frac{1}{2}\frac{gL^2}{V_0^2}. \end{aligned}$$

We here want $y|_{x=L} \geq h$; so, choosing $\tan\theta = \frac{V_0^2}{gL}$, we may require that

$$\frac{1}{2}\frac{V_0^2}{g} - \frac{1}{2}\frac{gL^2}{V_0^2} \geq h.$$

The equality should hold for $V_0 = V$ (: the minimum speed necessary), and accordingly

$$V^4 - 2ghV^2 - g^2L^2 = 0 \quad \longrightarrow \quad V^2 = gh + \sqrt{g^2h^2 + g^2L^2}.$$

Therefore, for the minimum speed, we find $V = [g(h+\sqrt{h^2+L^2})]^{1/2}$.

1-14 The water "bell" is cylindrically symmetrical about the vertical and so it is sufficient to solve the problem by considering a cross section. Let the small (i.e., more or less point-like) rose be at the origin of an x-y coordinate plane. The jets of water then follow parabolic paths starting from the origin, and our task is to find the "envelope" (see the figure) to this set of parabolas.

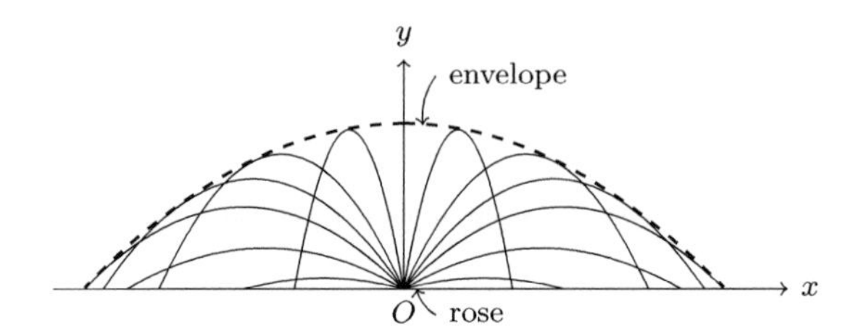

The equation of the path of a body projected with initial speed v at an angle α to the horizontal is

$$y = x\tan\alpha - \frac{g}{2v^2\cos^2\alpha}x^2,$$

which can also be written as ($\frac{1}{\cos^2\alpha} = 1 + \tan^2\alpha$)

$$\frac{gx^2}{2v^2}u^2 - xu + \left(y + \frac{gx^2}{2v^2}\right) = 0, \tag{1}$$

where $u = \tan\alpha$. Now, for some given values of (x, y), (1) — a quadratic equation in u — admits a real solution if its discriminant is nonnegative, i.e.,

$$x^2 - 4\frac{gx^2}{2v^2}\left(y + \frac{gx^2}{2v^2}\right) \geq 0.$$

This is equivalent to

$$y \leq \frac{v^2}{2g} - \frac{g}{2v^2}x^2.$$

This inequality divides the xy-plane into tow regions separated by a parabola. Water can reach the points under the parabola (a paraboloid of revolution in three dimensions), but not those above it. Hence the equation $y = \frac{v^2}{2g} - \frac{g}{2v^2}x^2$ describes the sought-for envelope in the xy-plane.
[Comment: From the equation of the envelope, the height of the bell is $\frac{v^2}{2g}$, as one would expect from considering an object thrown vertically upwards. Also the water "bell" defines a circle on the surface of the basin water, the radius of which can be found using the condition $y = 0$; it is $r = \frac{v^2}{g}$. This means that the radius of the basin should be at least two times the height of the water "bell" if no water is to be lost.]

1-15 (i) A vector normal to the surface at $P = (2,1,1)$ is determined by $\vec{N} = \vec{\nabla}(xy - 1 - z)|_P = (y, x, -1)|_P = (1, 2, -1)$. Take an arbitrary point $\vec{r}$ on the tangent plane. Then the vector $\vec{r} - \vec{r}_P$, where $\vec{r}_P$ denotes the point P, should be orthogonal to the normal vector $\vec{N}$. This condition

$$(\vec{r} - \vec{r}_P) \cdot \vec{N} = (x - 2) + 2(y - 1) - (z - 1) = 0$$

is the equation for the tangent plane, i.e., $x + 2y - z - 3 = 0$.

(ii) If (x, y) denotes any point P on the ellipse, we have, with $A = (f, 0)$ and $B = (-f, 0)$,

$$r_1 + r_2 = \sqrt{(x - f)^2 + y^2} + \sqrt{(x + f)^2 + y^2} = \text{const.}$$

Then, for any displacement $d\vec{r}$ along the ellipse, we should have $d\vec{r} \cdot \vec{\nabla}(r_1 + r_2) = 0$; hence, denoting $\overrightarrow{AP} = \vec{r}_1$ and $\overrightarrow{BP} = \vec{r}_2$.

$$\vec{\nabla}(r_1 + r_2) = \frac{(x - f)\mathbf{i} + y\mathbf{j}}{\sqrt{(x - f)^2 + y^2}} + \frac{(x + f)\mathbf{i} + y\mathbf{j}}{\sqrt{(x + f)^2 + y^2}}$$

$$= \hat{r}_1 + \hat{r}_2 \qquad \left(\text{where } \hat{r}_1 \equiv \frac{\overrightarrow{AP}}{|\overrightarrow{AP}|} \text{ and } \hat{r}_2 \equiv \frac{\overrightarrow{BP}}{|\overrightarrow{BP}|}\right)$$

is normal to the ellipse at $P = (x, y)$. The unit normal is thus given by $\mathbf{n} = \frac{1}{|\hat{r}_1+\hat{r}_2|}(\hat{r}_1 + \hat{r}_2)$.

Now let $\mathbf{t}$ be the unit tangent to the ellipse at $P = (x, y)$. Then, $\mathbf{t} \cdot \mathbf{n} = 0$, and thus

$$\mathbf{t} \cdot \hat{r}_1 = -\mathbf{t} \cdot \hat{r}_2.$$

Since $\hat{r}_1$ ($\hat{r}_2$) is a unit vector parallel to $\overrightarrow{AP}$ ($\overrightarrow{BP}$), this equation shows that $\overrightarrow{AP}$ and $\overrightarrow{BP}$ make equal angles with the tangent to the ellipse.

1-16 On the xy-plane, we find

$$\vec{V} \cdot d\vec{r}\Big|_{z=0} = [(A_y z - A_z y)dx + (A_z x - A_x z)dy]\Big|_{z=0} = A_z(xdy - ydx).$$

For the path C_1, $y = 0$ and thus we evaluate the line integral as

$$\int_{C_1} \vec{V} \cdot d\vec{r} = \int_{C_1} A_z(xdy - ydx)\Big|_{dy=0,y=0} = 0.$$

For the path C_2, we may set $x = a\cos\theta$ and $y = a\sin\theta$ to obtain

$$\int_{C_2} \vec{V} \cdot d\vec{r} = \int_{C_2} A_z(xdy - ydx)\Big|_{x=a\cos\theta, y=a\sin\theta}$$
$$= A_z \int_0^{\pi} a^2 d\theta = A_z \pi a^2.$$

1-17 Similarly to the previous problem, $(\mathbf{k} \times \vec{r}) \cdot d\vec{r} = (xdy - ydx)$ on the xy-plane. You can find that, after some explicit evaluations with simple closed curves, the line integral over a closed curve equals twice the area enclosed by the closed curve. (The integration path in the counterclockwise direction is assumed.)

For a general argument, first note that $(\mathbf{k} \times \vec{r}) \cdot d\vec{r} = \vec{r} \times d\vec{r} \cdot \mathbf{k}$. Here, $\frac{1}{2}\vec{r} \times d\vec{r}$ corresponds to the infinitesimal area vector of the triangle composed by $\vec{r}$ and $d\vec{r}$. Therefore, the integral $\int \vec{r} \times d\vec{r}$ over a closed curve is equal to twice the area enclosed by the curve. (This also follows from Stoke's theorem, discussed in the appendix to Chapter 8.)

2 Evolution in Time: Basic Elements of Newtonian Mechanics

2-1 Solving the three first-order differential equations given, we find

$$x'(t) = c_1 t + d_1, \quad y'(t) = c_2 t + d_2, \quad z'(t) = c_3 t + d_3$$

with three integration constants, d_1, d_2 and d_3. The coordinates x', y' and z' are non-inertial ones, but related to an inertial system $R = (x, y, z)$ by a uniform rotation about its z-axis. Explicitly, from the given relations

$$x = x' \cos \omega t + y' \sin \omega t, \quad y = -x' \sin \omega t + y' \cos \omega t, \quad z = z',$$

and hence the trajectory in terms of inertial coordinates reads

$$\begin{aligned} x(t) &= (c_1 t + d_1) \cos \omega t + (c_2 t + d_2) \sin \omega t, \\ y(t) &= -(c_1 t + d_1) \sin \omega t + (c_2 t + d_2) \cos \omega t, \\ z(t) &= c_3 t + d_3. \end{aligned} \tag{1}$$

Based on (1), the velocity (as seen in the inertial system) becomes

$$\begin{aligned} v_x(t) &= c_1 \cos \omega t + c_2 \sin \omega t - \omega(c_1 t + d_1) \sin \omega t + \omega(c_2 t + d_2) \cos \omega t, \\ v_y(t) &= -c_1 \sin \omega t + c_2 \cos \omega t - \omega(c_1 t + d_1) \cos \omega t - \omega(c_2 t + d_2) \sin \omega t, \\ v_z(t) &= c_3 \end{aligned}$$

and so, for the acceleration, we find

$$\begin{aligned} a_x(t) &= 2\omega(-c_1 \sin \omega t + c_2 \cos \omega t) - \omega^2\big((c_1 t + d_1) \cos \omega t \\ &\quad + (c_2 t + d_2) \sin \omega t\big), \\ a_y(t) &= -2\omega(c_1 \cos \omega t + c_2 \sin \omega t) + \omega^2\big((c_1 t + d_1) \sin \omega t \\ &\quad - (c_2 t + d_2) \cos \omega t\big), \\ a_z(t) &= 0. \end{aligned} \tag{2}$$

From (1), we see that the object moves along a curved trajectory in space, as its translational motion in the xy-plane is accompanied by a rotation. For a specific case of $c_1 = 1$, $c_2 = 2$, $c_3 = 1$ and $\omega = 1$ (with $d_1 = d_2 = d_3 = 0$), we obtain the trajectory as shown in the figure.

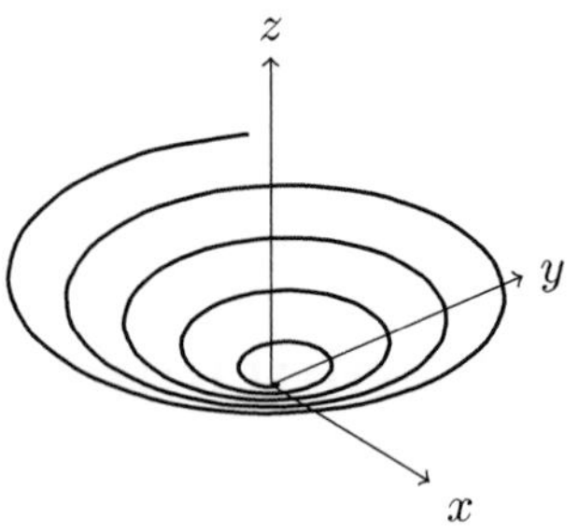

[Comment: Note that, if we introduce the rotating axes $(\mathbf{e}'_1, \mathbf{e}'_2, \mathbf{e}'_3)$ by

$$\mathbf{e}'_1 = \cos\omega t\mathbf{e}_1 - \sin\omega t\mathbf{e}_2, \quad \mathbf{e}'_2 = \sin\omega t\mathbf{e}_1 + \cos\omega t\mathbf{e}_2, \quad \mathbf{e}'_3 = \mathbf{e}_3,$$

it is easy to show that $\frac{d}{dt}\mathbf{e}'_i = \vec{\omega} \times \mathbf{e}'_i$ $(i = 1, 2, 3)$ with $\vec{\omega} = -\omega\mathbf{e}'_3$. Then our formula (2) for the acceleration is tantamount to the relation (derived in Chapter 9)

$$\vec{a}(t) = \underbrace{2\vec{\omega} \times \vec{v}'(t)}_{\text{(Coriolis acceleration)}} + \underbrace{\vec{\omega} \times [\vec{\omega} \times \vec{r}(t)]}_{\text{(centripetal acceleration)}}$$

with $\vec{v}' = c_i\mathbf{e}'_i(t)$ and $\vec{r}(t) = x'_i(t)\mathbf{e}'_i(t) = (c_i t + d_i)\mathbf{e}'_i(t)$ in our case.]

2-2 Let the two ships pass by at distance d. Suppose that the second ship fires the gun in the direction $\mathbf{n} = \sin\varphi\mathbf{i} + \cos\varphi\mathbf{j}$ at the instant when the two ships are on the y-axis (see the figure). Then the shell acquires the velocity $v_0\mathbf{n} - v_2\mathbf{i}$. If it hits the first ship at time t, it must be the case that

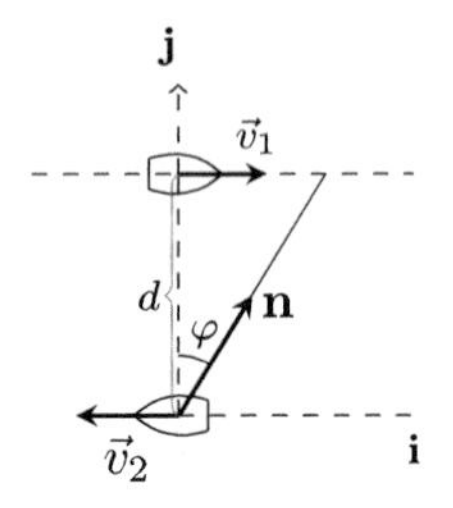

$$(v_0\mathbf{n} - v_2\mathbf{i})t = d\mathbf{j} + v_1 t\mathbf{i}$$

and so

$$(v_0\cos\varphi)t = d \quad \longrightarrow \quad t = d/(v_0\cos\varphi)$$
$$(v_0\sin\varphi - v_2)\not{t} = v_1\not{t} \quad \longrightarrow \quad v_0\sin\varphi = v_1 + v_2.$$

We thus see that the gun should be fired at angle $\varphi = \sin^{-1}\left(\frac{v_1+v_2}{v_0}\right)$ (assuming $v_0 > v_1 + v_2$).

2-3 Let $\mathbf{i}$ denote the direction of the current, and $\mathbf{j}$ the direction perpendicular to the current. Then the velocity of the boat can be expressed as

$$\vec{V} = v\mathbf{j} + u\left[1 - \frac{2}{d}|y - d/2|\right]\mathbf{i}.$$

Therefore, we can represent its trajectory by

$$\begin{aligned} y &= vt, \\ x &= \int_0^t u\left[1 - \frac{2}{d}|y(t') - d/2|\right]dt' \\ &= \int_0^y \frac{u}{v}\left[1 - \frac{2}{d}|y' - d/2|\right]dy'. \end{aligned}$$

From the second of these equations, we obtain the trajectory equation $x = x(y)$ of the form

$$x = \begin{cases} \frac{2u}{vd}\int_0^y y'dy' = \frac{u}{vd}y^2, \quad \text{for } y \le \frac{d}{2} \\ \frac{2u}{vd}\int_0^{d/2} y'dy' + \frac{2u}{v}\int_{d/2}^y (1 - \frac{y'}{d})dy' \\ \quad = -\frac{ud}{2v} + \frac{2u}{v}y - \frac{u}{vd}y^2, \quad \text{for } y > \frac{d}{2} \end{cases}.$$

Hence, when the boat arrives at the other bank, it has gone down along the current as much as the distance $x_0 = x(y = d) = \frac{ud}{2v}$. The sketch of the trajectory is given in the figure.

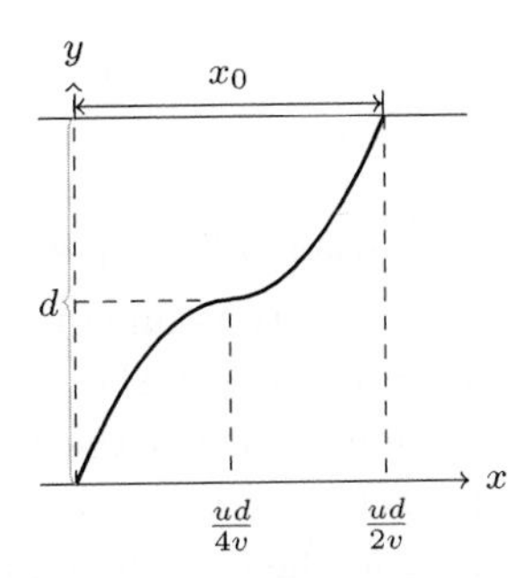

2-4 If a particle had a trajectory given as

$$x^i = x^i(t) \qquad \text{(with respect to } R\text{)},$$

then, in the primed system R', it should be possible to derive the trajectory equation $x'^i = x'^i(t')$ from the related Lorentz transformation

$$\begin{aligned} t' &= \frac{t - ux(t)/c^2}{\sqrt{1 - (u/c)^2}}, \quad x'(t') = \frac{x(t) - ut}{\sqrt{1 - (u/c)^2}}, \\ y'(t') &= y(t), \quad z'(t') = z(t). \end{aligned} \tag{1}$$

If $\frac{dx^i}{dt} = v^i(t)$, we find from the first relation of (1),

$$\frac{dt'}{dt} = \frac{1 - uv^1(t)/c^2}{\sqrt{1-(u/c)^2}}$$

and therefore

$$\begin{aligned}
v'^1(t') &\equiv \frac{dx'}{dt'} = \left(\frac{dt'}{dt}\right)^{-1}\frac{dx'}{dt} = \frac{v^1(t) - u}{1 - uv^1(t)/c^2},\\
v'^2(t') &\equiv \frac{dy'}{dt'} = \left(\frac{dt'}{dt}\right)^{-1}\frac{dy'}{dt} = \frac{\sqrt{1-u^2/c^2}}{1 - uv^1(t)/c^2}v^2(t),\\
v'^3(t') &\equiv \frac{dz'}{dt'} = \left(\frac{dt'}{dt}\right)^{-1}\frac{dz'}{dt} = \frac{\sqrt{1-u^2/c^2}}{1 - uv^1(t)/c^2}v^3(t). \qquad (2)
\end{aligned}$$

If the particle had the speed c with respect to R (i.e., $\sqrt{v^i v^i} = c$), then we find, from (2),

$$\begin{aligned}
v'^i v'^i &= \frac{(v^1-u)^2 + (1-u^2/c^2)[(v^2)^2 + (v^3)^2]}{(1-uv^1/c^2)^2}\\
&= \frac{c^2 - 2uv^1 + u^2 - \frac{u^2}{c^2}[(v^2)^2+(v^3)^2]}{(1-uv^1/c^2)^2} = c^2\frac{(1-uv^1/c^2)^2}{(1-uv^1/c^2)^2}\\
&= c^2,
\end{aligned}$$

where we used $(v^2)^2 + (v^3)^2 = c^2 - (v^1)^2$. So its speed with respect to R', $\sqrt{v'^i v'^i}$, is again equal to c.

2-5 (i) We are going to show that $d\tau'$ in the system R' equals $d\tau$ in the system R. From the previous problem, we know the relations between the velocity components of the particle with respect to R' and the corresponding ones in R:

$$v'_x = \frac{v_x - u}{1 - uv_x/c^2}, \quad v'_y = \frac{\sqrt{1-u^2/c^2}}{1-uv_x/c^2}v_y, \quad v'_z = \frac{\sqrt{1-u^2/c^2}}{1-uv_x/c^2}v_z. \qquad (1)$$

So the square of velocity vector in the system R' is given by

$$\begin{aligned}
(\vec{v}')^2 &= \frac{(v_x-u)^2 + (1-u^2/c^2)(v_y^2+v_z^2)}{(1-uv_x/c^2)^2}\\
&= \frac{(1-u^2/c^2)(\vec{v}^2 - c^2) + c^2(1-uv_x/c^2)^2}{(1-uv_x/c^2)^2}. \qquad (2)
\end{aligned}$$

Also $dt' = (dt'/dt)dt = \frac{1-uv_x/c^2}{\sqrt{1-u^2/c^2}}dt$. Plugging these results into the expression for $d\tau'$, we have

$$\begin{aligned}
d\tau' &= [1 - (\vec{v}')^2/c^2]^{1/2} dt' \\
&= \frac{[(1-u^2/c^2)(1-\vec{v}^2/c^2)]^{1/2}}{1-uv_x/c^2} \frac{1-uv_x/c^2}{\sqrt{1-u^2/c^2}} dt \\
&= \sqrt{1-\vec{v}^2/c^2}\, dt = d\tau.
\end{aligned}$$

This establishes that $d\tau$ behaves like a Lorentz scalar.

(ii) When the coordinates transform as $x'^\mu = \Lambda^\mu{}_\nu x^\nu$, dx^μ transforms in the same way, i.e., $dx'^\mu = \Lambda^\mu{}_\nu dx^\nu$. Since $U^\mu = \frac{dx^\mu}{d\tau}$ and $d\tau$ behaves like a Lorentz scalar, U^μ transforms as a Lorentz four-vector, i.e., $U'^\mu = \Lambda^\mu{}_\nu U^\nu$.

2-6 With two component stars moving in circles of radii r_1 and r_2 (drawn from the common center), the mutual gravitational attraction should be responsible for the (radially-directed) centripetal accelerations of individual stars. This can happen only when the line connecting the two component stars at any given time passes through the center of the circles; that is, in a double star, $m_1 r_1 \omega^2 = m_2 r_2 \omega^2 =$ (mutual gravitational force) if ω denotes the rotational angular speed. Hence $\frac{m_1}{m_2} = \frac{r_2}{r_1}$ (i.e., the center of the circles should coincide with the center of mass of the system).

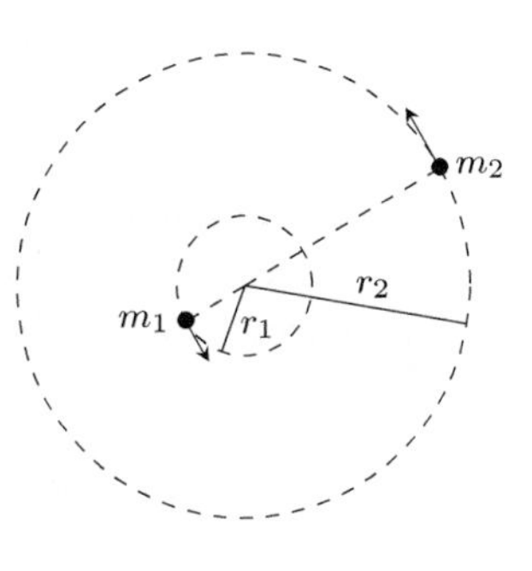

2-7 Here, we simply provide the answer:

on M_3: F to the right, $\frac{M_1+M_2}{M_1+M_2+M_3}F$ to the left,

on M_2: $\frac{M_1+M_2}{M_1+M_2+M_3}F$ to the right,

$\frac{M_1}{M_1+M_2+M_3}F$ to the left,

on M_1: $\frac{M_1}{M_1+M_2+M_3}F$ to the right, 0 to the left.

2-8 If $T(x)$ denotes the magnitude of the tension at the point x from the bottom of the rope, it must be able to balance the net weight of the man (with the mass M) and the part of the rope below the point x (with mass $m\frac{x}{l}$). Hence, $T(x) = Mg + m(\frac{x}{l})g$.

2-9 (i)

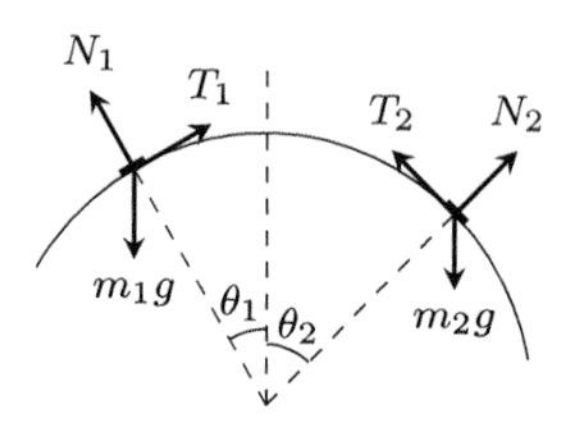

(ii) Consider a small segment AB of length ds on the cable. We may take the inclinations (to some fixed direction) of the tangents to the cable at A and B to be θ and $\theta + d\theta$. As the cable is very light, all forces acting on the cable segment should balance out. There are three forces, namely, the tension T at A, the tension $T + dT$ at B, and a normal force Nds (which may be supposed to act along the normal at A). [Clearly, N here represents the normal force per unit length.] Now, resolving forces along the tangent and normal at A, the balance equations imply

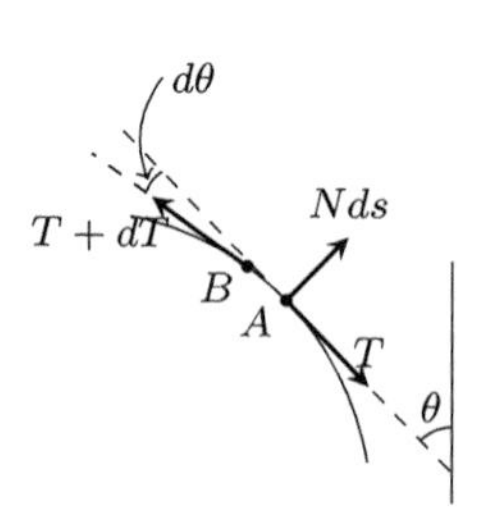

$$
\begin{aligned}
(T + dT)\left[1 + \cancel{O((d\theta)^2)}\right] &= T \quad \longrightarrow \quad dT = 0 \\
(T + \cancel{dT})d\theta &= Nds \quad \longrightarrow \quad Td\theta = Nds.
\end{aligned}
$$

Hence *the tension is constant along a light cable.* Also, since $ds = ad\theta$, we have the normal force per unit length $N = T/a$. [Remark: If one wishes, one can use the equations of motion for the angles θ_1 and θ_2 (defined in (i))

$$
\begin{aligned}
m_2 g \sin\theta_2 - T &= m_2\ddot{\theta}_2, \\
T - m_1 g \sin\theta_1 &= -m_1\ddot{\theta} = m_1\ddot{\theta}_2,
\end{aligned}
$$

to determine also the magnitude of the tension in the form $T = \frac{m_1 m_2}{m_1+m_2} g(\sin\theta_1 + \sin\theta_2)$.]

(iii) For the two masses to have zero tangential accelerations, the following condition should hold:

$$T = m_1 g \sin\theta_1 = m_2 g \sin\theta_2.$$

Hence, in equilibrium, θ_1 and θ_2 (with $\theta_1 + \theta_2$ held fixed) must satisfy the condition $\frac{\sin\theta_2}{\sin\theta_1} = \frac{m_1}{m_2}$.

2-10 For the segment in the infinitesimal interval $[r, r + dr]$, we can write the equation of motion as

$$T(r) - T(r + dr) = \lambda dr\, r\omega^2,$$

where $T(r)$ is the tension at r and λ $(= \frac{m}{L})$ is the line density of the rope. From this equation we obtain $\frac{dT}{dr} = -\lambda r\omega^2$. Integrating over r then gives

$$\begin{aligned} T(r) &= \int_L^r (-\lambda) r'\omega^2 dr' + T(L) \\ &= \frac{1}{2} m \left(L - \frac{r^2}{L} \right) \omega^2 + ML\omega^2, \end{aligned}$$

where the tension in the end of the rope, $T(L) = ML\omega^2$, has been used.

2-11 When T denotes the tension in the string, we have the equations of motion (a is the acceleration)

$$\begin{aligned} &5\,\text{kg weight}: \qquad (5\,\text{kg})\,a = (5\,\text{kg})\,g - T, \\ &1\,\text{kg weight}: \qquad (1\,\text{kg})\,a = T - (1\,\text{kg})\,g\,. \end{aligned}$$

Solving these, we find $a = \frac{2}{3}g$ and $T = (\frac{5}{3}\,\text{kg})\,g$. The spring balance reads $2T/g$, i.e., $\frac{10}{3}$ kg. Hence, during the motion, it reads a value *less* than 6 kg.

2-12 (i) For a flexible cable hanging under the influence of its own weight, the tension along the cable can be represented by

$$\vec{T}(s) = T(s) \left[\frac{dx}{ds}\mathbf{i} + \frac{dy}{ds}\mathbf{j} \right],$$

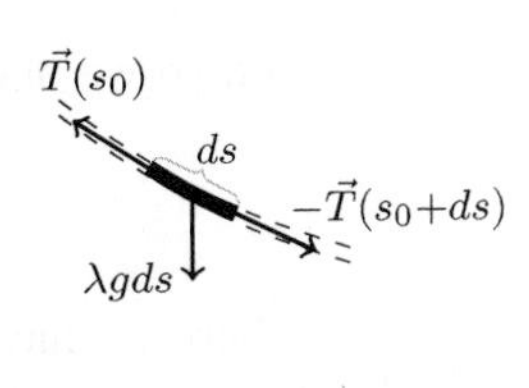

where s denotes the arc length along the cable. Hence, on the cable segment ds in equilibrium, the force balance equation will read (see the figure)

$$\text{horizontal:} \quad \left(T\frac{dx}{ds}\right)\bigg|_{s=s_0} = \left(T\frac{dx}{ds}\right)\bigg|_{s=s_0+ds}$$

$$\text{or} \quad \frac{d}{ds}\left(T\frac{dx}{ds}\right) = 0, \tag{1}$$

$$\text{vertical:} \quad \left(T\frac{dy}{ds}\right)\bigg|_{s=s_0+ds} - \left(T\frac{dy}{ds}\right)\bigg|_{s=s_0} = \lambda g ds$$

$$\text{or} \quad \frac{d}{ds}\left(T\frac{dy}{ds}\right) = \lambda g \quad (\text{: const.}). \tag{2}$$

From (1), we can set

$$T\frac{dx}{ds} = H_0 \quad (\text{: const.}) \tag{3}$$

and then, substituting this into (2) yields (with $T\frac{dy}{ds} = T\frac{dy}{dx}\frac{dx}{ds} = H_0\frac{dy}{dx}$)

$$\frac{d}{ds}\left(\frac{dy}{dx}\right) = \frac{\lambda g}{H_0} \quad \text{or} \quad \frac{d}{dx}\left(\frac{dy}{dx}\right) = \frac{\lambda g}{H_0}\sqrt{1+\left(\frac{dy}{dx}\right)^2}, \tag{4}$$

since $\frac{dx}{ds} = [1 + \left(\frac{dy}{dx}\right)^2]^{-1/2}$. This is the *differential equation for a hanging cable.* Writing $\frac{dy}{dx} = \sinh z$, this equation assumes the form

$$\cancel{\cosh z}\frac{dz}{dx} = \frac{\lambda g}{H_0}\cancel{\cosh z} \quad \longrightarrow \quad z = \frac{\lambda g}{H_0}x + C$$

with an integration constant C. Choosing the origin O at the lowest point of the cable (see the figure), the integration constant C can be taken to be zero; then

$$\frac{dy}{dx} = \sinh\left(\frac{\lambda g}{H_0}x\right), \tag{5}$$

and so we obtain the curve

$$y = \frac{H_0}{\lambda g}\left[\cosh\left(\frac{\lambda g}{H_0}x\right) - 1\right], \tag{6}$$

where we have fixed another integration constant by the condition $y(0) = 0$. (This curve is called the *common catenary*, the lowest point O being its vertex.) The constant H_0 can be fixed implicitly by the equation

$$l = \int_0^a \sqrt{1 + \left(\frac{dy}{dx}\right)^2}\, dx = \int_0^a \cosh\left(\frac{\lambda g}{H_0} x\right) dx$$
$$= \frac{H_0}{\lambda g} \sinh\left(\frac{\lambda g}{H_0} a\right). \qquad (7)$$

(ii) The tension T along the cable can now be determined from (3), i.e.,

$$T = H_0 \frac{ds}{dx} = H_0 \cosh\left(\frac{\lambda g}{H_0} x\right) = \lambda g y + H_0.$$

The maximum tension occurs when y assumes the largest value, i.e., *at points A and B* (with $x = \pm a$), where the tension has the magnitude

$$T_{\text{max}} = H_0 \left[\cosh\left(\frac{\lambda g}{H_0} a\right) - 1\right] + H_0$$
$$= H_0 \cosh\left(\frac{\lambda g}{H_0} a\right)$$

with H_0 determined by (7).

3 One-Dimensional Motion

3-1 The equation of motion in a first-order form reads

$$\dot{x}(t) = v(t),$$
$$\dot{v}(t) = -\omega^2 x(t) \qquad \left(\omega \equiv \sqrt{\frac{k}{m}}\right).$$

We may then introduce successive time steps $[t_i, t_{i+1} = t_i + \epsilon](i = 0, 1, 2, \ldots, N-1)$ with $\epsilon = t/N$, to write the approximate expressions

$$x(t_0{=}0) = x_0 = 0, \quad v(t_0{=}0) = v_0,$$
$$x_1 \equiv x(t_1) = x_0 + v_0\epsilon, \quad v_1 \equiv v(t_1) = v_0 - \omega^2 x_0\epsilon,$$
$$\vdots$$
$$x_i \equiv x(t_i) = x_{i-1} + v_{i-1}\epsilon, \quad v_i \equiv v(t_i) = v_{i-1} - \omega^2 x_{i-1}\epsilon,$$
$$\vdots$$
$$x_N \equiv x(t = t_N) = x_{N-1} + v_{N-1}\epsilon,$$
$$v_N \equiv v(t = t_N) = v_{N-1} - \omega^2 x_{N-1}\epsilon.$$

Exact results are expected to follow by the limiting procedure, i.e., $x(t) = \lim_{N\to\infty} x_N$ and $v(t) = \lim_{N\to\infty} v_N$.

To calculate the above limits, notice that

$$\begin{pmatrix} x_i \\ v_i \end{pmatrix} = \begin{pmatrix} 1 & \epsilon \\ -\omega^2\epsilon & 1 \end{pmatrix} \begin{pmatrix} x_{i-1} \\ v_{i-1} \end{pmatrix} \qquad (i = 1, \ldots, N).$$

Based on these and using $\epsilon = t/N$, we have

$$\begin{pmatrix} x_N \\ v_N \end{pmatrix} = \begin{pmatrix} 1 & t/N \\ -\omega^2 t/N & 1 \end{pmatrix}^N \begin{pmatrix} x_0 \\ v_0 \end{pmatrix}$$

and therefore

$$\begin{pmatrix} x(t) \\ v(t) \end{pmatrix} = \lim_{N\to\infty} \begin{pmatrix} 1 & t/N \\ -\omega^2 t/N & 1 \end{pmatrix}^N \begin{pmatrix} x_0 \\ v_0 \end{pmatrix} = e^{\omega t Q} \begin{pmatrix} x_0 \\ v_0 \end{pmatrix}$$

where $Q \equiv \begin{pmatrix} 0 & \frac{1}{\omega} \\ -\omega & 0 \end{pmatrix}$. [Note that $\begin{pmatrix} 1 & t/N \\ -\omega^2 t/N & 1 \end{pmatrix} = \begin{pmatrix} 1 & 0 \\ 0 & 1 \end{pmatrix} + \omega \frac{t}{N} Q = e^{\omega \frac{t}{N} Q} + O(\frac{1}{N^2})$, with the exponential matrix $e^A \equiv \sum_{n=0}^{\infty} \frac{1}{n!} A^n$.] Then, since $Q^2 = -\begin{pmatrix} 1 & 0 \\ 0 & 1 \end{pmatrix} = -I$, we have

$$\begin{aligned} e^{\omega t Q} &= \sum_{n=0}^{\infty} \frac{1}{n!} (\omega t)^n Q^n \\ &= \sum_{k=0}^{\infty} \frac{1}{(2k)!} (\omega t)^{2k} (-1)^k I + \sum_{k=0}^{\infty} \frac{1}{(2k+1)!} (\omega t)^{2k+1} (-1)^k Q \\ &= I \cos \omega t + Q \sin \omega t = \begin{pmatrix} \cos \omega t & \frac{1}{\omega} \sin \omega t \\ -\omega \sin \omega t & \cos \omega t \end{pmatrix} \end{aligned}$$

where we used the well-known Maclaurin series for the sine and cosine functions. We thus find

$$\begin{pmatrix} x(t) \\ v(t) \end{pmatrix} = \begin{pmatrix} \cos \omega t & \frac{1}{\omega} \sin \omega t \\ -\omega \sin \omega t & \cos \omega t \end{pmatrix} \begin{pmatrix} x_0 \\ v_0 \end{pmatrix}.$$

Hence, if $x_0 = 0$, we obtain the formula $x(t) = \frac{v_0}{\omega} \sin \omega t$ (and $v(t) = v_0 \cos \omega t$).

3-2 (i) The given equation of motion can be written as

$$\frac{dv(t)}{dt} + \frac{\kappa}{m} v = g - \frac{\kappa}{m} f'(t),$$

or, multiplying both sides by $e^{\frac{\kappa}{m} t}$, as

$$\frac{d}{dt}(e^{\frac{\kappa}{m} t} v) = \left[g - \frac{\kappa}{m} f'(t) \right] e^{\frac{\kappa}{m} t}.$$

Hence, with $v(t{=}0) = 0$, we find for $t > 0$

$$\begin{aligned} v(t) &= e^{-\frac{\kappa}{m} t} \left\{ \int_0^t \left[g - \frac{\kappa}{m} f'(t_1) \right] e^{\frac{\kappa}{m} t_1} \, dt_1 \right\} \\ &= \frac{mg}{\kappa} (1 - e^{-\frac{\kappa}{m} t}) - \frac{\kappa}{m} e^{-\frac{\kappa}{m} t} \int_0^t f'(t_1) e^{\frac{\kappa}{m} t_1} dt_1. \end{aligned}$$

(ii) Assuming $x(t{=}0) = 0$, the traversed distance $x(t)$ for $t > 0$ is given by

$$
\begin{aligned}
x(t) &= \int_0^t v(t_1)dt_1 \\
&= \frac{mg}{\kappa}t + \frac{m^2 g}{\kappa^2}(e^{-\frac{\kappa}{m}t} - 1) \\
&\quad - \frac{\kappa}{m}\int_0^t e^{-\frac{\kappa}{m}t_2}\left[\int_0^{t_2} f'(t_1)e^{\frac{\kappa}{m}t_1}\,dt_1\right]dt_2 .
\end{aligned}
$$

(iii) If $f'(t) = C\cos pt = C\mathrm{Re}(e^{ipt})$, we find for $t > 0$

$$
\begin{aligned}
v(t) &= \frac{mg}{\kappa}(1 - e^{-\frac{\kappa}{m}t}) - C\frac{\kappa}{m}e^{-\frac{\kappa}{m}t}\mathrm{Re}\left(\int_0^t e^{(\frac{\kappa}{m}+ip)t_1}dt_1\right) \\
&= \frac{mg}{\kappa}(1 - e^{-\frac{\kappa}{m}t}) - C\frac{e^{-\frac{\kappa}{m}t}}{1 + (\frac{mp}{\kappa})^2} \\
&\quad \times \mathrm{Re}\left\{\left(1 - i\frac{mp}{\kappa}\right)\left[e^{(\frac{\kappa}{m}+ip)t} - 1\right]\right\} \\
&= \frac{mg}{\kappa}(1 - e^{-\frac{\kappa}{m}t}) - \frac{C}{1 + (\frac{mp}{\kappa})^2}\left[\cos pt + \frac{mp}{\kappa}\sin pt - e^{-\frac{\kappa}{m}t}\right],
\end{aligned} \tag{1}
$$

and

$$
\begin{aligned}
x(t) &= \frac{mg}{\kappa}\left[t + \frac{m}{\kappa}\left(e^{-\frac{\kappa}{m}t} - 1\right)\right] \\
&\quad - \frac{C}{1 + (\frac{mp}{\kappa})^2}\left[\frac{1}{p}\sin pt - \frac{m}{\kappa}(\cos pt - 1) + \frac{m}{\kappa}(e^{-\frac{\kappa}{m}t} - 1)\right].
\end{aligned} \tag{2}
$$

From (1), we see that

$$
v(t) \longrightarrow \frac{mg}{\kappa} - \frac{C}{1 + (\frac{mp}{\kappa})^2}\left[\cos pt + \frac{mp}{\kappa}\sin pt\right], \quad \text{as } t \to \infty, \tag{3}
$$

i.e., $v(t)$ tends to oscillate about the value $\frac{mg}{\kappa}$ — the terminal velocity when there is no fluctuating air current — with the same frequency as that of the air current. For the asymptotic form of $x(t)$, we find from (2)

$$
x(t) \longrightarrow -\frac{m^2 g}{\kappa^2} + \left\{\frac{mg}{\kappa}t - \frac{C}{1 + (\frac{mp}{\kappa})^2}\left[\frac{1}{p}\sin pt - \frac{m}{\kappa}\cos pt\right]\right\},
$$

and this is entirely consistent with the asymptotic expression for $v(t)$ given above.

3-3 (i) Based on the given information, we have the initial velocity $v_0 = \sqrt{2gh}$. The equation of motion, when $v = \dot{x} > 0$ (i.e, going upward), reads

$$m\ddot{x} = -mg - kv^2.$$

But, $\ddot{x} = \frac{dv}{dt} = \frac{dx}{dt}\frac{dv}{dx} = v\frac{dv}{dx}$ and hence the equation of motion can be cast into the form

$$v\frac{dv}{dx} = -g - \frac{k}{m}v^2 \quad \text{or} \quad \frac{dv}{dx} = -\frac{g}{v} - \frac{k}{m}v.$$

From the last equation, we find $\mathcal{F}(v) = -\frac{g}{v} - \frac{k}{m}v$. (In the downward motion, one must use $F = kv^2$ and so find $\mathcal{F}(v) = -\frac{g}{v} + \frac{k}{m}v$.)

(ii) The equation obtained in (i) leads to

$$\frac{dv^2}{v^2 + \frac{mg}{k}} = -\frac{2k}{m}dx$$

and therefore

$$\ln\left(v^2 + \frac{mg}{k}\right) = -\frac{2k}{m}x + C \tag{1}$$

with an integration constant C. Since we have $x = 0$ and $v = v_0$ at $t = 0$, the constant C here is taken as $C = \ln(v_0^2 + \frac{mg}{k})$. Using this value in (1), we find

$$\ln\left(\frac{v^2 + \frac{mg}{k}}{v_0^2 + \frac{mg}{k}}\right) = -\frac{2k}{m}x \quad \text{or} \quad v^2 = \left(v_0^2 + \frac{mg}{k}\right)e^{-\frac{2k}{m}x} - \frac{mg}{k}.$$

Then, at the position of the maximum height $x = H$, we should have $v = 0$ and hence

$$\left(v_0^2 + \frac{mg}{k}\right)e^{-\frac{2k}{m}H} = \frac{mg}{k} \quad \longrightarrow \quad H = \frac{m}{2k}\ln\left(1 + \frac{2kh}{m}\right).$$

In the last equation, we have used $v_0^2 = 2gh$. Terminal velocity is obtained when the particle is in the downward motion. From the equation of motion

$$m\ddot{x} = -mg + kv^2$$

we find $\ddot{x} = 0$ when $kv^2 = mg$; hence, the terminal velocity is $v = -\sqrt{mg/k}$.

3-4 See the sketch of the potential given below.

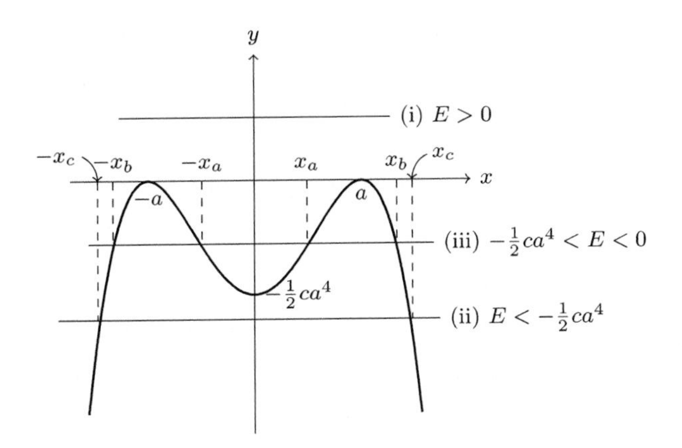

(i) $E > 0$. In this case the particle moves right or left without turning, depending on the initial value of velocity.

(ii) $E < -\frac{1}{2}ca^4$. Motion of the particle is unbounded but restricted to the region $x < -x_c$ or $x > x_c$.

(iii) $-\frac{1}{2}ca^4 < E < 0$. There are three possibilities: the particle oscillates in the finite region $-x_a < x < x_a$, or moves in the region $x < -x_b$, or moves in the region $x > x_b$. In the last two cases, its motions are bounded from one side.

3-5 (i) The force is

$$F(x) = -\frac{d}{dx}V(x) = -6\frac{a}{x^7} + 12\frac{b}{x^{13}}$$

(see the plot). There is one equilibrium point — the point where $F(x)$ vanishes — at $x_{\text{eq}} = (\frac{2b}{a})^{\frac{1}{6}}$.

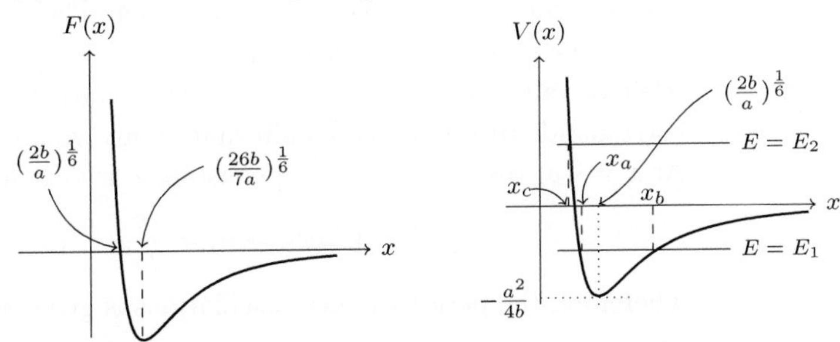

(ii) Let the heavy atom be at $x = 0$. Then we can describe the motion of the lighter atom with the help of the energy conservation equation

$$\frac{1}{2}m\dot{x}^2 + V(x) = E \quad (\text{: const.}),$$

where $V(x) = -\frac{a}{x^6} + \frac{b}{x^{12}}$ (see the figure for the plot). In detail,

a) if the energy is such that $-\frac{a^2}{4b} < E < 0$ (the case denoted as $E = E_1$), it undergoes an oscillatory motion in the region $x_a \leq x \leq x_b$ (with the turning points x_a and x_b determined by the equations $V(x_a) = V(x_b) = E$), and

b) if the energy value is positive (the case denoted as $E = E_2$), an unbounded motion (in the region $x > x_c$) follows.

(iii) Equilibrium distance is $x_{\text{eq}} = (\frac{2b}{a})^{\frac{1}{6}}$ (see (i)), and the period of small oscillations is given by

$$T = \frac{2\pi}{\omega} \left(\text{with } \omega = \sqrt{\frac{V''(x_{\text{eq}})}{m}} = 3a\left(\frac{4a}{m^3b^4}\right)^{\frac{1}{6}} \right)$$

$$= \frac{2\pi}{3a}\left(\frac{m^3b^4}{4a}\right)^{\frac{1}{6}}.$$

3-6 (i) See the figure.

(ii) To find equilibrium points, let us consider the derivative of the potential energy function

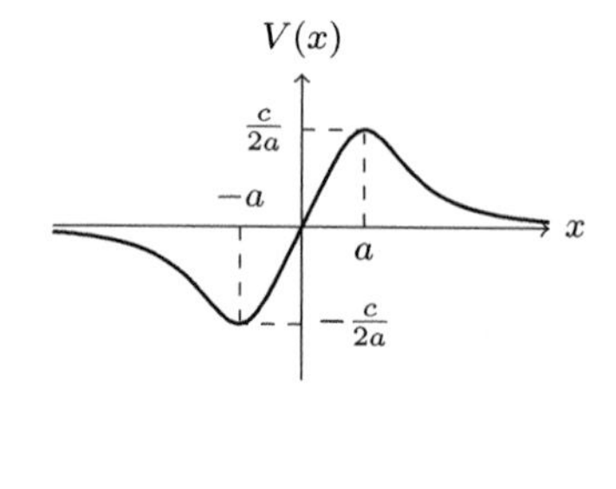

$$\frac{dV(x)}{dx} = \frac{c(x^2+a^2) - 2cx^2}{(x^2+a^2)^2}$$

$$= \frac{c(a^2 - x^2)}{(x^2+a^2)^2},$$

which becomes zero at two points $x = \pm a$. The second derivative of the potential energy function, $V''(x) = \frac{2cx(x^2-3a^2)}{(x^2+a^2)^3}$ has a positive value $\frac{c}{2a^3}$ at $x = -a$ ($-\frac{c}{2a^3}$ at $x = a$); hence, $x = -a$ corresponds to the stable equilibrium point we are looking for. At $x = -a$, the effective spring constant k is determined by

$$k = V''(x{=}-a) = \frac{c}{2a^3}.$$

Therefore, the period of small oscillations is given as

$$T = \frac{2\pi}{\sqrt{k/m}} = \frac{2\pi}{\sqrt{\frac{c}{2ma^3}}} = 2\pi a\sqrt{\frac{2ma}{c}}.$$

(iii) From the conditions given, we find that the energy of the particle is

$$E = \frac{1}{2}mv_0^2 + V(x=-a) = \frac{1}{2}mv_0^2 - \frac{c}{2a}.$$

Hence the particle oscillates in a finite region if

$$-\frac{c}{2a} < E < 0 \quad \longrightarrow \quad v_0^2 < \frac{\not{2}c}{\not{2}ma}, \quad \text{i.e.,} \quad |v_0| < \sqrt{\frac{c}{ma}}.$$

On the other hand, for the particle to escape to $+\infty$ in the positive x-direction, we must have

$$E > \frac{c}{2a} \quad \longrightarrow \quad v_0^2 > \frac{2c}{ma} \quad \text{i.e.,} \quad v_0 > \sqrt{\frac{2c}{ma}}.$$

3-7 With our formula

$$t - t_0 = \pm\sqrt{\frac{m}{2}}\int_{x_0}^{x} \frac{dx'}{\sqrt{E - \frac{1}{2}kx'^2}} \qquad (x(t_0) = x_0),$$

let us choose x_0 at one of the turning points. Then, inserting $E = \frac{1}{2}kx_0^2$ into this equation, we find

$$t - t_0 = \pm\sqrt{\frac{m}{k}}\int_{x_0}^{x} \frac{dx'}{\sqrt{x_0^2 - x'^2}}.$$

This becomes, after making the change of variable $x' = x_0 \sin\phi'$,

$$t - t_0 = \pm\sqrt{\frac{m}{k}}\int_{\pi/2}^{\phi} \frac{\cos\phi'}{\sqrt{1 - \sin^2\phi'}}d\phi' = \pm\frac{1}{\omega}\left(\phi - \frac{\pi}{2}\right),$$

where $x = x_0 \sin\phi$ and $\omega = \sqrt{\frac{k}{m}}$. Here the sign factor $\pm$ should be identified to $\alpha \equiv \text{sign}[\dot{x}] = \text{sign}[\cos\phi]$. From the last equation, we have $\phi = \alpha\omega(t - t_0) + \frac{\pi}{2}$ and so

$$x(t) = x_0 \sin\left[\alpha\omega(t - t_0) + \frac{\pi}{2}\right] = \sqrt{\frac{2E}{k}}\cos[\omega(t - t_0)],$$

where we used $E = \frac{1}{2}kx_0^2$.

3-8 (i) To find the energy conservation law, let us consider the time derivative of the potential energy function

$$\frac{dV(x)}{dt} = \frac{dV(x)}{dx}\frac{dx}{dt} = -F(x)v.$$

Applying the equation of motion then leads us to

$$\frac{dV(x)}{dt} = -m_0 v\left(\frac{\dot{v}}{(1-\frac{v^2}{c^2})^{\frac{1}{2}}} + \frac{\dot{v}\frac{v^2}{c^2}}{(1-\frac{v^2}{c^2})^{\frac{3}{2}}}\right) = -\frac{d}{dt}\left(\frac{m_0c^2}{(1-\frac{v^2}{c^2})^{\frac{1}{2}}}\right).$$

This implies the following relativistic energy conservation equation

$$\frac{m_0c^2}{\sqrt{1-v^2/c^2}} + V(x) = E_{\text{rel}} \quad (\text{: const.}).$$

(ii) Clearly, $\frac{m_0c^2}{\sqrt{1-v^2/c^2}}$ cannot be smaller m_0c^2 and hence the energy conservation equation shows that the motion of the particle can occur only in the spatial region $V(x) \leq E_{\text{rel}} - m_0c^2$. (The positions $x = \bar{x}$ satisfying $V(\bar{x}) = E_{\text{rel}} - m_0c^2$ correspond to turning points.) Also, from the energy conservation equation, we can determine the velocity at any given position x as

$$v \equiv \frac{dx}{dt} = \pm c\sqrt{1 - \frac{m_0^2c^4}{(E_{\text{rel}} - V(x))^2}}.$$

Integrating this equation, we also obtain

$$t - t_0 = \pm\int_{x_0}^{x} \frac{(E_{\text{rel}} - V(x))dx}{c\sqrt{(E_{\text{rel}} - V(x))^2 - m_0^2c^4}},$$

which corresponds to (3.18) in nonrelativistic cases.

3-9 Suppose that the length of the spring is x_0 when no weight is hung. After the weight is released, let us denote its position at time t by $x(t)$. Then the restoring force on the spring will be equal to $-k(x - x_0)$, while the weight of the mass is mg. We thus have the equation of motion

$$m\frac{d^2x}{dt^2} = mg - k(x - x_0).$$

If we introduce a new variable $y = x - x_0 - \frac{mg}{k}$, this can be simplified to

$$\frac{d^2y}{dt^2} = -\frac{k}{m}y$$

with the initial conditions $y(t=0) = -\frac{mg}{k}$ (because $x(t=0) = x_0$) and $\dot{y}(t=0) = 0$. Then it has a solution $y(t) = -\frac{mg}{k}\cos\omega t$ with $\omega = \sqrt{\frac{k}{m}}$, and so we obtain

$$x(t) = x_0 + \frac{mg}{k}\left(1 - \cos\left(\sqrt{\frac{k}{m}}\,t\right)\right).$$

3-10 Let the relaxed lengths of the two springs be $\bar{l}_1$ and $\bar{l}_2$, and let x denote the position of mass m as measured from the bottom. Then we have the equation of motion

$$\begin{aligned} m\ddot{x} &= -k(x - \bar{l}_1) + k[(2l - x) - \bar{l}_2] - mg \\ &= -2k\left(x - l - \frac{\bar{l}_1 - \bar{l}_2}{2} + \frac{mg}{2k}\right). \end{aligned}$$

In terms of the shifted variable $y = x - l - \frac{\bar{l}_1 - \bar{l}_2}{2} + \frac{mg}{2k}$, this takes the form

$$m\ddot{y} = -2ky,$$

which has the general solution $y(t) = A\cos\left(\sqrt{\frac{2k}{m}}\,t\right) + B\sin\left(\sqrt{\frac{2k}{m}}\,t\right)$ (A, B are arbitrary constants). Hence the general solution for $x(t)$ is provided by

$$x(t) = l + \frac{\bar{l}_1 - \bar{l}_2}{2} - \frac{mg}{2k} + A\cos\left(\sqrt{\frac{2k}{m}}\,t\right) + B\sin\left(\sqrt{\frac{2k}{m}}\,t\right).$$

The constants A and B can be fixed by the initial conditions.

3-11 The equation of motion satisfied by the height $y(t)$, including the effect of the buoyant force, takes the form

$$\begin{aligned} M\frac{d^2y}{dt^2} &= -Mg + \rho A(l - y)g + \text{const.} \\ &= -gA\rho(y - \bar{y}) \end{aligned}$$

where $\bar{y}$ is also a constant. From this equation we can infer that the hydrometer oscillates about its equilibrium position with the angular frequency $\omega = \sqrt{\frac{gA\rho}{M}}$. Hence the period

$$T = \frac{2\pi}{\omega} = 2\pi\sqrt{\frac{M}{A\rho g}}\,.$$

3-12 (i) The potential energy function is

$$V(x) = \frac{1}{2}m\omega_0^2\left(x^2 + \frac{a^4}{x^2}\right).$$

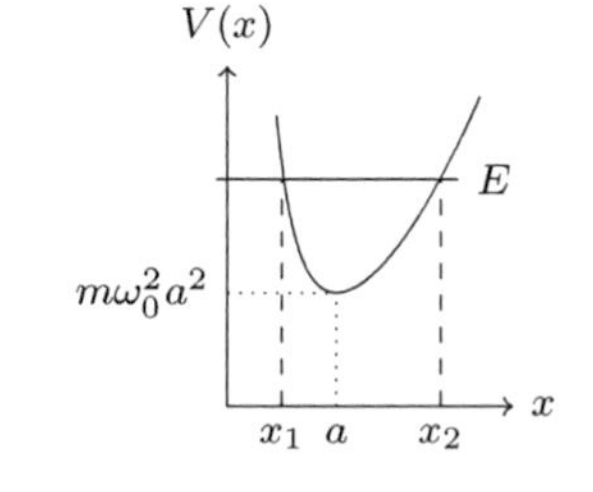

For the plot of this function, see the figure.

(ii) The equilibrium point is determined by the condition $F(x) = 0$, and obviously it is at $x = a$. The derivative of the force function at the equilibrium point determines the effective spring constant

$$k = -F'(x{=}a) = 4m\omega_0^2,$$

from which one may read off the period

$$T = 2\pi\sqrt{\frac{m}{k}} = \frac{\pi}{\omega_0}.$$

(iii) The result in (ii) is valid for small oscillations. For an arbitrary energy E, the period can be determined (see (3.18) in the text) by

$$T = 2\sqrt{\frac{m}{2}}\int_{x_1}^{x_2}\frac{dx}{\sqrt{E - V(x)}},$$

where x_1 and x_2 ($> x_1$) are two turning points satisfying $V(x) = E$. (See the figure in (i).) Inserting the potential energy function in (i) into this, we get

$$T = \sqrt{2m}\int_{x_1}^{x_2}\frac{dx}{[E - \frac{1}{2}m\omega_0^2(x^2 + \frac{a^4}{x^2})]^{\frac{1}{2}}}.$$

Changing the integration variable to $y = x^2$, this becomes

$$T = \frac{1}{\omega_0}\int_{y_1}^{y_2}\frac{dy}{\sqrt{\mathcal{E}y - y^2 - a^4}}$$

with $y_1 = x_1^2$, $y_2 = x_2^2$ and $\mathcal{E} = \frac{2E}{m\omega_0^2}$. The end points y_1 and y_2 are two zeros of the denominator and they have the form $\frac{\mathcal{E}}{2} \mp \sqrt{\frac{\mathcal{E}^2}{4} - a^4}$. Noting that the terms inside the square root can be written in the form $(\frac{\mathcal{E}^2}{4} - a^4) - (y - \frac{\mathcal{E}}{2})^2$, we change the integration variable one more time by writing $(y - \frac{\mathcal{E}}{2}) = \sqrt{\frac{\mathcal{E}^2}{4} - a^4}\sin\theta$. Then the end points y_1 and y_2 correspond to $\theta = -\frac{\pi}{2}$ and $\theta = \frac{\pi}{2}$, respectively. Hence, for the period, we find the result

$$T = \frac{1}{\omega_0}\int_{-\frac{\pi}{2}}^{\frac{\pi}{2}} \frac{\cancel{\sqrt{\frac{\mathcal{E}^2}{4} - a^4}\cos\theta}}{\cancel{\sqrt{\frac{\mathcal{E}^2}{4} - a^4}\cos\theta}}\, d\theta = \frac{\pi}{\omega_0},$$

which is independent of energy E.

3-13 (i) As there are many candidate functions, no specific forms will be suggested here.

(ii) For the potential $V_a(x)$, we have two equilibrium points, an elliptical fixed point at $x = \alpha$ and a hyperbolic fixed point at $x = \beta$. (See the figure.) On the other hand, for the potential $V_b(x)$, there are three equilibrium points, i.e., two elliptical fixed points at $x = a$ and $x = c$ and a hyperbolic fixed point at $x = b$. We can

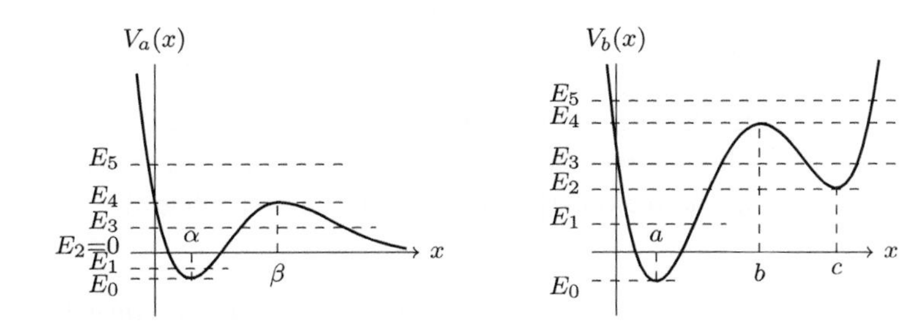

then draw local phase portraits in the vicinity of these equilibrium points. Also, on the basis of the energy conservation equation $\frac{1}{2}mv^2 + V(x) = E$ with appropriate potentials, we can infer the general motional characteristics for every possible value of E. From this information, the phase portraits for the given two systems should acquire, qualitatively, the following forms:

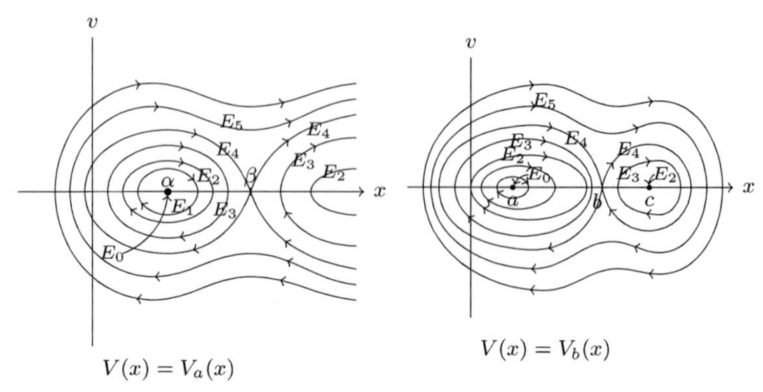

3-14 (i)

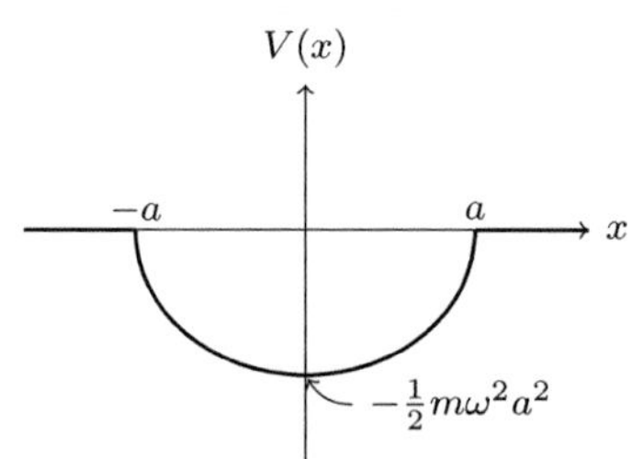

(ii) The equation of motion in a first-order form reads

$$\dot{x} = v,$$

$$\dot{v} = -V'(x) = \begin{cases} -m\omega^2 x, & |x| < a \\ 0, & |x| > a. \end{cases}$$

In the region $|x| > a$, we have $\dot{v} = 0$ (i.e., $v = \text{const.}$) and so phase curves take the form of straight lines. On the other hand, in the region $|x| < a$, we have the first integral

$$\frac{1}{2}mv^2 - \frac{m\omega^2}{2}(a^2 - x^2) = E \quad (\text{: const.})$$

which can be recast as

$$x^2 + \frac{v^2}{\omega^2} = a^2 + \frac{2E}{m\omega^2} \quad (\text{: ellipse}).$$

Based on these informations, we have the phase portrait given below.

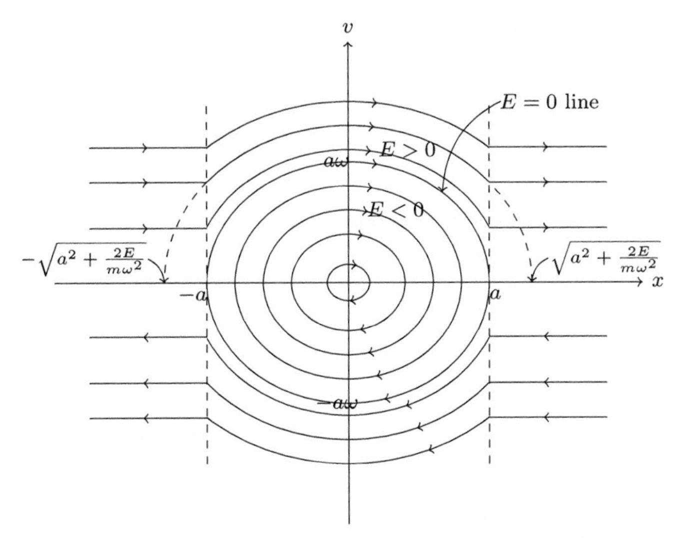

(iii) When the particle is in the region $-a < x < a$, the general solution of the equation of motion $m\ddot{x} = -m\omega^2 x$ is $x(t) = A\cos\omega t + B\sin\omega t$ with two arbitrary constants A and B. Then, from the initial conditions given, we find $A = -a$ and $B\omega = v_0$ and so

$$x(t) = -a\cos\omega t + \frac{v_0}{\omega}\sin\omega t.$$

Now, let t_0 denote the time this object reaches the point $x = a$, i.e.,

$$\begin{aligned} x(t_0) &= -a\cos\omega t_0 + \frac{v_0}{\omega}\sin\omega t_0 = a \\ \longrightarrow \quad & \frac{v_0}{\omega}\sin\omega t_0 = a(1+\cos\omega t_0) \\ \longrightarrow \quad & \frac{v_0}{\omega} 2\sin\frac{\omega t_0}{2}\cos\frac{\omega t_0}{2} = 2a\cos^2\frac{\omega t_0}{2}. \end{aligned}$$

The time t_0 is thus determined by

$$\tan\frac{\omega t_0}{2} = \frac{a\omega}{v_0}.$$

For $t > t_0 = \frac{2}{\omega}\tan^{-1}\frac{a\omega}{v_0}$, the particle moves with a constant velocity $v = v_0$ (because of the conservation of energy) and so we conclude that

$$x(t) = a + v_0(t - t_0).$$

3-15 General solutions to the equation of motion for the underdamped oscillator case are given by the form

$$x(t) = e^{-\gamma t}(c_1 \cos \omega t + c_2 \sin \omega t) \quad \left(\omega = \sqrt{\omega_0^2 - \gamma^2}\right).$$

Initial conditions $x(0) = 0$ and $\dot{x}(0) = v_0$ determine the two constants to have the values $c_1 = 0$ and $c_2 = \frac{v_0}{\omega}$, and thus we obtain a specific solution

$$x(t) = \frac{v_0}{\omega} e^{-\gamma t} \sin \omega t.$$

In the limit $\omega_0 \to \gamma$, ω approaches 0 and accordingly

$$x(t) = v_0 e^{-\gamma t} \lim_{\omega \to 0} \frac{\sin \omega t}{\omega} = v_0 t\, e^{-\omega_0 t},$$

which is precisely the solution corresponding to the critically damped case.

3-16 Let x denote the position of the weight measured from the unextended position of the spring. For a critically damped spring, $x(t)$ then satisfies the equation of motion

$$m\ddot{x} + 2m\omega_0 \dot{x} + m\omega_0^2 x = mg \quad \left(\omega_0 = \sqrt{\frac{k}{m}}\right).$$

We can then consider the shifted variable $\bar{x} = x - \frac{mg}{k}$ to recast this as

$$m\ddot{\bar{x}} + 2m\omega_0 \dot{\bar{x}} + m\omega_0^2 \bar{x} = 0.$$

The new equation has the general solution of the form

$$\bar{x}(t) = (A + Bt)e^{-\omega_0 t} \quad (A, B \text{ are constants}).$$

Then, with the initial conditions $\bar{x}(t{=}0) = -\frac{mg}{k}$ (corresponding to $x(t{=}0) = 0$) and $\dot{\bar{x}}(t{=}0) = 0$, we must choose $A = -\frac{mg}{k} = \frac{B}{\omega_0}$ to obtain the expression

$$\bar{x}(t) = -\frac{mg}{k}\left(1 + \sqrt{\frac{k}{m}}t\right) e^{-\sqrt{\frac{k}{m}}\,t} \quad (t > 0)$$

or

$$x(t) = \frac{mg}{k} - \frac{mg}{k}\left(1 + \sqrt{\frac{k}{m}}t\right)e^{-\sqrt{\frac{k}{m}}\,t} \quad (t > 0).$$

Now, if $\frac{mg}{k} = 0.4\,\text{m}$, we have $\omega_0 = \sqrt{\frac{k}{m}} = \sqrt{\frac{g}{mg/k}} = \sqrt{\frac{9.8}{0.4}}\,\text{s}^{-1} = 4.95\,\text{s}^{-1}$. Accordingly, after 1 sec, we find

$$|\bar{x}(t{=}1)| = 0.4(1 + 4.95)e^{-4.95}\ \text{m} = 0.017\ \text{m},$$

i.e., the particle is at a distance 0.017 m from the equilibrium position.

3-17 (i) First of all, since the new equilibrium position after the impact is 0.2 m below the original position, the spring constant must have the value given by

$$(0.2\,\text{m})k = 1000 \times 9.8\,\text{kg}\,\text{m}\,\text{s}^{-2},$$

i.e., $k = 49\,000\,\text{kg}\,\text{s}^{-2}$ and this implies that $\omega_0 \equiv \sqrt{\frac{k}{m}} = 7\,\text{s}^{-1}$. Let us take the $x = 0$ point at the new equilibrium position. Then, immediately after impact, the platform will start moving downward (from the position $x_0 = 0.2\,\text{m}$) with the velocity $v_0 = -\sqrt{2 \cdot (9.8\,\text{m}\,\text{s}^{-2}) \cdot 10\text{m}} = -14\,\text{m}\,\text{s}^{-1}$. Now, not to have overshooting, it must be overdamped ($\gamma > \omega_0$) with

$$|v_0| \leq x_0\left(\gamma + \sqrt{\gamma^2 - \omega_0^2}\right). \tag{1}$$

Inserting the values $|v_0| = 14\,\text{m}\,\text{s}^{-1}$, $\omega_0 = 7\,\text{s}^{-1}$ and $x_0 = 0.2\,\text{m}$ into the second condition, we find that $\gamma > 35.35\,\text{s}^{-1}$. For the shortest relaxation time (which is governed by $(\gamma - \sqrt{\gamma^2 - \omega_0^2})^{-1}$), we should take $\gamma = 35.35\,\text{s}^{-1}$. This corresponds to an overdamped oscillation, and leads to the damping constant $b = 2m\gamma = 70\,700\,\text{kg}\,\text{s}^{-1}$.

(ii) For an overdamped oscillation the general solution is

$$x(t) = c_1 e^{-\gamma_+ t} + c_2 e^{-\gamma_- t} \quad (c_1, c_2 : \text{ arbitrary constants}), \tag{2}$$

and, in our case,

$$\gamma_+ \equiv \gamma + \sqrt{\gamma^2 - \omega_0^2} = 70\,\text{s}^{-1}, \quad \gamma_- \equiv \gamma - \sqrt{\gamma^2 - \omega_0^2} = 0.7\,\text{s}^{-1}.$$

The constants c_1, c_2 are determined by the initial conditions, i.e.,

$$\begin{aligned} x_0 &= c_1 + c_2 = 0.2, \\ v_0 &= -\gamma_+ c_1 - \gamma_- c_2 = -14, \end{aligned} \tag{3}$$

and these lead to the values $c_1 = 0.2\,\text{m}$ and $c_2 = 0\,\text{m}$. Having $c_2 = 0$ is not surprising, since then the equality in the second equation in (1) holds. Now, the condition $|x(t)| \leq 0.001\,\text{m}$ becomes,

$$|c_1 e^{-\gamma_+ t}| \leq 0.001 \quad \longrightarrow \quad e^{70t} \geq \frac{0.2}{0.001} = 200,$$

since the decay is governed by γ_+. This yields $t \geq 0.076\,\text{s}$, i.e., it takes 0.076 seconds for the platform to settle down within 1 mm of the equilibrium position.

Now, suppose that the damping factor γ is increased by 20%, which means that it becomes $42.42\,\text{s}^{-1}$. (Then we may expect much longer relaxation time.) Two exponential factors $\gamma_\pm$ are now given by $\gamma_+ = 84.26\,\text{s}^{-1}$ and $\gamma_- = 0.5815\,\text{s}^{-1}$. The constants c_1, c_2 are again determined by (3) but with these new values of $\gamma_\pm$, and the results are $c_1 = 0.166\,\text{m}$, $c_2 = 0.034\,\text{m}$. Since the first term in (2) decays much faster, the condition $|x(t)| \leq 0.001\,\text{m}$ becomes, essentially,

$$|c_2 e^{-\gamma_- t}| \leq 0.001 \quad \longrightarrow \quad e^{0.5815t} \geq \frac{0.034}{0.001} = 34.$$

This implies the corresponding time of $t \geq 6.06$ seconds.

3-18 (i) In the case of underdamping, the motion of the body can be expressed by the general form

$$x(t) = e^{-\gamma t}\{A\cos\omega t + B\sin\omega t\} \quad \left(\omega = \sqrt{\omega_0^2 - \gamma^2}\right).$$

The constants A and B can be fixed by imposing the initial conditions

$$x(0) = A = a, \quad \dot{x}(0) = -\gamma A + \omega B = 0 \quad \left(\longrightarrow B = \frac{\gamma}{\omega}A\right).$$

Hence,

$$x(t) = ae^{-\gamma t}\left(\cos\omega t + \frac{\gamma}{\omega}\sin\omega t\right).$$

(For the velocity, we have $\dot{x}(t) = -a\frac{\omega_0^2}{\omega}e^{-\gamma t}\sin\omega t$.)

(ii) While the velocity of the body *remains negative* after its release at $t = 0$, the equation of motion reads

$$m\ddot{x} + b\dot{x} + kx = \mu mg \quad \longrightarrow \quad \ddot{x} + 2\gamma\dot{x} + \omega_0^2 x = \mu g, \qquad (1)$$

and this has the solution (satisfying the initial conditions $x(0) = a$ and $\dot{x}(0) = 0$)

$$x(t) = \frac{\mu g}{\omega_0^2} + \Delta_0 e^{-\gamma t}\big(\cos\omega t + \frac{\gamma}{\omega}\sin\omega t\big) \qquad (2)$$

with the constant $\Delta_0 = a - \frac{\mu g}{\omega_0^2}$. We assume that Δ_0 is positive, as it is required for the body to *undergo movement*. Note that, for the velocity, we find $\dot{x}(t) = -\Delta_0 e^{-\gamma t}\frac{\omega_0^2}{\omega}\sin\omega t \leq 0$, for $0 \leq t \leq \frac{\pi}{\omega}$. At $t = \frac{\pi}{\omega}$, the velocity becomes zero and after this time there are two possibilities, depending on which is larger between $|kx(t = \frac{\pi}{\omega})|$ and μmg. (We will write $x(t = \frac{\pi}{\omega}) = -\Delta_1$, $\Delta_1 \equiv \Delta_0 e^{-\gamma\frac{\pi}{\omega}} - \frac{\mu g}{\omega_0^2}$ below.)
a) If $|k\Delta_1| < \mu mg$ (this happens as long as $\Delta_1 < \frac{\mu g}{\omega_0^2}$, i.e., $\Delta_0 e^{-\gamma\frac{\pi}{\omega}} < \frac{2\mu g}{\omega_0^2}$, including all cases with $\Delta_1 < 0$), the body stops moving at $x = -\Delta_1$ after time $t = \frac{\pi}{\omega}$ as the spring restoring force gets completely balanced by the static friction.
b) If $|k\Delta_1| > \mu mg$ (this happens as long as $\Delta_0 e^{-\gamma\frac{\pi}{\omega}} > \frac{2\mu g}{\omega_0^2}$, for Δ_1 positive), the body changes the direction of motion at $t = \frac{\pi}{\omega}$ and continues the motion but now with the sliding friction acting in the other (i.e., negative) direction. Thus, after this time, the equation of motion becomes

$$m\ddot{x} + b\dot{x} + kx = -\mu mg \quad \longrightarrow \quad \ddot{x} + 2\gamma\dot{x} + \omega_0^2 x = -\mu g$$

with the "initial" conditions $x(t{=}\frac{\pi}{\omega}) = \frac{\mu g}{\omega_0^2} - \Delta_0 e^{-\gamma\frac{\pi}{\omega}} \equiv -\Delta_1$ (<0) and $\dot{x}(t{=}\frac{\pi}{\omega}) = 0$. The solution is

$$x(t) = -\frac{\mu g}{\omega_0^2} + \left[\Delta_0 - \frac{2\mu g}{\omega_0^2}e^{\gamma\frac{\pi}{\omega}}\right] e^{-\gamma t}\left(\cos\omega t + \frac{\gamma}{\omega}\sin\omega t\right)$$

$$\dot{x}(t) = -\left[\Delta_0 - \frac{2\mu g}{\omega_0^2}e^{\gamma\frac{\pi}{\omega}}\right]\frac{\omega_0^2}{\omega}e^{-\gamma t}\sin\omega t\,,$$

which should be valid for $\frac{\pi}{\omega} < t < \frac{2\pi}{\omega}$ (before the next velocity zero occurring at $t = \frac{2\pi}{\omega}$). [If $|k\Delta_2| < \mu mg$ with $\Delta_2 = [\Delta_0 - \frac{2\mu g}{\omega_0^2}e^{\gamma\frac{\pi}{\omega}}] - \frac{\mu g}{\omega_0^2}$, the body stops moving after time $t = \frac{2\pi}{\omega}$ at the position $x = \Delta_2$.]

(iii) See A. Ricchiuto and A. Tozzi, *Am. J. Phys.* **50**, 176 (1982), where related aspects are discussed.

3-19 In the case of overdamping ($\omega_0 < \gamma$), the eigenvalue equation allows two solutions: one eigenvector is $\mathbf{X}_{(1)} = \left(\begin{smallmatrix} 1 \\ -\gamma_+ \end{smallmatrix}\right)$ with eigenvalue $\lambda = -\gamma_+$, and the other $\mathbf{X}_{(2)} = \left(\begin{smallmatrix} 1 \\ -\gamma_- \end{smallmatrix}\right)$ with eigenvalue $\lambda = -\gamma_-$, where $\gamma_\pm = \gamma \pm \sqrt{\gamma^2 - \omega_0^2}$. Hence the general solution is

$$\mathbf{X}(t) = c_1 e^{-\gamma_+ t}\mathbf{X}_{(1)} + c_2 e^{-\gamma_- t}\mathbf{X}_{(2)},$$

or

$$\begin{pmatrix} x(t) \\ v(t) \end{pmatrix} = \begin{pmatrix} c_1 e^{-\gamma_+ t} + c_2 e^{-\gamma_- t} \\ -\gamma_+ c_1 e^{-\gamma_+ t} - \gamma_- c_2 e^{-\gamma_- t} \end{pmatrix}$$

with two constants c_1 and c_2. We may draw the phase portraits, when ω_0/γ has the values 1/3 and 2/3, as in the figure below.

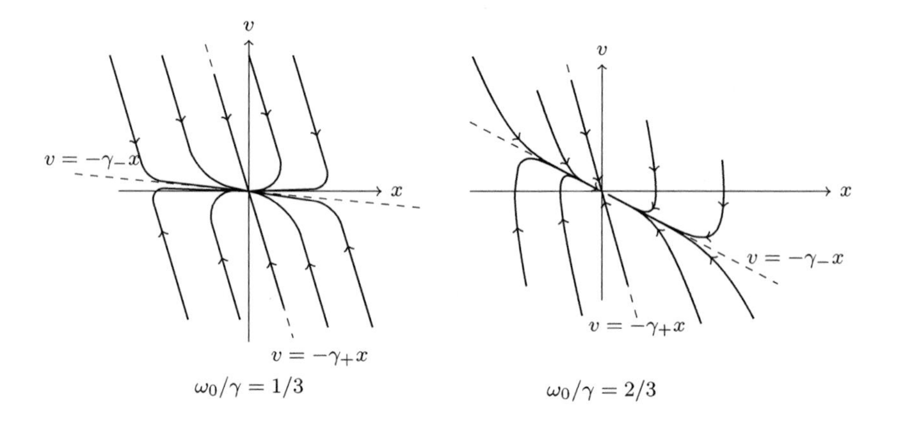

In the case of critical damping ($\omega_0 = \gamma$), the general solution is given by

$$\begin{aligned} \mathbf{X}(t) &= (C + Dt)e^{-\gamma t}\begin{pmatrix} 1 \\ -\gamma \end{pmatrix} + De^{-\gamma t}\begin{pmatrix} 0 \\ 1 \end{pmatrix} \\ &= \begin{pmatrix} (C + Dt)e^{-\gamma t} \\ \left[D - \gamma(C + Dt)\right]e^{-\gamma t} \end{pmatrix} \end{aligned}$$

with two arbitrary constants C and D. [See the discussions in the text (Sec. 3.3.5).] The phase portrait for this case is exhibited in the figure below.

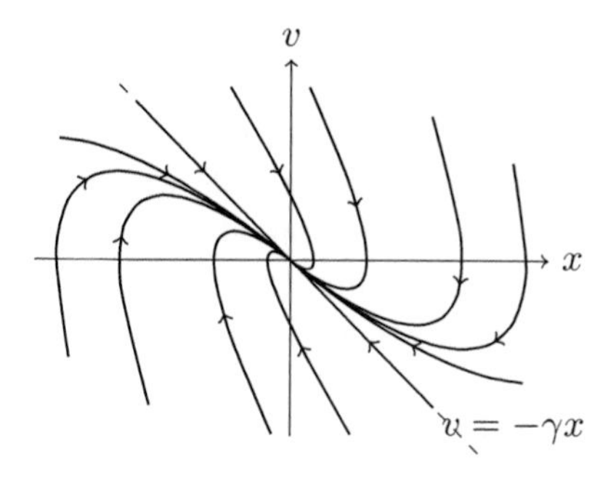

3-20 With material oscillators of the real world, it is a general fact that oscillations, or any regular motions, die out with time (due to dissipative effects such as frictional damping or electrical resistance) unless there is some external driving force acting. If the driving force oscillates with some given frequency, then the oscillator system in its steady state tends to oscillate with the same frequency as the driving force system (this includes the case of resonance); i.e., the frequency is transferred from one physical system to another. It is the mechanism that operates when we hear sound for instance. Take a string instrument such as a violin. The vibrating string produces sound wave, i.e., oscillating pressure waves in the air with the same frequency as the string (and governed by the three-dimensional wave equation). Then, the ear canal channels the sound to the tympanic membrane (eardrum) which transforms the air pressure oscillations into mechanical vibrations of the eardrum. These are then transmitted to the inner part of the ear (consisting of some stuffs that move in response to the vibrations), to be eventually picked up by the receptor cells that inform the brain of a sound. We can hear sound because we have an auditory system that makes the transfer of one oscillatory motion to another (with the same frequency) possible.

Other examples involving frequency transfer are as follows. If an electromagnetic wave is incident on a dielectric medium, it induces electric dipole oscillations in the medium with the same frequency as the incident wave. (This was discussed in the text.) To the electromagnetic field oscillator system filling the space, these oscillating dipoles play a role similar to an external driving force that impels the field system to oscillate with the frequency of oscillating dipoles. As a result, additional electromagnetic waves with the same frequency as the incident wave are generated. These waves interfere with the incident wave to produce the known wave motion in the presence of a dielectric medium. It is also possible to have the frequency of mechanical oscillation (including rotation) transferred to an electrical circuit oscillation. Take loops of stacked coils which are rotating, say, with angular frequency ω in the region where a uniform magnetic field is present. Since the magnetic flux through the loops of coil varies with time, the law of electromagnetic induction tells us that there will be a periodically-varying induced emf in the coils, which becomes a source for an alternating electric current with frequency ω.

The ocean tides, the periodic rise and fall of sea levels as observed at a fixed location on the Earth's surface, are caused by the combined effect of the gravitational (tidal) forces exerted by the Moon and the Sun and the rotation of the Earth. Their twice-daily periodicity is related to the fact that the tidal forces (on the body of sea water) due to the gravitational pull of the Moon and the Sun oscillate approximately with the period of 12 hours, if viewed in the reference frame which is rigidly fixed to the Earth and rotates with it. In this sense, one may say that the observed frequency of ocean tides is largely due to the frequency transfer from the apparently oscillating tidal forces (for an observer at a fixed location on the Earth and so rotating with it).

3-21 If the relaxed length of the spring is l, the equation of motion should read

$$m\frac{d^2x}{dt^2} = -b\frac{dx}{dt} - k(x - A\sin\omega t - l).$$

Using the shifted variable $\bar{x} = x - l$, we can recast this equation of motion as

$$\ddot{\bar{x}} + 2\gamma\dot{\bar{x}} + \omega_0^2\bar{x} = \omega_0^2 A\sin\omega t,$$

with $\gamma = \frac{b}{2m}$ and $\omega_0 = \sqrt{\frac{k}{m}}$. This corresponds to the driven damped oscillator discussed in Sec. 3.4. When $\frac{b}{2m} < \sqrt{\frac{k}{m}}$ (i.e., $\gamma < \omega_0$), we can now write down the general solution in the form

$$\bar{x} = Ce^{-\gamma t}\cos(\omega_1 t - \theta_0) + \frac{A\omega_0^2}{\sqrt{(\omega_0^2 - \omega^2)^2 + 4\gamma^2\omega^2}}\sin(\omega t - \alpha) \quad \left(\alpha = \tan^{-1}\frac{2\gamma\omega}{\omega_0^2 - \omega^2}\right)$$

where the first term corresponds to the general solution of the related homogeneous equation with $\omega_1 = \sqrt{\omega_0^2 - \gamma^2}$ and the second corresponds to a particular solution.

3-22 (i) The equation given is

$$m\ddot{x} + m\omega_0^2 x = F_1 \cos(\omega_1 t + \theta_1).$$

If $\omega_1^2 \neq \omega_0^2$, $x_p(t) = \frac{F_1/m}{\omega_0^2-\omega_1^2}\cos(\omega_1 t + \theta_1)$ provides a particular solution and so we have the general solution (containing two arbitrary constants A and θ_0)

$$x(t) = A\sin(\omega_0 t + \theta_0) + \frac{F_1/m}{\omega_0^2 - \omega_1^2}\cos(\omega_1 t + \theta_1).$$

(ii) Since the particular solution found in (i) diverges as $\omega_1 \to \pm\omega_0$, we cannot use it if $\omega_1^2 = \omega_0^2$. But, if we combine it with a specific form of the corresponding homogeneous equation in the form $\frac{F_1/m}{\omega_0^2-\omega_1^2}[\cos(\omega_1 t + \theta_1) - \cos(\pm\omega_0 t + \theta_1)]$, this has a finite limit as $\omega_1 \to \pm\omega_0$, which is $\frac{F_1/m}{(\pm 2\omega_0)} t\sin(\pm\omega_0 t + \theta_1)$ according to the L'Hospital's rule. This limiting form provides a good particular solution when $\omega_1 = \pm\omega_0$. Hence, for $\omega_1 = \pm\omega_0$, we can write the general solution as

$$x(t) = A\sin(\omega_0 t + \theta_0) \pm \frac{F_1}{2m\omega_0} t\sin(\pm\omega_0 t + \theta_1).$$

In this case, the system is in resonance and so the driving force can induce a *large* oscillation (with its amplitude increasing with time).

3-23 To find a particular solution $x_p(t)$ with a driving force $F(t) = F_0 e^{-\alpha t}\cos\omega_1 t$, let us consider the related complexified equation

$$\ddot{z} + 2\gamma\dot{z} + \omega_0^2 z = (F_0/m)e^{(-\alpha+i\omega_1)t}.$$

If $z_p(t)$ is a particular solution of this equation, we can set $x_p(t) = \mathrm{Re}[z_p(t)]$. Substituting a trial solution $z_p(t) = Ae^{(-\alpha+i\omega_1)t}$, we then find

$$\left[(-\alpha + i\omega_1)^2 + 2\gamma(-\alpha + i\omega_1) + \omega_0^2\right]Ae^{(-\alpha+i\omega_1)t} = (F_0/m)e^{(-\alpha+i\omega_1)t},$$

i.e., the constant A must be chosen as

$$A = \frac{F_0/m}{(\alpha^2 - 2\gamma\alpha + \omega_0^2 - \omega_1^2) + 2i(\gamma - \alpha)\omega_1}$$
$$= \frac{F_0/m}{\sqrt{(\alpha^2 - 2\gamma\alpha + \omega_0^2 - \omega_1^2)^2 + 4(\gamma - \alpha)^2\omega_1^2}}e^{-i\beta}$$

with $\tan\beta = 2(\gamma-\alpha)\omega_1/(\alpha^2 - 2\gamma\alpha + \omega_0^2 - \omega_1^2)$. Hence the particular solution we are looking for is

$$x_p(t) = \mathrm{Re}[z_p(t)] = \frac{(F_0/m)e^{-\alpha t}\cos(\omega_1 t - \beta)}{\sqrt{(\alpha^2 - 2\gamma\alpha + \omega_0^2 - \omega_1^2)^2 + 4(\gamma-\alpha)^2\omega_1^2}}.$$

The general solution of the problem can thus be written as

$$x(t) = Ce^{-\gamma t}\cos(\omega t + \theta_0) + \frac{(F_0/m)e^{-\alpha t}\cos(\omega_1 t - \beta)}{\sqrt{(\alpha^2 - 2\gamma\alpha + \omega_0^2 - \omega_1^2)^2 + 4(\gamma-\alpha)^2\omega_1^2}},$$

where $\omega = \sqrt{\omega_0^2 - \gamma^2}$ and two arbitrary constants C, θ_0 can be fixed by initial conditions.

3-24 As the given "saw-tooth" driving force (see the figure) is an odd function in t, its Fourier series involves only sine functions, i.e.,

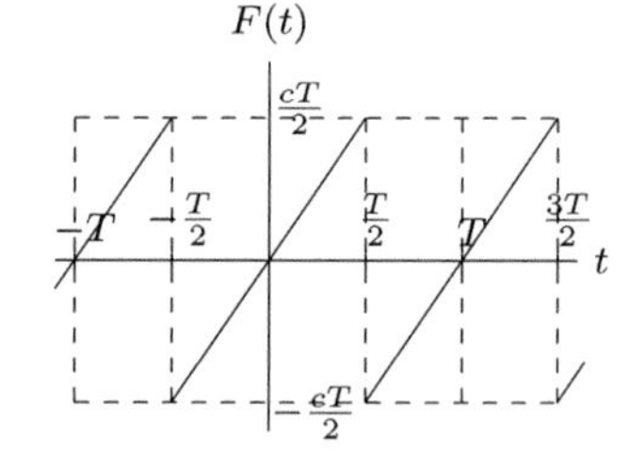

$$F(t) = \sum_{l=1}^{\infty} B_l \sin(\omega l t) \quad \left(\omega \equiv \frac{2\pi}{T}\right)$$

with coefficients

$$\begin{aligned} B_l &= \frac{2}{T}\left[\int_0^{\frac{T}{2}} ct\sin(\omega l t)dt + \int_{\frac{T}{2}}^{T} c(t-T)\sin(\omega l t)dt\right] \\ &= \frac{2c}{\omega l}(-1)^{l+1}. \end{aligned}$$

Then we can present the steady-state solution as (see (3.158))

$$\begin{aligned} x_s(t) &= \frac{2c}{m\omega}\sum_{l=1}^{\infty}\frac{(-1)^{l+1}}{2li}\left\{\frac{e^{i\omega l t}}{\omega_0^2 - \omega^2 l^2 + 2i\omega l\gamma} - \frac{e^{-i\omega l t}}{\omega_0^2 - \omega^2 l^2 - 2i\omega l\gamma}\right\} \\ &= \frac{2c}{m\omega}\sum_{l=1}^{\infty}\frac{(-1)^{l+1}}{l[(\omega_0^2 - \omega^2 l^2)^2 + 4\gamma^2\omega^2 l^2]^{\frac{1}{2}}}\sin(\omega l t - \alpha_l) \end{aligned}$$

with $\alpha_l = \tan^{-1}\frac{2\gamma\omega l}{\omega_0^2 - \omega^2 l^2}$.

3-25 (i) Clearly, for $z = \dot{x} + i\omega_0 x$, we find

$$\begin{aligned}\dot{z} - i\omega_0 z &= \ddot{x} + i\omega_0\dot{x} - i\omega_0(\dot{x} + i\omega_0 x)\\ &= \ddot{x} + \omega_0^2 x\\ &= \frac{1}{m}F(t),\end{aligned}$$

as claimed.

(ii) With both sides of the equation $\dot{z} - i\omega_0 z = \frac{1}{m}F(t)$ multiplied by $e^{-i\omega_0 t}$, the result can be written as

$$\frac{d}{dt}\left(e^{-i\omega_0 t}z\right) = \frac{1}{m}e^{-i\omega_0 t}F(t).$$

Hence, assuming $z(t_0) = v_0 + i\omega_0 x_0$, we may integrate this equation to obtain

$$z(t) = e^{i\omega_0(t-t_0)}z_0 + \frac{1}{m}e^{i\omega_0 t}\int_{t_0}^{t} e^{-i\omega_0 t'}F(t')dt'.$$

Taking the imaginary part of this equation then yields

$$\begin{aligned}x(t) &= \frac{1}{\omega_0}\mathrm{Im}\Big\{\big[\cos\omega_0(t-t_0) + i\sin\omega_0(t-t_0)\big](v_0 + i\omega_0 x_0)\Big\}\\ &\quad + \frac{1}{m\omega_0}\int_{t_0}^{t}\sin\omega_0(t-t')F(t')dt'\\ &= x_0\cos\omega_0(t-t_0) + \frac{v_0}{\omega_0}\sin\omega_0(t-t_0)\\ &\quad + \frac{1}{m\omega_0}\int_{t_0}^{t}\sin\omega_0(t-t')F(t')dt',\end{aligned}$$

which is consistent with the expected solution from the Green's function method.

3-26 Using the result of Problem 3-25, the quantity $z(t) \equiv \dot{x}(t) + i\omega_0 x(t)$ corresponding to the given situation can be expressed by

$$z(t) = \frac{1}{m}e^{i\omega_0 t}\int_{-\infty}^{t} e^{-i\omega_0 t}F(t')dt',$$

or, performing the integration by parts, by

$$\begin{aligned}z(t) &= \frac{1}{m}e^{i\omega_0 t}\left\{\left[\frac{i}{\omega_0}e^{-i\omega_0 t'}F(t')\right]\Bigg|_{t'=-\infty}^{t'=t} - \frac{i}{\omega_0}\int_{-\infty}^{t} e^{-i\omega_0 t'}\frac{dF(t')}{dt'}dt'\right\}\\ &= \frac{i}{m\omega_0}F(t) - \frac{i}{m\omega_0}e^{i\omega_0 t}\int_{-\infty}^{t} e^{-i\omega_0 t'}\frac{dF(t')}{dt'}dt'. \qquad (1)\end{aligned}$$

The last expression is convenient to find the large-(positive) t behavior of $z(t)$ as the integral $\int_{-\infty}^{t} e^{-i\omega_0 t'} \frac{dF(t')}{dt'} dt'$, even when $F(t) \to F_0$ ($\neq 0$) for t large, can have the well-defined $t \to \infty$ limit. Let $\mathcal{F} = \int_{-\infty}^{\infty} e^{-i\omega_0 t} \frac{dF(t)}{dt} dt$, a complex number in general. Then, from (1), we can clearly describe $z(t)$, for $t\,(> 0)$ very large, by the expression

$$z(t) = \frac{i}{m\omega_0} F_0 - \frac{i\mathcal{F}}{m\omega_0} e^{i\omega_0 t}.$$

Taking the imaginary part, we now see that $x(t)$ for large t would acquire the form

$$x(t) = \frac{F_0}{m\omega_0^2} - \frac{1}{m\omega_0^2} \mathrm{Re}\big[\mathcal{F} e^{i\omega_0 t}\big]. \tag{2}$$

Hence the desired amplitude of oscillation, about the new equilibrium position $x = \frac{1}{m\omega_0^2} F_0$, is $\frac{1}{m\omega_0^2}|\mathcal{F}| = \frac{1}{m\omega_0^2}\left|\int_{-\infty}^{\infty} e^{-i\omega_0 t} \frac{dF(t)}{dt} dt\right|$.

As for the total energy gain, notice that, for large positive t,

$$\begin{aligned}\int_{-\infty}^{t} F(t')\dot{x}(t')dt' &= \int_{-\infty}^{t} [m\ddot{x}(t') + m\omega_0^2 x(t')]\dot{x}(t')dt' \\ &= \frac{1}{2} m\dot{x}^2(t) + \frac{1}{2} m\omega_0^2 x^2(t),\end{aligned} \tag{3}$$

since $x(t{=}-\infty) = \dot{x}(t{=}-\infty) = 0$. From (2) and (3), we can now conclude that the driving force caused the energy gain equal to

$$\frac{1}{2} m\omega_0^2 \left(\frac{|\mathcal{F}|}{m\omega_0^2}\right)^2 = \frac{1}{2m\omega_0^2}\left|\int_{-\infty}^{\infty} e^{-i\omega_0 t} \frac{dF(t)}{dt} dt\right|^2.$$

3-27 Recall the formula

$$\begin{aligned}\frac{d}{dt}\left[\int_{h_1(t)}^{h_2(t)} \mathcal{G}(t,t')dt'\right] &= h_2'(t) \lim_{t' \to h_2(t)} \mathcal{G}(t,t') - h_1'(t) \lim_{t' \to h_1(t)} \mathcal{G}(t,t') \\ &\quad + \int_{h_1(t)}^{h_2(t)} \frac{\partial}{\partial t} \mathcal{G}(t,t')dt'\end{aligned}$$

for any differentiable functions $h_1(t)$, $h_2(t)$, and $\mathcal{G}(t,t')$. Hence, with our expression

$$x_s(t) = \int_{t_0}^{t} \left[\frac{e^{-\gamma(t-t')}}{m\omega} \sin\omega(t-t')\right] F(t')dt' \equiv \int_{t_0}^{t} G(t,t')F(t')dt', \tag{1}$$

we find

$$\frac{dx_s(t)}{dt} = \lim_{t'\to t-} G(t,t')F(t') + \int_{t_0}^{t} \left[\frac{\partial}{\partial t} G(t,t')\right] F(t')dt'$$
$$= \int_{t_0}^{t} \frac{\partial}{\partial t}\left[\frac{e^{-\gamma(t-t')}}{m\omega}\sin\omega(t-t')\right] F(t')dt', \qquad (2)$$

since $\lim_{t'\to t-} G(t,t') = 0$, and

$$\frac{d^2x_s(t)}{dt^2} = \lim_{t'\to t-} \frac{\partial}{\partial t}\left[\frac{e^{-\gamma(t-t')}}{m\omega}\sin\omega(t-t')\right] F(t')$$
$$+ \int_{t_0}^{t} \frac{\partial^2}{\partial t^2}\left[\frac{e^{-\gamma(t-t')}}{m\omega}\sin\omega(t-t')\right] F(t')dt'$$
$$= \frac{F(t)}{m} + \int_{t_0}^{t} \frac{\partial^2}{\partial t^2}\left[\frac{e^{-\gamma(t-t')}}{m\omega}\sin\omega(t-t')\right] F(t')dt'. \qquad (3)$$

Therefore, from these,

$$m\ddot{x}_s + b\dot{x}_s + kx_s$$
$$= F(t) + \int_{t_0}^{t}\left(m\frac{\partial^2}{\partial t^2} + b\frac{\partial}{\partial t} + k\right)\left[\frac{e^{-\gamma(t-t')}}{m\omega}\sin\omega(t-t')\right] F(t')dt'$$
$$= F(t),$$

where we used the fact that $\left(m\frac{\partial^2}{\partial t^2} + b\frac{\partial}{\partial t} + k\right)\left[\frac{e^{-\gamma(t-t')}}{m\omega}\sin\omega(t-t')\right] = 0$. We have thus shown that our expression $x_s(t)$ corresponds to a particular solution of the problem, satisfying the conditions $x_s(t{=}t_0) = 0$ and $\dot{x}_s(t{=}t_0) = 0$ from (1) and (2) above.

3-28 From the parameter values given, we have $\gamma = \frac{b}{2m} = \frac{a}{2}$, $\omega_0 = \sqrt{\frac{k}{m}} = \sqrt{\frac{4ma^2}{m}} = 2a$ and $\omega = \sqrt{\omega_0^2 - \gamma^2} = \sqrt{15}a/2$. Then, when there is a driving force $F(t) = F_0(1 - e^{-at})$ $(t > 0)$, the solution satisfying the initial conditions $x(t{=}0) = \dot{x}(t{=}0) = 0$ can be obtained as

$$x(t) = \int_0^t \left[\frac{e^{-\gamma(t-t')}}{m\omega}\sin\omega(t-t')\right] F(t')dt'$$
$$= \frac{F_0}{m\omega}\mathrm{Im}\left[\int_0^t e^{(-\gamma+i\omega)(t-t')}(1 - e^{-at'})dt'\right]$$

$$
\begin{aligned}
&= \frac{F_0}{m\omega}\text{Im}\left[\frac{1-e^{(-\gamma+i\omega)t}}{\gamma-i\omega} - \frac{e^{-at}-e^{(-\gamma+i\omega)t}}{\gamma-a-i\omega}\right] \\
&= \frac{2F_0}{m\omega}\text{Im}\left[\frac{1-e^{(-\gamma+i\omega)t}}{a-2i\omega} + \frac{e^{-at}-e^{(-\gamma+i\omega)t}}{a+2i\omega}\right] \\
&= \frac{2F_0}{m\omega}\left[\frac{-2ae^{-\gamma t}\sin\omega t + 2\omega(1-e^{-at})}{a^2+4\omega^2}\right] \\
&= \frac{F_0}{4a^2 m}\left[1-e^{-at} - \frac{2}{\sqrt{15}}e^{-\frac{a}{2}t}\sin\frac{\sqrt{15}}{2}at\right].
\end{aligned}
$$

3-29 An overdamped oscillator is described by the equation

$$\ddot{x} + 2\gamma\dot{x} + \omega_0^2 x = 0 \quad (\gamma > \omega_0),$$

which has the general solution

$$x(t) = c_1 e^{-\gamma_+ t} + c_2 e^{-\gamma_- t} \quad \left(\gamma_\pm = \gamma \pm \sqrt{\gamma^2 - \omega_0^2}\right)$$

with two arbitrary real constants c_1 and c_2. To find the Green's function for this system, let us consider an impulse given during a small interval $[t_0, t_0 + \delta t]$. Before $t = t_0$ the solution was $x(t) = 0$. However, just after the impulse, the particle is still there but picks up the velocity $v = \frac{\delta p}{m}$ with $\delta p = \int_{t_0}^{t_0+\delta t} F dt$. So, for the subsequent motion, we must use c_1 and c_2 determined by

$$
\begin{aligned}
&c_1 + c_2 = 0, \quad -c_1\gamma_+ - c_2\gamma_- = \frac{\delta p}{m} \\
&\longrightarrow c_2 = -c_1 = \frac{\delta p}{2m\sqrt{\gamma^2 - \omega_0^2}},
\end{aligned}
$$

thus finding the motion

$$x(t) = \frac{\delta p}{2m\sqrt{\gamma^2-\omega_0^2}}\left(e^{-\gamma_-(t-t_0)} - e^{-\gamma_+(t-t_0)}\right) \quad (t > t_0).$$

Based on this result, we can then write down the Green's function appropriate to this case as

$$G(t-t') = \begin{cases} 0, & t < t' \\ \frac{e^{-\gamma_-(t-t')} - e^{-\gamma_+(t-t')}}{2m\sqrt{\gamma^2-\omega_0^2}}, & t > t' \end{cases}.$$

Now consider an overdamped oscillator which is initially at rest in its equilibrium position. If it is subject to a driving force $F(t) = ct$

for $t > 0$, we may then use the Green's function method to obtain the subsequent motion:

$$\begin{aligned}x(t) &= \int_0^t G(t-t')F(t')dt' \\ &= \frac{c}{2m\sqrt{\gamma^2-\omega_0^2}}\int_0^t t'\left(e^{-\gamma_-(t-t')} - e^{-\gamma_+(t-t')}\right)dt' \\ &= \frac{c}{2m\sqrt{\gamma^2-\omega_0^2}}\left(-\frac{t}{\gamma_+} + \frac{t}{\gamma_-} + \frac{(1-e^{-\gamma_+ t})}{\gamma_+^2} - \frac{(1-e^{-\gamma_- t})}{\gamma_-^2}\right).\end{aligned}$$

3-30 (i) For $0 < t < \tau$, we have the motion described by

$$x(t) = \frac{v_0}{\omega}e^{-\gamma t}\sin\omega t \quad \left(\text{with } \omega = \sqrt{\omega_0^2 - \gamma^2}\right),$$

since the initial conditions $x(0) = 0$ and $\dot{x}(0) = v_0$ are given. At $t = \tau$, an impulsive force is applied and it changes velocity. Accordingly, for $t > \tau$, the motion is given by

$$x(t) = \frac{v_0}{\omega}e^{-\gamma t}\sin\omega t + \frac{\delta p}{m\omega}e^{-\gamma(t-\tau)}\sin\omega(t-\tau).$$

(ii) For large t, we can ignore the transient term to write

$$\begin{aligned}x(t) &= \sum_{n=1}^{N(t)=\lfloor t/\tau\rfloor}\frac{\delta p}{m\omega}e^{-\gamma(t-n\tau)}\sin\omega(t-n\tau) \\ &= \frac{\delta p}{m\omega}e^{-\gamma t}\mathrm{Im}\left[\sum_{n=1}^{N(t)} e^{i\omega t}e^{n(\gamma-i\omega)\tau)}\right] \\ &= \frac{\delta p}{m\omega}e^{-\gamma t}\mathrm{Im}\left[e^{i\omega t}\frac{e^{(\gamma-i\omega)\tau} - e^{(N(t)+1)(\gamma-i\omega)\tau}}{1-e^{(\gamma-i\omega)\tau}}\right], \qquad (1)\end{aligned}$$

where $N(t) \equiv \lfloor t/\tau\rfloor$ denotes the largest integer not exceeding t/τ. Then, after some manipulations, one can show that the above expression is equivalent to

$$x(t) = -\frac{\delta p}{m\omega}\mathrm{Im}\left[Ae^{-(\gamma+i\omega)t}\right] + \frac{\delta p}{m\omega}\mathrm{Im}\left[Ae^{-(\gamma+i\omega)\{t-N(t)\tau\}}\right], \qquad (2)$$

where $A = \frac{e^{(\gamma+i\omega)\tau}}{1-e^{(\gamma+i\omega)\tau}}$, a t-independent complex number. For large t, the first term on the right hand side of (2) can be neglected, and we are thus left with the second term which is a periodic function of t with period τ (and does not depend on the initial conditions). [Notice that the function $t - N(t)\tau$ is of the saw-tooth wave form.] This corresponds to a limit cycle behavior.

For sufficiently small γ so that $\omega \approx \omega_0$, the complex number A becomes very large if

$$e^{i\omega\tau} \approx 1 \quad \longrightarrow \quad \omega\tau \approx \omega_0\tau = 2l\pi \quad (l = 1, 2, \ldots).$$

This implies that, if the impulses are provided at the time interval τ approximately equal to an integer multiple of the natural oscillator period $T_0 = \frac{2\pi}{\omega_0}$, we see the resonance phenomenon.

3-31 When a is relatively small, the integrand may be expanded in a power series,

$$\int_0^{\varphi} \left[1 + \frac{a^2}{2}\sin^2\varphi' + \cdots\right] d\varphi' = \omega_0(t - t_0).$$

This can be integrated to

$$\varphi - \frac{a^2}{8}\sin 2\varphi + O(a^4) = \omega(t - t_0)$$

where $\omega = \omega_0[1 - \frac{a^2}{4} + O(a^4)]$. This can be solved for $\varphi(t)$ in the form

$$\varphi(t) = \omega(t - t_0) + \frac{a^2}{8}\sin[2\omega(t - t_0)] + O(a^4). \tag{1}$$

On the other hand, from $\sin\varphi = \frac{1}{a}\sin\frac{\theta}{2}$, it follows that $\theta = 2a\sin\varphi + O(a^3)$. This can also be extended to the next order by writing $\sin\frac{\theta}{2} = \frac{\theta}{2} - \frac{1}{3!}\left(\frac{\theta}{2}\right)^3 + O(a^5)$ in the formula $\sin\frac{\theta}{2} = a\sin\varphi$, then,

$$\begin{aligned}\theta &= 2a\sin\varphi + \frac{1}{24}(2a\sin\varphi)^3 + O(a^5)\\ &= \left(2a + \frac{1}{4}a^3\right)\sin\varphi - \frac{1}{12}a^3\sin 3\varphi + O(a^5).\end{aligned} \tag{2}$$

In (2), we have used the iteration method and $\sin^3\varphi = \frac{3}{4}\sin\varphi - \frac{1}{4}\sin 3\varphi$. Now use (1) with the expression (2) and the addition theorem for sine to find

$$\theta(t) = \left(2a + \frac{1}{4}a^3\right) \sin\left\{\omega(t - t_0) + \frac{a^2}{8}\sin[2\omega(t - t_0)]\right\}$$
$$- \frac{1}{12}a^3 \sin[3\omega(t - t_0)] + O(a^5)$$
$$= \left(2a + \frac{1}{4}a^3\right)\left\{\sin\omega(t - t_0) + \frac{a^2}{8}\sin[2\omega(t - t_0)]\cos\omega(t - t_0)\right\}$$
$$- \frac{1}{12}a^3 \sin[3\omega(t - t_0)] + O(a^5)$$
$$= \left(2a + \frac{3}{8}a^3\right)\sin\omega(t - t_0) + \frac{a^3}{24}\sin[3\omega(t - t_0)] + O(a^5).$$

3-32 Noting the periodicity of motion, we may express the solution $x(t)$ by a Fourier series

$$x(t) = \frac{A_0}{2} + \sum_{l=1}^{\infty}(A_l \cos l\omega t + B_l \sin l\omega t), \tag{1}$$

where the angular frequency ω is not known yet. Suppose that $x(t) = \bar{A}\cos\omega_0 t$ when the perturbation is off (i.e., $\beta = 0$). Then, in the presence of the perturbation, we would have that $A_1 = \bar{A} + O(\beta)$, all of the other coefficients in the series are $O(\beta)$ or higher, and ω differs from ω_0 at most by an $O(\beta)$ term. To make these explicit, let us write

$$A_l = \bar{A}\delta_{l1} + \sum_{n=1}^{\infty}\beta^n A_l^{(n)} \quad (l = 0, 1, 2, \ldots)$$

$$B_l = \sum_{n=1}^{\infty}\beta^n B_l^{(n)} \quad (l = 1, 2, \ldots)$$

and assume that the angular frequency ω is the zeroth order value of ω_0, i.e., ω_0 admits the expansion

$$\omega_0 = \omega - \sum_{n=1}^{\infty}\beta^n \Delta^{(n)}.$$

Then, to apply the perturbation method, we may write our equation of motion in the form

$$\ddot{x} + \omega_0^2 x = -\beta x^2 \tag{2}$$

and compare both sides in each order of β after inserting the series development for $x(t)$.

The right hand side of (2) has the perturbative development

$$-\beta\bar{A}^2\cos^2\omega t - 2\beta^2\bar{A}\cos\omega t\left[\frac{1}{2}A_0^{(1)} + \sum_{l=1}^{\infty}(A_l^{(1)}\cos l\omega t + B_l^{(1)}\sin l\omega t)\right]$$
$$+ O(\beta^3), \tag{3}$$

while the corresponding development for its left hand side takes the form

$$\begin{aligned}
&(\omega_0^2 - \omega^2)\bar{A}\cos\omega t \\
&\quad + \beta\left[\frac{\omega_0^2}{2}A_0^{(1)} + \sum_{l=1}^{\infty}(\omega_0^2 - l^2\omega^2)(A_l^{(1)}\cos l\omega t + B_l^{(1)}\sin l\omega t)\right] \\
&\quad + \beta^2\left[\frac{\omega_0^2}{2}A_0^{(2)} + \sum_{l=1}^{\infty}(\omega_0^2 - l^2\omega^2)(A_l^{(2)}\cos l\omega t + B_l^{(2)}\sin l\omega t)\right] \\
&\quad + O(\beta^3) \\
&\quad = \beta\Bigg[- 2\omega_0\Delta^{(1)}\bar{A}\cos\omega t + \frac{\omega_0^2}{2}A_0^{(1)} + \sum_{l=2}^{\infty}(1 - l^2)\omega_0^2(A_l^{(1)}\cos l\omega t \\
&\quad + B_l^{(1)}\sin l\omega t)\Bigg] \\
&\quad + \beta^2\Bigg[- ((\Delta^{(1)})^2 + 2\omega_0\Delta^{(2)})\bar{A}\cos\omega t - 2\omega_0\Delta^{(1)}(A_1^{(1)}\cos\omega t \\
&\quad + B_1^{(1)}\sin\omega t) + \sum_{l=2}^{\infty}(1 - l^2)\omega_0^2(A_l^{(2)}\cos l\omega t + B_l^{(2)}\sin l\omega t)\Bigg] \\
&\quad + O(\beta^3).
\end{aligned} \tag{4}$$

(For the second expression of (4), we have used $\omega^2 = (\omega_0 + \sum_{n=1}^{\infty}\beta^n\Delta^{(n)})^2 = \omega_0^2 + 2\beta\omega_0\Delta^{(1)} + \beta^2((\Delta^{(1)})^2 + 2\omega_0\Delta^{(2)}) + O(\beta^3)$.) The $O(\beta)$ term of (3), i.e., $-\beta\bar{A}^2\cos^2\omega t = -\frac{\beta}{2}\bar{A}^2(1 + \cos 2\omega t)$ can match the $O(\beta)$ term of (4) only when

$$-\frac{\bar{A}^2}{2} = \frac{\omega_0^2}{2}A_0^{(1)}, \quad -2\omega_0\Delta^{(1)} = 0, \quad -\frac{\bar{A}^2}{2} = -3\omega_0^2A_2^{(1)},$$
$$A_3^{(1)} = A_4^{(1)} = \cdots = B_2^{(1)} = B_3^{(1)} = \cdots = 0. \tag{5}$$

The values of $A_1^{(1)}$ and $B_1^{(1)}$ are left undetermined, and we here take them to vanish as they are related to what choice is made for the zeroth order solution. (Also note that the general solution to our equation of motion should contain two adjustable parameters.) We now have $A_0^{(1)} = -\frac{\bar{A}^2}{\omega_0^2}$, $A_2^{(1)} = -\frac{\bar{A}^2}{6\omega_0^2}$, $\Delta^{(1)} = 0$ and accordingly the expression

$$x(t) = \bar{A}\cos\omega t + \beta\left(\frac{\bar{A}^2}{6\omega_0^2}\right)[-3 + \cos 2\omega t] + O(\beta^2) \tag{6}$$

with $\omega = \omega_0 + O(\beta^2)$.

Let us look at the $O(\beta^2)$ terms of (3) and (4). As the values obtained in the above $O(\beta)$-analysis are used, the $O(\beta^2)$ terms of the two expressions match each other only when the following relation holds:

$$\begin{aligned} &-2\bar{A}\cos\omega t\left[-\frac{1}{2}\frac{\bar{A}^2}{\omega_0^2} + \frac{\bar{A}^2}{6\omega_0^2}\cos 2\omega t\right] \\ &\quad = -2\omega_0\Delta^{(2)}\bar{A}\cos\omega t + \sum_{l=2}^{\infty}(1-l^2)\omega_0^2(A_l^{(2)}\cos l\omega t + B_l^{(2)}\sin l\omega t). \end{aligned} \tag{7}$$

As the left hand side of (7) can be rewritten as

$$\frac{\bar{A}^3}{\omega_0^2}\cos\omega t - \frac{\bar{A}^3}{6\omega_0^2}[\cos 3\omega t + \cos\omega t] = \frac{5}{6}\frac{\bar{A}^3}{\omega_0^2}\cos\omega t - \frac{\bar{A}^3}{6\omega_0^2}\cos 3\omega t, \tag{8}$$

we obtain from (7) the following relationships:

$$\begin{aligned} &-2\omega_0\Delta^{(2)}\bar{A} = \frac{5}{6}\frac{\bar{A}^3}{\omega_0^2}, \quad A_2^{(2)} = 0, \quad -8\omega_0^2 A_3^{(2)} = -\frac{\bar{A}^3}{6\omega_0^2}, \\ &A_4^{(2)} = A_5^{(2)} = \cdots = B_2^{(2)} = B_3^{(2)} = \cdots = 0. \end{aligned} \tag{9}$$

Hence, $\Delta^{(2)} = -\frac{5}{12}\frac{\bar{A}^2}{\omega_0^3}$ and $A_3^{(2)} = \frac{\bar{A}^3}{48\omega_0^4}$. We again take $A_1^{(2)} = B_1^{(2)} = 0$, so that we can write $\bar{A} = A_1$. Based on these findings, we secure the following result:

$$\begin{aligned} x(t) &= A_1\cos\omega t + \beta\left(\frac{A_1^2}{6\omega_0^2}\right)[-3 + \cos 2\omega t] \\ &\quad + \beta^2\left(\frac{A_1^3}{48\omega_0^4}\right)\cos 3\omega t + O(\beta^3) \end{aligned}$$

with $\omega = \omega_0[1 - \beta^2\frac{5A_1^2}{12\omega_0^4}]$.

3-33 Let us consider a general case when a perturbation δV is added to the given potential V. Then the change of period is determined by

$$\delta T = \sqrt{2m}\left[\int_{x_1+\delta x_1}^{x_2+\delta x_2} \frac{dx}{\sqrt{E - V(x) - \delta V(x)}} - \int_{x_1}^{x_2} \frac{dx}{\sqrt{E - V(x)}}\right].$$

It is not possible to expand the integrand in this equation in terms of δV since the resulting integral diverges. However, it becomes possible to expand the integrand (up to the first order of δV) if we write δT in the form

$$\delta T = 2\sqrt{2m}\frac{\partial}{\partial E}\left[\int_{x_1+\delta x_1}^{x_2+\delta x_2} \sqrt{E - V(x) - \delta V(x)}dx - \int_{x_1}^{x_2} \sqrt{E - V(x)}dx\right].$$

We are thus led to the following formula:

$$\delta T = -\sqrt{2m}\frac{\partial}{\partial E}\int_{x_1}^{x_2} \frac{\delta V}{\sqrt{E - V(x)}}dx. \tag{1}$$

Since $\frac{dx}{dt} = \sqrt{\frac{2}{m}(E - V(x))}$, one can rewrite this formula in the form (given in footnote 50 of Chapter 3 in the text) $\delta T = -2\frac{d}{dE}\int_0^T \delta V(\bar{x}_0(t))dt$, where $\bar{x}_0(t)$ denotes the unperturbed solution with given energy E. Higher-order corrections to the period can be obtained by similar means:

$$T = \sqrt{2m}\sum_{n=0}^{\infty}\frac{(-1)^n}{n!}\frac{\partial^n}{\partial E^n}\int_{x_1}^{x_2}\frac{[\delta V]^n}{\sqrt{E - V(x)}}dx. \tag{2}$$

We apply this to the two cases given (with $V = \frac{1}{2}\omega_0^2 x^2$ and $m = 1$):

(i) With the perturbation $\delta V = \frac{1}{4}\beta x^4$, we evaluate the change in the period as

$$\begin{aligned}\delta T &= -\sqrt{2}\frac{\partial}{\partial E}\int_{x_1=-\sqrt{2E}/\omega_0}^{x_2=\sqrt{2E}/\omega_0}\frac{\frac{1}{4}\beta x^4}{\sqrt{E - \frac{1}{2}\omega_0^2 x^2}}dx \\ &= -\frac{\beta}{\omega_0}\frac{\partial}{\partial E}\left(\frac{2E}{\omega_0^2}\right)^2\int_0^1\frac{t^4}{\sqrt{1-t^2}}dt \\ &= -\beta\frac{8E}{\omega_0^5}\frac{3\pi}{16} = -\beta\frac{3\pi E}{2\omega_0^5}.\end{aligned} \tag{3}$$

(ii) With the perturbation $\delta V = \frac{1}{3}\beta x^3$, the first-order correction vanishes identically. We thus evaluate the second-order correction based on our formula (2):

$$\delta T = \frac{\sqrt{2}}{2}\frac{\partial^2}{\partial E^2}\int_{x_1=-\sqrt{2E}/\omega_0}^{x_2=\sqrt{2E}/\omega_0} \frac{\frac{1}{9}\beta^2 x^6}{\sqrt{E-\frac{1}{2}\omega_0^2 x^2}}dx$$

$$= \frac{2}{9}\frac{\beta^2}{\omega_0}\frac{\partial^2}{\partial E^2}\left(\frac{2E}{\omega_0^2}\right)^3 \int_0^1 \frac{t^6}{\sqrt{1-t^2}}dt$$

$$= \beta^2 \frac{5\pi E}{3\omega_0^7}. \tag{4}$$

For the case (i), the perturbation theory provides us with the result $\omega = \omega_0 + \frac{3\beta A^2}{8\omega_0}$ (see Sec. 3.5.3 in the text) and this corresponds to the value of δT in (3). For the case (ii) the result $\omega = \omega_0\left(1 - \beta^2 \frac{5A^2}{12\omega_0^4}\right)$ (obtained in Problem 3-32) is consistent with the value of δT found in (4).

3-34 From the result of Problem 3-8, the period of oscillation is given by

$$T = 2\int_{-x_0}^{x_0} \frac{E_{\text{rel}} - \frac{1}{2}kx^2}{c\sqrt{(E_{\text{rel}} - \frac{1}{2}kx^2)^2 - m_0^2 c^4}}dx$$

with the turning points $\pm x_0 = \pm\sqrt{\frac{2}{k}(E_{\text{rel}} - m_0 c^2)}$. Subtracting the rest energy from the relativistic energy, we introduce $\bar{E} = E_{\text{rel}} - m_0 c^2$ and consider the nonrelativistic limit by taking a small value of $\bar{E}$ (or, more properly, of $(\bar{E} - \frac{1}{2}kx^2)$). We then find

$$T \approx 2\int_{-x_0=-\sqrt{\frac{2\bar{E}}{k}}}^{x_0=\sqrt{\frac{2\bar{E}}{k}}} \frac{\sqrt{m_0/2}}{\sqrt{\bar{E} - \frac{1}{2}kx^2}}\left\{1 + \frac{3}{4}\frac{(\bar{E} - \frac{1}{2}kx^2)}{m_0 c^2}\right\}dx$$

$$= T_0 + \frac{3}{2}\frac{\sqrt{m_0/2}}{m_0 c^2}\sqrt{\frac{k}{2}}\int_{-x_0}^{x_0}\sqrt{x_0^2 - x^2}\,dx$$

$$= T_0 + \frac{3}{4}\frac{\sqrt{k}}{\sqrt{m_0}c^2}\frac{\pi}{2}x_0^2$$

$$= T_0\left[1 + \frac{3}{16}\left(\frac{x_0\omega_0}{c}\right)^2\right],$$

where we used the nonrelativistic expression for the period, $T_0 \equiv 2\pi\sqrt{m_0/k} = \frac{2\pi}{\omega_0}$, and the fact that the value of the last integral in the second line corresponds to the area of a half circle with radius x_0. Therefore the angular frequency with the relativistic correction is

$$\omega = \frac{2\pi}{T} \approx \omega_0 \left[1 - \frac{3}{16}\left(\frac{x_0\omega_0}{c}\right)^2\right].$$

[Remark: The same result can be obtained by perturbation theory. First note that, if $|\dot{x}/c| \ll 1$, we can approximate $\frac{m\dot{x}}{\sqrt{1-\dot{x}^2/c^2}}$ by $m\dot{x}(1+\frac{1}{2c^2}\dot{x}^2)$ and so replace the given equation of motion by the form

$$\ddot{x} + \omega_0^2 x = -\beta\dot{x}^2\ddot{x}, \tag{1}$$

where $\omega_0 = \sqrt{\frac{k}{m}}$ and $\beta = \frac{3}{2c^2}$. (β is the perturbation parameter.) Then write $x(t) = \sum_{n=1}^{\infty} A_n \cos n\omega t$, with $A_1 = O(1)$, $A_n = O(\beta)$ if $n \neq 1$, and $\omega = \omega_0 + O(\beta)$. Using this form, the left and right hand sides of (1) can be written as

$$\begin{aligned}
\ddot{x} + \omega_0^2 x &= \sum_{n=1}^{\infty}(\omega_0^2 - n^2\omega^2)A_n \cos n\omega t,\\
-\beta\dot{x}^2\ddot{x} &= -\beta A_1^2\omega^2 \sin^2\omega t\,(-A_1\omega^2\cos\omega t) + O(\beta^2)\\
&= \beta\omega^4 A_1^3(\cos\omega t - \cos^3\omega t) + O(\beta^2)\\
&= \beta\omega^4 A_1^3\left(\frac{1}{4}\cos\omega t - \frac{1}{4}\cos 3\omega t\right) + O(\beta^2),
\end{aligned}$$

and hence

$$(\omega_0^2 - \omega^2)A_1 = \frac{1}{4}\beta\omega^4 A_1^3, \quad A_2 = 0.$$

So, using $\omega_0^2 - \omega^2 \approx 2\omega_0(\omega_0 - \omega)$, we find that $\omega_0 - \omega = \frac{1}{8}\beta\omega_0^3 A_1^2 = \frac{3}{16c^2}\omega_0^3 A_1^2$, i.e., $\omega = \omega_0\{1 - \frac{3}{16}(\frac{A_1\omega_0}{c})^2\}$.]

3-35 We are here given the equation of motion

$$\ddot{x} + 2\gamma\dot{x} + \omega_0^2 x = \frac{\epsilon}{m}x^3 + \frac{f}{m}\cos\omega t \tag{1}$$

with $\gamma = \frac{b}{2m}$ and $\omega_0 = \sqrt{\frac{k}{m}}$. When there is no nonlinear term (i.e., $\epsilon = 0$), a steady-state solution is of the form

$$x_0(t) = A\cos(\omega t - \alpha).$$

Inserting this into (1), the left hand side of (1) becomes

$$\begin{aligned} & A[(\omega_0^2-\omega^2)\cos(\omega t-\alpha)-2\gamma\omega\sin(\omega t-\alpha)] \\ & = A\sqrt{(\omega_0^2-\omega^2)^2+4\gamma^2\omega^2}\,\cos(\omega t-\alpha+\beta_0) \end{aligned}$$

with $\tan\beta_0 = \frac{2\gamma\omega}{(\omega_0^2-\omega^2)}$. If we choose the two constants A and α as

$$A = \frac{f/m}{\sqrt{(\omega_0^2-\omega^2)^2+4\gamma^2\omega^2}} \equiv A_0, \quad \alpha = \beta_0 = \tan^{-1}\frac{2\gamma\omega}{(\omega_0^2-\omega^2)},$$

$x_0(t)$ satisfies (1) for $\epsilon = 0$. When $\epsilon \neq 0$, noting that the nonlinear term generates higher harmonics

$$\begin{aligned} \frac{\epsilon}{m}x_0^3 &= \frac{\epsilon}{m}A_0^3\cos^3(\omega t-\beta_0) \\ &= \frac{1}{4}\frac{\epsilon}{m}A_0^3\{3\cos(\omega t-\beta_0)+\cos[3(\omega t-\beta_0)]\}, \end{aligned}$$

we may set the approximate solution in the form

$$x(t) = A\cos(\omega t-\alpha)+B\cos[3(\omega t-\alpha)]+C\sin[3(\omega t-\alpha)]. \quad (2)$$

Obviously the two constants B and C are $O(\epsilon)$. But, as A and α take values A_0 and β_0 respectively when $\epsilon = 0$, we may set $A = A_0 + A_1$ and $\alpha = \beta_0 + \beta_1$ where A_1 and β_1 are $O(\epsilon)$. Substituting (2) into (1) and keeping terms up to $O(\epsilon)$, we obtain

$$\begin{aligned} & (\omega_0^2-\omega^2)A\cos(\omega t-\alpha)+(\omega_0^2-9\omega^2)\{B\cos[3(\omega t-\alpha)] \\ & \quad + C\sin[3(\omega t-\alpha)]\}+2\gamma\omega\{-A\sin(\omega t-\alpha) \\ & \quad - 3B\sin[3(\omega t-\alpha)]+3C\cos[3(\omega t-\alpha)]\} \\ & = \frac{1}{4}\frac{\epsilon}{m}A_0^3\{3\cos(\omega t-\alpha)+\cos[3(\omega t-\alpha)]\}+\frac{f}{m}\cos\omega t. \quad (3) \end{aligned}$$

Matching the higher harmonic terms in (3) gives us the equations

$$\begin{aligned} & (\omega_0^2-9\omega^2)B+6\gamma\omega C = \frac{1}{4}\frac{\epsilon}{m}A_0^3, \\ & (\omega_0^2-9\omega^2)C-6\gamma\omega B = 0, \end{aligned}$$

which determine

$$B = \frac{(\omega_0^2-9\omega^2)\frac{1}{4}\frac{\epsilon}{m}A_0^3}{(\omega_0^2-9\omega^2)^2+(6\gamma\omega)^2}, \qquad C = \frac{(6\gamma\omega)\frac{1}{4}\frac{\epsilon}{m}A_0^3}{(\omega_0^2-9\omega^2)^2+(6\gamma\omega)^2}.$$

The lower harmonic terms in (3) also give rise to the equation

$$\begin{aligned}(A_0 + A_1)\Big\{(\omega_0^2 - \omega^2)[\cos(\omega t - \beta_0)\cos\beta_1 + \sin(\omega t - \beta_0)\sin\beta_1] \\ - 2\gamma\omega[\sin(\omega t - \beta_0)\cos\beta_1 - \cos(\omega t - \beta_0)\sin\beta_1]\Big\} \\ = \frac{3}{4}\frac{\epsilon}{m}A_0^3\cos(\omega t - \beta_0) + \frac{f}{m}\cos\omega t, \qquad (4)\end{aligned}$$

which reduces, upon setting $\cos\beta_1 \sim 1$ and $\sin\beta_1 \sim \beta_1$, to

$$\begin{aligned}A_0\beta_1(\omega_0^2 - \omega^2)\sin(\omega t - \beta_0) + A_1(\omega_0^2 - \omega^2)\cos(\omega t - \beta_0) \\ - 2A_1\gamma\omega\sin(\omega t - \beta_0) + 2\beta_1\gamma\omega A_0\cos(\omega t - \beta_0) \\ = \frac{3}{4}\frac{\epsilon}{m}A_0^3\cos(\omega t - \beta_0).\end{aligned}$$

This is valid only when

$$\begin{aligned}\beta_1 A_0(\omega_0^2 - \omega^2) - 2A_1\gamma\omega = 0, \\ A_1(\omega_0^2 - \omega^2) + 2\beta_1\gamma\omega A_0 = \frac{3}{4}\frac{\epsilon}{m}A_0^3, \qquad (5)\end{aligned}$$

and we can use these equations to determine A_1 and β_1:

$$A_1 = \frac{\frac{3}{4}\frac{\epsilon}{m}(\omega_0^2 - \omega^2)A_0^3}{(\omega_0^2 - \omega^2)^2 + 4\gamma^2\omega^2}, \quad \beta_1 = \frac{3}{4}\frac{\epsilon}{m}\frac{(2\gamma\omega)A_0^2}{(\omega_0^2 - \omega^2)^2 + 4\gamma^2\omega^2}.$$

This completes the construction of the approximate solution of the form (2) to first order in ϵ.

4 Motion of a Particle in Two or Three Dimensions

4-1 (i) Conservative, the given force being of the form $-\vec{\nabla}V$ if one takes $V = -6abxyz^3 + 5bx^4y^2$.

(ii) Nonconservative.

(iii) Conservative, the given force being of the form $-\vec{\nabla}V$ if one takes $V = -a(x^2y^3 + x^2z^3 + y^2z^3) - ay^5$.

(iv) In Cartesian coordinates the force acquires the expression

$$\begin{aligned}\vec{F} &= -3a\rho^2\cos\varphi(\cos\varphi\mathbf{i} + \sin\varphi\mathbf{j}) + a\rho^2\sin\varphi(-\sin\varphi\mathbf{i} \\ &\quad + \cos\varphi\mathbf{j}) + 2az^2\mathbf{k} \\ &= a(-3x^2 - y^2)\mathbf{i} - 2axy\mathbf{j} + 2az^2\mathbf{k}.\end{aligned}$$

This is a conservative force, and can be written as $\vec{F} = -\vec{\nabla}V$ if one takes $V = ax(x^2 + y^2) - \frac{2}{3}az^3 = a\rho^3\cos\varphi - \frac{2}{3}az^3$.

(v) One can rewrite this force using Cartesian coordinates to obtain

$$\vec{F} = -a\left(\frac{x^2}{r} + r\right)\mathbf{i} - a\frac{xy}{r}\mathbf{j} - a\frac{xz}{r}\mathbf{k} \quad (r \equiv \sqrt{x^2 + y^2 + z^2}).$$

This is a conservative force, and can be written as $\vec{F} = -\vec{\nabla}V$ if one takes $V = arx = ar^2\sin\theta\cos\varphi$.

4-2 Given a force field of the form $\vec{F}(\vec{r}) = -\frac{\vec{p}}{r^3} + c\frac{(\vec{p}\cdot\vec{r})\vec{r}}{r^5}$, we can calculate its curl as

$$\begin{aligned}(\vec{\nabla} \times \vec{F})_i &= \epsilon_{ijk}\partial_j\left(-\frac{p_k}{r^3} + c\frac{p_l x_l x_k}{r^5}\right) \\ &= \epsilon_{ijk}\left[\frac{3}{r^5}p_k x_j + c\left(\frac{p_j x_k + \cancel{p_l x_l \delta_{jk}}}{r^5} - \cancel{5\frac{p_l x_l x_k x_j}{r^7}}\right)\right] \\ &= \epsilon_{ijk}\left(-\frac{3}{r^5} + \frac{c}{r^5}\right)p_j x_k.\end{aligned}$$

Hence we have $\vec{\nabla} \times \vec{F} = 0$ for $c = 3$, and $\vec{\nabla} \times \vec{F} \neq 0$ otherwise. This shows that the given force form can be conservative only when $c = 3$. To find the potential energy function leading to the force $\vec{F} = -\vec{\nabla} V = -\frac{\vec{p}}{r^3} + 3\frac{(\vec{p}\cdot\vec{r})\vec{r}}{r^5}$, we here note that $V(\vec{r})$ should be a rotational scalar (if $\vec{p}$ is regarded as a vector quantity) while being linear in $\vec{p}$. The only candidate we can write for such scalar is

$$V(\vec{r}) = h(r)\vec{p}\cdot\vec{r} \quad (h(r): \text{ a function of } r).$$

Now, for this form, we find

$$-\vec{\nabla} V = -h'(r)\frac{\vec{r}}{r}\vec{p}\cdot\vec{r} - h(r)\vec{p}.$$

As the given force follows for $h(r) = \frac{1}{r^3}$, the related potential energy function is $V(r) = \frac{\vec{p}\cdot\vec{r}}{r^3}$.

[Remark: The given form with $c = 3$ can be used to describe the electric field due to an electric dipole (located at the origin).]

4-3 For the given force field, we find

$$\begin{aligned}\frac{\partial F_y}{\partial x} &= \frac{\partial}{\partial x}\left(\gamma\frac{x}{x^2+y^2}\right) = \gamma\left(\frac{1}{x^2+y^2} - 2\frac{x^2}{(x^2+y^2)^2}\right)\\ &= \gamma\frac{-x^2+y^2}{(x^2+y^2)^2},\\ \frac{\partial F_x}{\partial y} &= \frac{\partial}{\partial y}\left(\gamma\frac{-y}{x^2+y^2}\right) = \gamma\left(\frac{-1}{x^2+y^2} + 2\frac{y^2}{(x^2+y^2)^2}\right)\\ &= \gamma\frac{-x^2+y^2}{(x^2+y^2)^2},\end{aligned}$$

and hence $\frac{\partial F_y}{\partial x} = \frac{\partial F_x}{\partial y}$ away from the origin — this force is conservative in a simply-connected region not including the origin. On the other hand, if one calculates the line integral $\oint_C \vec{F}\cdot d\vec{r}$ over a closed curve C that encloses the origin, it does not vanish. (Take a circle C: $r = \sqrt{x^2+y^2} = a$, then $\oint_C \vec{F}\cdot d\vec{r} = 2\pi$.) So, in a region containing the origin as an interior point, no everywhere-single-valued potential exists. But there may exist a "multi-valued potential" V, yielding $-\vec{\nabla} V = \vec{F}$ for the given force. Indeed, as polar coordinates (r, θ) (with $x = r\cos\theta$, $y = r\sin\theta$) are used, the given force field can be written as

$$\vec{F} = \gamma\frac{1}{r}(-\sin\theta\mathbf{i} + \cos\theta\mathbf{j}) = \vec{\nabla}(\gamma\theta) \tag{1}$$

where the last form follows from

$$\frac{\partial\theta}{\partial x} = \frac{\partial}{\partial x}\tan^{-1}\frac{y}{x} = \frac{-y/x^2}{1+(y/x)^2} = \frac{-y}{x^2+y^2},$$
$$\frac{\partial\theta}{\partial y} = \frac{\partial}{\partial y}\tan^{-1}\frac{y}{x} = \frac{1/x}{1+(y/x)^2} = \frac{x}{x^2+y^2}.$$

Thus $V = -\gamma\theta = -\gamma\tan^{-1}\frac{y}{x}$, with the angle θ allowed to increase continuously in counterclockwise rotations, can serve as a potential function for the force field $\vec{F} = \gamma\frac{-y\mathbf{i}+x\mathbf{j}}{x^2+y^2}$ (although it is not a single-valued function in any shell-like region enclosing the origin).

In polar coordinates, the above force is represented by $\vec{F} = \gamma\frac{1}{r}\hat{\theta}$ (i.e., $F_r = 0$ and $F_\theta = \gamma\frac{1}{r}$) and so we have equations of motion (see Sec. 4.2)

$$m(\ddot{r} - r\dot{\theta}^2) = 0$$
$$m(r\ddot{\theta} + 2\dot{r}\dot{\theta}) = \gamma\frac{1}{r}. \tag{2}$$

For this system we have the energy conservation law

$$\frac{1}{2}m(\dot{r}^2 + r^2\dot{\theta}^2) - \gamma\theta = E \quad (\text{: const.}), \tag{3}$$

and the angular momentum equation $\frac{dL_z}{dt} = \Gamma_z$ simplifies to

$$\frac{d}{dt}(mr^2\dot{\theta}) = \gamma \quad \longrightarrow \quad mr^2\dot{\theta} = \gamma t + C \quad (C: \text{ cosnt.}) \tag{4}$$

(since $\Gamma_z = rF_\theta = \gamma$). (3) and (4) provide us with the equations of the form $\dot{r} = f(r,\theta,t)$, $\dot{\theta} = g(r,\theta,t)$; in principle, given the initial conditions, these two equations may be integrated (numerically) to determine the motion explicitly.

An alternative way to characterize the above system is through the following two equations:

$$mr^2\dot{\theta} = l(t), \qquad m\ddot{r} = \frac{l^2(t)}{mr^3} \tag{5}$$

where $l(t) \equiv \gamma t + C$. (The second of these, with the centrifugal force term $\frac{l^2(t)}{mr^3}$ follows from the first of (2).) If $l(t)$ were time-independent, these would have inertial (hence straight-line) trajectories as their solutions. In our case where the angular momentum value increases linearly with time, we expect the motion to exhibit following general patterns:

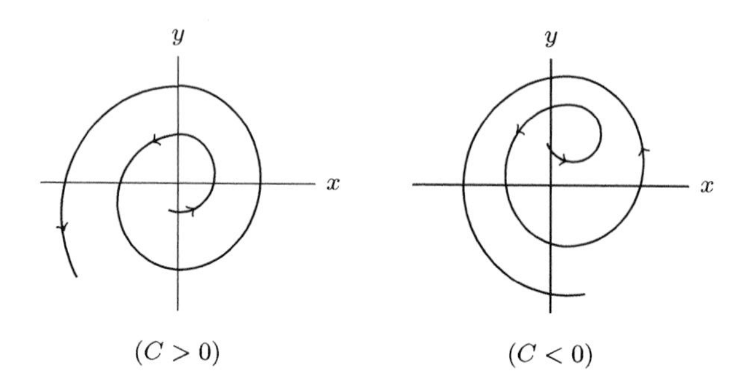

$(C > 0)$ $(C < 0)$

4-4 Schematically, equipotential curves for the respective cases are of following forms (we assume $Q > 0$ and solid (dashed) lines represent equipotential curves (electric field lines)):

(i)

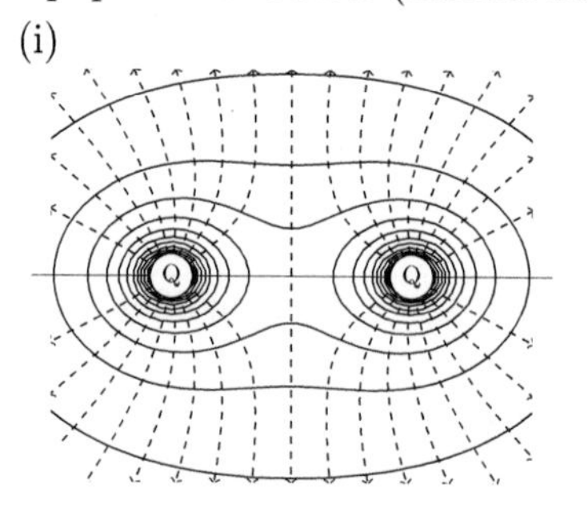

(ii)

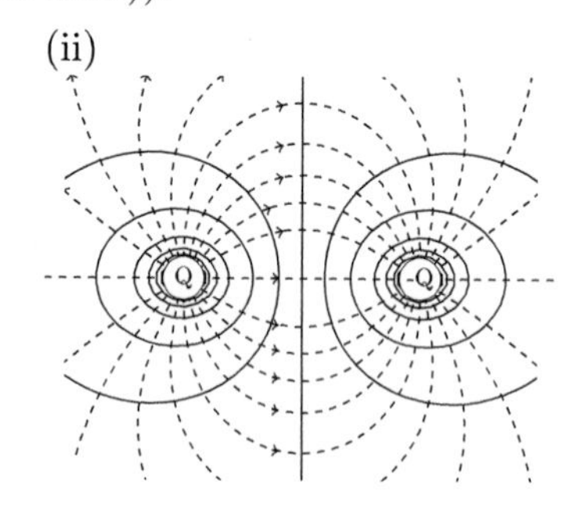

4-5 From the angular momentum conservation,

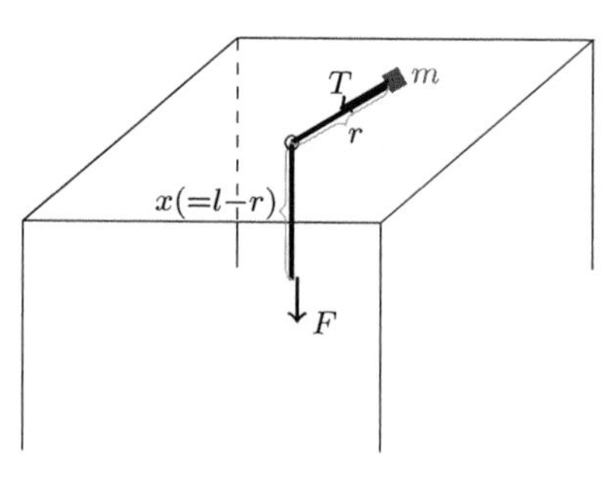

$$mr^2\dot{\theta} = ml^2\omega \quad \longrightarrow \quad \dot{\theta} = \frac{l^2}{r^2}\omega,$$

and so the angular velocity when $r = \frac{1}{2}l$ is equal to 4ω. Also the tension T in the string, which has the same magnitude as the force F pulling the string, is given from the radial equation of motion by

$$-T = m(\ddot{r} - r\dot{\theta}^2) = m\left(\ddot{r} - \frac{l^4}{r^3}\omega^2\right).$$

When the string is pulled *slowly*, we can here take $\ddot{r} \approx 0$ to obtain $T = m\frac{l^4\omega^2}{r^3}$. Now suppose that the length of the string below the hole, $x\ (= l - r)$, has been increased by $\bar{x}\ (= l - \bar{r})$. Then the net work W done by the force F is

$$\begin{aligned} W &= \int_0^{\bar{x}} F dx = -\int_l^{\bar{r}} T dr = -\int_l^{\bar{r}} m\frac{l^4\omega^2}{r^3} dr \\ &= \frac{1}{2}m\frac{l^4\omega^2}{\bar{r}^2} - \frac{1}{2}ml^2\omega^2 \\ &= [\text{kinetic energy at } r = \bar{r}] \\ &\quad - [\text{kinetic energy at } r = l], \end{aligned}$$

when the kinetic energy expression $T = \frac{1}{2}m\dot{r}^2 + \frac{1}{2}mr^2\dot{\theta}^2 \approx \frac{1}{2}mr^2\dot{\theta}^2$ is used.

4-6 (i) Since $\dot{\theta} = \omega$ is a constant, it follows that $\theta = \omega t + \delta$. On the other hand, the radial equation for the bead (using polar coordinates) reads

$$m\ddot{r} = -mg\sin\theta + mr\omega^2,$$

including the centrifugal force term $mr\omega^2$. Hence we obtain

$$\ddot{r} - \omega^2 r = -g\sin(\omega t + \delta).$$

As $r_p(t) = \frac{g}{2\omega^2}\sin(\omega t + \delta)$ provides a particular solution to this equation, we can write its general solution as

$$r = ae^{\omega t} + be^{-\omega t} + \frac{g}{2\omega^2}\sin(\omega t + \delta) \tag{1}$$

with arbitrary constants a, b.

(ii) Suppose the rod is horizontal at $t = 0$, then $\delta = 0$. At this time, we have from the given initial conditions $r = 0$ and $\dot{r} = v_0$. Hence the constants a and b in (1) must be chosen such that

$$a + b = 0, \quad \omega(a - b) + \frac{g}{2\omega} = v_0. \tag{2}$$

The motion of the bead along the rod will be simple harmonic if and only if the constants a and b are both zero (see (1)); clearly, from (2), this is realized only when $v_0 = \frac{g}{2\omega}$.

4-7 (i) Let us take the Cartesian z-axis along the downward vertical. Then we can express the gravitational force on the mass by

$$\begin{aligned}\vec{F}_g &= mg\mathbf{k} = mg(\hat{r}\cdot\mathbf{k})\hat{r} + mg(\hat{\Theta}\cdot\mathbf{k})\hat{\Theta} \\ &= mg\cos\Theta\,\hat{r} - mg\sin\Theta\,\hat{\Theta},\end{aligned}$$

and the elastic force due to the spring by

$$\vec{F}_e = -k(r-l)\hat{r}.$$

Thus the net force acting on the mass is

$$\vec{F} = [mg\cos\Theta - k(r-l)]\hat{r} - mg\sin\Theta\hat{\Theta},$$

and accordingly we have the equations of motion (see (4.59))

$$\begin{aligned} m(\ddot{r} - r\dot{\Theta}^2 - r\dot{\varphi}^2\sin^2\Theta) &= mg\cos\Theta - k(r-l), \\ m(r\ddot{\Theta} + 2\dot{r}\dot{\Theta} - r\dot{\varphi}^2\sin\Theta\cos\Theta) &= -mg\sin\Theta, \\ m\left\{\frac{d}{dt}(r\dot{\varphi}\sin\Theta) + \dot{r}\dot{\varphi}\sin\Theta + r\dot{\Theta}\dot{\varphi}\cos\Theta\right\} &= 0. \end{aligned}$$

(ii) For a circular orbit we may take $\dot{r} = \dot{\Theta} = 0$ and $\dot{\varphi} = \omega$ (: const.) in the above equations of motion: then,

$$\begin{aligned} -r\omega^2\sin^2\Theta &= g\cos\Theta - \frac{k}{m}(r-l), \\ r\omega^2\cancel{\sin\Theta}\cos\Theta &= g\cancel{\sin\Theta} \quad\longrightarrow\quad r = \frac{g}{\omega^2\cos\Theta} \end{aligned}$$

and so we find

$$\begin{aligned} -g\frac{\sin^2\Theta}{\cos\Theta} &= g\cos\Theta - \frac{k}{m}\left(\frac{g}{\omega^2\cos\Theta} - l\right) \quad\longrightarrow\quad \cos\Theta \\ &= \frac{g}{l}\left(\frac{1}{\omega^2} - \frac{m}{k}\right). \end{aligned}$$

4-8 (i) Given a central force of the form $\vec{F} = -cr^n\hat{r}$ $(c > 0)$, the related effective radial potential is

$$V_{\text{eff}}(r) = \frac{c}{n+1}r^{n+1} + \frac{L^2}{2mr^2},$$

where L is the angular momentum of the particle about the origin. Then, for a circular orbit of radius r_0, we must have

$$0 = V'_{\text{eff}}(r_0) = cr_0^n - \frac{L^2}{mr_0^3} \quad \rightarrow \quad r_0 = \left(\frac{L^2}{mc}\right)^{\frac{1}{n+3}}$$
$$(\text{with } n \neq -3).$$

(ii) To check the stability of this circular orbit, let us calculate the second derivative $V''_{\text{eff}}(r_0)$:

$$V''_{\text{eff}}(r_0) = cnr_0^{n-1} + \frac{3L^2}{mr_0^4}$$
$$= \frac{L^2}{mr_0^4}(n+3),$$

where we used $r_0^{n+3} = \frac{L^2}{mc}$ from (i). *We thus have $V''_{\text{eff}}(r_0) < 0$ if $n < -3$ (and the corresponding circular orbit is unstable)*; on the other hand, if $n > -3$, we find $V''_{\text{eff}}(r_0) > 0$ (and so a stable circular orbit).
[Remark: Notice that, depending on the value of n, the graph of the effective potential $V_{\text{eff}}(r)$ changes its character:

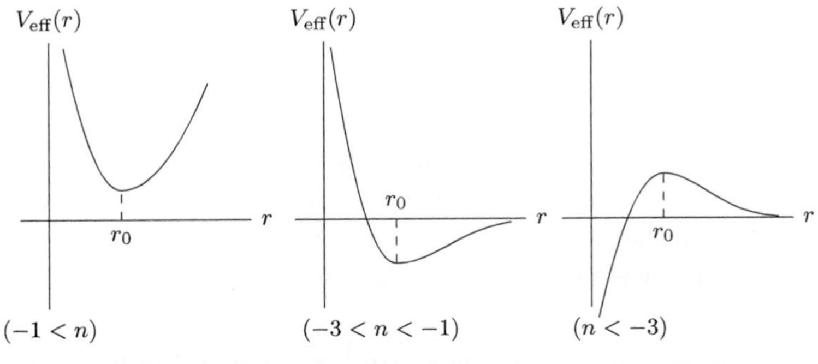

.]

4-9 (i) We can investigate the motion in terms of the effective radial potential

$$V_{\text{eff}}(r) = V(r) + \frac{L^2}{2mr^2},$$

where $-V'(r) = -\frac{K}{r^2}e^{-r/a}$, and L is the angular momentum of the particle about the origin. For a stable circular orbit of radius

R, the following two conditions must be satisfied:

$$V'_{\text{eff}}(r{=}R) = 0, \quad V''_{\text{eff}}(r{=}R) > 0.$$

From the first condition, we obtain

$$\frac{K}{R^2}e^{-R/a} - \frac{L^2}{mR^3} = 0 \quad \to KRe^{-R/a} = \frac{L^2}{m}.$$

Then the second condition yields

$$\begin{aligned} V''_{\text{eff}}(r{=}R) &= -\frac{2K}{R^3}e^{-R/a} - \frac{K}{aR^2}e^{-R/a} + \frac{3L^2}{mR^4} \\ &= \frac{L^2}{mR^4}\left(1 - \frac{R}{a}\right) > 0 \end{aligned}$$

where we used $KRe^{-R/a} = \frac{L^2}{m}$. Thus, for stability, we must have $R < a$.

(ii) As we write $\xi \equiv r - R$, the equation for small radial oscillations reads

$$m\ddot{\xi} + V''(r{=}R)\xi = 0.$$

Therefore, the angular frequency for small oscillations is given by

$$\omega = \sqrt{\frac{V''(r{=}R)}{m}} = \frac{L}{mR^2}\sqrt{1 - \frac{R}{a}},$$

where we used the result of (i).

4-10 (i) If $\vec{N} = N(s)\,\mathbf{n}$ denotes the normal force that the wire exerts on the particle, we may apply Newton's second law to write the particle's equation of motion in the form

$$m\left(\ddot{s}\mathbf{t} - \frac{\dot{s}^2}{\rho}\mathbf{n}\right) = -mg\mathbf{k} + N(s)\mathbf{n},$$

where we used the representation of the acceleration given in Sec. 4.2. Taking the scalar product with $\mathbf{t} = \frac{dx}{ds}\mathbf{i} + \frac{dz}{ds}\mathbf{k}$ on this equation then yields

$$\frac{d^2s}{dt^2} = -g\frac{dz}{ds}, \tag{1}$$

since $\mathbf{t} \cdot \mathbf{k} = \frac{dz}{ds}$. To write this as a differential equation for $x = x(t)$, note that

$$\frac{dz}{ds} = f'(x)\frac{dx}{ds} = \frac{f'(x)}{\sqrt{1+[f'(x)]^2}},$$

$$\frac{d^2s}{dt^2} = \frac{d}{dt}\left(\frac{ds}{dx}\frac{dx}{dt}\right) = \frac{d}{dt}\left(\sqrt{1+[f'(x)]^2}\frac{dx}{dt}\right)$$

$$= \sqrt{1+[f'(x)]^2}\frac{d^2x}{dt^2} + \frac{f'(x)f''(x)}{\sqrt{1+[f'(x)]^2}}\left(\frac{dx}{dt}\right)^2.$$

If we insert these in (1), we obtain the following equation:

$$\frac{d^2x}{dt^2} + \frac{f'(x)f''(x)}{1+[f'(x)]^2}\left(\frac{dx}{dt}\right)^2 + \frac{gf'(x)}{1+[f'(x)]^2} = 0. \qquad (2)$$

(ii) If we multiply both sides of (1) by $\frac{ds}{dt}$, the result can be presented as

$$\frac{d}{dt}\left[\frac{1}{2}\left(\frac{ds}{dt}\right)^2 + gz\right] = 0.$$

Since $\frac{ds}{dt} = \sqrt{\left(\frac{dx}{dt}\right)^2 + \left(\frac{dz}{dt}\right)^2} = \sqrt{1+[f'(x)]^2}\frac{dx}{dt}$ $(\equiv v(t))$, we are thus led to

$$\frac{1}{2}v^2 + gf(x) = \frac{1}{2}(1+[f'(x)]^2)\left(\frac{dx}{dt}\right)^2 + gf(x) = \frac{E}{m} \quad (\text{: const}).$$

As one can check easily, this also provides the first integral of the second-order differential equation in (2).

4-11 (i) To find the equation of motion along the cycloidal path (see the figure), we may use the energy conservation equation (see Problem 4-10 (ii))

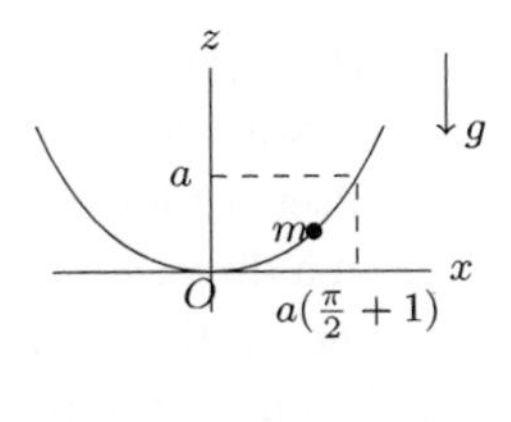

$$\frac{1}{2}mv^2 + mgz = \frac{m}{2}(\dot{x}^2 + \dot{z}^2) + mgz$$
$$= E \quad (\text{: const}). \qquad (1)$$

Since $\dot{x} = a\dot{\phi}(1+\cos\phi)$ and $\dot{z} = a\dot{\phi}\sin\phi$, we are then led to the following expression for the energy in terms of ϕ:

$$E = ma^2\dot{\phi}^2(1+\cos\phi) + mga(1-\cos\phi)$$
$$= 2ma^2\dot{\phi}^2\cos^2\frac{\phi}{2} + 2mga\sin^2\frac{\phi}{2}. \tag{2}$$

In terms of the variable $w = \sin\frac{\phi}{2}$, this energy equation can be rewritten as

$$E = 8ma^2\dot{w}^2 + 2mgaw^2, \tag{3}$$

which also implies that $\ddot{w} = -\frac{g}{4a}w$.

(ii) In (i) we saw that the variable w satisfies the equation of motion corresponding to a harmonic oscillator. From this fact we can immediately conclude that the particle undergoes periodic motion whose frequency is *independent of the amplitude of oscillator*, unlike the case of the simple pendulum for which the frequency depends on the amplitude.

4-12 We have two conservation laws

$$\frac{1}{2}m\dot{r}^2 + V_{\text{eff}}(r) = E \quad \left(\text{with } V_{\text{eff}}(r) = \frac{1}{2}kr^2 + \frac{L^2}{2mr^2}\right),$$
$$mr^2\dot{\theta} = L.$$

With the particle moving in a circle of radius a, we have $ma^2\dot{\theta} = L$ and

$$0 = V'_{\text{eff}}(r=a) = ka - \frac{L^2}{ma^3}$$
$$\rightarrow \quad L = \sqrt{mk}a^2.$$

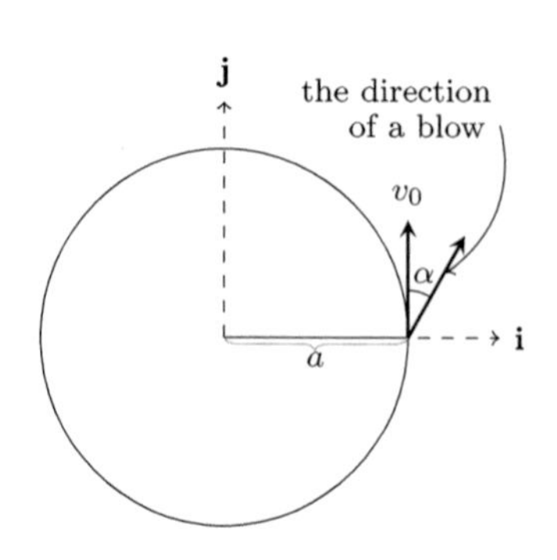

Hence the orbital velocity is $v_0 = a\dot{\theta} = a\left(\frac{L}{ma^2}\right) = a\sqrt{\frac{k}{m}}$.

Now let us consider the motion after a blow of impulse mv_0 is given in a direction making an angle α with the original velocity (see the figure). Then, as the particle's initial velocity can be taken as

$$\begin{aligned}\vec{v}_0 &= v_0\mathbf{j} + (v_0 \sin\alpha\mathbf{i} + v_0\cos\alpha\mathbf{j})\\ &= v_0(1+\cos\alpha)\mathbf{j} + v_0\sin\alpha\mathbf{i},\end{aligned}$$

we have

$$\begin{aligned}\textit{energy}:\quad E &= \frac{1}{2}m[v_0^2\sin^2\alpha + v_0^2(1+\cos\alpha)^2] + \frac{1}{2}ka^2\\ &= ka^2\left(\frac{3}{2}+\cos\alpha\right),\\ \textit{angular momentum}:\quad L &= mav_0(1+\cos\alpha) = a^2\sqrt{mk}(1+\cos\alpha),\end{aligned}$$

where we have used $v_0 = a\sqrt{\frac{k}{m}}$. The minimum and maximum radial distances are given by the roots of the equation $V_{\text{eff}}(r) = \frac{1}{2}kr^2 + \frac{L^2}{2mr^2} = E$. In the present case,

$$\begin{aligned}r_{\min}^2 &= \frac{E}{k} - \sqrt{\frac{E^2}{k^2} - \frac{L^2}{mk}}\\ &= a^2\left(\frac{3}{2}+\cos\alpha\right) - a^2\sqrt{\left(\frac{3}{2}+\cos\alpha\right)^2 - (1+\cos\alpha)^2}\\ &= a^2\left[\left(\frac{3}{2}+\cos\alpha\right) - \sqrt{\frac{5}{4}+\cos\alpha}\right],\\ r_{\max}^2 &= \frac{E}{k} + \sqrt{\frac{E^2}{k^2} - \frac{L^2}{mk}} = a^2\left[\left(\frac{3}{2}+\cos\alpha\right) + \sqrt{\frac{5}{4}+\cos\alpha}\right].\end{aligned}$$

With $\alpha = 0$, we obtain from the above expressions the values $r_{\min} = a$ and $r_{\max} = 2a$ — this corresponds to an elliptical orbit with the semi-minor axis a and the semi-major axis $2a$. On the other hand, with $\alpha = \pi$, we find the values $r_{\min} = 0$ and $r_{\max} = a$ (with $L = 0$) — this corresponds to a straight-line oscillatory motion passing through the origin. These two cases are illustrated as below.

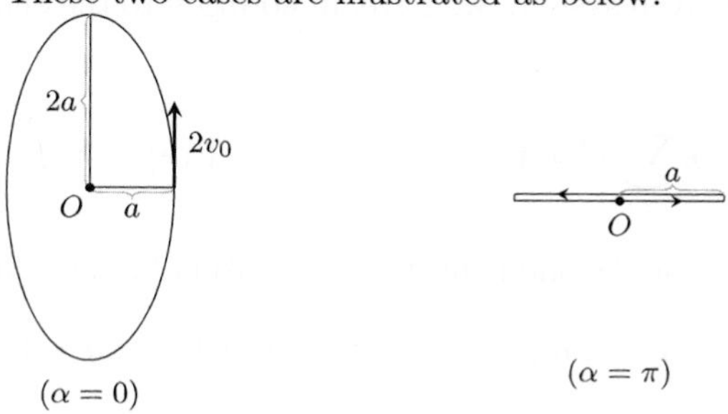

4-13 Here are Lissajous figures for a few values of the phase $\theta_y = 0$, $\pi/4$, $\pi/3$, $\pi/2$:

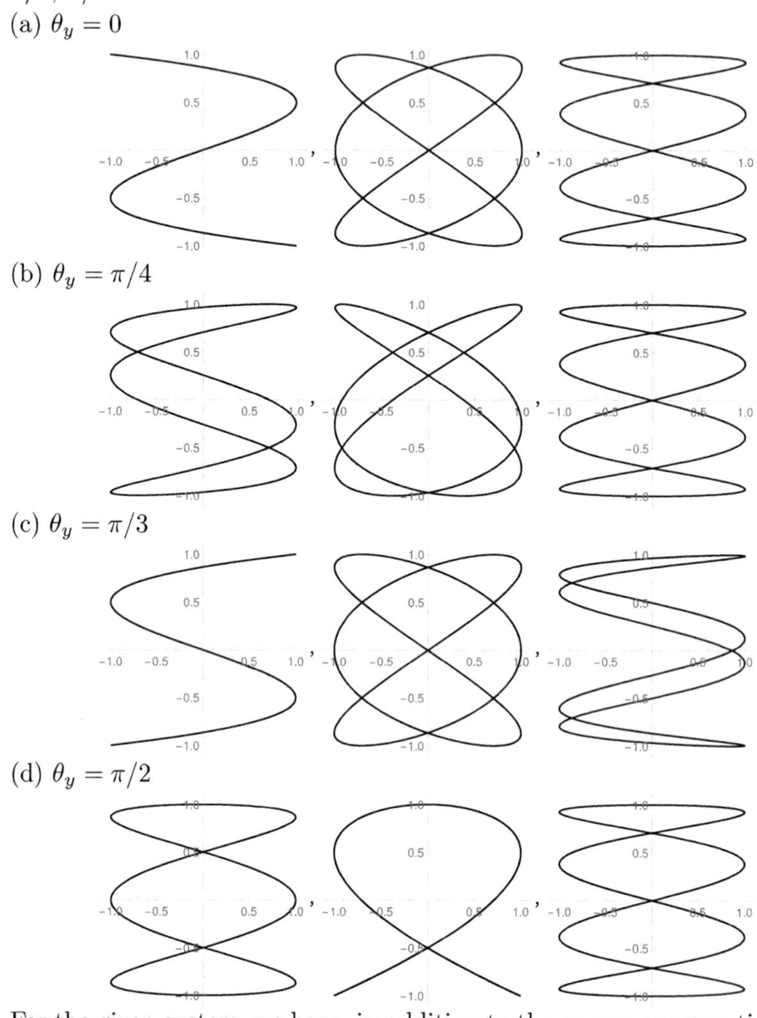

4-14 For the given system, we have, in addition to the energy conservation equation

$$\frac{1}{2}m\sum_{i=1}^{n}\dot{x}_i^2 + V(r) = E \quad \left(\text{with } V(r) = -\int^r f(r')dr'\right), \tag{1}$$

also the angular momentum conservation laws in the form

$$L_{ij} \equiv m(x_i\dot{x}_j - x_j\dot{x}_i) = \text{const.} \quad (i, j = 1, \ldots, n) \tag{2}$$

(note that $L_{ji} = -L_{ij}$), since

$$\frac{d}{dt}L_{ij} = m(x_i\ddot{x}_j - x_j\ddot{x}_i) = \frac{f(r)}{r}(x_ix_i - x_jx_i) = 0.$$

Now consider a particle with the initial conditions

$$\mathbf{x}(t{=}0) = (x_{01}, x_{02}, \ldots, x_{0n}), \quad \dot{\mathbf{x}}(t{=}0) = (v_{01}, v_{02}, \ldots, v_{0n}). \tag{3}$$

We may then choose a new set of Cartesian axes, $(\bar{\mathbf{e}}_1, \ldots, \bar{\mathbf{e}}_n)$, such that $\mathbf{x}(t{=}0)$ and $\dot{\mathbf{x}}(t{=}0)$ may lie on the $\bar{1}\bar{2}$-plane, i.e.,

$$\mathbf{x}(t{=}0) = \bar{x}_{01}\bar{\mathbf{e}}_1 + \bar{x}_{02}\bar{\mathbf{e}}_2, \quad \dot{\mathbf{x}}(t{=}0) = \bar{v}_{01}\bar{\mathbf{e}}_1 + \bar{v}_{02}\bar{\mathbf{e}}_2. \tag{4}$$

The equations of motion, which are clearly rotation-symmetric, would still have the same form, that is

$$m\frac{d^2\bar{x}_i}{dt^2} = f(r)\frac{\bar{x}_i}{r} \quad \left(\text{with } r = \sqrt{\bar{x}_1^2 + \cdots + \bar{x}_n^2}\right), \tag{5}$$

so that we may express the energy and angular momentum conservation equations as

$$\frac{1}{2}m\sum_{i=1}^{n}\dot{\bar{x}}_i^2 + V(r) = E, \tag{6a}$$

$$\bar{L}_{ij} \equiv m(\bar{x}_i\dot{\bar{x}}_j - \bar{x}_j\dot{\bar{x}}_i) = \text{const.} \quad (i, j = 1, \ldots, n). \tag{6b}$$

In terms of the new coordinates, however, the only nonvanishing angular momentum is $\bar{L}_{12}$. Then, for any $k \neq 1$ or 2, one can use the conditions like

$$\begin{aligned}\bar{L}_{1k} &= m(\bar{x}_1\dot{\bar{x}}_k - \bar{x}_k\dot{\bar{x}}_1) = 0,\\ \bar{L}_{2k} &= m(\bar{x}_2\dot{\bar{x}}_k - \bar{x}_k\dot{\bar{x}}_2) = 0\end{aligned} \tag{7}$$

to show that

$$\bar{x}_k(t{=}0) = \left.\frac{d\bar{x}_k}{dt}\right|_{t=0} = \left.\frac{d^2\bar{x}_k}{dt^2}\right|_{t=0} = \left.\frac{d^3\bar{x}_k}{dt^3}\right|_{t=0} \cdots = 0. \tag{8}$$

(For these, differentiate (7) with respect to t successively before setting $t = 0$.) This implies that the particle remains in the $\bar{1}\bar{2}$-plane for all $t \geq 0$. Hence the equations of motion can always be reduced to those of the two-dimensional motion (regardlessly of the spatial dimension n (≥ 2)), with the conservation laws of the form

$$\frac{1}{2}m(\dot{\bar{x}}_1^2 + \dot{\bar{x}}_2^2) + V(r) = E \quad \left(r = \sqrt{\bar{x}_1^2 + \bar{x}_2^2}\right),$$
$$m(\bar{x}_1\dot{\bar{x}}_2 - \bar{x}_2\dot{\bar{x}}_1) = \bar{L}_{12} \quad (\text{: const.}).$$

These systems can be analyzed most conveniently in terms of the polar coordinates introduced in the plane of motion (as we have done explicitly when there are three spatial dimensions).

4-15 (i) Observe that, for $\vec{p} \equiv \frac{m_0\vec{v}}{\sqrt{1-\frac{\vec{v}^2}{c^2}}}$,

$$\vec{v}\cdot\frac{d\vec{p}}{dt} = \vec{v}\cdot\left[\frac{m_0\dot{\vec{v}}}{\sqrt{1-\frac{\vec{v}^2}{c^2}}} + \frac{m_0\vec{v}(\vec{v}\cdot\dot{\vec{v}}/c^2)}{(1-\frac{\vec{v}^2}{c^2})^{3/2}}\right] = \frac{m_0\vec{v}\cdot\dot{\vec{v}}}{(1-\frac{\vec{v}^2}{c^2})^{3/2}}$$

while

$$\frac{d}{dt}\left(\frac{m_0c^2}{\sqrt{1-\frac{\vec{v}^2}{c^2}}}\right) = \frac{m_0\vec{v}\cdot\dot{\vec{v}}}{(1-\frac{\vec{v}^2}{c^2})^{3/2}}.$$

Hence

$$\frac{d}{dt}\left(\frac{m_0c^2}{\sqrt{1-\frac{\vec{v}^2}{c^2}}}\right) = \vec{v}\cdot\frac{d\vec{p}}{dt} = -\vec{v}\cdot\vec{\nabla}V(r) = -\frac{d}{dt}V(r),$$

so that we have the conservation law

$$\frac{m_0c^2}{\sqrt{1-\frac{\vec{v}^2}{c^2}}} + V(r) = E_{\text{rel}} \quad (\text{: const.}). \tag{1}$$

(ii) In the presence of a central force, the (relativistic) angular momentum defined by

$$\vec{L} = \vec{r}\times\vec{p} = m_0\vec{r}\times\frac{\vec{v}}{\sqrt{1-\frac{\vec{v}^2}{c^2}}} \tag{2}$$

is a conserved quantity. In terms of the proper time variable τ defined by $\frac{d\tau}{dt} = \sqrt{1-\frac{\vec{v}^2}{c^2}}$ (see Problem 2-5), we can express the energy and angular momentum conservation laws by the forms

$$m_0 c^2 \frac{dt}{d\tau} = E_{\text{rel}} - V(r) \tag{3a}$$

$$m_0 \vec{r} \times \frac{d\vec{r}}{d\tau} = \vec{L}, \tag{3b}$$

while, from the very definition of the variable τ, we should have

$$\left(\frac{dt}{d\tau}\right)^2 - \frac{1}{c^2}\frac{d\vec{r}}{d\tau} \cdot \frac{d\vec{r}}{d\tau} = 1. \tag{4}$$

Because of the angular momentum conservation law, it would be sufficient to consider a two-dimensional motion restricted to the plane perpendicular to $\vec{L}$. Using polar coordinates (r, θ) in the plane of motion, we can write

$$\frac{d\vec{r}}{d\tau} \cdot \frac{d\vec{r}}{d\tau} = \left(\frac{dr}{d\tau}\right)^2 + r^2 \left(\frac{d\theta}{d\tau}\right)^2 \tag{5}$$

and, from (3b),

$$m_0 r^2 \frac{d\theta}{d\tau} = L \quad (L = \pm|\vec{L}|). \tag{6}$$

Then, substituting (3a), (5) and (6) into (4), we obtain the equation

$$\left(\frac{dr}{d\tau}\right)^2 + \frac{L^2}{m_0 r^2} = \frac{1}{m_0^2 c^2}[E_{\text{rel}} - V(r)]^2 - c^2. \tag{7}$$

(This can be regarded as the generalization of Sec. 4.3 to the relativistic case.)

To obtain the equation for the orbit $r = r(\theta)$, it is convenient to work with the variable $u = \frac{1}{r}$. Now,

$$\frac{dr(\theta)}{d\tau} = \frac{d\theta}{d\tau}\frac{dr}{d\theta} = -r^2 \frac{d\theta}{d\tau}\frac{du}{d\theta} = -\frac{L}{m_0}\frac{du}{d\theta}, \tag{8}$$

and hence (7) can be recast into the form

$$\left(\frac{du}{d\theta}\right)^2 + u^2 = \frac{1}{L^2 c^2}\left[E_{\text{rel}} - V\left(\frac{1}{u}\right)\right]^2 - \frac{m_0^2 c^2}{L^2}. \tag{9}$$

This is the desired relativistic orbit equation.

4-16 Kepler's first law is equivalent to the statement that, as the Sun's position is taken as the coordinate origin, planetary orbits are described by the equation of the form

$$r = \frac{\alpha}{1 + e\cos(\theta - \theta_0)}.$$

For these orbits, we have $u = \frac{1}{r} = \frac{1}{\alpha}[1 + e\cos(\theta - \theta_0)]$ and so

$$\begin{aligned}\left(\frac{du}{d\theta}\right)^2 + u^2 &= \frac{1}{\alpha^2}\{e^2\sin^2(\theta - \theta_0) + [1 + e\cos(\theta - \theta_0)]^2\} \\ &= \frac{1}{\alpha^2}\{1 + e^2 + 2e\cos(\theta - \theta_0)\}.\end{aligned}$$

According to the orbit equation given, this must be equal to $\frac{2m}{L^2}[E - V(\frac{1}{u})]$, when $u = \frac{1}{\alpha}[1 + e\cos(\theta - \theta_0)]$, i.e., $e\cos(\theta - \theta_0) = \alpha u - 1$. From this consideration,

$$\frac{2m}{L^2}\left[E - V\left(\frac{1}{u}\right)\right] = \frac{1}{\alpha^2}\{-1 + e^2 + 2\alpha u\}.$$

This can be true only when $V(\frac{1}{u}) = -\frac{L^2}{m\alpha}u$ or $V(r) = -\frac{L^2}{m\alpha}\frac{1}{r}$ (i.e., an inverse square force).

4-17 Let r_p (r_a) denote the distance to the perihelion (aphelion). Then, from the given data for the Pluto, we have

$$\begin{aligned} r_p &= (1 - e)a = 4.425 \times 10^{12}\,\text{m}, \\ r_a &= (1 + e)a = 7.375 \times 10^{12}\,\text{m}, \end{aligned}$$

so that

$$a \text{ (: the semi-major axis of the orbit)} = \frac{r_a + r_p}{2} = 5.9 \times 10^{12}\,\text{m} \quad (\approx 39.5\,\text{AU})$$

$$e \text{ (: the eccentricity)} = \frac{r_a - r_p}{r_a + r_p} = 0.25.$$

Using Kepler's 3rd law, we can then estimate the length of Pluto's planetary year, T (in Earth's year), to be

$$T \approx (39.5)^{3/2} \approx 249\,\text{years}.$$

4-18 (i) To calculate the shaded area S we write the equation of the Kepler elliptic orbit in the form

$$\frac{1}{r} = c(1 + e\cos\theta) \quad \text{with} \quad c = \frac{1}{a(1-e^2)}, \tag{1}$$

where r and θ are the polar coordinates of an orbital point with the Sun as the origin and OW as the reference line, e is the orbital eccentricity, and a is the semi-major axis. Then, because of the symmetry about the line OW, we have

$$S = 2\int_0^{\pi/2} \frac{1}{2} r^2 d\theta = \frac{1}{c^2}\int_0^{\pi/2} \frac{d\theta}{(1+e\cos\theta)^2}. \tag{2}$$

The last integral can be evaluated with the help of following relations:

$$\int \frac{d\theta}{(1+e\cos\theta)^2} = -\frac{e}{1-e^2}\frac{\sin\theta}{1+e\cos\theta} + \frac{1}{1-e^2}\int \frac{d\theta}{1+e\cos\theta},$$

$$\int \frac{d\theta}{1+e\cos\theta} = \frac{2}{\sqrt{1-e^2}}\tan^{-1}\left[\frac{\sqrt{1-e^2}\tan(\theta/2)}{1+e}\right]. \tag{3}$$

Using (3) to evaluate (2), we get

$$\begin{aligned} S &= \frac{1}{c^2(1-e^2)}\left(-e + \frac{2}{\sqrt{1-e^2}}\tan^{-1}\sqrt{\frac{1-e}{1+e}}\right), \\ &= a^2(1-e^2)\left(-e + \frac{\cos^{-1} e}{\sqrt{1-e^2}}\right), \end{aligned}$$

where the simplification effected in the second line follows from using (1) and a trigonometric identity.

(ii) The total orbital area is $A = \pi a^2\sqrt{1-e^2}$. Putting this and the above expression for S into the equation $S/A = 179/365$, we obtain

$$\cos^{-1} e - e\sqrt{1-e^2} = \frac{179}{365}\pi, \tag{4}$$

which is the transcendental equation for the eccentricity. For $e \ll 1$ (as is the case for the Earth's orbit), we may use the following approximations:

$$\cos^{-1} e = \pi/2 - e - e^3/6 + O(e^5),$$
$$e\sqrt{1-e^2} = e - e^3/2 + O(e^5). \tag{5}$$

Using (5) in (4) and ignoring terms of $O(e^5)$ and higher, we find the following cubic equation for the eccentricity:

$$\frac{1}{3}e^3 - 2e + \frac{7}{730}\pi = 0.$$

This can readily be solved by Cardano's formula. Only one of the roots is less than unity and it has the value 0.0151, which is about 10% off the true value of 0.0167.
[Remark: The source for this problem is P. K. Aravind, *Am. J. Phys.* **55**, 1144 (1987). We also remark that the true perihelion passage occurs on 3 January and not on the winter solstice (as we assumed above. But, this does not introduce a serious discrepancy because of the smallness of the Earth's eccentricity.]

4-19 The radial coordinate of a planet (with mass m), $r = r(t)$, should satisfy the radial energy equation

$$\frac{1}{2}m\dot{r}^2 + \frac{L^2}{2mr^2} - \frac{|k|}{r} = E \quad (|k| = GMm;\ M \text{ is the Sun's mass}) \tag{1}$$

and accordingly also the radial equation of motion

$$m\ddot{r} - \frac{L^2}{mr^3} = -\frac{|k|}{r^2}. \tag{2}$$

For a bounded orbit (which is an ellipse), we have $E = -\frac{|k|}{2a}$ if a is the semi-major axis of the orbit. Now, after multiplying both sides of (2) by r and then integrating over t for the full period, we find from the left hand side

$$\int_0^T \left(mr\ddot{r} - \frac{L^2}{mr^2}\right) dt = -\int_0^T \left(m\dot{r}^2 + \frac{L^2}{mr^2}\right) dt$$
$$= -2ET - 2|k| \int_0^T \frac{dt}{r(t)},$$

where we performed integration by parts and used (1). On the other hand, from the right hand side of the equation, this must equal

$-|k|\int_0^T \frac{dt}{r(t)}$. Therefore,

$$-2ET - 2|k|\int_0^T \frac{dt}{r(t)} = -|k|\int_0^T \frac{dt}{r(t)}$$

or

$$\int_0^T \frac{dt}{r(t)} = -\frac{2E}{|k|}T = \frac{T}{a}. \tag{3}$$

Based on (3), the time-averaged potential energy becomes

$$\overline{V} \equiv \frac{1}{T}\int_0^T \left(-\frac{|k|}{r}\right) dt = 2E;$$

while we have, for the time-averaged kinetic energy,

$$\overline{T} = E - \overline{V} = -E = |E|.$$

Hence, $\overline{V} = -2\overline{T}$, for a bounded Kepler orbit.
[Remark: This corresponds to a specialized application of a more general observation called the virial theorem.]

4-20 (i) Using polar coordinates, the Kepler elliptic orbit with the pericenter at $\theta = 0$ is represented by

$$r = \frac{L^2/m|k|}{1 + e\cos\theta} \quad \left(k = -GmM < 0;\ e = \sqrt{1 + \frac{2EL^2}{mk^2}}\right). \tag{1}$$

In our problem, L and E are given as (see the figure)

$$L = mVR\cos\alpha,$$
$$E = \frac{1}{2}mV^2 - \frac{GmM}{R} = \frac{1}{2}m(V^2 - 2gR)$$

(here $g \equiv \frac{GM}{R^2}$), while we have

$$e = \sqrt{1 + \frac{V^2}{g^2R^2}\cos^2\alpha(V^2 - 2gR)}$$
$$= \sqrt{1 - C^2(2 - C^2)\cos^2\alpha},$$

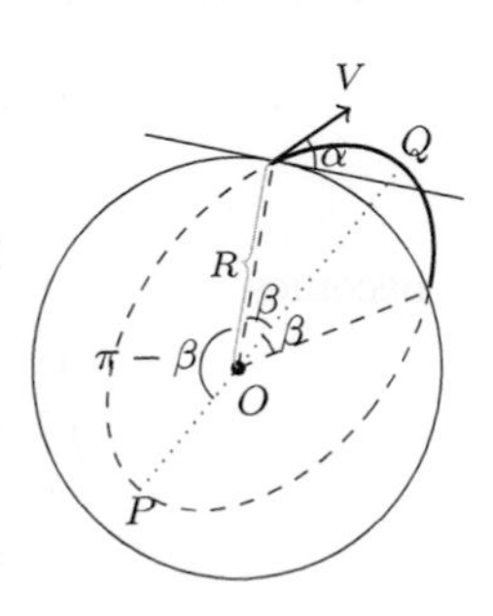

where we have set $\frac{V^2}{gR} \equiv C^2$. Also the point Q in the figure, where $\frac{dr}{d\theta} = 0$, must correspond to the apocenter of the orbit (since the distance $\overline{OQ}$ is supposed to be larger than R while we had $r = R$ initially): hence, the pericenter is at the point P of the figure.

At $\theta = \pi \pm \beta$, the projectile touches the surface of the Earth and thus $r = R$. Inserting this into the orbit equation (1), we have

$$R = \frac{m^2V^2R^2\cos^2\alpha/m^2GM}{1 - e\cos\beta}.$$

This gives us

$$e\cos\beta = 1 - C^2\cos^2\alpha,$$

and so

$$\begin{aligned} e\sin\beta &= \sqrt{e^2 - (1 - C^2\cos^2\alpha)^2} \\ &= \sqrt{1 - C^2(2 - C^2)\cos^2\alpha - (1 - C^2\cos^2\alpha)^2} \\ &= C^2\sin\alpha\cos\alpha. \end{aligned}$$

Hence the angular range 2β is determined through $\tan\beta = \frac{\frac{1}{2}C^2\sin 2\alpha}{1 - C^2\cos^2\alpha}$.

(ii) If $C \ll 1$, we have $\tan\beta \approx \frac{1}{2}C^2\sin 2\alpha \approx \beta$. In this limit, we thus have the approximation

$$\begin{aligned} (\text{horizontal distance}) &\approx R\cdot 2\beta \approx RC^2\sin 2\alpha \\ &= \frac{V^2}{g}\sin 2\alpha, \end{aligned}$$

which is the familiar result from the flat-Earth-constant-gravity approach.

4-21 In the limit $e \to 1-$ with the elliptic orbit given, we note that the quantity (the energy $E \to 0-$)

$$\frac{x + ea}{a} = \frac{2|E|}{|k|}\left(x + \sqrt{1 - \frac{2|E|L^2}{mk^2}}\frac{|k|}{2|E|}\right)$$

becomes

$$\begin{aligned} &\frac{2|E|}{|k|}\left(x + \frac{|k|}{2|E|}\left[1 - \frac{|E|L^2}{mk^2}\right] + O(|E|)\right) \\ &= 1 + \frac{2|E|}{|k|}\left[x - \frac{L^2}{2m|k|}\right] + O(|E|^2). \end{aligned}$$

Hence the elliptic orbit equation smoothly goes over to

$$\not{1} + \frac{4|E|}{|k|}\left[x - \frac{L^2}{2m|k|}\right] + \frac{y^2}{b^2} = \not{1},$$

or, using $b^2 = L^2/(2m|E|)$, to the form (corresponding to a parabolic orbit)

$$x - \frac{L^2}{2m|k|} = -\frac{m|k|}{2L^2}y^2.$$

4-22 Before we answer specific questions in this problem, some general discussions on basic Kepler orbit theory should be useful. As the related potential is $-\frac{k}{r} = -ku$, we here have the orbit equation for $u = \frac{1}{r}$

$$\frac{L^2}{2m}\left(\frac{du}{d\theta}\right)^2 + V_{\text{eff}}\left(\frac{1}{u}\right) = E \quad \left(V_{\text{eff}}\left(\frac{1}{u}\right) = \frac{L^2}{2m}u^2 - ku\right), \tag{1}$$

where the constants E, L are given from the initial configuration by

$$E = -\frac{k}{r_0} + \frac{1}{2}mV^2, \quad L = mVr_0. \tag{2}$$

Then, casting (1) into the form

$$\left(\frac{du}{d\theta}\right)^2 + \left(u - \frac{mk}{L^2}\right)^2 = \frac{2mE}{L^2} + \left(\frac{mk}{L^2}\right)^2,$$

we can immediately recognize the solution:

$$u = \frac{mk}{L^2} + \sqrt{\frac{2mE}{L^2} + \left(\frac{mk}{L^2}\right)^2}\cos(\theta - \theta_0)$$

or

$$r = \frac{(L^2/mk)}{1 + e\cos(\theta - \theta_0)} \quad \left(e \equiv \sqrt{1 + \frac{2EL^2}{mk^2}} < 1\right). \tag{3}$$

Here note that the given energy E (see (2) above) is equal to $V_{\text{eff}}(u = 1/r_0)$ and hence, at our initial point (taken to be the angle $\theta = 0$), $\frac{du}{d\theta} = 0$ or $\frac{dr}{d\theta} = 0$ because of (1). This implies that the given initial point corresponds to the pericenter or apocenter of the orbit.

In accordance with this observation, we may choose θ_0 in (3) to be 0 or π — this gives rise to the orbit equation of the form

$$r = \frac{(L^2/mk)}{1+e\cos\theta} \quad \text{(for an initial point corresponding to the pericenter of the orbit)},$$
$$r = \frac{(L^2/mk)}{1-e\cos\theta} \quad \text{(for an initial point corresponding to the apocenter of the orbit)}.$$

We are now ready to answer specific questions in the problem.

(i) From $E = -\frac{k}{r_0} + \frac{1}{2}mV_e^2 = 0$, we find $V_e = \sqrt{\frac{2k}{mr_0}}$ and $e = 1$. The related orbit is thus given by $r = \frac{(L^2/ml)}{1+\cos\theta}$, which corresponds to a parabola. (Note that the initial point in this case cannot be the apocenter.) Now, depending on $V \gtrless V_e = \sqrt{\frac{2k}{mr_0}}$, we can say the followings:

a) If $V > V_e$ (and so $E > 0$), we have $e > 1$ and the corresponding orbit $r = \frac{(L^2/mk)}{1+e\cos\theta}$ is a hyperbola.

b) If $V < V_e$ (and so $E < 0$), we have $e < 1$ and the corresponding orbit $r = \frac{(L^2/mk)}{1+e\cos\theta}$ is a bounded closed orbit, i.e., an *ellipse with the given initial point as its pericenter.*

Schematic plots of related orbits are given after (ii).

(ii) For a circular orbit (i.e., $r = r_0 = \frac{L^2}{mk}$), we should have $e = 0$; then, $E = -\frac{mk^2}{2L^2}$ and this demands that

$$-\frac{k}{r_0} + \frac{1}{2}mV_c^2 = -\frac{mk^2}{2m^2V_c^2r_0^2} \quad \rightarrow \quad \left(mV_c^2 - \frac{k}{r_0}\right)^2 = 0,$$

i.e., $V_c = \sqrt{\frac{k}{mr_0}}$ (and the energy value $E = -\frac{k}{2r_0}$). Now, depending on $V \gtrless V_c = \sqrt{\frac{k}{mr_0}}$, we can say the followings:

a) If $V > V_c$, we have $0 < e$ (< 1), appropriate to an elliptical orbit with $-\frac{k}{2r_0} < E < 0$. Since we have $|E| = \frac{k}{2a}$ if a is the semi-major axis, we infer that $a > r_0$ and so the initial point in this case corresponds to the pericenter of the orbit. The orbit is described by the equation $r = \frac{(L^2/mk)}{1+e\cos\theta}$.

b) If $V < V_c$, we have $e > 0$, an ellipse but with $E < -\frac{k}{2r_0}$. Here, since $|E| = \frac{k}{2a} > \frac{k}{2r_0}$, we infer that $a < r_0$ and so the initial point in this case corresponds to the apocenter of the orbit. The orbit is described by the equation $r = \frac{(L^2/mk)}{1-e\cos\theta}$.

Schematic plots of corresponding orbits are as follows:

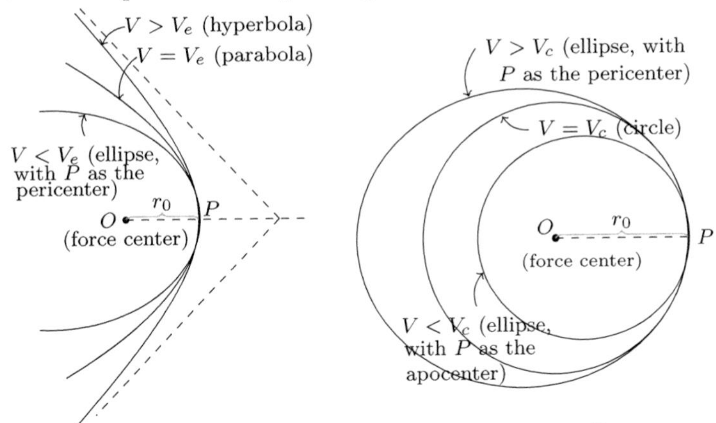

(iii) With a very small initial speed V, e $(= 1 - \frac{mr_0}{k}V^2)$ becomes very closed to 1 from below and the elliptical orbit found in (ii) (b) above can be written as

$$r = \frac{(L^2/mk)}{1 - e\cos\theta} = r_0 \frac{\left(\frac{mr_0}{k}\right)V^2}{(1-\cos\theta) + \frac{mr_0}{k}V^2\cos\theta}.$$

From this form we can see that the ellipse has the apocenter at $(r = r_0,\ \theta = 0)$, and away from this point its orbit can be approximated closely by $r(1-\cos\theta) \approx r_0(\frac{mr_0}{k})V^2 \approx 0$ (i.e., the orbit traces the part of the x-axis closely, and $r \approx 0$ if $\theta \neq 0$). This leads to an orbit resembling the one like

r_0

x

O (force center) apocenter

4-23 We will here utilize the relativistic orbit equation derived in Problem 4-15, i.e.,

$$\left(\frac{du}{d\theta}\right)^2 + u^2 = \frac{1}{L^2c^2}\left[E_{\text{rel}} - V\left(\frac{1}{u}\right)\right]^2 - \frac{m_0^2c^2}{L^2} \tag{1}$$

with $V(\frac{1}{u}) = -\kappa u$. As this can be rearranged as

$$\left(\frac{du}{d\theta}\right)^2 + \left(1 - \frac{\kappa^2}{L^2c^2}\right)\left(u - \frac{\kappa E_{\text{rel}}}{L^2c^2 - \kappa^2}\right)^2 = \frac{L^2(E_{\text{rel}}^2 - m_0^2c^4) + m_0^2c^2\kappa^2}{L^2(L^2c^2 - \kappa^2)}, \tag{2}$$

we can write the solution in the form

$$u = \frac{1}{r} = \frac{\kappa E_{\rm rel}}{L^2c^2 - \kappa^2} + \frac{\sqrt{c^2L^2(E_{\rm rel}^2 - m_0^2c^4) + m_0^2c^4\kappa^2}}{(L^2c^2 - \kappa^2)} \cos\left(\sqrt{1 - \frac{\kappa^2}{L^2c^2}}\theta + \text{const.}\right)$$
$$\equiv \frac{\kappa E_{\rm rel}}{L^2c^2 - \kappa^2}\left[1 + \tilde{e}\cos\left(\sqrt{1 - \frac{\kappa^2}{L^2c^2}}\theta + \text{const.}\right)\right], \quad (3)$$

where we defined $\tilde{e} = \sqrt{\frac{m_0^2c^4\kappa^2 + c^2L^2(E_{\rm rel}^2 - m_0^2c^4)}{\kappa^2E_{\rm rel}^2}}$. [Note that, in the nonrelativistic limit of $E_{\rm nonrel} \equiv E_{\rm rel} - m_0c^2 \ll m_0c^2$, $\tilde{e}$ reduces to the expression $\sqrt{1 + \frac{2E_{\rm nonrel}L^2}{m_0\kappa^2}}$, representing the eccentricity of the Kepler orbit.]

For bounded orbits, we should have $E_{\rm rel} < m_0c^2$ and so $\tilde{e} < 1$. In this case the locus described by (3) corresponds to an ellipse whose axes rotate around the origin. In particular, for a single period of radial oscillation (i.e., from one perihelion to the next perihelion), (3) tells us that the orbital angle θ changes by

$$\Delta\theta = \frac{2\pi}{\sqrt{1 - \frac{\kappa^2}{L^2c^2}}} \equiv \frac{2\pi}{1 - \sigma},$$

leading to the perihelion advance $\sigma \approx \frac{\kappa^2}{2L^2c^2}$ when $\frac{\kappa^2}{L^2c^2} \ll 1$.
[Remark: For a test mass m in the gravitational field due to a massive localized body (having mass M), General relativity predicts the perihelion advance of $\sigma \approx \frac{3\kappa^2}{L^2c^2}$, where $\kappa = GmM$. That this formula (rather than the above special-relativity-based result) can account for the observed perihelion advance of Mercury indicates that $\frac{d}{dt}\left(\frac{m_0\vec{r}}{\sqrt{1-\vec{v}^2/c^2}}\right) = -\frac{\kappa}{r^3}\vec{r}$ is a *wrong* relativistic generalization as far as gravity is concerned. It can still be a valid generalization for the dynamics of a charged particle in an electrostatic field.]

4-24 (i) The effective radial potential for this system is

$$V_{\rm eff}(r) = \frac{L^2}{2mr^2} - \frac{GMm}{r} + \frac{1}{2}kr^2,$$

and the radial equation of motion is

$$m\ddot{r} = -\frac{d}{dr}V_{\rm eff}(r) = \frac{L^2}{mr^3} - \frac{GMm}{r^2} - kr. \quad (1)$$

When the planet makes a circular orbit, we have $\ddot{r} = 0$ and the radius r_0 is determined by

$$\frac{L^2}{mr_0^3} - \frac{GMm}{r_0^2} - kr_0 = 0. \tag{2}$$

(ii) Let ξ express a small radial deviation around r_0, i.e., $\xi = r - r_0$. For $\xi \ll r_0$, (1) becomes

$$\begin{aligned} m\ddot{\xi} &= \frac{L^2}{m(r_0+\xi)^3} - \frac{GMm}{(r_0+\xi)^2} - k(r_0+\xi) \\ &\approx \frac{L^2}{mr_0^3}\left(1 - \frac{3\xi}{r_0}\right) - \frac{GMm}{r_0^2}\left(1 - \frac{2\xi}{r_0}\right) - kr_0\left(1 + \frac{\xi}{r_0}\right). \end{aligned}$$

Using (2), we can rewrite this as

$$\begin{aligned} m\ddot{\xi} &= -\xi\left[\frac{3L^2}{mr_0^4} - \frac{2GMm}{r_0^3} + k\right] \\ &= -\xi\left[\frac{L^2}{mr_0^4} + \frac{2}{r_0}\left(\frac{L^2}{mr_0^3} - \frac{GMm}{r_0^2}\right) + k\right] \\ &= -\xi\left[\frac{L^2}{mr_0^4} + 3k\right], \end{aligned}$$

or

$$\ddot{\xi} = -\left[\frac{L^2}{m^2r_0^4} + 3\frac{k}{m}\right]\xi.$$

This corresponds to the equation for a harmonic oscillator with angular frequency

$$\omega_r = \sqrt{\frac{L^2}{m^2r_0^4} + 3\frac{k}{m}}.$$

As the radial frequency ω_r is slightly larger than the azimuthal frequency $\omega_0 = \frac{L}{mr_0^2}$, the orbit is a precessing ellipse. [Note that, to first order in k, the azimuthal frequency is not affected by the presence of dust.] The precessional frequency is

$$\begin{aligned} \omega_p = \omega_r - \omega_0 &= \sqrt{\frac{L^2}{m^2r_0^4} + 3\frac{k}{m}} - \frac{L}{mr_0^2} \\ &\approx \frac{3}{2}\frac{kr_0^2}{L}. \end{aligned}$$

(iii) Since the radial oscillation is faster than the orbital revolution, the axis of the ellipse precesses in the direction opposite to the orbital angular velocity.

4-25 The time of flight for the spaceship by the Hohmann orbit will be $T/2$, where

$$T^2 = \frac{4\pi^2}{GM_\odot} a^3 = \frac{\pi^2}{2GM_\odot}(r_1 + r_2)^3.$$

On the other hand, for the destination planet (with mass m) moving in a circular orbit, we know that its radius r_2 should be equal to $\frac{L^2}{m(GmM_\odot)}$ if L denotes the related angular momentum, that is

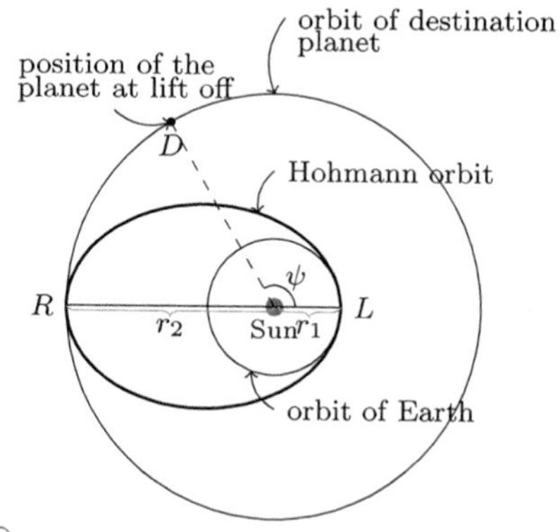

$$mr_2^2\dot{\theta} = L.$$

From this we can readily determine the angular velocity $\dot{\theta}$ of the destination planet:

$$\dot{\theta}^2 = \frac{L^2}{m^2 r_2^4} = \frac{Gm^2 M_\odot r_2}{m^2 r_2^4} = \frac{GM_\odot}{r_2^3}.$$

In the time $\frac{T}{2}$, the destination planet will have travelled through an angle $\dot{\theta}\frac{T}{2} = \sqrt{\frac{GM_\odot}{r_2^3}}\frac{1}{2}\sqrt{\frac{\pi^2(r_1+r_2)^3}{2GM_\odot}}$. This angle should equal $\pi - \psi$ and hence, for the angular separation, we obtain

$$\psi = \pi - \frac{\pi}{2\sqrt{2}}\left(1 + \frac{r_1}{r_2}\right)^{3/2}.$$

4-26 (i) Scattering occurs only when the impact parameter b satisfies the condition $b \le a$. Schematically, we can represent the scattering trajectory as in the figure. Then, the velocity inside the potential well should be given by

$$v_1 = \sqrt{v_0^2 + \frac{2}{m}V_0}, \quad (1)$$

due to the conservation of energy, $\frac{1}{2}mv_1^2 - V_0 = \frac{1}{2}mv_0^2$. Because of the angular momentum conservation (valid with any central potential), it must also be the case that

$$v_0 \sin\alpha = v_1 \sin\beta,$$

and therefore we may set

$$\frac{\sin\alpha}{\sin\beta} = n \quad \left(\text{with } n = \frac{v_1}{v_0} = \sqrt{1 + \frac{2V_0}{mv_0^2}} > 1\right). \tag{2}$$

From the figure, the deflection angle can now be identified as

$$\Theta = 2(\alpha - \beta). \tag{3}$$

Clearly, $\sin\alpha = \frac{b}{a}$ and, from (3),

$$\begin{aligned}\sin\beta &= \sin\left(\alpha - \frac{\Theta}{2}\right) \\ &= \sin\alpha\cos\frac{\Theta}{2} - \cos\alpha\sin\frac{\Theta}{2},\end{aligned}$$

so that we have (see (2))

$$\frac{1}{n} = \cos\frac{\Theta}{2} - \cot\alpha\sin\frac{\Theta}{2}.$$

This implies that the angle α is related to the deflection angle Θ by

$$\cot\alpha = \frac{\cos\frac{\Theta}{2} - \frac{1}{n}}{\sin\frac{\Theta}{2}}. \tag{5}$$

Using this with $1 + \cot^2\alpha = \frac{1}{\sin^2\alpha} = \frac{a^2}{b^2}$, we obtain

$$b^2 = \frac{a^2}{1 + \left(\frac{\cos\frac{\Theta}{2} - \frac{1}{n}}{\sin\frac{\Theta}{2}}\right)^2} = \frac{a^2\sin^2\frac{\Theta}{2}}{1 - \frac{2}{n}\cos\frac{\Theta}{2} + \frac{1}{n^2}}.$$

Accordingly, replacing Θ by the scattering angle θ, we have

$$b(\theta) = a\frac{n\sin\frac{\theta}{2}}{(n^2 - 2n\cos\frac{\theta}{2} + 1)^{1/2}} \quad \left(n = \sqrt{1 + \frac{2V_0}{mv_0^2}}\right). \tag{6}$$

The angle θ here can take the value from zero (for $b = 0$) up to the value $\theta_{\max}$ (for $b = a$) given by

$$\cos\frac{\theta_{\max}}{2} = \frac{1}{n}, \tag{7}$$

from the requirement that the right hand side of (6) should not exceed a (or from the condition that $\cot\alpha > 0$ with (5)).

(ii) To get the differential cross section, we calculate $\frac{db}{d\theta}$ for $b = b(\theta)$ given in (6):

$$\begin{aligned}\frac{db}{d\theta} &= an\left\{\frac{\frac{1}{2}\cos\frac{\theta}{2}}{(n^2 - 2n\cos\frac{\theta}{2} + 1)^{1/2}} - \frac{\frac{1}{2}n\sin^2\frac{\theta}{2}}{(n^2 - 2n\cos\frac{\theta}{2} + 1)^{3/2}}\right\}\\ &= \frac{an}{2}\frac{n^2\cos\frac{\theta}{2} - n\cos^2\frac{\theta}{2} + \cos\frac{\theta}{2} - n}{(n^2 - 2n\cos\frac{\theta}{2} + 1)^{3/2}}\\ &= \frac{an}{2}\frac{(n\cos\frac{\theta}{2} - 1)(n - \cos\frac{\theta}{2})}{(n^2 - 2n\cos\frac{\theta}{2} + 1)^{3/2}}.\end{aligned} \tag{8}$$

Now, from (6) and (8), we obtain the differential cross section

$$\begin{aligned}\frac{d\sigma}{d\Omega} &= \frac{b(\theta)}{\sin\theta}\left|\frac{db(\theta)}{d\theta}\right|\\ &= \begin{cases}\frac{a^2n^2}{4}\frac{1}{\cos\frac{\theta}{2}}\frac{(n\cos\frac{\theta}{2}-1)(n-\cos\frac{\theta}{2})}{(n^2-2n\cos\frac{\theta}{2}+1)^2}, & \text{for } \theta \le \theta_{\max} = 2\cos^{-1}\frac{1}{n}\\ 0, & \text{for } \theta > \theta_{\max}.\end{cases}\end{aligned} \tag{9}$$

Based on (9), the total cross section is

$$\begin{aligned}\sigma_{\text{tot}} &= \frac{a^2n^2}{4}\int_0^{\theta_{\max}}\frac{1}{\cos\frac{\theta}{2}}\frac{(n\cos\frac{\theta}{2} - 1)(n - \cos\frac{\theta}{2})}{(n^2 - 2n\cos\frac{\theta}{2} + 1)^2}2\pi\sin\theta d\theta\\ &= \pi a^2n^2\int_0^{\theta_{\max}}\frac{(n\cos\frac{\theta}{2} - 1)(n - \cos\frac{\theta}{2})}{(n^2 - 2n\cos\frac{\theta}{2} + 1)^2}\sin\frac{\theta}{2}d\theta\\ &= 2\pi a^2n^2\int_{\frac{1}{n}}^1\frac{(nu - 1)(n - u)}{(n^2 + 1 - 2nu)^2}du,\end{aligned}$$

where, in the last expression, we made the substitution $\cos\frac{\theta}{2} = u$. Setting $n^2 + 1 - 2nu = -x$ so that $u = \frac{x+n^2+1}{2n}$, this integral becomes

$$
\begin{aligned}
\sigma_{\text{tot}} &= \pi a^2 \cancel{n} \int_{1-n^2}^{-(n-1)^2} \frac{(x+n^2-1)(n^2-x-1)}{4\cancel{n}x^2}\,dx \\
&= \frac{\pi a^2}{4}\Bigg[(n-1)^2 + (1-n^2) + (n^2-1)^2 \\
&\quad \times \left\{\frac{1}{(n-1)^2} + \frac{1}{1-n^2}\right\}\Bigg] \\
&= \pi a^2,
\end{aligned}
$$

thus confirming that σ_{tot} in this problem is equal to the geometrical cross section πa^2.

4-27 (i) We will first find the formula relating the scattering angle θ to the impact parameter b. With the angular momentum given by $L = mv_0 b$, the effective one-dimensional radial potential becomes

$$
V_{\text{eff}}(r) = \begin{cases} \frac{(mbv_0)^2}{2mr^2}, & r > a \\ V_0 + \frac{(mbv_0)^2}{2mr^2}, & r < a \end{cases}.
$$

If the impact parameter b is larger than a, the particle will see only the centrifugal barrier and move along a straight line, i.e., no scattering if $b > a$. On the other hand, if $b < a$, the particle may penetrate the barrier at $r = a$ if its energy $E = \frac{1}{2}mv_0^2$ (assumed to be larger than V_0) happens to exceed $V_0 + \frac{1}{2}m(\frac{b}{a})^2v_0^2$, where the term $\frac{1}{2}m(\frac{b}{a})^2v_0^2$ represents the centrifugal barrier; if E is less than $V_0 + \frac{1}{2}m(\frac{b}{a})^2v_0^2$, the particle is reflected from the spherical barrier at $r = a$ instead (see the figure, where respective cases are shown). Translating this as a statement involving the magnitude of the impact parameter b (for given energy $E = \frac{1}{2}mv_0^2$), we may say that

(a) if $b \le a\sqrt{1 - \frac{V_0}{E}}$, the particle can enter the region $r < a$, and

(b) if $a\sqrt{1 - \frac{V_0}{E}} < b \le a$, the particle gets reflected at the surface $r = a$ (as in a hard-sphere scattering).

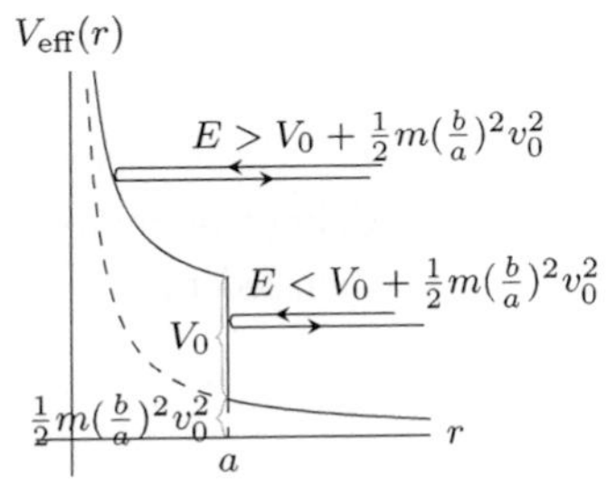

Let us denote $\sqrt{1-\frac{V_0}{E}} \equiv n\ (<1)$. Then, if $na < b \le a$ (the case (b) above), the result for a hard-sphere scattering (see the figure) may be used to write

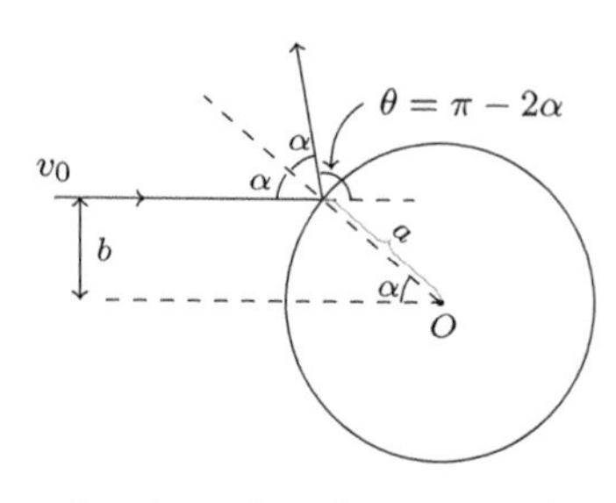

$$b(\theta) = a\sin\alpha = a\cos\frac{\theta}{2}, \quad (1)$$

where the scattering angle, θ, can take the value from zero (for $b = a$) up to the value $\theta_{\max} = 2\cos^{-1} n$ (for $b \to n\,a+$). For $b \le an$, the scattering trajectory can be drawn schematically as in the figure below. In this case we find from the angular momentum conservation

$$v_0 \sin\alpha = v_1 \sin\beta, \quad (2)$$

where $v_1 = \sqrt{\frac{2}{m}(E - V_0)}$. This means that

$$\frac{\sin\alpha}{\sin\beta} = \frac{v}{v_0} = n. \quad (3)$$

From the figure, we can relate the scattering angle to $\theta = 2(\beta - \alpha)$. Now, to find the relation $b = b(\theta)$, note that

$$\begin{aligned} \sin\beta &= \sin\left(\frac{\theta}{2} + \alpha\right) \\ &= \sin\alpha\cos\frac{\theta}{2} + \cos\alpha\sin\frac{\theta}{2}. \end{aligned} \quad (4)$$

Form (3) and (4), we thus obtain

$$\frac{1}{n} = \cos\frac{\theta}{2} + \cot\alpha \sin\frac{\theta}{2}$$

or

$$\cot\alpha = \frac{\frac{1}{n} - \cos\frac{\theta}{2}}{\sin\frac{\theta}{2}}. \quad (5)$$

Using this relation with $1 + \cot^2\alpha = \frac{1}{\sin^2\alpha}$ and $\sin\alpha = \frac{b}{a}$, we are then led to

$$b^2 = \frac{a^2\sin^2\frac{\theta}{2}}{1 - \frac{2}{n}\cos\frac{\theta}{2} + \frac{1}{n^2}}. \quad (6)$$

Accordingly, we obtain the formula

$$b(\theta) = \frac{an \sin\frac{\theta}{2}}{(n^2 - 2n\cos\frac{\theta}{2} + 1)^{1/2}} \qquad \left(n \equiv \sqrt{1 - \frac{V_0}{E}} < 1\right) \tag{7}$$

where the angle θ here can take again the value from zero (for $b = 0$) up to the value $\theta_{\max} = 2\cos^{-1} n$ (for $b = an$).

We see that scattering occurs entirely into the cone $\theta \le \theta_{\max} = 2\cos^{-1} n$, and there are *two* possible impact parameter values — the values given by (1) and (7) — for each given scattering angle $\theta \in (0, \theta_{\max})$.

(ii) From the above analysis, the related differential cross section can be found by *adding* the contribution obtained using the formula (1) to that obtained using the formula (7). Then, after some calculations (which are very similar to the ones given in Problem 4-26 (ii)), we obtain the expression

$$\frac{d\sigma}{d\Omega} = \begin{cases} \frac{a^2n^2}{4}\frac{1}{\cos\frac{\theta}{2}}\frac{(n\cos\frac{\theta}{2}-1)(n-\cos\frac{\theta}{2})}{(1+n^2-2n\cos\frac{\theta}{2})^2} + \frac{1}{4}a^2, & \text{for } \theta < \theta_{\max} = 2\cos^{-1} n \\ 0, & \text{for } \theta_{\max} < \theta < \pi, \end{cases}$$

where $n = \sqrt{1 - \frac{V_0}{E}}$. Based on this expression, one can also verify that the total cross section is equal to the geometrical cross section πa^2.

4-28 (i) The equation governing the radial motion is

$$\frac{1}{2}m\dot{r}^2 + V_{\text{eff}}(r) = E \qquad \left(E = \frac{1}{2}mv_0^2 > \frac{k}{a}\right)$$

with the effective radial potential

$$V_{\text{eff}}(r) = \frac{L^2}{2mr^2} + V(r) = \begin{cases} E\left(\frac{b}{r}\right)^2 + \frac{k}{r}, & r > a \\ E\left(\frac{b}{r}\right)^2 + \frac{k}{a}, & r < a \end{cases}$$

where we used $L = mv_0 b$. Only particles incident with the impact parameter b less than a certain value $\bar{b}$ will penetrate into the region $r < a$, the value $\bar{b}$ here being the impact parameter for which the condition $V_{\rm eff}(r{=}a) = E$ holds, i.e.,

$$E\left(\frac{\bar{b}}{a}\right)^2 + \frac{k}{a} = E \quad \longrightarrow \quad \bar{b} = a\sqrt{1 - \frac{1}{\epsilon}},$$

where we defined the dimensionless energy $\epsilon = E/\left(\frac{k}{a}\right)$. This means that, if the particle's impact parameter is larger than $\bar{b}$ (i.e., $s \equiv \frac{b}{\bar{b}} > 1$), the corresponding scattering angle Θ is given strictly by the Rutherford formula (see Sec. 4.4.3 in text)

$$\Theta = \pi - 2\cos^{-1}\frac{1}{\sqrt{1 + \frac{2EL^2}{mk^2}}} = 2\cot^{-1}\sqrt{\frac{2EL^2}{mk^2}}$$
$$= 2\cot^{-1}\left(2s\sqrt{\epsilon^2 - \epsilon}\right) \tag{1}$$

(or $b = \frac{k}{2E}\cot\frac{\Theta}{2}$). For particle with $s > 1$ (and $\epsilon > 1$), the scattering angle Θ cannot exceed the value $\bar{\Theta} = 2\cot^{-1}(2\sqrt{\epsilon^2 - \epsilon})$.

On the other hand, if the particle's impact parameter b is less than $\bar{b}$ (i.e., $s < 1$), we will have the scattering orbit that can be represented schematically as

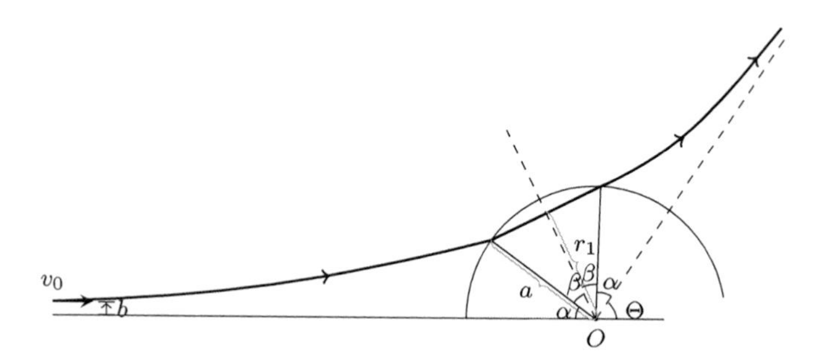

(Note that, in the region $r < a$, this particle moves along a straight line since it is subject to no force.) Here, if r_1 denotes the distance of closest approach to the origin for a particle with the impact parameter $b = s\bar{b}$, it can be found by solving the equation

$$E\left(\frac{b}{r_1}\right)^2 + \frac{k}{a} = E \quad \longrightarrow \quad r_1 = \frac{b}{1 - \frac{1}{\epsilon}} = as.$$

Hence the angle β, shown in the figure, can be identified as $\beta = \cos^{-1}\frac{r_1}{a} = \cos^{-1}s$. To determine the angle α (see the figure), we can utilize the Kepler orbit equation $r = \frac{(\frac{2\epsilon}{a})b^2}{e\cos(\theta-\theta_0)-1}$ where $e = \sqrt{1+\frac{2EL^2}{mk^2}} = \sqrt{1+(\frac{2\epsilon b}{a})^2}$. Evidently, $\alpha = \theta_{r=\infty} - \theta_{r=a} = (\theta_{r=\infty}-\theta_0) - (\theta_{r=a}-\theta_0)$ while we have

$$\theta_{r=\infty} - \theta_0 = \cos^{-1}\frac{1}{e}, \qquad \theta_{r=a} - \theta_0 = \cos^{-1}\left[\frac{1+2\epsilon(\frac{b}{a})^2}{e}\right].$$

This gives, writing $b = \bar{b}s$,

$$\begin{aligned}\alpha &= \cos^{-1}\frac{1}{e} - \cos^{-1}\left(\frac{1+2\epsilon(\frac{b}{a})^2}{e}\right) \\ &= \tan^{-1}\left(2s\sqrt{\epsilon^2-\epsilon}\right) - \tan^{-1}\left(\frac{2(\epsilon-1)s\sqrt{1-s^2}}{1+2(\epsilon-1)s^2}\right). \qquad (2)\end{aligned}$$

From the figure the scattering angle Θ equals $\pi - 2(\alpha+\beta)$, and accordingly we have, when $b \equiv \bar{b}s < \bar{b}$,

$$\begin{aligned}\Theta = \pi - 2\Bigg[&\tan^{-1}\left(2s\sqrt{\epsilon^2-\epsilon}\right) - \tan^{-1}\left(\frac{2(\epsilon-1)s\sqrt{1-s^2}}{1+2(\epsilon-1)s^2}\right) \\ &+ \cos^{-1}s\Bigg]. \qquad (3)\end{aligned}$$

(ii) If the incident energy E were less than $\frac{k}{a}$, the particle could not enter the region $r < a$ and so Rutherford cross section formula $\frac{d\sigma}{d\Omega} = \frac{k^2}{16E^2}\frac{1}{\sin^4\frac{\theta}{2}}$ would remain strictly valid. But, with energy $E = \epsilon\frac{k}{a}$ where $\epsilon > 1$, the Rutherford relation $b = \frac{k}{2E}\cot\frac{\Theta}{2}$ between the scattering angle and the impact parameter cannot be used for a particle incident with the impact parameter less than the value $\bar{b} = a\sqrt{1-1/\epsilon}$; for these particles, the relation (3) holds instead. This means that, as the incident energy E is raised to have the value larger than $\frac{k}{a}$, the scattering cross section tends to show deviations from the Rutherford cross section formula.

Although our formula (3) is rather complicated, it should not be too difficult to guess its general character: as the impact parameter is further reduced from the value $\bar{b}$, the scattering angle Θ does not increase (as would be the case if the Rutherford relation $b = \frac{k}{2E}\cot\frac{\Theta}{2}$ were valid), but decreases (to the zero scattering angle at $b = 0$). The scattering cross section is thus practically zero beyond the angle $\theta = \bar{\Theta} = 2\cot^{-1}(2\sqrt{\epsilon^2 - \epsilon})$, while there will be *two* impact parameter values for the given scattering angle $\theta < \bar{\Theta}$ — one based on $b = \frac{k}{2E}\cot\frac{\theta}{2}$ and the other based on (3). [Accordingly, at angle $\theta < \bar{\Theta}$, the scattering cross section is given by the sum of the cross sections obtained using the formula (3) and the Rutherford cross section.] It is the high-energy large-angle scattering that is most sensitive to possible deviations in the potential at small distance.

4-29 (i) With the potential $V = \frac{k}{2r^2}$, the orbit equation in Sec. 4.3 becomes

$$\left(\frac{du}{d\theta}\right)^2 + \left[1 + \frac{km}{L^2}\right]u^2 = \frac{2m}{L^2}E \quad (\text{: const.}), \tag{1}$$

where $u = 1/r$ and L and E denote the angular momentum and the energy. From the resemblance between (1) and the energy equation for the harmonic oscillator, one may readily recognize that the orbit is described by

$$u(\theta) = A\sin(n\theta + \phi_0) \tag{2}$$

with $n = \sqrt{1 + \frac{km}{L^2}}$, $A = \sqrt{\frac{2mE}{L^2+km}}$ and the constant of integration ϕ_0 to be determined from the initial conditions. From the figure, we have $\theta \to 0$ as $r \to \infty$ (i.e., $u \to 0$); hence, we set $\phi_0 = 0$. Then, (2) can be written in the desired form

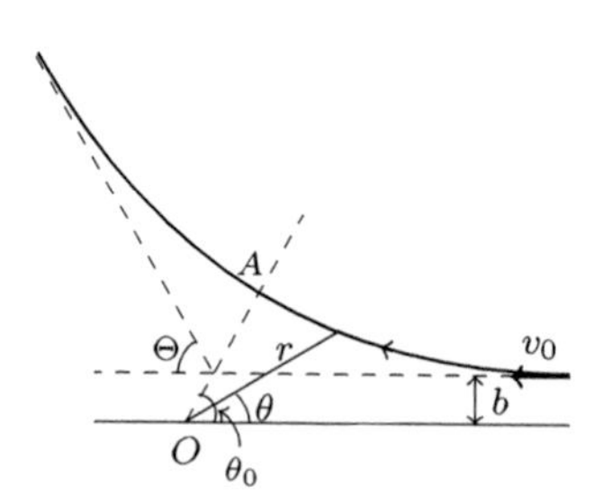

$$r\cos(n[\theta - \theta_0]) = C,$$

where $\theta_0 = \frac{\pi}{2}\frac{1}{n}$ and $C = \frac{1}{A} = \sqrt{\frac{L^2+km}{2mE}}$.

(ii) From the result in (i) it is seen that r has a minimum value when $\theta = \theta_0 = \frac{\pi}{2n}$. This is the distance of closest approach OA (see the figure). Then the scattering angle is determined as (see the figure)

$$\Theta = \left|\pi - 2\theta_0\right| = \pi\left(1 - \frac{1}{n}\right). \tag{3}$$

As we have $E = \frac{1}{2}mv_0^2$ and $L^2 = (mbv_0)^2 = 2b^2mE$, we find from (3)

$$1 - \frac{\Theta}{\pi} = \left(1 + \frac{mk}{L^2}\right)^{-\frac{1}{2}} = \left(1 + \frac{k}{b^2mv_0^2}\right)^{-\frac{1}{2}}$$

or

$$\frac{k}{mb^2v_0^2 + k} = -\left(\frac{\Theta^2}{\pi^2} - \frac{2\Theta}{\pi}\right).$$

This gives

$$b^2 = \frac{k}{mv_0^2}\frac{(\pi - \Theta)^2}{(2\pi - \Theta)\Theta}$$

as the relation between the scattering angle Θ and the impact parameter b. The differential cross section is then given as

$$\frac{d\sigma}{d\Omega} = \left|\frac{b}{\sin\Theta}\frac{db}{d\Theta}\right| = \frac{k}{mv_0^2 \sin\Theta}\frac{\pi^2(\pi - \Theta)}{(2\pi - \Theta)^2\Theta^2}.$$

4-30 (i) The effective radial potential is

$$V_{\text{eff}}(r) = -\frac{\alpha}{r^4} + \frac{L^2}{2mr^2} \quad (\alpha > 0),$$

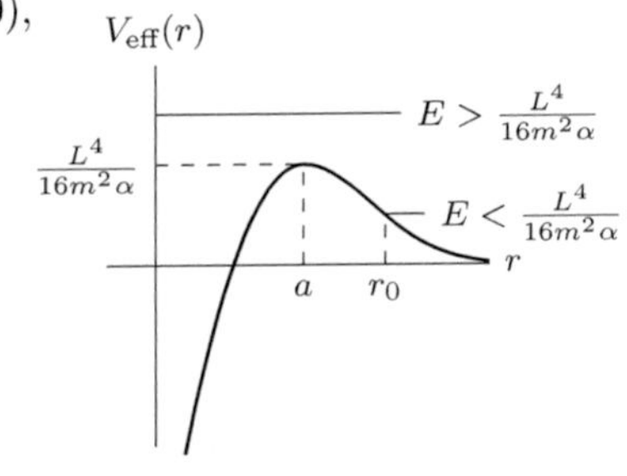

where L denotes the angular momentum. As shown in the figure, V_{eff} attains the maximum value, $\frac{L^4}{16m^2\alpha}$, at $r = \frac{\sqrt{4\alpha m}}{L}$ $(\equiv a)$, assuming $L > 0$. There are two possibilities for the motion, depending on the value of energy E:

(a) When $E < \frac{L^4}{16m^2\alpha}$, there is a minimum distance r_0 (where $V_{\text{eff}}(r_0) = E$) from the potential center that the incident particle can approach. The particle moves toward the center and makes a turn at $r = r_0$.

(b) When $E > \frac{L^4}{16m^2\alpha}$, the particle has sufficient energy to overcome the potential barrier and will fall to the center (i.e., be captured by the potential).

(ii) Suppose that a particle approaches from far with speed v_0 and impact parameter b. Then its kinetic energy is $E = \frac{1}{2}mv_0^2$ and it carries angular momentum $L = mv_0 b$. The particle will fall to the center if the impact parameter b is such that

$$\frac{1}{2}mv_0^2 > \frac{(mv_0 b)^4}{16m^2\alpha} \quad \rightarrow \quad b < b_0 \equiv \left(\frac{8\alpha}{mv_0^2}\right)^{1/4}.$$

Accordingly, the total capture cross section is given by

$$\sigma_{\text{capture}} = \pi b_0^2 = \pi\sqrt{\frac{8\alpha}{mv_0^2}}\,.$$

4-31 To determine the horizontal range, we may set $z = 0$ in the trajectory equation, i.e., consider the (positive) root of

$$0 = \frac{\gamma v_{0z} + g}{\gamma v_{0x}}x + \frac{g}{\gamma^2}\ln\left(1 - \frac{\gamma x}{v_{0x}}\right). \tag{1}$$

We may here write $v_{0x} = v_0\cos\alpha$ and $v_{0z} = v_0\sin\alpha$, α being the angle of launch. Now, for small γ, it is possible to use the approximation

$$\begin{aligned}\frac{g}{\gamma^2}\ln\left(1 - \frac{\gamma x}{v_{0x}}\right) &= -\frac{g}{\gamma^2}\left\{\frac{\gamma x}{v_{0x}} + \frac{1}{2}\left(\frac{\gamma x}{v_{0x}}\right)^2 + \frac{1}{3}\left(\frac{\gamma x}{v_{0x}}\right)^3\right\} + O(\gamma^2)\\ &= -\frac{g}{\gamma v_{0x}}x - \frac{g}{2v_{0x}^2}x^2 - \frac{g}{3v_{0x}^3}\gamma x^3 + O(\gamma^2).\end{aligned}$$

Hence, the above equation for the range becomes

$$0 = \tan\alpha - \frac{g}{2v_0^2\cos^2\alpha}x - \frac{g}{3v_0^2\cos^3\alpha}\gamma x^2 + O(\gamma^2)$$

and so we find the formula

$$\begin{aligned}x &= \frac{v_0^2}{g}\sin 2\alpha - \frac{2}{3v_0\cos\alpha}\gamma x^2 + O(\gamma^2)\\ &= \frac{v_0^2}{g}\sin 2\alpha - \frac{8v_0^3\gamma}{3g^2}\cos\alpha\sin^2\alpha + O(\gamma^2),\end{aligned} \tag{2}$$

which expresses the range for the given launch angle α. To find the launch angle which maximizes this range, we may consider, with the expression given in (2),

$$0 = \frac{dx}{d\alpha} = \frac{2v_0^2}{g}\cos 2\alpha - \frac{8v_0^3\gamma}{3g^2}\left[-\sin^3\alpha + 2\cos^2\alpha\sin\alpha\right] + O(\gamma^2).$$

Then, writing $\alpha = \frac{\pi}{4} - \delta$, we find from (2),

$$0 = \frac{2v_0^2}{g}\sin 2\delta - \frac{8v_0^3\gamma}{3g^2}\frac{1}{2\sqrt{2}} + O(\gamma^2) \quad \rightarrow \quad \delta = \frac{v_0\gamma}{3\sqrt{2}g}.$$

Hence, for the launch angle which maximizes the range, we find the value $\alpha = \frac{\pi}{4} - \frac{v_0\gamma}{3\sqrt{2}g}$ (to first order in γ).

4-32 The equation of motion for the bomb is

$$\ddot{x} = -\gamma\dot{x}, \qquad \ddot{z} = -\gamma\dot{z} - g.$$

Here, z corresponds to the vertical height of the bomb and x represents the horizontal distance traversed. The solution satisfying the initial conditions $x(t{=}0) = 0$, $\dot{x}(t{=}0) = V_0$, $z(t{=}0) = h$ and $\dot{z}(t{=}0) = 0$ is easily found:

$$x(t) = \frac{V_0}{\gamma}(1 - e^{-\gamma t}),$$
$$z(t) = h + \frac{g}{\gamma^2}(1 - \gamma t - e^{-\gamma t}).$$

For a small value of γ, this solution can be approximated as

$$x(t) = \frac{V_0}{\gamma}\left(\gamma t - \frac{1}{2}\gamma^2 t^2 + \cdots\right) = V_0 t - \frac{1}{2}V_0\gamma t^2 + \cdots, \tag{1}$$

$$z(t) = h + \frac{g}{\gamma^2}\left(-\frac{1}{2}\gamma^2 t^2 + \frac{1}{6}\gamma^3 t^3 + \cdots\right)$$
$$= h - \frac{1}{2}gt^2 + \frac{1}{6}g\gamma t^3 + \cdots. \tag{2}$$

The time of fall, say t_0, is determined by setting $z(t_0) = 0$. It has a solution $\sqrt{\frac{2h}{g}}$ when $\gamma = 0$. So we may find an approximate solution by setting $t_0 = \sqrt{\frac{2h}{g}}(1 + A\gamma)$ for some small nonzero value of γ. Inserting this into (2), we get

$$0 = h - \frac{1}{2}g\frac{2h}{g}(1 + 2A\gamma) + \frac{1}{6}g\gamma\frac{2h}{g}\sqrt{\frac{2h}{g}} + O(\gamma^2),$$

from which we obtain $A = \frac{1}{6}\sqrt{\frac{2h}{g}}$. Hence, for the time of fall, we find

$$t_0 = \sqrt{\frac{2h}{g}}\left(1 + \frac{1}{6}\gamma\sqrt{\frac{2h}{g}}\right). \tag{3}$$

Inserting (3) into (1) gives the approximate horizontal distance travelled:

$$x(t_0) \approx V_0 \left[\sqrt{\frac{2h}{g}} \left(1 + \frac{1}{6}\gamma \sqrt{\frac{2h}{g}} \right) - \frac{1}{2}\gamma \frac{2h}{g} \right]$$
$$= V_0 \sqrt{\frac{2h}{g}} \left(1 - \frac{1}{3}\gamma \sqrt{\frac{2h}{g}} \right).$$

4-33 To obtain the short-time expansion, we may use the following differential relations

$$\frac{dx}{dt} = v\cos\theta, \quad \frac{dz}{dt} = v\sin\theta, \quad \frac{dt}{d\theta} = -\frac{v}{g\cos\theta}$$

together with

$$\frac{1}{v}\frac{dv}{d\theta} = \frac{Cv^2 + \sin\theta}{\cos\theta}.$$

Then, with the initial conditions $v(t{=}0) = v_0$ and $\theta(t{=}0) = \theta_0$, we may use

$$\frac{d\theta}{dt} = -\frac{g\cos\theta}{v}, \qquad \frac{dv}{dt} = \frac{d\theta}{dt}\frac{dv}{d\theta} = -g(Cv^2 + \sin\theta),$$

to obtain

$$\left.\frac{dx}{dt}\right|_{t=0} = v_0\cos\theta_0, \qquad \left.\frac{dz}{dt}\right|_{t=0} = v_0\sin\theta_0,$$
$$\left.\frac{d}{dt}\frac{dx}{dt}\right|_{t=0} = \left.\left[\frac{dv}{dt}\cos\theta - v\sin\theta\frac{d\theta}{dt}\right]\right|_{t=0} = -gCv_0^2\cos\theta_0,$$
$$\left.\frac{d}{dt}\frac{dz}{dt}\right|_{t=0} = \left.\left[\frac{dv}{dt}\sin\theta + v\cos\theta\frac{d\theta}{dt}\right]\right|_{t=0} = -g(Cv_0^2\sin\theta_0 + 1),$$
$$\left.\frac{d^2}{dt^2}\left(\frac{dx}{dt}\right)\right|_{t=0} = g^2Cv_0(2Cv_0^2 + \sin\theta_0)\cos\theta_0,$$
$$\left.\frac{d^2}{dt^2}\left(\frac{dz}{dt}\right)\right|_{t=0} = g^2Cv_0[(2Cv_0^2 + \sin\theta_0)\sin\theta_0 + 1].$$

Based on these, we can write the power series

$$\dot{x}(t) = v_0\cos\theta_0\left\{1 - gCv_0t + \frac{1}{2}g^2C(2Cv_0^2 + \sin\theta_0)t^2 + \cdots\right\}$$

$$
\begin{aligned}
&= \frac{v_{0x}}{1 + gCv_0 t - \frac{1}{2}g^2 C \sin\theta_0 t^2 + \cdots}, \\
\dot{z}(t) &= v_0 \sin\theta_0 \left\{1 - gCv_0 t + \frac{1}{2}g^2 C(2Cv_0^2 + \sin\theta_0)t^2 + \cdots\right\} \\
&\quad - gt\left\{1 - \frac{1}{2}gCv_0 t + \cdots\right\} \\
&= \frac{v_{0z} - gt[1 + \frac{1}{2}gCv_0 t + \cdots]}{1 + gCv_0 t - \frac{1}{2}g^2 C \sin\theta_0 t^2 + \cdots},
\end{aligned}
$$

which should be useful for small t.

For $t \to \infty$ we expect the projectile to be in the state in which the gravity force is balanced by the resistive force. As this happens if $Cv^2 = 1$, the terminal speed is $\frac{1}{\sqrt{C}}$. Also, from $\frac{dv}{dt} = -g(Cv^2 + \sin\theta)$, we must have $\theta \to -\frac{\pi}{2}$ for $t \to \infty$, and so it must be the case that

$$
t \to \infty: \qquad \dot{x} \to 0, \quad \dot{z} \to -\frac{1}{\sqrt{C}}.
$$

To study this asymptotic approach, it is convenient to use the variable ψ defined by $\tanh\psi = \sin\theta$, since we have $\frac{d\psi}{dt} = -\frac{g}{v}$ thanks to $\frac{d\theta}{dt} = -\frac{g\cos\theta}{v}$. This means that, when t is sufficiently large, we can set $\psi \sim -g\sqrt{C}t + \text{const}$. Now, for $\psi \to -\infty$,

$$
\begin{aligned}
\sin\theta &= \tanh\psi = -[1 - 2e^{2\psi} + \cdots], \\
\cos\theta &= \operatorname{sech}\psi = -2e^{\psi}[1 - e^{2\psi}],
\end{aligned}
$$

and then, it also follows that

$$
\psi \to \infty: \qquad \frac{1}{v^2} \to C + O(e^{2\psi}).
$$

Hence, using $\psi \sim -g\sqrt{C}t$ for large t, we are led to the behaviors

$$
\begin{aligned}
t \to \infty: \qquad & \dot{x} = v\cos\theta = O(e^{-g\sqrt{C}t}), \\
& \dot{z} = v\sin\theta = -\frac{1}{\sqrt{C}}[1 + O(e^{-2g\sqrt{C}t})].
\end{aligned}
$$

4-34 (i) $\frac{m_0 c^2}{\sqrt{1-\frac{\vec{v}^2}{c^2}}} - qEz = \mathcal{E}$ (: const.).

(ii) Using the equation of motion together with the above energy conservation law, we here have

$$p_x \equiv \frac{m_0 \frac{dx}{dt}}{\sqrt{1-\frac{\vec{v}^2}{c^2}}} = \frac{1}{c^2}(\mathcal{E}+qEz)\frac{dx}{dt} = p_{0x}, \tag{1}$$

$$p_z \equiv \frac{m_0 \frac{dz}{dt}}{\sqrt{1-\frac{\vec{v}^2}{c^2}}} = \frac{1}{c^2}(\mathcal{E}+qEz)\frac{dz}{dt} = qEt, \tag{2}$$

where $p_{0x} = \frac{m_0 v_{0x}}{\sqrt{1-\frac{\vec{v}^2}{c^2}}}$. Then, from (2), we find that

$$\frac{1}{2c^2}\frac{1}{qE}\frac{d}{dt}(\mathcal{E}+qEz)^2 = qEt,$$

so that it becomes possible to write

$$(\mathcal{E}+qEz)^2 - (qE)^2(ct)^2 = C \quad (\text{: const.}). \tag{3}$$

From the given initial conditions, we can set $C = (\mathcal{E}+qEz_0)^2$ and then, from (3), obtain

$$\mathcal{E}+qEz = \sqrt{(\mathcal{E}+qEz_0)^2 + (qE)^2(ct)^2}, \tag{4}$$

assuming $\mathcal{E}+qEz > 0$. Now, inserting this into (1), we obtain the equation

$$\frac{dx}{dt} = \frac{c^2 p_{0x}}{\sqrt{(\mathcal{E}+qEz_0)^2 + (qE)^2(ct)^2}}. \tag{5}$$

This can readily be integrated to yield

$$x(t) = \frac{cp_{0x}}{qE}\sinh^{-1}\frac{ct}{z_0+(\mathcal{E}/qE)} + x_0, \tag{6}$$

while the expression for $z = z(t)$ is available from (4):

$$z(t) = \frac{1}{qE}\{\sqrt{(\mathcal{E}+qEz_0)^2 + (qE)^2(ct)^2} - \mathcal{E}\}. \tag{7}$$

(iii) For $t \to \infty$, $\frac{dx}{dt} \to 0$ from (5). Also, based on (7), we have

$$\frac{dz}{dt} \to \pm ct, \quad \text{as } t \to \infty.$$

Hence the speed of the particle approaches c as $t \to \infty$.

(iv) We eliminate ct from (6) and (7) to obtain the orbit represented by

$$\begin{aligned}\mathcal{E} + qEz &= (\mathcal{E} + qEz_0)\sqrt{1 + \sinh^2\left[\frac{qE(x - x_0)}{cp_{0x}}\right]} \\ &= (\mathcal{E} + qEz_0)\cosh\left[\frac{qE(x - x_0)}{cp_{0x}}\right]. \qquad (8)\end{aligned}$$

[This corresponds to the orbit equation describing a catenary.] If the energy $\mathcal{E}$ is dominated by the particle's rest energy m_0c^2 and the speed of the particle is small compared to c, we may replace the (relativistic) orbit equation (8) by the approximate from

$$\not{\mathcal{E}} + qEz \approx \not{\mathcal{E}} + qEz_0 + m_0c^2\frac{(qE)^2(x - x_0)^2}{2c^2(m_0v_{0x})^2}$$

or

$$z - z_0 \approx \frac{qE}{2m_0}\left(\frac{x - x_0}{v_{0x}}\right)^2, \qquad (9)$$

i.e., we recover the nonrelativistic parabolic orbit. [In this limit, (6) and (7) reduce to the corresponding nonrelativistic equations, $x(t) \approx x_0 + vt$ and $z(t) \approx z_0 + \frac{qE}{2m_0}t^2$.] But the orbit deviates significantly from the nonrelativistic trajectory if the total energy is not dominated by the particle's rest energy or if the particle's speed becomes comparable to c.

4-35 Let us represent the initial velocity of the particles by

$$\begin{aligned}\vec{v}_0 &= (v_0 \sin\alpha\cos\phi_0, v_0 \sin\alpha\sin\phi_0, v_0\cos\alpha) \\ &\approx (v_0\alpha\cos\phi_0, v_0\alpha\sin\phi_0, v_0),\end{aligned}$$

assuming that $\alpha \approx 0$. Then, in the presence of uniform $\vec{B} = B\mathbf{k}$, particles emitted with different angles α and ϕ_0, but all assumed to have the same initial position of $x(t = 0) = y(t = 0) = 0$ and $z(t = 0) = z_0$, will follow somewhat different helical trajectories given by

$$x = \frac{mv_0\alpha}{qB}\left[\sin\left(\frac{qB}{m}t - \phi_0\right) + \sin\phi_0\right],$$

$$y = \frac{mv_0\alpha}{qB}\left[\cos\left(\frac{qB}{m}t - \phi_0\right) - \cos\phi_0\right],$$

$$z = v_0 t + z_0.$$

In this case, regardless of α and ϕ_0, we find $x(\bar{t}) = y(\bar{t}) = 0$ for $t = \bar{t} = \frac{2\pi m}{|q|B}$. This shows that the beam is brought to a focus at the point which is at a distance $l = z(\bar{t}) - z_0 = \frac{2\pi m v_0}{|q|B}$. [It is thus possible to use the measurement of l as a way of measuring the charge-to-mass ratio q/m.]

4-36 Assuming that the motion lies entirely in the xy-plane, we have the equations of motion

$$m\ddot{x} = -m\omega_0^2 x + qB\dot{y}$$
$$m\ddot{y} = -m\omega_0^2 y - qB\dot{x}$$

with $\omega_0 = \sqrt{\frac{k}{m}}$. These equations may be combined into a single equation for the (complex) variable $X = x + iy$,

$$\ddot{X} = -\omega_0^2 X - i\frac{qB}{m}\dot{X}.$$

The solutions to this homogeneous linear differential equation may be found in the form

$$X = Ce^{i(\alpha t + \phi)},$$

where α is a root of the quadratic equation

$$-\alpha^2 - \frac{qB}{m}\alpha + \omega_0^2 = 0.$$

The two roots of this equation are

$$\alpha = -\frac{qB}{2m} \pm \sqrt{\omega_0^2 + \frac{q^2B^2}{4m^2}} \equiv \alpha_{\pm}. \tag{1}$$

We thus obtain two linearly independent solutions

$$X_1 = C_1 e^{i(\alpha_+ t + \phi_1)}, \quad X_2 = C_2 e^{i(\alpha_- t + \phi_2)},$$

and the general solution is a linear combination of these two. Taking the real and imaginary parts of the first solution, we obtain a real solution of the form

$$x_1 = |C_1| \cos(\alpha_+ t + \theta_1), \qquad y_1 = |C_1| \sin(\alpha_+ t + \theta_1)$$

which represent a circular motion of the particle with the radius $|C_1|$ and angular velocity $\omega_1 = \alpha_+$. We also have a second real solution given by

$$x_2 = |C_2| \cos(\alpha_- t + \theta_2), \qquad y_2 = |C_2| \sin(\alpha_- t + \theta_2),$$

which represents a circular motion with radius $|C_2|$ and angular velocity $\omega_2 = \alpha_-$. General motion may be described in terms of juxtaposition of these two modes.

Let us consider a case when the applied magnetic field is relatively small and satisfies the condition

$$\left| \frac{qB}{2m} \right| \ll \omega_0.$$

Then, $\alpha_\pm$ in (1) may be approximated by $\alpha_\pm \approx \pm\omega_0 - \omega_L$ where $\omega_L = \frac{qB}{2m}$. Hence, for the above two angular frequencies, we find $\omega_1 \approx \omega_0 - \omega_L$ and $\omega_2 = -\omega_0 - \omega_L$. This implies that the introduction of a magnetic field causes the unperturbed orbit of the isotropic oscillator (i.e., with $\vec{B} = 0$) to precess with the additional angular velocity equal to $\vec{\omega}_L = -\frac{qB}{2m}\mathbf{k}$. This is called the *Larmor precession* (see Sec. 9.3 for discussions on this phenomenon).

5 The Two-Body Problem, Collision and Many-Particle System

5-1 In the center-of-mass frame, we have ($M = m_1 + m_2$)

$$\vec{r}_1 = \frac{m_2}{M}\vec{r}, \quad \vec{r}_2 = -\frac{m_1}{M}\vec{r}\left(= -\frac{m_1}{m_2}\vec{r}_1\right)$$

with $\vec{r}$ $(= \vec{r}_1 - \vec{r}_2)$ satisfying the following equation of motion

$$\mu\ddot{\vec{r}} = -k\vec{r} \quad \left(\mu = \frac{m_1 m_2}{m_1 + m_2}\right). \tag{1}$$

The solution of (1) is

$$\vec{r}(t) = \vec{a}\cos\omega t + \vec{b}\sin\omega t \quad \left(\omega = \sqrt{\frac{k}{\mu}}\right)$$

which describes an elliptical motion ($\vec{r} = 0$ corresponds to the center of the ellipse). When $\vec{r}(t)$ corresponds to an ellipse, both $\vec{r}_1(t) = \frac{m_2}{M}\vec{r}(t)$ and $\vec{r}_2(t) = -\frac{m_1}{M}\vec{r}(t)$ describe ellipses of their own. Choosing the x-axis (y-axis) in direction of the semi-major (semi-minor) axis of an ellipse defined by $\vec{r}_1(t) = \frac{m_2}{M}\vec{r}(t)$, we can draw these orbits as

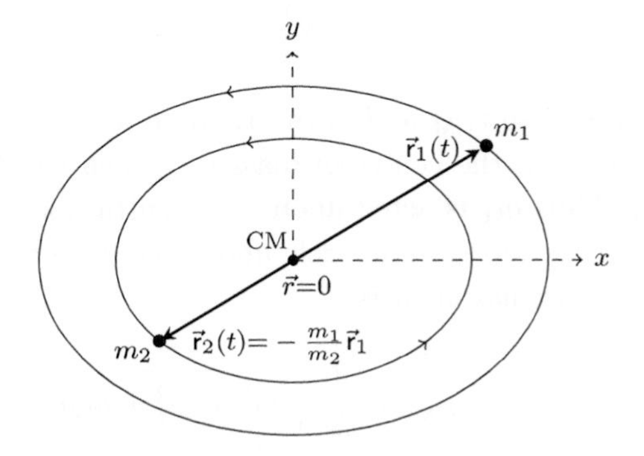

To have circular orbits both for $\vec{r}_1(t) = \frac{m_2}{M}\vec{r}(t)$ and $\vec{r}_2(t) = -\frac{m_1}{M}\vec{r}(t)$, it must be the case that $r = |\vec{r}| = \left|\frac{M}{m_2}\vec{r}_1\right| =$ const. From (1), this happens when the total energy in the CM frame is such that the relative distance r may have a fixed value satisfying the condition $kr = \frac{\vec{L}^2}{\mu r^3}$ (i.e., $r = \left(\frac{\vec{L}^2}{\mu k}\right)^{1/4}$), where $\vec{L} \equiv \mu\vec{r} \times \dot{\vec{r}}$ is the constant of motion representing the relative angular momentum. Then the circular radius of mass m_1 is $\frac{m_2}{M}\left(\frac{\vec{L}^2}{\mu k}\right)^{\frac{1}{4}}$ and that of mass m_2 is $\frac{m_1}{M}\left(\frac{\vec{L}^2}{\mu k}\right)^{\frac{1}{4}}$.

5-2 In Cartesian frame shown in the figure, the three masses m_1, m_2, m_3 have coordinates $(0,0)$, $(L,0)$, and $(\frac{1}{2}L, \frac{\sqrt{3}}{2}L)$ respectively. The position of the center of mass C is

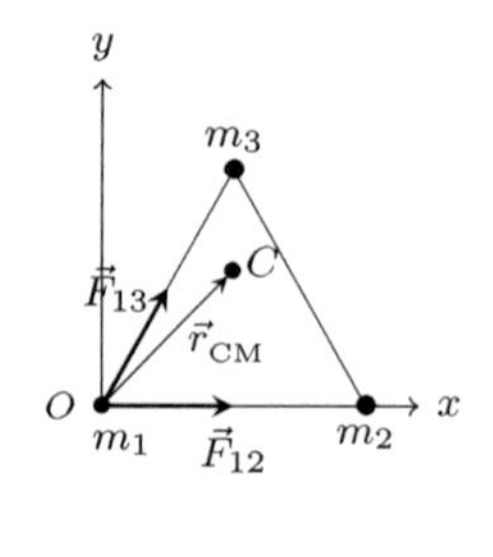

$$\vec{r}_{\text{CM}} = \frac{\sum_i m_i\vec{r}_i}{\sum_i m_i}$$
$$= \frac{L}{M}\left[\left(m_2 + \frac{1}{2}m_3\right)\mathbf{i} + \frac{\sqrt{3}}{2}m_3\mathbf{j}\right]$$

where $M = m_1 + m_2 + m_3$. Now consider the forces acting on m_1. There are two attractive forces

$$\vec{F}_{12} = \frac{Gm_1m_2}{L^2}\mathbf{i} \quad \text{and} \quad \vec{F}_{13} = \frac{Gm_1m_3}{L^2}\left(\frac{1}{2}\mathbf{i} + \frac{\sqrt{3}}{2}\mathbf{j}\right)$$

due to m_2 and m_3 respectively. Their resultant is

$$\vec{F}_1 = \vec{F}_{12} + \vec{F}_{13} = \frac{Gm_1}{L^2}\left[\left(m_2 + \frac{1}{2}m_3\right)\mathbf{i} + \frac{\sqrt{3}}{2}m_3\mathbf{j}\right].$$

As $\vec{F}_1$ is parallel to $\vec{r}_{\text{CM}}$ and both originate from the same point O, $\vec{F}_1$ passes through the center of mass C (which can be assumed to be at rest). Thus m_1 is acted upon by a central force with center at C and hence m_1 moves in a circle about the center of mass C. The radius of this circular orbit is

$$R_1 = |\vec{r}_{\text{CM}}| = \frac{L}{M}\sqrt{m_2^2 + m_3^2 + m_2m_3}\,.$$

The velocity of m_1, v_1, is given by $\frac{m_1 v_1^2}{R_1} = |\vec{F}_1|$, or

$$v_1^2 = \frac{R_1}{m_1}|\vec{F}_1| = \frac{G}{L}\left(\frac{m_2^2 + m_3^2 + m_2 m_3}{M}\right).$$

By permutation of the indices, the above result applies also to m_2 and m_3. Hence the rotational motion which leaves the separation of each pair of masses unchanged is a circular motion of period

$$T = \frac{2\pi R_1}{v_1} = 2\pi L\sqrt{\frac{L}{GM}}$$

which is the same for all three masses.

5-3 This problem can be viewed as a 2-body collision under the interaction potential $V(x)$ (here $x = x_2 - x_1$) drawn in the figure.

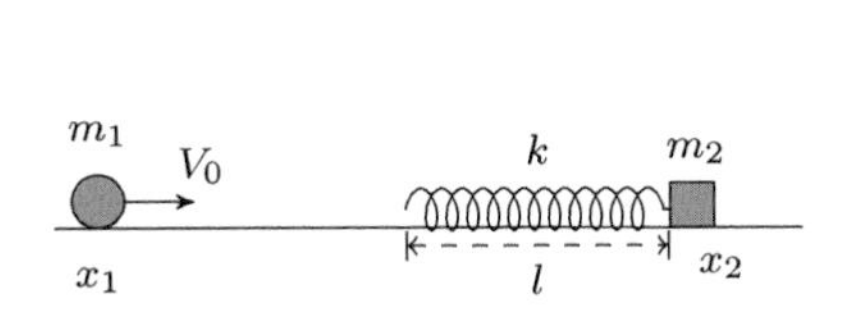

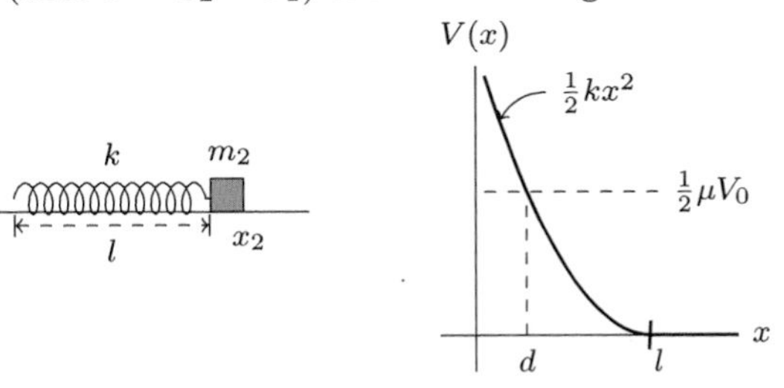

(i) We can here use the energy conservation equation, $\frac{1}{2}\mu\dot{x}^2 + \frac{1}{2}kx^2 = \text{const.}$ ($\mu = \frac{m_1 m_2}{m_1+m_2}$ is the reduced mass). Hence, if d denotes the *maximum compression* of the spring, we should have $\frac{1}{2}kd^2 = \frac{1}{2}\mu V_0^2$ since $\dot{x} = 0$, i.e., the two masses m_1 and m_2 have the same velocity when the spring is compressed maximally. From this relation, we obtain $d = \sqrt{\frac{\mu}{k}}V_0$.

(ii) Let us denote initial (final) velocities of the two masses m_1 and m_2 by v_{1i} and v_{2i} (v_{1f} and v_{2f}). Then, from the conservation laws of energy and momentum, we have

$$\frac{1}{2}m_1(v_{1f}^2 - v_{1i}^2) = -\frac{1}{2}m_2(v_{2f}^2 - v_{2i}^2)$$
$$m_1(v_{1f} - v_{1i}) = -m_2(v_{2f} - v_{2i}).$$

These imply

$$\begin{pmatrix} v_{1f} \\ v_{2f} \end{pmatrix} = \begin{pmatrix} \dfrac{m_1 - m_2}{m_1 + m_2} & \dfrac{2m_2}{m_1 + m_2} \\ \dfrac{2m_1}{m_1 + m_2} & \dfrac{m_2 - m_1}{m_1 + m_2} \end{pmatrix} \begin{pmatrix} v_{1i} \\ v_{2i} \end{pmatrix}. \tag{1}$$

Because we have $v_{1i} = V_0$ and $v_{2i} = 0$, we thus find

$$v_{1f} = \frac{m_1 - m_2}{m_1 + m_2} V_0 \quad \text{and} \quad v_{2f} = \frac{2m_1}{m_1 + m_2} V_0.$$

5-4 Note that the energy loss is

$$\begin{aligned} Q &= \frac{1}{2} m_1 v_{1i}^2 + \frac{1}{2} m_2 v_{2i}^2 - \frac{1}{2} m_1 v_{1f}^2 - \frac{1}{2} m_2 v_{2f}^2 \\ &= \frac{1}{2} (v_{1i},\, v_{2i}) \begin{pmatrix} m_1 & 0 \\ 0 & m_2 \end{pmatrix} \begin{pmatrix} v_{1i} \\ v_{2i} \end{pmatrix} - \frac{1}{2} (v_{1f},\, v_{2f}) \begin{pmatrix} m_1 & 0 \\ 0 & m_2 \end{pmatrix} \begin{pmatrix} v_{1f} \\ v_{2f} \end{pmatrix}. \end{aligned} \tag{1}$$

Then use (see (5.34))

$$\begin{pmatrix} v_{1f} \\ v_{2f} \end{pmatrix} = \frac{1}{M} \begin{pmatrix} m_1 - em_2 & (1+e)m_2 \\ (1+e)m_1 & m_2 - em_1 \end{pmatrix} \begin{pmatrix} v_{1i} \\ v_{2i} \end{pmatrix} \qquad (M \equiv m_1 + m_2)$$

to evaluate the second term in (1):

$$\begin{aligned} &\frac{1}{2} (v_{1f}, v_{2f}) \begin{pmatrix} m_1 & 0 \\ 0 & m_2 \end{pmatrix} \begin{pmatrix} v_{1f} \\ v_{2f} \end{pmatrix} \\ &\quad = \frac{1}{2M^2} (v_{1i}, v_{2i}) \begin{pmatrix} m_1 - em_2 & (1+e)m_1 \\ (1+e)m_2 & m_2 - em_1 \end{pmatrix} \begin{pmatrix} m_1 & 0 \\ 0 & m_2 \end{pmatrix} \\ &\qquad \times \begin{pmatrix} m_1 - em_2 & (1+e)m_2 \\ (1+e)m_1 & m_2 - em_1 \end{pmatrix} \begin{pmatrix} v_{1i} \\ v_{2i} \end{pmatrix} \\ &\quad = \frac{1}{2M^2} (v_{1i}, v_{2i}) \\ &\qquad \times \begin{pmatrix} m_1[m_1^2 + e^2 m_2^2 + m_1 m_2 (1+e^2)] & (1-e^2) m_1 m_2 M \\ (1-e^2) m_1 m_2 M & m_2[m_2^2 + e^2 m_1^2 + m_1 m_2 (1+e^2)] \end{pmatrix} \begin{pmatrix} v_{1i} \\ v_{2i} \end{pmatrix}. \end{aligned}$$

Inserting this result in (1) yields

$$\begin{aligned} Q &= \frac{1}{2} \frac{m_1 m_2}{M} (1 - e^2) (v_{1i},\, v_{2i}) \begin{pmatrix} 1 & -1 \\ -1 & 1 \end{pmatrix} \begin{pmatrix} v_{1i} \\ v_{2i} \end{pmatrix} \\ &= \frac{1}{2} \frac{m_1 m_2}{(m_1 + m_2)} (v_{2i} - v_{1i})^2 (1 - e^2). \end{aligned}$$

This agrees with the expression given.

5-5 We can apply the definition of coefficient of elasticity to each bounce, regarding the ball as the 1st object and the floor as the 2nd. Then the velocity of the ball after the first bounce,

$$\begin{aligned} v_{1f} &= -ev_{1i} \\ &= e\sqrt{2gh}, \end{aligned} \tag{1}$$

where we used $v_{1i} = -\sqrt{2gh}$, the value corresponding to the ball dropped from height h (after time $\Delta t = \sqrt{\frac{2h}{g}}$). Then the ball goes up and comes down (in the time span $\Delta t = \frac{2v_{1f}}{g}$) before bouncing against the floor; in this case, we have $v_{2i} = -e\sqrt{2gh}$ and so the velocity after this second bounce is

$$v_{2f} = -ev_{2i} = e^2\sqrt{2gh}\ . \tag{2}$$

After the nth bounce (after time $\Delta t = \frac{2v_{n-1,f}}{g}$ from the $(n-1)$-th collision), the velocity of the ball will be

$$v_{nf} = -ev_{ni} = e^n\sqrt{2gh}\ .$$

The ball will come to rest after a time

$$\begin{aligned} \sqrt{\frac{2h}{g}} + \frac{2v_{1f}}{g} + \frac{2v_{2f}}{g} + \cdots &= \sqrt{\frac{2h}{g}} + \frac{2}{g}\sqrt{2gh}\, e(1 + e + e^2 + \cdots) \\ &= \sqrt{\frac{2h}{g}}\left(1 + \frac{2e}{1-e}\right) \\ &= \sqrt{\frac{2h}{g}}\frac{1+e}{1-e}. \end{aligned}$$

5-6 Using the formula (with $E_1 = \frac{|\vec{p}_{1i}|^2}{2m_1}$, $E_1' = \frac{|\vec{p}_{1f}|^2}{2m_1}$), we have

$$\cos\theta_1 = \frac{\frac{m_1-m_2}{2m_1}|\vec{p}_{1i}|^2 + \frac{m_1+m_2}{2m_1}|\vec{p}_{1f}|^2}{|\vec{p}_{1i}||\vec{p}_{1f}|}.$$

Let $X \equiv \frac{|\vec{p}_{1f}|}{|\vec{p}_{1i}|}$, then this gives us the quadratic equation

$$(m_1 + m_2)X^2 - 2m_1\cos\theta_1 X + (m_1 - m_2) = 0,$$

the roots of which are given as

$$X = \frac{m_1 \cos\theta_1 \pm \sqrt{m_2^2 - m_1^2 \sin^2\theta_1}}{m_1 + m_2}. \tag{1}$$

To have real X, $m_2^2 - m_1^2 \sin^2\theta_1 \geq 0$; for $m_1 > m_2$, this determines the allowed range for θ_1. And X should be positive also. When $m_1^2 \cos\theta_1^2 > m_2^2 - m_1^2 \sin\theta_1^2$ (which implies $m_1 > m_2$), both signs in (1) are allowed, i.e., there are two possible values for X. (Look for the second value of $|\vec{p}_{1f}|$ in Fig. 5-7 (b)!) On the other hand, when $m_2 \geq m_1$, the negative sign in (1) must be dropped and thus we obtain a unique value for X.

5-7 Recall that the lab scattering angle θ_1 is related to the CM scattering angle θ^* by

$$\tan\theta_1 = \frac{\sin\theta^*}{\cos\theta^* + \frac{m_1}{m_2}}. \tag{1}$$

If we combine the vectors of lab and CM frames in the same diagram, we see that $\theta_2 = \frac{1}{2}(\pi - \theta^*)$ so that

$$\tan\theta_2 = \cot\frac{\theta^*}{2}. \tag{2}$$

Using (1), (2) and the identity

$$\tan(\theta_1 + \theta_2) = \frac{\tan\theta_1 + \tan\theta_2}{1 - \tan\theta_1 \tan\theta_2},$$

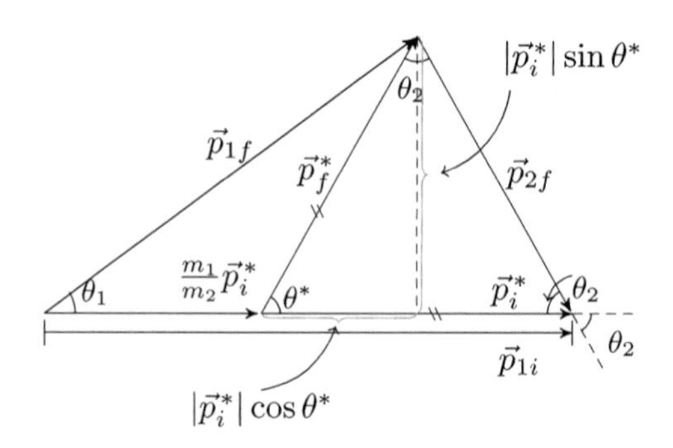

we obtain

$$\begin{aligned}\tan(\theta_1+\theta_2) &= \frac{\left(\frac{\sin\theta^*}{\cos\theta^*+\frac{m_1}{m_2}}\right)+\cot\frac{\theta^*}{2}}{1-\cot\frac{\theta^*}{2}\left(\frac{\sin\theta^*}{\cos\theta^*+\frac{m_1}{m_2}}\right)}\\ &= \frac{\sin\theta^*+\cot\frac{\theta^*}{2}(\cos\theta^*+\frac{m_1}{m_2})}{\cos\theta^*+\frac{m_1}{m_2}-\cot\frac{\theta^*}{2}\sin\theta^*}\\ &= \left(\cot\frac{\theta^*}{2}\right)\frac{2\sin^2\frac{\theta^*}{2}+\cos\theta^*+\frac{m_1}{m_2}}{\cos\theta^*+\frac{m_1}{m_2}-2\cos^2\frac{\theta^*}{2}}\\ &= \tan\theta_2\frac{1+\frac{m_1}{m_2}}{\frac{m_1}{m_2}-1}.\end{aligned}$$

Hence,

$$\frac{\tan(\theta_1+\theta_2)}{\tan\theta_2}=\frac{m_1+m_2}{m_1-m_2}.$$

5-8 In a 2-body elastic collision involving relativistic particles, let $\vec{p}_i^*$ and $-\vec{p}_i^*$ ($\vec{p}_f^*$ and $-\vec{p}_f^*$) denote the initial (final) momenta of two particles of rest masses m_{10} and m_{20}, respectively. (Here, $|\vec{p}_i^*|=|\vec{p}_f^*|$.) Then, to go to the lab frame, perform the *Lorentz transformation* with boost velocity $\vec{v}=c^2\frac{(-\vec{p}_i^*)}{E_2^*}$; then, noting that $E_{1i}^*=E_{1f}^*=E_1^*$ and $E_{2i}^*=E_{2f}^*=E_2^*$ (and $\frac{1}{\sqrt{1-\vec{v}^2/c^2}}=E_2^*/m_{20}c^2$), we find

$$\begin{aligned}\vec{p}_{1i} &= \frac{\vec{p}_i^*-\frac{\vec{v}}{c}\frac{E_1^*}{c}}{\sqrt{1-\vec{v}^2/c^2}}=\frac{E_2^*}{m_{20}c^2}\left(1+\frac{E_1^*}{E_2^*}\right)\vec{p}_i^*,\\ \vec{p}_{2i} &= \frac{(-\vec{p}_i^*)-\frac{\vec{v}}{c}\frac{E_2^*}{c}}{\sqrt{1-\vec{v}^2/c^2}}=0,\end{aligned}$$

and, when $\theta^*=\angle(p_f^*,\vec{p}_i^*)$,

$$\begin{aligned}\vec{p}_{1f\|} &= \frac{\vec{p}_{f\|}^*-\frac{\vec{v}}{c}\frac{E_1^*}{c}}{\sqrt{1-\vec{v}^2/c^2}}=\frac{E_2^*}{m_{20}c^2}\left(\cos\theta^*+\frac{E_1^*}{E_2^*}\right)\vec{p}_i^*,\\ \vec{p}_{1f\perp} &= \vec{p}_{f\perp}^*=|\vec{p}_i^*|\sin\theta^*\hat{p}_{i\perp}^*\quad(\hat{p}_{i\perp}^*\equiv\vec{p}_{i\perp}^*/|\vec{p}_{i\perp}^*|),\\ \vec{p}_{2f\|} &= \frac{-\vec{p}_{f\|}^*-\frac{\vec{v}}{c}\frac{E_2^*}{c}}{\sqrt{1-\vec{v}^2/c^2}}=\frac{E_2^*}{m_{20}c^2}\left(-\cos\theta^*+1\right)\vec{p}_i^*,\\ \vec{p}_{2f\perp} &= -\vec{p}_{f\perp}^*=-|\vec{p}_i^*|\sin\theta^*\hat{p}_{i\perp}^*,\end{aligned}$$

where $\vec{p}_{1f\parallel}$ ($\vec{p}_{1f\perp}$) denote the part that is parallel (perpendicular) to $\vec{p}_i^{\,*}$, etc. Therefore, from $\vec{p}_{1i} \mathbin{//} \vec{p}_i^{\,*}$ and

$$\vec{p}_{1f} = \frac{E_2^*}{m_{20}c^2}\left(\cos\theta^* + \frac{E_1^*}{E_2^*}\right)\vec{p}_i^{\,*} + |\vec{p}_i^{\,*}|\sin\theta^*\hat{p}_{i\perp}^{\,*},$$

we find

$$\tan\theta_1 = \frac{\sin\theta^*}{\frac{E_2^*}{m_{20}c^2}\left(\frac{E_1^*}{E_2^*} + \cos\theta^*\right)} = \frac{m_{20}c^2\sin\theta^*}{E_1^* + E_2^*\cos\theta^*}.$$

Then, using this relativistic connection formula, it is straightforward to show that the lab scattering cross section $\frac{d\sigma}{d\Omega_{\rm Lab}}(\theta_1, \varphi_1)$ is related to the CM scattering cross section $\frac{d\sigma}{d\Omega_{\rm CM}}(\theta^*, \varphi^*)$ by

$$\frac{d\sigma}{d\Omega_{\rm Lab}}(\theta_1, \varphi_1) = \frac{\left[(m_{01}^2 + m_{02}^2)c^4 + \vec{p}_i^{\,*2}c^2(1+\cos^2\theta^*) + 2E_1^*E_2^*\cos\theta^*\right]^{3/2}}{m_{02}^2c^4|E_2^* + E_1^*\cos\theta^*|} \times \frac{d\sigma}{d\Omega_{\rm CM}}(\theta^*, \varphi^*)\,.$$

5-9 In the case of Compton scattering (or elastic $e\gamma$ scattering in the lab frame) depicted schematically as

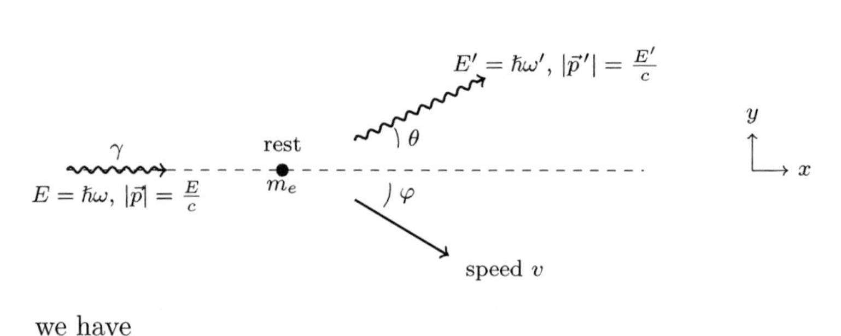

we have

(i) (energy conservation)

$$E + m_ec^2 = E' + \frac{m_ec^2}{\sqrt{1 - v^2/c^2}}, \tag{1}$$

(ii) (conservation of momentum)

$$x\text{-comp.}: \quad E/c = (E'/c)\cos\theta + \frac{m_e v}{\sqrt{1-v^2/c^2}}\cos\varphi, \tag{2a}$$

$$y\text{-comp.}: \quad 0 = (E'/c)\sin\theta - \frac{m_e v}{\sqrt{1-v^2/c^2}}\sin\varphi. \tag{2b}$$

Given $E = \hbar\omega$ (here, $\hbar = \frac{h}{2\pi}$) and the scattering angle θ, these equations may be used to determine v, $E' = \hbar\omega'$ and the recoil angle φ. We are particularly interested in E' as a function of the scattering angle θ, and so we eliminate φ and v from these equations. Eliminating φ from (2a) and (2b) gives

$$\begin{aligned}\frac{m_e v^2}{1-v^2/c^2} &= \frac{\hbar^2}{c^2}(\omega - \omega'\cos\theta)^2 + \frac{\hbar^2}{c^2}\omega'^2\sin^2\theta \\ &= \frac{\hbar^2}{c^2}(\omega^2 - 2\omega\omega'\cos\theta + \omega'^2),\end{aligned}$$

while we find from (1)

$$\frac{m_e^2 c^2}{1-v^2/c^2} = \frac{\hbar^2}{c^2}\left(\omega - \omega' + \frac{m_e c^2}{\hbar}\right)^2.$$

Finally we may eliminate v from these equations by subtracting them from each other to obtain

$$\begin{aligned}m_e^2 c^2 &= \frac{\hbar^2}{c^2}\left(\omega - \omega' + \frac{m_e c^2}{\hbar}\right)^2 - \frac{\hbar^2}{c^2}(\omega^2 - 2\omega\omega'\cos\theta + \omega'^2) \\ &= m_e^2 c^2 + \frac{\hbar^2}{c^2}\left\{-2\omega\omega' + 2\frac{m_e c^2}{\hbar}(\omega - \omega') + 2\omega\omega'\cos\theta\right\},\end{aligned}$$

viz.,

$$m_e(\omega - \omega') = \frac{\hbar}{c^2}(1 - \cos\theta)\omega\omega'$$

or, dividing both sides by $\omega\omega'$ and using $E = \hbar\omega$ and $E' = \hbar\omega'$,

$$m_e c^2\left(\frac{1}{E'} - \frac{1}{E}\right) = 1 - \cos\theta. \tag{3}$$

Introducing the wavelengths $\lambda = \frac{hc}{E}$, $\lambda' = \frac{hc}{E'}$, we therefore obtain the *Compton relation*

$$\lambda' - \lambda = 2\frac{h}{m_e c}\sin^2\frac{\theta}{2}.$$

Note that the greatest shift in the wavelength takes place when the photon is back scattered, i.e., for $\theta = \pi$; here, $\Delta\lambda = 2\frac{h}{m_e c}$.

5-10 Let m_1 and m_2 denote the masses of the two hard spheres (with radii r_1 and r_2). Then, for the relative coordinate $\vec{r}$ $(= \vec{r}_1 - \vec{r}_2)$ connecting the centers of the two spheres, we can here write the equation of motion in the form

$$m\ddot{\vec{r}} = -\vec{\nabla}V(r) \quad \left(\mu = \frac{m_1 m_2}{m_1 + m_2}\right) \tag{1}$$

where $V(r)$ is a central potential specified as

$$V(r) = \begin{cases} 0, & r > r_1 + r_2 \\ \infty, & r \leq r_1 + r_2. \end{cases}$$

As explained in Sec. 5.2 in the text, the CM scattering cross section can be found by studying the "fixed-center scattering problem" based on (1). In this context, the impact parameter b can be related to the scattering angle Θ by

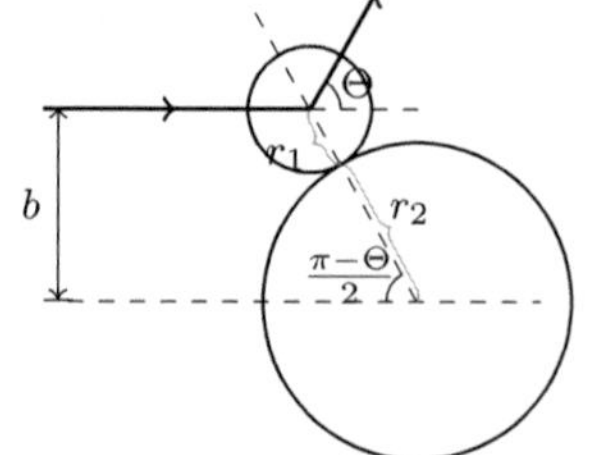

$$\begin{aligned} b(\Theta) &= (r_1 + r_2)\sin\left(\frac{\pi - \Theta}{2}\right) \\ &= (r_1 + r_2)\cos\frac{\Theta}{2}. \end{aligned}$$

(See the figure.) Using this result, the CM scattering cross section is immediately found as

$$\begin{aligned} \frac{d\sigma}{d\Omega_{\text{CM}}}(\theta^*, \varphi^*) &= \frac{b(\theta^*)}{\sin\theta^*}\left|\frac{db(\theta^*)}{d\theta^*}\right| = \frac{(r_1 + r_2)\cos\frac{\theta^*}{2} \cdot \frac{1}{2}(r_1 + r_2)\sin\frac{\theta^*}{2}}{\sin\theta^*} \\ &= \frac{1}{4}(r_1 + r_2)^2, \end{aligned}$$

i.e., the CM scattering cross section is isotropic. For the lab scattering cross section,

$$\begin{aligned} \frac{d\sigma}{d\Omega_{\text{Lab}}}(\theta_1, \varphi_1) &= \left|\frac{\sin\theta^*}{\sin\theta_1}\frac{d\theta^*}{d\theta_1}\right| \frac{d\sigma}{d\Omega_{\text{CM}}}(\theta^*, \varphi^*) \\ &= \frac{\left\{1 + \left(\frac{m_1}{m_2}\right)^2 + 2\frac{m_1}{m_2}\cos\theta^*\right\}^{3/2}}{\left|1 + \frac{m_1}{m_2}\cos\theta^*\right|}\frac{1}{4}(r_1 + r_2)^2, \end{aligned} \tag{3}$$

where the lab scattering angles θ_1, φ_1 are related to the CM scattering angles θ^*, φ^* by $\tan\theta_1 = \frac{\sin\theta^*}{\cos\theta^* + \frac{m_1}{m_2}}$ and $\varphi_1 = \varphi^*$. [In writing (3), we have assumed $m_1 \leq m_2$; if $m_1 > m_2$, there is a complication arising from the fact that for each given value of θ_1 less than the maximum scattering angle $\theta_{\max} = \sin^{-1}\frac{m_2}{m_1}$ (find the extremum of $\tan\theta_1 = \frac{\sin\theta^*}{\cos\theta^* + \frac{m_1}{m_2}}$), there are two possible values of θ^*.]

The total cross section can be calculated using (2) or (3), to obtain

$$\sigma_t = \int \frac{d\sigma}{d\Omega_{\text{CM}}} d\Omega_{\text{CM}} = \int \frac{d\sigma}{d\Omega_{\text{Lab}}} d\Omega_{\text{Lab}} = \pi(r_1 + r_2)^2.$$

5-11 (i) Since there exists a reflection symmetry about the x-axis, it is sufficient to determine the x-coordinate of the CM. Let us consider a uniform disk with no hole cut. Combining a contribution form the hole and that from the rest, we must have (see the figure)

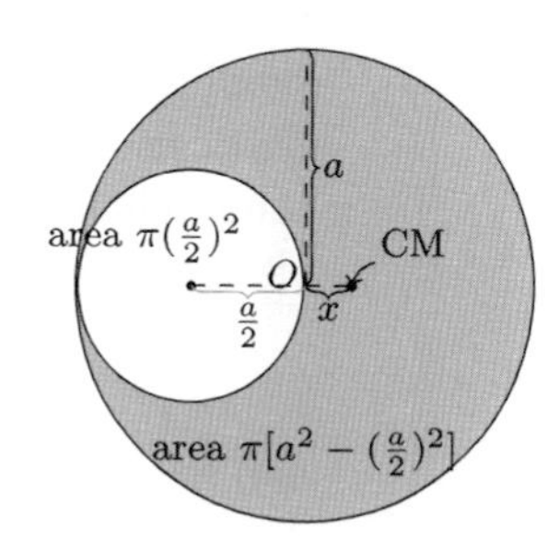

$$\left(-\frac{a}{2}\right)\pi\left(\frac{a}{2}\right)^2 + x \cdot \pi[a^2 - (a/2)^2]$$
$$= 0 \cdot \pi a^2.$$

Hence the x-coordinate of the CM is $x = \frac{\frac{1}{8}\pi a^3}{\frac{3}{4}\pi a^2} = \frac{a}{6}$.

(ii) <u>x-coordinate of the CM:</u>

$$\frac{1}{7a}[la + 2l \cdot 2a + la + 3l \cdot 3a] = \frac{15}{7}l.$$

<u>y-coordinate of the CM:</u>

$$\frac{1}{7a}[la + 3l \cdot 2a + 5l \cdot a + 7l \cdot 3a] = \frac{33}{7}l.$$

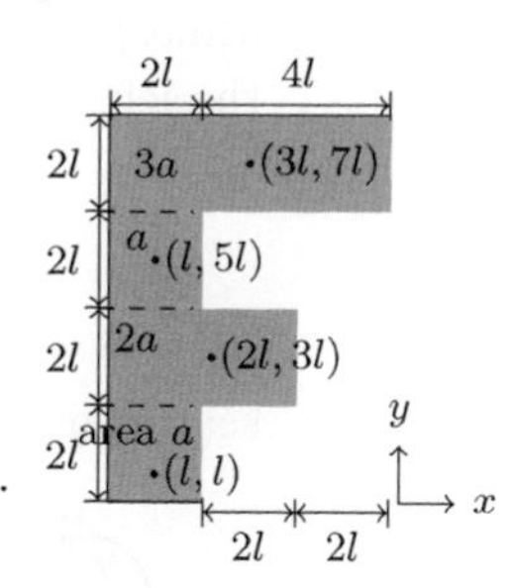

(iii) A uniform solid tetrahedron $ABCD$: in this case, the CM is the point of intersection of the lines joining the vertices to the centroids of the opposite triangular faces.

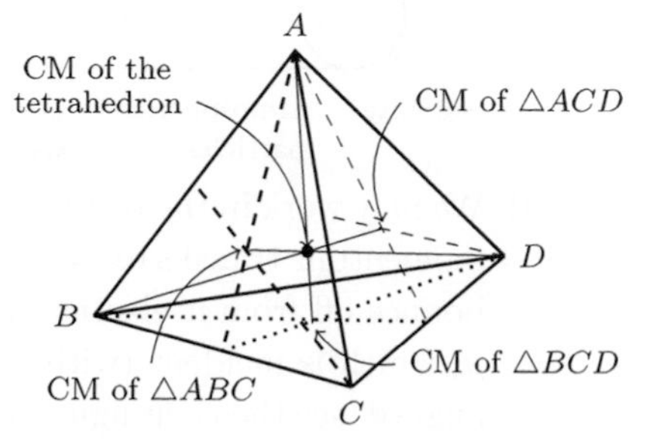

(iv) A homogeneous circular pyramid:

$$\bar{y} = \frac{\int_0^h y \cdot y^2\, dy}{\int_0^h y^2\, dy} = \frac{3}{4}h.$$

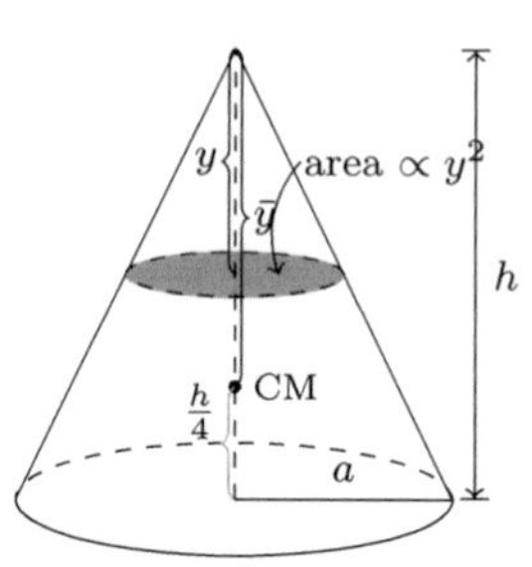

5-12 (i) Let $m(t)$, $v(t)$ denote the mass and velocity of the satellite at time t. When every atmospheric particle striking the satellite adheres to its surface, we can *define* the retarding force on the satellite to be given by $F_{\text{ret}} = m\frac{dv}{dt}$. We assume that atmospheric particles are *initially at rest.* Then, if the mass and velocity of the satellite at time $t+dt$ are denoted as $m(t)+dm$ and $v(t)+dv$, we should have, because of the conservation of total momentum

$$(m + dm)(v + dv) = mv$$

and so

$$mdv + vdm = 0 \qquad \text{or} \qquad \frac{d(mv)}{dt} = 0.$$

But $dm = \rho_{\text{atm}}(\pi r^2)vdt$ since this must equal the total mass of atmospheric particles striking the satellite during $(t,\, t+dt)$ (see the left figure below). Hence

$$F_{\text{ret}} = m\frac{dv}{dt} = -v\frac{dm}{dt} = -\rho_{\text{atm}}(\pi r^2)v^2,$$

i.e., the retarding force is equal to $\rho_{\text{atm}}v^2$ times the hard-sphere scattering cross section πr^2.

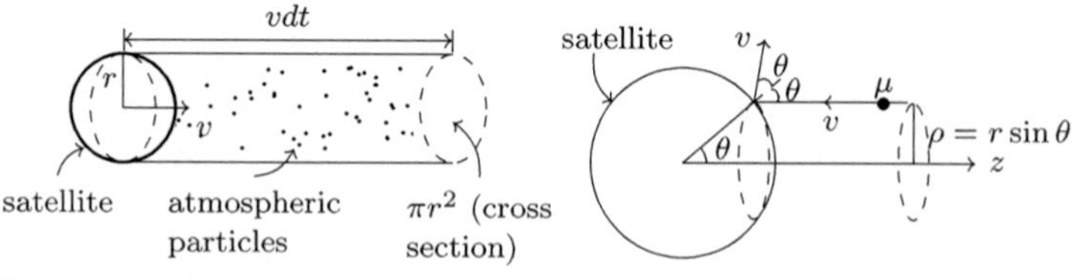

(ii) We may work in the rest frame of the satellite to calculate the net momentum transferred to the satellite as atmospheric particles bounce off from it elastically. If an atmospheric particle of mass μ, which is incident (with speed v) on the spherical satellite at angle θ (see the right figure above), bounces off from the satellite, its momentum in the z-direction will get changed by $\mu v \cos 2\theta + \mu v$. On the other hand, during the time interval $(t,\ t + \Delta t)$, the number of atmospheric particles hitting the satellite at the angular range $(\theta,\ \theta + d\theta)$ will be equal to $\frac{\rho_{\text{atm}}}{\mu}(v\Delta t)2\pi\rho\, d\rho$ where $\rho = r\sin\theta$. Then, due to the conservation of total momentum, the net momentum transferred to the satellite during the time interval $(t,\ t + \Delta t)$ should be equal to

$$\begin{aligned}\Delta p_z &= \left[-\int_{\theta=0}^{\theta=\frac{\pi}{2}} \mu v(\cos 2\theta + 1)\frac{\rho_{\text{atm}}}{\mu} v \cdot 2\pi r \sin\theta\, d(r\sin\theta)\right]\Delta t \\ &= -2\pi r^2 \rho_{\text{atm}} v^2 \left[\int_{\sin\theta=0}^{\sin\theta=1} 2(1-\sin^2\theta)\sin\theta\, d(\sin\theta)\right]\Delta t \\ &= -\pi r^2 \rho_{\text{atm}} v^2 \Delta t.\end{aligned}$$

Hence, for the retarding force, we find $F_{\text{ret}} = \frac{\Delta p_z}{\Delta t} = -\rho_{\text{atm}} v^2(\pi r^2)$ again.

5-13 (i) From footnote 25 in Chapter 5 of the text, the equation of motion is

$$M\frac{dv}{dt} = -Mg - u_0\frac{dM}{dt}. \tag{1}$$

Since the rate of exhaust $A = -\frac{dM}{dt}$ is constant, the mass of the rocket is $M(t) = M_0 - At$. Hence (1) becomes

$$\frac{dv}{dt} = -g + u_0\frac{A}{M(t)}.$$

This has the solution

$$v(t) = -gt + u_0 \ln\frac{M_0}{M(t)},$$

before the fuel gets exhausted, i.e., for $t < t_1 \equiv \Delta M/A$. Integrating the velocity with respect to time gives the height of the rocket:

$$\begin{aligned}h(t) &= \int_0^t v(t')dt' \\ &= -\frac{1}{2}gt^2 + \frac{u_0}{A}\left(M(t)\ln\frac{M(t)}{M_0} + At\right).\end{aligned}$$

(ii) At $t = t_1$, the velocity and height of the rocket become, respectively,

$$v_1 = v(t_1) = -g\frac{\Delta M}{A} + u_0 \ln \frac{M_0}{M_0 - \Delta M}$$

and

$$H = h(t_1) = -\frac{g}{2}\left(\frac{\Delta M}{A}\right)^2 + \frac{u_0}{A}\left((M_0 - \Delta M)\ln\frac{M_0 - \Delta M}{M_0} + \Delta M\right).$$

When $t > t_1$, the height of the rocket is

$$h(t) = H + v_1(t - t_1) - \frac{1}{2}g(t - t_1)^2, \qquad t > t_1 .$$

Hence, the rocket arrives at the highest altitude at the time $t = t_2 \equiv t_1 + \frac{v_1}{g}$ and the height is

$$\begin{aligned} h(t_2) &= H + \frac{v_1^2}{2g} \\ &= \frac{u_0^2}{2g}\left(\ln\frac{M_0 - \Delta M}{M_0}\right)^2 + \frac{u_0}{A}\left(\Delta M + M_0 \ln\frac{M_0 - \Delta M}{M_0}\right) \\ &= \frac{u_0^2}{2g}\left(\ln\frac{M_0 - \Delta M}{M_0}\right)^2 - \frac{u_0}{A}M_0\left[\sum_{k=2}^{\infty}\frac{1}{k}\left(\frac{\Delta M}{M_0}\right)^k\right]. \end{aligned}$$

Therefore the rocket reaches a higher altitude when the rate of exhaust A has a larger value.

5-14 (i) When the longer portion of the chain is of length x, the potential energies of the two portions of the chain are $-(\frac{M}{2l}x)g\frac{x}{2}$ and $-\left[\frac{M}{2l}(2l - x)\right]g\frac{2l-x}{2}$, while the kinetic energy of the chain is $\frac{1}{2}M\dot{x}^2$. Hence, from the conservation of total energy, we should have

$$\frac{1}{2}M\dot{x}^2 - \frac{Mg}{4l}[x^2 + (2l - x)^2] = -Mg\frac{l}{2}.$$

After algebraic simplification this gives

$$\dot{x}^2 = \frac{g}{l}(x - l)^2. \tag{1}$$

(ii) To find the tension T in the chain it is necessary to write down the equation of motion, say, for the right hand portion of the chain. The mass of this portion is changing (as in the rocket problem) — here, chain is being added on to the right hand portion with a *zero relative velocity.* Then, based on the same reasoning as given in footnote 26 of this chapter, the right hand portion (with instantaneous mass $\frac{M}{2l}x$ and velocity $v = \dot{x}$) will satisfy the equation

$$\left(\frac{M}{2l}x\right)\ddot{x} = \left(\frac{M}{2l}x\right)g - T. \tag{2}$$

On the other hand, differentiating (1) gives $\ddot{x} = \frac{g}{l}(x - l)$ and using this result in (2) gives

$$T = \frac{Mx}{2l}\left[g - \frac{g}{l}(x - l)\right] = \frac{Mg}{2l^2}x(2l - x).$$

5-15 Clearly, the CM of the triangle, S, has to be directly below the point of suspension. Denote the vectors pointing from the CM S to the vertices of the triangle by $\vec{r}_1$, $\vec{r}_2$ and $\vec{r}_3$, and that to the suspension point by $\vec{m}$. Then the forces $\vec{F}_1$, $\vec{F}_2$ and $\vec{F}_3$ exerted on the plate by the threads can be represented as

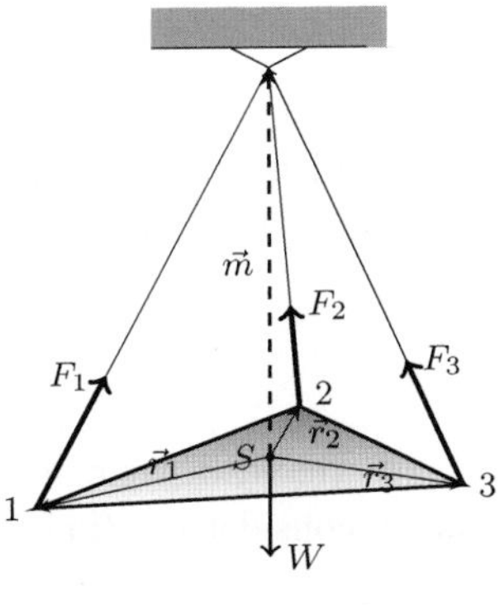

$$\vec{F}_i = \lambda_i(\vec{m} - \vec{r}_i) \quad (i = 1, 2, 3)$$

and, as the plate is in equilibrium,

$$\vec{F}_1 + \vec{F}_2 + \vec{F}_3 + \vec{W} = 0.$$

But, $\vec{r}_1 + \vec{r}_2 + \vec{r}_3 = 0$ (why?) and $\vec{W}//\vec{m}$ (so we are allowed to write $\vec{W} = -k\vec{m}$). Eliminating $\vec{r}_3$ from the above equations gives

$$\begin{aligned} 0 &= \lambda_1(\vec{m} - \vec{r}_1) + \lambda_2(\vec{m} - \vec{r}_2) + \lambda_3(\vec{m} + \vec{r}_1 + \vec{r}_2) - k\vec{m} \\ &= (\lambda_3 - \lambda_1)\vec{r}_1 + (\lambda_3 - \lambda_2)\vec{r}_2 + (\lambda_1 + \lambda_2 + \lambda_3 - k)\vec{m}. \end{aligned} \tag{1}$$

Since $\vec{r}_1$, $\vec{r}_2$ and $\vec{m}$ are linearly independent, (1) can be true only when the coefficient of each vector is zero, i.e.,

$$\lambda_3 = \lambda_1 = \lambda_2, \quad \lambda_1 + \lambda_2 + \lambda_3 = k.$$

Accordingly, the tensions in the threads are proportional to their lengths (i.e., $|\vec{F}_i| \propto |\vec{m} - \vec{r}_i| \equiv h_i$). Also, with $\lambda_1 = \lambda_2 = \lambda_3 \equiv \lambda$

(and so $\vec{F}_i = \lambda(\vec{m} - \vec{r}_i)$) and $|\vec{m}| = l$ (: the distance from the point of suspension to the CM of the plate),

$$3\lambda = k \quad \rightarrow \quad \lambda = \frac{1}{3}k = \frac{1}{3}\frac{|-\vec{W}|}{|\vec{m}|} = \frac{W}{3l}.$$

We can thus write

$$|\vec{F}_i| = \frac{W}{3l}h_i.$$

(Note that, for a given triangular plate, l can be determined if h_1, h_2 and h_3 are known.)

5-16 Let the energy of the N-body system be described by the general expression

$$E = \frac{1}{2}\sum_{i=1}^{N} m_i\vec{v}_i^2 + V(\vec{r}_1, \ldots, \vec{r}_N),$$

so that we have the equation of motion $m_i\dot{\vec{v}}_i = -\vec{\nabla}_{\vec{r}_i}V$ $(= \vec{F}_i)$. The problem is to find the configuration leading to minimal total energy, under the condition that

$$L_z = \sum_{i=1}^{N} m_i(\vec{r}_i \times \vec{v}_i)_z = \text{fixed}.$$

To this end, we can apply the method of Lagrange multipliers. That is, consider the function (of the $\vec{r}$'s and $\vec{v}$'s) given by

$$W = \frac{1}{2}\sum_{i=1}^{N} m_i\vec{v}_i^2 + V(\vec{r}_1, \ldots, \vec{r}_N) - \lambda\left(\sum_{i=1}^{N} m_i(x_i v_{yi} - y_i v_{xi})\right)$$

(λ is the Lagrange multiplier), and demand that, at the minimal energy configuration,

$$\left.\begin{aligned} \frac{\partial W}{\partial v_{zi}} &= m_i v_{zi} = 0 \\ \frac{\partial W}{\partial v_{xi}} &= m_i v_{xi} + \lambda m_i y_i = 0 \\ \frac{\partial W}{\partial v_{yi}} &= m_i v_{yi} - \lambda m_i x_i = 0 \end{aligned}\right\} \longrightarrow \quad \vec{v}_i = \lambda\mathbf{k} \times \vec{r}_i \quad (i = 1, \ldots, N)$$

$$\left.\begin{aligned} \frac{\partial W}{\partial z_i} &= \frac{\partial V}{\partial z_i} = 0 \\ \frac{\partial W}{\partial x_i} &= \frac{\partial V}{\partial x_i} - \lambda m_i v_{yi} = 0 \\ \frac{\partial W}{\partial y_i} &= \frac{\partial V}{\partial y_i} + \lambda m_i v_{xi} = 0 \end{aligned}\right\} \longrightarrow \quad -\frac{1}{m_i}\vec{\nabla}_{\vec{r}_i}V \ (= \dot{\vec{v}}_i) = \lambda\mathbf{k} \times \vec{v}_i \quad (i = 1, \ldots, N).$$

Clearly, this corresponds to a state in which the system is rotating as a rigid body about the z-axis with angular velocity λ (whose value can be determined in terms of the value of L_z).

5-17 (i) From the symmetry of the layout and initial conditions, we deduce that all the bodies fall towards the center O of the n-gon with the same non-uniform acceleration. The formation keeps its original shape, but the distance r from the center decreases at a non-uniformly accelerating rate. The resultant force acting on one (say the nth) body when it is at distance r from the center O is, after some calculations,

$$F(r) = G\frac{m^2}{r^2}\sum_{k=1}^{n-1}\frac{1}{4\sin\frac{\pi k}{n}}.$$

This force, made up of the gravitational forces exerted by all the other bodies, or, more precisely, of those components of these forces which are directed towards the center, is identical to the gravitational attraction of a fixed body situated at the center of the mass

$$M_n = \frac{m}{4}\sum_{k=1}^{n-1}\frac{1}{\sin\frac{\pi k}{n}}.$$

The values of the masses M_n (in units of m) can be calculated numerically for all values of n as

$$M_2 = 0.25, \quad M_3 = 0.58, \quad M_4 = 0.96, \cdots, M_{10} = 3.86.$$

(ii) The time T of the collapse from an initial distance R onto a central mass M can be considered as half of the period T_e for a severely flattened (degenerate) elliptical orbit of semi-major axis $\frac{R}{2}$. (By this we mean an elliptical orbit of the limiting shape shown in the figure.) The period T_c of a circular orbit of radius R can be calculated directly from the dynamical equation for circular motion

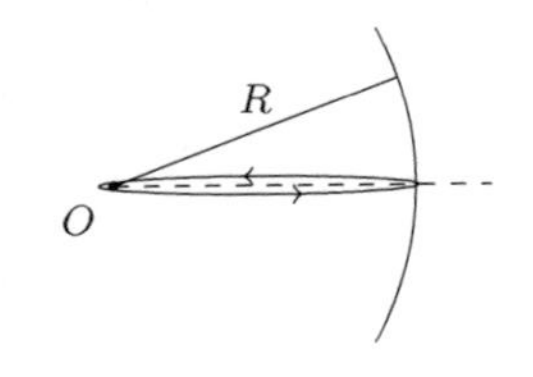

$$G\frac{Mm}{R^2} = mR\left(\frac{2\pi}{T_c}\right)^2, \qquad \text{giving } T_c = 2\pi\sqrt{\frac{R^3}{GM}}.$$

But, according to Kepler's third law,

$$\left(\frac{T_e}{T_c}\right)^2 = \left(\frac{R/2}{R}\right)^3.$$

Thus, finally, we obtain $T = \frac{1}{2}T_e = \frac{1}{2}\left(\frac{1}{2}\right)^{3/2}T_c = \pi\sqrt{R^3/8GM_n}$ for the time.

The limiting case $n \gg 1$ is interesting. As the number n of bodies increases, M_n increases even if the total mass of the system is fixed at M_0, i.e., $m = M_0/n$. The more finely a given amount of matter is spread around a circle, the shorter the time it takes for it to collapse under its own gravitational attraction. (However, there is no point in examining a continuous matter distribution spread along an arbitrarily thin line; the extent of the matter in the transverse, i.e., radial direction cannot be neglected.)

5-18 Because of the energy–momentum conservation, the reaction

$$p_{\text{incident}} + p_{\text{target}} = p + p + p + \bar{p}$$

cannot go unless the kinetic energy of the incident proton (in the lab frame, i.e., when the target proton is at rest) exceeds a certain threshold value, T_{th}. To find this threshold value, let

$$P^\mu_{(1)} = (P^0_{(1)} \equiv E_1/c, P^1_{(1)}, P^2_{(1)}, P^3_{(1)}) : \text{4-momentum of the incident proton,}$$

$$P^\mu_{(2)} = (P^0_{(2)} \equiv E_2/c, P^1_{(2)}, P^2_{(2)}, P^3_{(2)}) : \text{4-momentum of the target proton}$$

in the CM frame (i.e., when $\vec{P}_{(1)} + \vec{P}_{(2)} = 0$), and let

$$\bar{P}^\mu_{(1)} = (\bar{P}^0_{(1)} \equiv \bar{E}_1/c, \bar{P}^1_{(1)}, \bar{P}^2_{(1)}, \bar{P}^3_{(1)}) : \text{4-momentum of the incident proton,}$$

$$\bar{P}^\mu_{(2)} = (\bar{P}^0_{(2)} \equiv m_p c, 0, 0, 0) : \text{4-momentum of the target proton}$$

in the lab frame. Now, if we consider the given reaction in the CM frame, it is clear from the energy conservation that we must have

$$E_1 + E_2 \geq 4m_p c^2. \tag{1}$$

But it follows from the Lorentz scalar nature that

$$(P^0_{(1)} + P^0_{(2)})^2 - \sum_{i=1}^{3}(P^i_{(1)} + P^i_{(2)})^2 = (\bar{P}^0_{(1)} + \bar{P}^0_{(2)})^2 - \sum_{i=1}^{3}(\bar{P}^i_{(1)} + \bar{P}^i_{(2)})^2,$$

and hence, using $\vec{P}_{(1)} + \vec{P}_{(2)} = 0$, $\bar{P}^i_{(2)} = 0$ and $\bar{P}^0_{(2)} = m_p c$,

$$\begin{aligned}(E_1 + E_2)^2/c^2 &= (\bar{P}^0_{(1)})^2 - \sum_{i=1}^{3}(\bar{P}^i_{(1)})^2 + 2m_p c\bar{P}^0_{(1)} + m_p^2 c^2 \\ &= 2m_p^2 c^2 + 2m_p \bar{E}_1\,. \end{aligned} \qquad (2)$$

Then, from (1) and (2), we infer that this reaction can go only when the energy of the incident proton (in the lab frame) satisfies the condition

$$\bar{E}_1 \geq 7m_p c^2 \quad \longrightarrow \quad \bar{E}_{\text{th}} = 7m_p c^2.$$

Therefore the threshold value for the relativistic kinetic energy is

$$T_{\text{th}} = \bar{E}_{\text{th}} - m_p c^2 = 6m_p c^2 = 5.628\,\text{GeV}.$$

5-19 This problem is relevant when an excited atom at rest in an inertial frame emits a photon and falls into a lower energy state. Let m_0, m_0' denote the rest masses of the atom before and after the emission, and $-v$ the velocity of the recoiling atom (after emitting a photon of energy $h\nu$). Then, from the laws of energy–momentum conservation,

$$\begin{aligned} m_0 c^2 &= \frac{m_0' c^2}{\sqrt{1 - v^2/c^2}} + h\nu, \\ 0 &= \frac{-m_0' v}{\sqrt{1 - v^2/c^2}} + \frac{h\nu}{c}. \end{aligned}$$

From the second of these, we find $\sqrt{1 - v^2/c^2} = \frac{m_0' c}{\sqrt{m_0'^2 c^2 + (\frac{h\nu}{c})^2}}$. Inserting this result into the first, we then find

$$\left(m_0 c - \frac{h\nu}{c}\right)^2 - \left(\frac{h\nu}{c}\right)^2 = m_0'^2 c^2. \qquad (1)$$

We can rewrite this result in terms of the rest energy difference, $\Delta E = (m_0 - m_0')c^2$, between the two energy levels of the atom involved in the transition. Then,

$$(m_0c)^2 - 2\left(\frac{h\nu}{c}\right)m_0c = \frac{1}{c^2}(m_0c^2 - \Delta E)^2$$

or

$$h\nu = \Delta E\left(1 - \frac{\Delta E}{2m_0c^2}\right).$$

(In most atomic transitions, the term $\frac{\Delta E}{2m_0c^2}$ may be neglected in comparison with unity to a high degree of approximation.)

5-20 We shall here discuss the case of a general relativistic gas involving particles of rest mass m_0 (and energy $\mathcal{E} = \sqrt{m_0^2c^4 + c^2\vec{p}^2}$), following closely the approach given in Sec. 5.4. Let $\Psi(p)$, where $p \equiv |\vec{p}|$ with $\vec{p} = \frac{m_0\vec{v}}{\sqrt{1-\vec{v}^2/c^2}}$, denote the (relativistic) momentum distribution in the gas which is normalized according to $\int_0^\infty \Psi(p)dp = 1$. Then, we can consider the quantity

$$n\Psi(p)dp\frac{1}{2}\sin\theta d\theta,$$

which represents the number of molecules per unit volume having momentum between p and $p + dp$ and travelling at angles between θ and $\theta + d\theta$ to the chosen direction. Now, the number of such molecules hitting unit area of wall in unit time is

$$v\cos\theta\, n\Psi(p)dp\frac{1}{2}\sin\theta d\theta,$$

where $v = |\vec{v}| = \frac{c^2p}{\mathcal{E}}$ is the speed of a relativistic particle. Based on these, we can integrate over θ and v to get the pressure

$$\begin{aligned} P &= \int_0^\infty dp \int_0^{\frac{\pi}{2}} d\theta(2p\cos\theta)\frac{c^2p}{\mathcal{E}}\cos\theta\; n\Psi(p)dp\frac{1}{2}\sin\theta \\ &= n\int_0^\infty dp\frac{c^2p^2}{\mathcal{E}}\Psi(p)\frac{1}{3} \\ &= \frac{1}{3}\frac{V}{N}\left(\overline{\frac{c^2p^2}{\mathcal{E}}}\right) \end{aligned}$$

where $\left(\overline{\frac{c^2p^2}{\mathcal{E}}}\right) = \int_0^\infty dp\left(\frac{c^2p^2}{\mathcal{E}}\right)\Psi(p)$. This is the expression for the pressure. For an ultrarelativistic gas with $\mathcal{E} = pc$, this leads to the formula $P = \frac{1}{3}\frac{V}{N}\bar{\mathcal{E}}$.

5-21 Aside from $\int_0^\infty e^{-p^2x^2}dx = \frac{\sqrt{\pi}}{2p}$ $(p > 0)$, following integral formulas are useful: for $p > 0$ and $n = 0, 1, 2, \ldots,$

$$\int_0^\infty x^{2n}e^{-px^2}dx = \frac{(2n-1)!!}{2(2p)^n}\sqrt{\frac{\pi}{p}},$$

where $(2n-1)!! = 1 \cdot 3 \cdot 5 \cdots (2n-1)$,

$$\int_0^\infty x^{2n+1}e^{-px^2}dx = \frac{n!}{2p^{n+1}}.$$

Now, with $\Phi(v) = \frac{4}{\sqrt{\pi}}\left(\frac{m}{2k_BT}\right)^{3/2} v^2 e^{-\frac{mv^2}{2k_BT}}$ (normalized according to $\int_0^\infty dv\Phi(v) = 1$), we find

$$\begin{aligned}
\bar{v} &= \int_0^\infty v\Phi(v)dv = \sqrt{\frac{2}{\pi}}\left(\frac{m}{k_BT}\right)^{3/2}\int_0^\infty v^3 e^{-\frac{mv^2}{k_BT}}dv \\
&= \sqrt{\frac{2}{\pi}}\left(\frac{m}{k_BT}\right)^{3/2}\frac{1}{2(m/2k_BT)^2} \\
&= \sqrt{\frac{8k_BT}{\pi m}},
\end{aligned}$$

$$\begin{aligned}
\overline{(v^2)} &= \int_0^\infty v^2\Phi(v)dv = \sqrt{\frac{2}{\pi}}\left(\frac{m}{k_BT}\right)^{3/2}\int_0^\infty v^4 e^{-\frac{mv^2}{2k_BT}}dv \\
&= \frac{3k_BT}{m}.
\end{aligned}$$

Also, if $\Phi_E(E)$ denotes the corresponding energy distribution, we must have from $E = \frac{1}{2}mv^2$

$$\begin{aligned}
\Phi_E(E)dE &= \Phi(v)dv\Big|_{v=\sqrt{\frac{2E}{m}}} \\
&= \Phi\left(v = \sqrt{\frac{2E}{m}}\right)\frac{dE}{\sqrt{2mE}} \\
&= \sqrt{\frac{2}{\pi}}\left(\frac{m}{k_BT}\right)^{3/2}\frac{2E}{m}e^{-\frac{E}{k_BT}}\frac{dE}{\sqrt{2mE}} \\
&= 2\sqrt{\frac{E}{\pi}}\left(\frac{1}{k_BT}\right)^{3/2}e^{-\frac{E}{k_BT}}dE.
\end{aligned}$$

6 Gravitational Field Equations

6-1 As the mass density is a function of r only, we can take the gravitational field to have a radially symmetric form, i.e., $\vec{g}(\vec{r}) = h(r)\frac{\vec{r}}{r}$. The function $h(r)$ can then be determined using Gauss's law, viz., choosing for the Gauss's surface a surface of a sphere of radius r, we have

$$h(r) \cdot 4\pi r^2 = -4\pi G \int_0^r \frac{Ma^2}{2\pi r(r^2+a^2)^2} 4\pi r^2 dr$$
$$= -4\pi GM \frac{r^2}{r^2+a^2}.$$

Hence $h(r) = -\frac{GM}{r^2+a^2}$, giving

$$\vec{g}(\vec{r}) = -\frac{GM\vec{r}}{r(r^2+a^2)}.$$

If $\mathcal{G}(r)$ is the corresponding gravitational potential, we must have

$$\frac{d\mathcal{G}(r)}{dr} = \frac{GM}{r^2+a^2}.$$

Integrating this yields

$$\mathcal{G}(r) = \frac{GM}{a} \tan^{-1}\left(\frac{r}{a}\right) + \text{const.}$$

The constant here may be chosen such that we find the behavior $G(r) \sim -\frac{GM}{r}$ as $r \to \infty$. We then obtain the expression

$$\mathcal{G}(r) = \frac{GM}{a}\left[\tan^{-1}\left(\frac{r}{a}\right) - \frac{\pi}{2}\right].$$

6-2 If $\vec{g}(\vec{r})$ denotes the gravitational field due to the infinite sheet (occupying the xy-plane), the force on the sphere will be

$$\vec{F} = \int_{\text{sphere}} \vec{g}(\vec{r})\rho_m(\vec{r})d^3\vec{r}$$

$$= -\mathbf{k}\int_{\text{sphere}} g(|z|)\rho_m(\vec{r})d^3\vec{r},$$

where we used $\vec{g}(\vec{r}) = -g(|z|)\mathbf{k}$ (for $z > 0$) by symmetry. But, applying Gauss's theorem over the rectangular surface shown in the figure yields

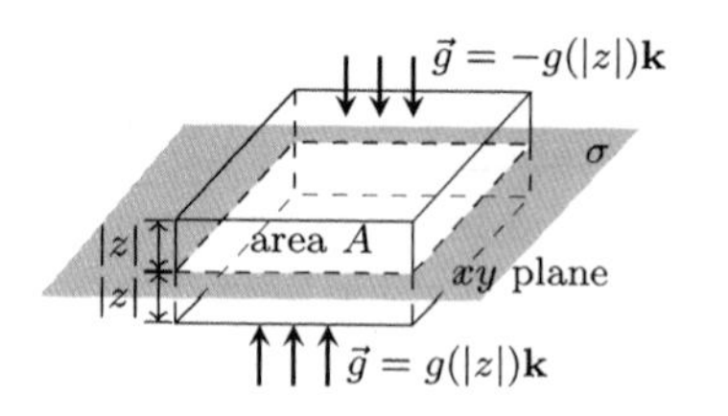

$$-2g(|z|)\cancel{A} = -4\pi G\sigma\cancel{A}$$

$$\to\ g(|z|) = 2\pi G\sigma$$

(: independent of $|z|$).

Hence we find

$$\vec{F} = -\mathbf{k}\,2\pi G\sigma\int_{\text{sphere}} \rho_m(\vec{r})d^3\vec{r} = -(2\pi G\sigma M)\mathbf{k},$$

i.e., the sphere is subject to the attractive force of strength $2\pi G\sigma M$.

6-3 From the principle of superposition, the gravitational field in the cavity should be equal to the gravitational field of the completely filled sphere minus the gravitational field of the material which has been removed. The gravitational field inside a completely filled sphere (with radius R) of density ρ is (see Sec. 6.1),

$$\vec{g} = -\frac{G(\frac{4}{3}\pi\rho R^3)}{R^3}\vec{r} = -\frac{4}{3}\pi G\rho\,\vec{r}.$$

Now consider the sphere with a spherical cavity (see the figure). Let X denote a point inside the cavity such that the displacement of X from the center of the cavity is $\vec{\Gamma}$. Then the displacement of X from the center of the original sphere is $\vec{a}+\vec{\Gamma}$, and the gravitational field at X due to the original (i.e., completely filled) sphere is thus equal to

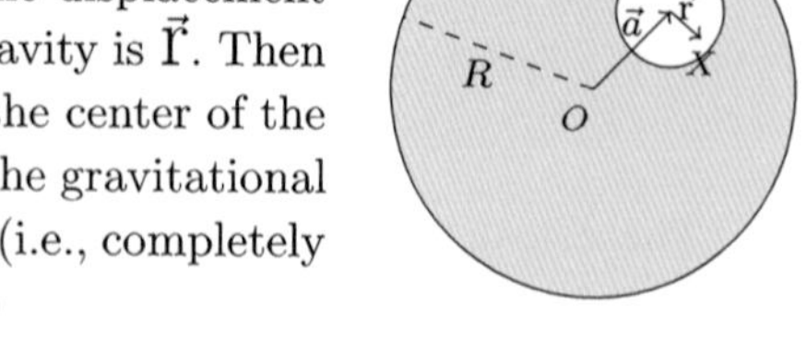

$$-\frac{4}{3}\pi G\rho(\vec{a}+\vec{\Gamma}).$$

As the gravitational field due to the material removed is $-\frac{4}{3}\pi G\rho\vec{\mathrm{r}}$, the net gravitational field at X becomes

$$\vec{g} = -\frac{4}{3}\pi G\rho(\vec{a} + \vec{\mathrm{r}}) + \frac{4}{3}\pi G\rho\vec{\mathrm{r}} = -\frac{4}{3}\pi G\rho\vec{a}.$$

The field within the cavity is thus *uniform*. Its magnitude is $\frac{4\pi}{3}G\rho a$ (i.e., depends only on the position of the cavity, and *not on its size*) and its direction is parallel to the line joining the center of the cavity to the center of the original sphere.

6-4 With the mass m_1 at a given position $\vec{r}_1$, imagine bringing up the mass m_2 from infinity to the position $\vec{r}_2$: the required work, to be done *against* the gravitational force from the mass m_1, is

$$W = -\int_{\infty}^{\vec{r}_2} m_2 \left[-G\frac{m_1(\vec{r} - \vec{r}_1)}{|\vec{r} - \vec{r}_1|^3}\right] \cdot d\vec{r} = -G\frac{m_1 m_2}{|\vec{r}_2 - \vec{r}_1|}.$$

Then, with the masses m_1 and m_2 at the positions $\vec{r}_1$ and $\vec{r}_2$, let us calculate the work required to bring up the mass m_3 from infinity:

$$\begin{aligned} W &= -\int_{\infty}^{\vec{r}_3} m_3 \left[-G\frac{m_1(\vec{r} - \vec{r}_1)}{|\vec{r} - \vec{r}_1|^3} - G\frac{m_2(\vec{r} - \vec{r}_2)}{|\vec{r} - \vec{r}_2|^3}\right] \cdot d\vec{r} \\ &= -G\frac{m_3 m_1}{|\vec{r}_3 - \vec{r}_1|} - G\frac{m_3 m_2}{|\vec{r}_3 - \vec{r}_2|}. \end{aligned}$$

If one brings up all N masses from infinity to the given positions in this manner, the total work required will evidently be

$$\begin{aligned} &-G\frac{m_2 m_1}{|\vec{r}_2 - \vec{r}_1|} - G\left[\frac{m_3 m_1}{|\vec{r}_3 - \vec{r}_1|} + \frac{m_3 m_2}{|\vec{r}_3 - \vec{r}_2|}\right] - \cdots \\ &\qquad - G\left[\frac{m_N m_1}{|\vec{r}_N - \vec{r}_1|} + \cdots + \frac{m_N m_{N-1}}{|\vec{r}_N - \vec{r}_{N-1}|}\right] \\ &= -\frac{G}{2}\sum_{i,j(\neq i)} \frac{m_i m_j}{|\vec{r}_i - \vec{r}_j|} = \frac{1}{2}\sum_{i=1}^{N} m_i \mathcal{G}(\vec{r}_i), \end{aligned}$$

where we used $\mathcal{G}(\vec{r}_i) = -G\sum_{j\neq i}\frac{m_j}{|\vec{r}_i - \vec{r}_j|}$. With a continuous mass distribution with mass density $\rho_m(\vec{r})$, the corresponding expression for the total required work should read

$$W = \frac{1}{2} \int \rho(\vec{r})\, \mathcal{G}(\vec{r})\, d^3\vec{r} = -\frac{1}{2} \iint \frac{\rho(\vec{r})\rho(\vec{r}\,')}{|\vec{r} - \vec{r}\,'|}\, d^3\vec{r}\, d^3\vec{r}\,'.$$

6-5 We can take the straight line from Seoul to Paris to be oriented as shown in the figure. Then consider the situation when the ball bearing is at the point P, a distance x from the mid-point of the tunnel. We will denote the perpendicular distance from the mid-point of the tunnel to the center of the Earth by a, and the radial distance of P from the Earth's center by r.

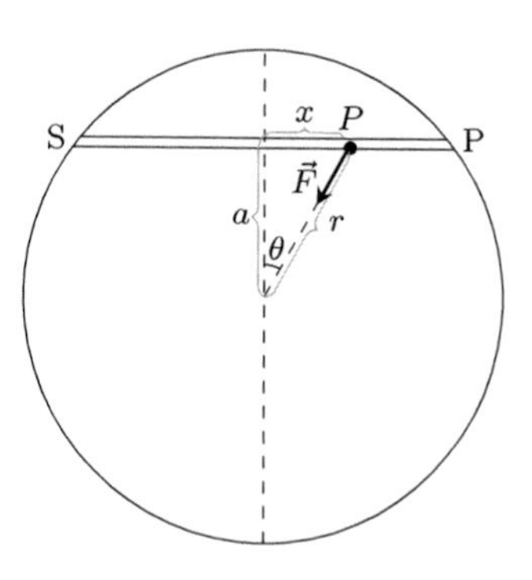

The gravitational force at a point P inside a uniform sphere is the same as the force that would be produced by all matter of the sphere closer to the center than P, and hence the force F acting on the ball bearing is given by $F = \frac{mG(\frac{4}{3}\pi r^3 \rho)}{r^2}$, where ρ is the Earth's density and m the mass of the ball bearing. The component of this force acting parallel to the tunnel in the direction of increasing x is $-F \sin\theta$, so that the acceleration of the ball bearing is

$$\begin{aligned} \frac{d^2x}{dt^2} &= -\frac{4}{3}\pi\rho G\, r \sin\theta \\ &= -\frac{4}{3}\pi\rho G x, \end{aligned}$$

i.e., we find the equation of simple harmonic motion (*independent of the distance* a!). The period of this motion is $2\pi\sqrt{\frac{3}{4\pi\rho G}}$, and the time T for the ball bearing to get from Seoul to Paris will be half a cycle; so the net time $T = \pi\sqrt{\frac{3}{4\pi\rho G}}$.

Now we may use the information we have been given about the Earth's radius and surface gravity to find the density ρ. Since g (: surface gravity) $= \frac{\frac{4}{3}\pi R^3 \rho G}{R^2} = \frac{4\pi R\rho G}{3}$, we have $\rho = \frac{3g}{4\pi R G}$ and substituting this into our expression for T gives

$$T = \pi\sqrt{\frac{R}{g}} = \pi\sqrt{\frac{6400 \times 10^3\,\mathrm{m}}{9.8\,\mathrm{m\,s^{-2}}}} = 2.5 \times 10^3\,\mathrm{s}\ (\sim 42\,\mathrm{minutes}).$$

6-6 Obviously, the formula

$$\vec{g}(\vec{r}) = -\kappa \int \frac{\rho(\vec{r}\,')(\vec{r}-\vec{r}\,')}{|\vec{r}-\vec{r}\,'|^{p+1}} d^n\vec{r}\,'$$
$$= -\vec{\nabla}\mathcal{G}(\vec{r}) \quad \left(\mathcal{G}(\vec{r}) = \frac{-\kappa}{(p-1)} \int \frac{\rho(\vec{r}\,')}{|\vec{r}-\vec{r}\,'|^{p-1}} d^n\vec{r}\,'\right) \tag{1}$$

for the gravitational field is an appropriate one when the gravitational force between two masses m_1, m_2 (located at the positions $\vec{r}_1$ and $\vec{r}_2$, respectively) is given by $\vec{F}_{12} = -\kappa m_1 m_2 \frac{\vec{r}_1-\vec{r}_2}{|\vec{r}-\vec{r}\,'|^{p+1}}$. If $\rho(\vec{r}) = \rho(r)$ where $r = \sqrt{(x_1)^2 + \cdots + (x_n)^2}$, i.e., we have a hyperspherically-symmetric mass distribution about the origin, the corresponding gravitational field will be radially symmetric (because of symmetry), that is,

$$\vec{g}(\vec{r}) = h(r)\frac{\vec{r}}{r} \tag{2}$$

with a certain radial function $h(r)$ (which in general depends on the details of the mass distribution $\rho(r)$). But, for some special value of p, we may have that the gravitational field (outside the mass distribution) is simply

$$\vec{g}(\vec{r}) = -\kappa \frac{M\vec{r}}{|\vec{r}|^{p+1}}, \tag{3}$$

i.e., given by the expression representing the field due to the net mass $M = \int \rho(|\vec{r}\,'|) d^n\vec{r}\,'$ accumulated at the origin, irrespective of the way the mass is distributed. What is that value for p?

It is not difficult to convince oneself that, for the above-mentioned fact to be true, we must have the n-dimensional analogue of Gauss's law — viz., for the "correct" value of p, one must find

$$\oint_S \left(-\kappa \frac{m\vec{r}}{|\vec{r}|^{p+1}}\right) \cdot \hat{n}\, da = \text{“a fixed (nonzero) constant, for } any \text{ } (n-1)\text{-dimensional closed hypersurface } S \text{ enclosing the origin”}, \tag{4}$$

where $\hat{n}$ denotes the (outward) unit normal to the hypersurface S. For this to be true, the left hand side of (4) should have the length dimension equal to zero: as the length dimension of da is $n-1$, we now find that p *must be equal to* $n-1$, i.e., $|\vec{F}_{12}| \propto \frac{1}{|\vec{r}_1-\vec{r}_2|^{n-1}}$. Now, choosing the value $p = n - 1$, we can write the above Gauss's law in a precise form:

$$\oint_S \left(\frac{-\kappa m\vec{r}}{|\vec{r}|^n}\right)\cdot\hat{n}\,da = -\kappa m\oint_S d\Omega_n = -\kappa m C_n \tag{5}$$

($d\Omega_m$ denotes the solid angle in n dimensions), with the constant C_n — the area of the unit hypersphere $(x_1)^2 + \cdots + (x_n)^2 = 1$ — given by

$$C_n = 2\frac{\pi^{\frac{n}{2}}}{\Gamma(\frac{n}{2})}. \tag{6}$$

[Here $\Gamma(x)$ is the gamma function with values $\Gamma(\frac{1}{2}) = \sqrt{\pi}$, $\Gamma(1) = 1$, $\Gamma(\frac{3}{2}) = \frac{1}{2}\sqrt{\pi}$; for the values of $\Gamma(\frac{n}{2})$ with other (half-) integer $\frac{n}{2}$, use the relation $\Gamma(x+1) = x\Gamma(x)$.] Then, the gravitational field (1) under the choice $p = n - 1$ will satisfy the relation

$$\oint_{S=\partial V}\vec{g}\cdot\hat{n}\,da = -C_n\kappa\int_V \rho(\vec{r})d^n\vec{r}, \tag{7}$$

which is the fully general form of Gauss's law. We may then invoke this Gauss's law with the radially symmetric field form (2) (with Gauss's surface, in the region outside the mass distribution, taken to be hyperspherical surfaces of radius r, for which we have $da = r^{n-1}d\Omega_n$ and $\hat{n} = \vec{r}/|\vec{r}|$), to have that $h(r) = -\kappa\frac{M}{|\vec{r}|^{n-1}}$ (with $M = \int\rho(|\vec{r}|)d^n\vec{r}\,'$), i.e., $\vec{g}(\vec{r}) = -\kappa\frac{M\vec{r}}{|\vec{r}|^n}$ as desired.

Then, using the divergence theorem in n dimensions, the left hand side of (7) can be equated to $\int_V \vec{\nabla}\cdot\vec{g}\,d^n\vec{r}$ (with $\vec{\nabla}\cdot\vec{g} \equiv \sum_{i=1}^n \frac{\partial g_i}{\partial x_i}$, if $\vec{g}(\vec{r}) = \sum_{i=1}^n g_i(\vec{r})\mathbf{e}_i$). As the volume V is really arbitrary, the resulting integral relation can be turned into a differential equation

$$\vec{\nabla}\cdot\vec{g}(\vec{r}) = -C_n\kappa\rho(\vec{r}). \tag{8}$$

If we insert $\vec{g}(\vec{r}) = -\vec{\nabla}\mathcal{G}(\vec{r})$ (with $\mathcal{G}(\vec{r}) = \frac{-\kappa}{(n-2)}\int\frac{\rho(\vec{r}\,')}{|\vec{r}-\vec{r}\,'|^{n-2}}d^n\vec{r}\,'$, for $n \neq 2$) into (8), we obtain the *n-dimensional Poisson's equation*

$$\vec{\nabla}^2\mathcal{G}(\vec{r}) \equiv \left(\frac{\partial^2}{\partial x_1^2} + \cdots + \frac{\partial^2}{\partial x_n^2}\right)\mathcal{G}(\vec{r}) = C_n\kappa\rho(\vec{r}) \tag{9}$$

with C_n in (6). Especially, in the region where $\rho(\vec{r}) = 0$, the potential should satisfy the (n-dimensional) Laplace equation $\vec{\nabla}^2\mathcal{G}(\vec{r}) = 0$.

[Remark 1: Given a hyperspherically symmetric distribution $\rho = \rho(r)$, the corresponding potential $G(r)$ — also a function of r only — should satisfy, in a region where $\rho(r) = 0$, the equation

$$0 = \sum_{i=1}^{n} \frac{\partial}{\partial x_i}\frac{\partial}{\partial x_i}\mathcal{G}(\vec{r}) = \sum_{i=1}^{n} \frac{\partial}{\partial x_i}\left[\frac{x_i}{r}\frac{d\mathcal{G}(r)}{dr}\right]$$
$$= \frac{n-1}{r}\frac{d}{dr}\mathcal{G}(r) + \frac{d^2}{dr^2}\mathcal{G}(r).$$

As this can be rewritten as

$$\frac{1}{r^{n-1}}\frac{d}{dr}\left(r^{n-1}\frac{d\mathcal{G}(r)}{dr}\right) = 0,$$

the potential should have the form $\mathcal{G}(r) = \frac{\alpha}{r^{n-2}} + \beta$, where α, β are constants. This result is consistent with our general finding presented above.

Remark 2: To demonstrate that the choice of the value $p = n - 1$ ensures the requested property, one may try to perform the related integral expression for $\mathcal{G}(\vec{r})$ (or that for $\vec{g}(\vec{r})$) explicitly, as we have done in the text for $n = 3$. But, when $n \neq 3$, it is not an easy thing to do. (If you want to see how this can be done, see for instance *Am. J. Phys.* **50**, 179 (1982).)]

6-7 Poisson's equation applies not only to gravitational potential but also to electrostatic potential ϕ (with $\vec{\nabla}^2\phi = -\frac{\rho}{\epsilon_0}$, if ρ represents the electric charge density). For other cases in which the Laplacian $\vec{\nabla}^2$ enters, we have:

(i) The velocity potential ϕ of a potential flow in fluid mechanics satisfies the Laplace equation $\vec{\nabla}^2\phi = 0$ (see Chapter 8);

(ii) The wave equation, for sounds or for electromagnetic waves, is of the form

$$\vec{\nabla}^2\phi - \frac{1}{c^2}\frac{\partial^2\phi}{\partial t^2} = 0,$$

where ϕ can be the pressure field (for sounds) or any particular components of electric or magnetic fields (for electromagnetic waves);

(iii) Propagation of "heat" can be described by the phenomenological equation of the form

$$\frac{\partial T}{\partial t} = \kappa \vec{\nabla}^2 T,$$

where T is the temperature, and κ the heat conductivity;

(iv) The one-particle Schrödinger equation in quantum mechanics takes the form

$$i\hbar \frac{\partial \psi}{\partial t} = -\frac{\hbar^2}{2m}\vec{\nabla}^2 \psi + V(\vec{r})\,\psi,$$

to name a few. The reason why the Laplacian operator appears so often in physical equations can be found above all from its rotational invariance — under arbitrary orthogonal coordinate transformations $x_i \to x_i' = O_{ij}x_j$, we have $\sum_{i=1}^{3} \frac{\partial}{\partial x_i'}\frac{\partial}{\partial x_i'} = \sum_{i=1}^{3} \frac{\partial}{\partial x_i}\frac{\partial}{\partial x_i}$. Therefore, when a local field equation approach is used to study any physical system which is intrinsically *isotropic*, the Laplacian operator is bound to play a prominent role. Furthermore, in view of the mean value theorem valid for any function $\phi(\vec{r})$ satisfying the Laplace equation, a nonvanishing value for $\vec{\nabla}^2\phi$ at a given point P can be seen as representing a (rotationally invariant) measure for how much "curvature" the function $\phi(\vec{r})$ has in the neighborhood of P; naturally, one may relate $\vec{\nabla}^2\phi$ to some kind of local physical source (or disturbance), which makes the value of ϕ at P deviate from the average ϕ-value in the region surrounding P.

6-8 (i) Let us make a Fourier expansion of the potential in the variable x

$$\mathcal{G}(x,z) = \int_{-\infty}^{\infty} \widetilde{\mathcal{G}}(k,z)e^{ikx}dk, \tag{1}$$

and then

$$\widetilde{\mathcal{G}}(k,z) = \frac{1}{2\pi}\int_{-\infty}^{\infty} \mathcal{G}(x,z)e^{-ikx}dx. \tag{2}$$

Now, when $\mathcal{G}(x,z)$ satisfies the Laplace equation $\left(\frac{\partial^2}{\partial x^2} + \frac{\partial^2}{\partial z^2}\right)\mathcal{G}(x,z) = 0$ in the region $z > 0$, we find, for $z > 0$,

$$\left(\frac{\partial^2}{\partial z^2} - k^2\right)\widetilde{\mathcal{G}}(k,z) = \frac{1}{2\pi}\int_{-\infty}^{\infty}\left(\frac{\partial^2}{\partial z^2} - k^2\right)\mathcal{G}(x,z)e^{-ikx}\,dx$$

$$= \frac{1}{2\pi}\int_{-\infty}^{\infty}\left(\frac{\partial^2}{\partial z^2}\mathcal{G}(x,z)\right)e^{-ikx}$$
$$+\left(\frac{\partial^2}{\partial x^2}e^{-ikx}\right)\mathcal{G}(x,z)\,dx$$
$$= \frac{1}{2\pi}\int_{-\infty}^{\infty}\left\{\left(\frac{\partial^2}{\partial z^2}+\frac{\partial^2}{\partial x^2}\right)\mathcal{G}(x,z)\right\}e^{-ikx}\,dx$$
$$= 0,$$

where we performed the integration by parts to change $\left(\frac{\partial^2}{\partial x^2}e^{-ikx}\right)\mathcal{G}(x,z)$ into $e^{-ikx}\frac{\partial^2}{\partial x^2}\mathcal{G}(x,z)$.

(ii) The general solution of

$$\frac{\partial^2\widetilde{\mathcal{G}}(k,z)}{\partial z^2} - k^2\widetilde{\mathcal{G}}(k,z) = 0$$

can be written as

$$\widetilde{\mathcal{G}}(k,z) = C_+(k)e^{kz} + C_-(k)e^{-kz}.$$

But, to have $\mathcal{G}(x,z)\to 0$ for $z\to\infty$, we must choose $C_+(k)=0$ for $k>0$ and $C_-(k)=0$ for $k<0$. This leads to

$$\widetilde{\mathcal{G}}(k,z) = C(k)e^{-|k|z}$$

with $C(k) = \widetilde{\mathcal{G}}(k,z{=}0)$. Hence, $\widetilde{\mathcal{G}}(k,z) = \widetilde{\mathcal{G}}(k,z{=}0)e^{-|k|z}$ (with $\widetilde{\mathcal{G}}(k,z{=}0) = \frac{1}{2\pi}\int_{-\infty}^{\infty}\mathcal{G}(x,z{=}0)e^{-ikx}\,dx$ from (2)) and inserting this result into (1) gives the expression

$$\mathcal{G}(k,z) = \int_{-\infty}^{\infty}\widetilde{\mathcal{G}}(k,z{=}0)e^{ikx}e^{-|k|z}\,dk.$$

6-9 With the Earth taken as a homogeneous oblate spheroid (of mass density ρ_0)

$$\frac{x^2}{a^2}+\frac{y^2}{a^2}+\frac{z^2}{c^2}\le 1 \quad (c = a(1-\eta))$$

the azimuthal symmetry condition $\rho(\vec{r}\,') = \rho(r',\theta')$ is clearly satisfied. Then, based on the discussions given in Sec. 6.2, the gravitational potential $\mathcal{G}(r,\theta)$ for $r>a$ can approximately be represented by the form

$$\mathcal{G}(r,\theta) = -\frac{GM}{r} - \frac{GQ}{4r^2}(3\cos^2\theta - 1)$$

with M (the total mass) and Q (see Sec. 6.2) given explicitly by the expressions

$$M = \iiint\limits_{\frac{x^2}{a^2}+\frac{y^2}{a^2}+\frac{z^2}{c^2}\leq 1} \rho_0\, dxdydz = \rho_0 a^2 c \iiint\limits_{\bar{x}^2+\bar{y}^2+\bar{z}^2\leq 1} d\bar{x}d\bar{y}d\bar{z} = \frac{4\pi}{3}\rho_0 a^2 c,$$

$$\begin{aligned} Q &= \iiint\limits_{\frac{x^2}{a^2}+\frac{y^2}{a^2}+\frac{z^2}{c^2}\leq 1} (2z^2 - x^2 - y^2)\rho_0\, dxdydz \\ &= \rho_0 a^2 c \iiint\limits_{\bar{x}^2+\bar{y}^2+\bar{z}^2\leq 1} (2c^2\bar{z}^2 - a^2\bar{x}^2 - a^2\bar{y}^2)\, d\bar{x}d\bar{y}d\bar{z}. \end{aligned}$$

(To simplify the integrals, we have performed the change of variables according to $x = a\bar{x}$, $y = a\bar{y}$ and $z = c\bar{z}$.) Then, since

$$\begin{aligned} \iiint\limits_{\bar{x}^2+\bar{y}^2+\bar{z}^2\leq 1} \bar{x}^2 d\bar{x}d\bar{y}d\bar{z} &= \iiint\limits_{\bar{x}^2+\bar{y}^2+\bar{z}^2\leq 1} \bar{y}^2 d\bar{x}d\bar{y}d\bar{z} = \iiint\limits_{\bar{x}^2+\bar{y}^2+\bar{z}^2\leq 1} \bar{z}^2 d\bar{x}d\bar{y}d\bar{z} \\ &= \frac{1}{3}\int_0^1 \bar{r}^2 \cdot 4\pi\bar{r}^2 d\bar{r} = \frac{4\pi}{15}, \end{aligned}$$

the above quadrupole moment Q gets the value

$$Q = \frac{8\pi}{15}\rho_0 a^2 c(c^2 - a^2) \quad \left(= \frac{2}{5}M(c^2 - a^2)\right).$$

This is an exact result. With $c = a(1-\eta)$ and η small, M and Q can be taken as

$$\begin{aligned} M &= \frac{4\pi}{3}\rho_0 a^3(1-\eta), \\ Q &\approx -\frac{4}{5}Ma^2\eta \approx -\frac{16\pi}{15}\rho_0 a^5\eta. \end{aligned}$$

6-10 As the gravitational potential found in Problem 6-9 can be rewritten in the form

$$\mathcal{G} = -\frac{GM}{r} - \frac{GQ}{4r^3}\left(\frac{3z^2}{r^2} - 1\right),$$

we find, for the related gravitational field,

$$\vec{g} = -\vec{\nabla}\mathcal{G} = -\frac{GM}{r^3}\vec{r} + \frac{3GQ}{4r^5}\left[\left(1 - \frac{5z^2}{r^2}\right)\vec{r} + 2z\mathbf{k}\right].$$

In the equatorial plane, i.e., with $z = 0$, this gives rise to the following force on an Earth satellite of mass m:

$$\vec{F} = \left(-\frac{GmM}{r^2} + \frac{3GmQ}{4r^4}\right)\hat{r}.$$

We see that the quadrupole term adds a radial perturbing force $\frac{3GmQ}{4r^4}\hat{r}$ ($\equiv \bar{f}(r)\hat{r}$) to the Keplerian force. The potential energy corresponding to this force is

$$V(r) = -\frac{GmM}{r} + \frac{GmQ}{4r^3}.$$

Using polar coordinates (r, θ) in the equatorial plane (with the Earth's center at the origin), the corresponding orbit equation for $u(\theta) \equiv \frac{1}{r(\theta)}$ reads (see Sec. 4.3)

$$\frac{L^2}{2m}\left(\frac{du}{d\theta}\right)^2 + \frac{L^2}{2m}u^2 + V\left(\frac{1}{u}\right) = E$$
$$\left(V\left(\frac{1}{u}\right) = -GmMu + \frac{GmQ}{4}u^3\right)$$

($L = mr^2\dot{\theta}$ denotes the conserved angular momentum), or

$$\frac{d^2u}{d\theta^2} + u = \frac{Gm^2M}{L^2} - \frac{m}{L^2}\frac{1}{u^2}\bar{f}\left(\frac{1}{u}\right) \quad \left(\bar{f}\left(\frac{1}{u}\right) = \frac{3}{4}GmQu^4\right). \quad (1)$$

The unperturbed solution (i.e., with $\bar{f}(\frac{1}{u})$ set to zero in (1)) is given by

$$u_0(\theta) = \frac{1}{r_0}[1 + e\cos(\theta - \theta_0)] \quad \left(r_0 \equiv \frac{L^2}{Gm^2M}\right), \quad (2)$$

where e is the eccentricity. Then, assuming that e is small (i.e., for a nearly circular orbit), we may approximate the perturbation term in

(1) by

$$\frac{m}{L^2}\frac{1}{u^2}\bar{f}\left(\frac{1}{u}\right) = \frac{3}{4}\frac{Gm^2Q}{L^2r_0^2}[1 + 2e\cos(\theta - \theta_0) + O(e^2)]. \quad (3)$$

Now, to first order in e, the orbit equation (1) can be written as

$$\frac{d^2u}{d\theta^2} + u - A = -\frac{3}{2}\frac{Gm^2Q}{L^2r_0^2}e\cos(\theta - \theta_0) \quad (4)$$

where $A = \frac{1}{r_0}\left(1 - \frac{3}{4}\frac{Gm^2Q}{L^2r_0}\right)$. We then try a solution of the form

$$u(\theta) = A\{1 + e\cos[(1 + B)(\theta - \theta_0)]\}$$

with (4), to find (to leading order in the perturbation term) that

$$B = \frac{1}{2A}\left(\frac{3}{2}\frac{Gm^2Q}{L^2r_0^2}\right) \approx \frac{3}{4}\frac{Gm^2Q}{L^2r_0}.$$

This gives the precession angle (per revolution)

$$\Delta\theta_p = -2\pi B = -\frac{3\pi}{2}\frac{Gm^2Q}{L^2r_0}$$

for the satellite.

6-11 From Problem 6-10, we know that the additional force on the satellite of mass m due to the Earth's quadrupole term is given by (with $Q < 0$)

$$\vec{F}_q = \frac{3GmQ}{4r^5}\left[\left(1 - \frac{5z^2}{r^2}\right)\vec{r} + 2z\mathbf{k}\right].$$

The first part in the square bracket, being a central force, does not change the magnitude and direction of the angular momentum about the center of the Earth; hence, it has no effect on the plane of orbit. The second part

$$\vec{F}' = \frac{3GmQ}{2r^5}z\mathbf{k}, \quad (1)$$

is not a central force and produces the torque (with respect to the Earth's center)

$$\vec{\Gamma} = \vec{r} \times \vec{F}' = \frac{3GmQ}{2r^5}(yz\mathbf{i} - xz\mathbf{j}), \quad (2)$$

which causes the orbital plane to precess about the z-axis. Here, $\vec{r} = (x, y, z)$ denotes the position of the satellite.

Let the line of intersection of the orbital plane and the equatorial plane of the Earth at time t be directed along (see the figure)

$$\widetilde{\mathbf{e}}_1(t) = \cos\varphi(t)\mathbf{i} + \sin\varphi(t)\mathbf{j}.$$

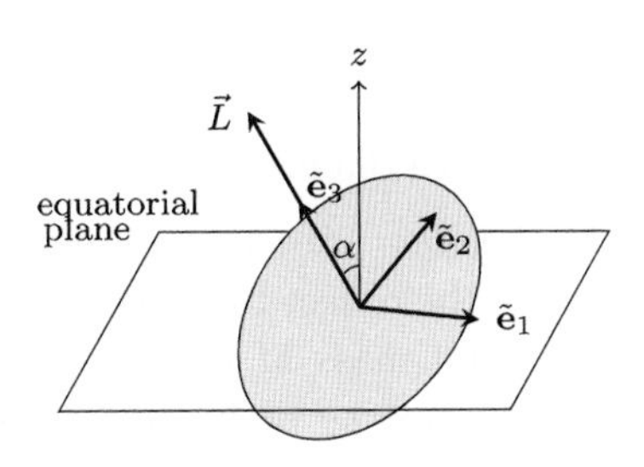

(Note that without the above $\vec{F}'$ term we must have $\dot{\varphi} = 0$.) Then the unit vector perpendicular to the orbital plane can be represented as

$$\widetilde{\mathbf{e}}_3(t) = -\sin\alpha[-\sin\varphi(t)\mathbf{i} + \cos\varphi(t)\mathbf{j}] + \cos\alpha\mathbf{k},$$

and another unit vector (in the orbital plane) $\widetilde{\mathbf{e}}_2(t)$, for an orthonormal triplet, as

$$\begin{aligned}\widetilde{\mathbf{e}}_2(t) &= \widetilde{\mathbf{e}}_3(t) \times \widetilde{\mathbf{e}}_1(t)\\ &= \cos\alpha[-\sin\varphi(t)\mathbf{i} + \cos\varphi(t)\mathbf{j}] + \sin\alpha\mathbf{k}.\end{aligned}$$

In terms of $(\widetilde{\mathbf{e}}_1, \widetilde{\mathbf{e}}_2, \widetilde{\mathbf{e}}_3)$, we can represent the location of the satellite at time t by

$$\vec{r}(t) = a_0[\cos(\omega t + \theta_0)\widetilde{\mathbf{e}}_1(t) + \sin(\omega t + \theta_0)\widetilde{\mathbf{e}}_2(t)],$$

where ω ($\gg |\dot{\varphi}|$) denotes the angular velocity of the satellite. Using this representation of $\vec{r}(t)$, the torque (2) is given as

$$\begin{aligned}\vec{\Gamma} = \frac{3GmQ}{2a_0^3}(\sin\alpha\sin(\omega t + \theta_0))\Big[&\cos(\omega t + \theta_0)\big(\sin\varphi(t)\mathbf{i} - \cos\varphi(t)\mathbf{j}\big)\\ &+ \sin(\omega t + \theta_0)\cos\alpha\big(\cos\varphi(t)\mathbf{i} + \sin\varphi(t)\mathbf{j}\big)\Big].\end{aligned} \tag{3}$$

In view of $\omega \gg |\dot{\varphi}|$, we may here take the time average as regards the rotational motion of the satellite; then, (3) can be simplified as

$$\vec{\Gamma} \sim \frac{3GmQ}{4a_0^3}\sin\alpha\cos\alpha\left(\cos\varphi(t)\mathbf{i} + \sin\varphi(t)\mathbf{j}\right). \tag{4}$$

Then, writing $\vec{L}(t) = L\,\widetilde{\mathbf{e}}_3(t)$ (with $L = ma_0^2\omega$ and $\widetilde{\mathbf{e}}_3(t)$ given above) for the angular momentum vector and using

$$\frac{d\vec{L}(t)}{dt} = \vec{\Gamma},$$

we find, for the precessional angular velocity, the expression

$$\dot{\varphi} = \frac{3GQ}{4a_0^5\omega}\cos\alpha.$$

Hence the precession angle per revolution is $\dot{\varphi}\frac{2\pi}{\omega} = \frac{6\pi GQ}{4a_0^5\omega^2}\cos\alpha$.

6-12 Let $\vec{R} = X\mathbf{i}+Y\mathbf{j}+Z\mathbf{k}$ denote the CM of the object from the location of the point mass M (see the figure). Relative to the body-fixed axes $(\mathbf{i}', \mathbf{j}', \mathbf{k}')$ where $\mathbf{k}'$ coincides with the symmetry axis of the object, we may also represent this vector $\vec{R}$ by the form

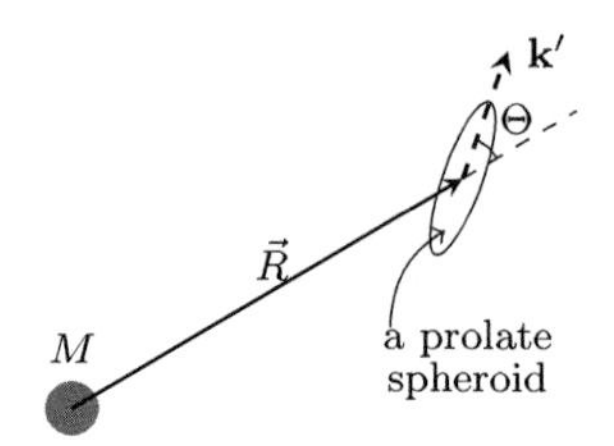

$$\vec{R} = X'\mathbf{i}' + Y'\mathbf{j}' + Z'\mathbf{k}'$$

with $Z' = \vec{R}\cdot\mathbf{k}' = R\cos\Theta$ and $X'^2 + Y'^2 = R^2\sin^2\Theta$. Then, using the formula from Sec. 6.2, we have the quadrupole-dependent part of the gravitational potential given by

$$\begin{aligned} V_{\text{ex}}(\vec{R},\alpha)\Big|_{\text{quad}} &= \frac{1}{6}Q_{ij}\frac{\partial}{\partial X_i}\frac{\partial}{\partial X_j}\left(-\frac{GM}{R}\right) \\ &= \frac{1}{6}Q'_{ij}\frac{\partial}{\partial X'_i}\frac{\partial}{\partial X'_j}\left(-\frac{GM}{R}\right), \end{aligned}$$

where Q'_{ij} denote quadrupole moments defined relative to the body axes $(\mathbf{i}', \mathbf{j}', \mathbf{k}')$. Then note that

$$Q'_{11} = Q'_{22} = -\frac{1}{2}Q, \quad Q'_{33} = Q \quad (Q'_{ij} = 0 \text{ if } i \neq j)$$

and

$$\frac{\partial}{\partial X'_i}\frac{\partial}{\partial X'_j}\left(-\frac{GM}{R}\right) = -GM\frac{3X'_iX'_j - R^2\delta_{ij}}{R^5}.$$

Hence we find

$$\begin{aligned} V_{\text{ex}}(\vec{R},\alpha)|_{\text{quad}} &= -\frac{GM}{6R^5}\Big[-\frac{1}{2}Q\{2(X')^2 - (Y')^2 - (Z')^2\} \\ &\quad -\frac{1}{2}Q\{2(Y')^2 - (X')^2 - (Z')^2\} \\ &\quad + Q\{2(Z')^2 - (X')^2 - (Y')^2\}\Big] \\ &= -\frac{GM}{4R^5}Q\left[-(X')^2 - (Y')^2 + 2(Z')^2\right] \end{aligned}$$

$$= -\frac{GM}{4R^5}Q\left[-R^2\sin^2\Theta + 2R^2\cos^2\Theta\right]$$

$$= -\frac{GM}{4R^3}Q\left[3\cos^2\Theta - 1\right].$$

Clearly, for $Q > 0$, this acquires the minimum for $\Theta = 0$ or π, i.e., when the symmetry axis points in the direction of the point mass M.

6-13 For a thin spherical shell of radius r_0 with an axially symmetric surface mass density $\sigma(\theta) = \sigma_0(\frac{3}{2}\cos^2\theta - \frac{1}{2})$, we find the gravitational quadrupole moment Q $(= Q_{33})$ given as (see Sec. 6.2)

$$Q = 2\pi r_0^2 \int_0^\pi d\theta \sin\theta\, r_0^2(2\cos^2\theta - \sin^2\theta)\left[\sigma_0\left(\frac{3}{2}\cos^2\theta - \frac{1}{2}\right)\right]$$

$$= 2\pi r_0^4\sigma_0 \int_{-1}^{1} dx\, \frac{1}{2}(3x^2 - 1)^2$$

$$= \frac{8\pi}{5}r_0^4\sigma_0\,.$$

Hence we can represent the related gravitational potential by the form (see Sec. 6.2)

$$\mathcal{G}_{\text{quad}}(\vec{r}) = -\frac{4\pi r_0^4}{5r^3}G\sigma_0\left(\frac{3}{2}\cos^2\theta - \frac{1}{2}\right) \quad (r > r_0). \tag{1}$$

Actually this potential is an exact one, valid as long as $r > r_0$, since all other gravitational multipole moments vanish for the given surface mass distribution. Notice that, as far as the angular dependences are concerned, the gravitational potential contains an identical factor as the surface mass density, i.e., $\frac{3}{2}\cos^2\theta - \frac{1}{2}$.

In the text we got, for the height of ocean tides, the expression $h(\theta) = h_0(\frac{3}{2}\cos^2\theta - \frac{1}{2})$ with $h_0 = \frac{mr_0^4}{MR^3}$, where r_0 (M) denotes the Earth's radius (mass). This was deduced by demanding that (on the ocean surface) the change in the Earth's gravitational potential, represented approximately by $\frac{GM}{r_0^2}h(\theta)$, becomes balanced by the tidal potential $\Phi_{\text{tidal}} = -Gm\frac{r_0^2}{R^3}(\frac{3}{2}\cos^2\theta - \frac{1}{2})$ due to the attraction from the Moon or Sun. This neglects the fact that the sea water, responsible for the tides, also feels the gravitational potential due to the tidal bulge itself — there is an effective gravitational quadrupole potential generated by the tidal bulge. To incorporate this effect in the tidal height calculation in a self-consistent manner, let us approximate the presence of the tidal bulge by introducing a surface mass density of the form $\sigma(\theta) = \rho_w \bar{h}(\theta)$. Here, ρ_w is the mass density of sea water and $\bar{h}(\theta)$ — the new formula for the height of ocean tide — is taken to have the form

$$\bar{h}(\theta) = \bar{h}_0 \left(\frac{3}{2} \cos^2\theta - \frac{1}{2} \right) \tag{2}$$

with an adjustable constant $\bar{h}_0$. This implies the surface mass density $\sigma(\theta) = \sigma_0\left(\frac{3}{2}\cos^2\theta - \frac{1}{2}\right)$ with $\sigma_0 = \rho_w \bar{h}_0$. Then, using the related gravitational potential in (1), the requirement that the surface of the ocean should be an equipotential can now be expressed by

$$\begin{aligned} 0 &= \frac{GM}{r_0^2}\bar{h}(\theta) + \mathcal{G}_{\text{quad}}(\vec{r})\Big|_{r=r_0} + \Phi_{\text{tidal}} \\ &= \frac{GM}{r_0^2}\bar{h}_0 \left(\frac{3}{2} \cos^2\theta - \frac{1}{2} \right) - \frac{4\pi}{5} r_0 G \rho_w \bar{h}_0 \left(\frac{3}{2} \cos^2\theta - \frac{1}{2} \right) \\ &\quad - Gm\frac{r_0^2}{R^3} \left(\frac{3}{2} \cos^2\theta - \frac{1}{2} \right). \end{aligned}$$

This leads to the value

$$\bar{h}_0 = \frac{1}{(1 - \frac{3}{5}\frac{\rho_w}{\rho_e})} \frac{m r_0^4}{MR^3} = \frac{h_0}{(1 - \frac{3}{5}\frac{\rho_w}{\rho_e})}, \tag{3}$$

where we defined $\rho_e \equiv M/\frac{4}{3}\pi r_0^3$, the average mass density of the Earth. [Given that $\rho_e \approx 5.5 \times 10^3\,\text{kg}\,\text{m}^{-3}$ and $\rho_w \approx 1.0 \times 10^3\,\text{kg}\,\text{m}^{-3}$, one finds $\bar{h}_0 \approx 1.12\, h_0$, i.e., the height of the tides is increased by about 12 percent with the gravitational self-interaction of the tidal bulge taken into account.]

6-14 Let $\hat{u}$ denote the (instantaneous) direction to the centers of the two small spheres from the Earth's center. Then the gravitational forces due to the Earth on the two spheres are given, respectively, by

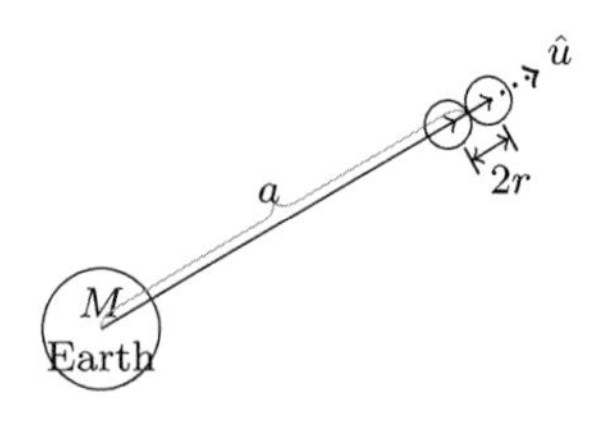

on the outer sphere:

$$-\frac{GM(\frac{4}{3}\pi r^3 \rho)}{(a+r)^2}\hat{u} = -\frac{GM(\frac{4}{3}\pi r^3 \rho)}{a^2}\left(1 - \frac{2r}{a} + \cdots\right)\hat{u},$$

on the inner sphere:

$$-\frac{GM(\frac{4}{3}\pi r^3 \rho)}{(a-r)^2}\hat{u} = -\frac{GM(\frac{4}{3}\pi r^3 \rho)}{a^2}\left(1 + \frac{2r}{a} + \cdots\right)\hat{u},$$

where $M = \frac{4}{3}\pi R_E^3 \rho_E$ is the Earth's mass. This shows that, aside from the term giving the overall inward acceleration $-\frac{GM}{a^2}\hat{u}$, the Earth also exerts a tidal force in the form of *mutual repulsive force* of the strength $\frac{GM(\frac{4}{3}\pi r^3 \rho)}{a^2}\left(\frac{2r}{a}\right)$. At the same time the two spheres are subject to the mutual attractive gravitational force of the strength $\frac{G(\frac{4}{3}\pi r^3\rho)^2}{(2r)^2}$. These two forces are balanced if a is equal to a_c, with a_c determined by

$$\frac{GM(\frac{4}{3}\pi r^3 \rho)}{a_c^2}\left(\frac{2r}{a_c}\right) = \frac{G(\frac{4}{3}\pi r^3 \rho)^2}{(2r)^2}.$$

This gives $a_c^3 = 8\frac{M}{(\frac{4}{3}\pi r^3 \rho)}$ or $a_c = 2\left(\frac{\rho_E}{\rho}\right)R_E$, using $M = \frac{4}{3}\pi R_E^3 \rho_E$. If $a < a_c$, the tidal force becomes dominant, and as a result the two spheres will be pulled apart.

6-15 (i) If the observer fixed to the ceiling sees the photon with $P^0 = P^z = \frac{h\omega_e}{2\pi} = h\nu_e$, the measured energy value by the outside observer (moving in the $+z$-direction with relative speed $v = \frac{gt}{c}$) should be

$$\bar{P}^0 \equiv h\nu_o = \frac{P^0 - \frac{v}{c}P^z}{\sqrt{1-(\frac{v}{c})^2}} = h\nu_e\sqrt{\frac{1-v/c}{1+v/c}}.$$

That is, $\nu_o = \nu_e\sqrt{\frac{1-v/c}{1+v/c}}$ (this is the Doppler-shift formula). Hence, for $\Delta\nu \equiv \nu_o - \nu_e$, one finds

$$\frac{\Delta\nu}{\nu_e} = \sqrt{\frac{1-v/c}{1+v/c}} - 1 \overset{v \ll c}{\simeq} -\frac{v}{c} = -\frac{gh}{c^2}.$$

(ii) In the case of (i), one has $\nu_o < \nu_e$; this corresponds to a red-shift. On the other hand, for a photon *falling down* (and hence $P^0 = -P^z = h\nu_e$), its energy seen by the outside observer becomes

$$\bar{P}^0 \equiv h\nu_o = h\nu_e\frac{1+\frac{v}{c}}{\sqrt{1-(\frac{v}{c})^2}} = h\nu_e\sqrt{\frac{1+v/c}{1-v/c}}.$$

So one now finds $\nu_o = \nu_e\sqrt{\frac{1+v/c}{1-v/c}}$ and $\frac{\Delta\nu}{\nu_e} \sim \frac{v}{c} = \frac{gh}{c^2}$. In this case, $\nu_o > \nu_e$ and this corresponds to a blue-shift.

7 Rigid Body Dynamics I

7-1 If one has some idea on how this kind of game works, it will not take much time to realize that "for the pile not to topple over, the CM of all the bricks lying above the point P_0 (shown in the figure) should be located on *the left of* P_0." Why? Otherwise, the torque balance equation about the point P_0, applied to the "rigid body" consisting of all the bricks above the point P_0, cannot be satisfied because the upward normal force from the bottom brick may act on the body only at the left of the point P_0. So, taking the origin (in the horizontal direction) at the position of P_0, we have the condition

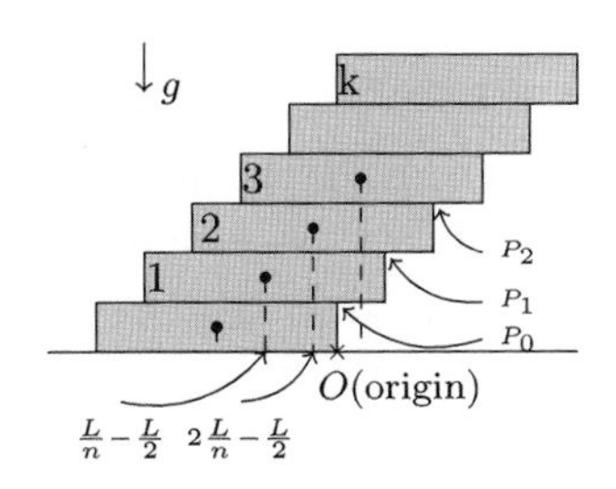

$$
\begin{aligned}
&(\text{the CM position of the } k \text{ bricks above } P_0)\\
&\quad = \frac{(\frac{L}{n}-\frac{L}{2})+(2\frac{L}{n}-\frac{L}{2})+\cdots+(k\frac{L}{n}-\frac{L}{2})}{k} \leq 0\\
&\quad \longrightarrow \quad \frac{L}{n}(1+2+\cdots+k)-\frac{L}{2}k \leq 0\\
&\quad \longrightarrow \quad k \leq n-1.
\end{aligned}
$$

Hence the desired answer is at most n bricks, including the one at the bottom.

7-2 Let N and F be the normal force and frictional force at the foot of the ladder, and N' the normal force at the wall (with zero frictional force, i.e., $F' = 0$ for the smooth wall). Then the force balance gives two conditions,

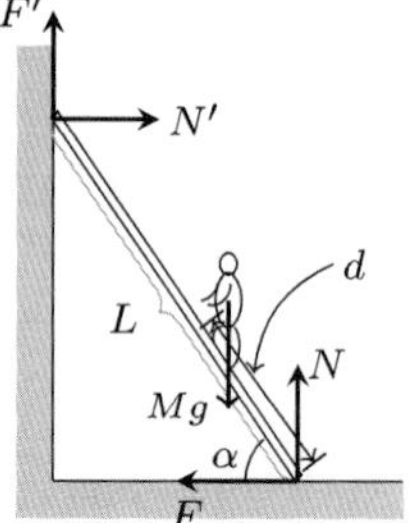

$$N = Mg, \tag{1}$$

$$N' = F. \tag{2}$$

It is convenient to consider the torque balance about the foot of the ladder, which gives the condition

$$N'L\sin\alpha = Mgd\cos\alpha, \tag{3}$$

where d is the distance that the man has climbed up the ladder. From the last two conditions, we find

$$F = \frac{Mgd\cos\alpha}{L\sin\alpha},$$

while we must have $F = \mu N$ ($= \mu Mg$, from (1)) on the verge of slipping. Hence a man of mass M can climb a distance $d = \mu L\tan\alpha$ up the ladder without it slipping.

Now consider the case when the wall is also rough with the same coefficient of friction μ. On the verge of slipping, the frictions at both contacts must again correspond to limiting friction, i.e., $F = \mu N$ and $F' = \mu N'$. In this case, the following three conditions should hold at the point of slipping:

$$\mu N' + N = Mg,$$
$$N' = \mu N$$
$$N'L\sin\alpha + \mu N'L\cos\alpha = Mgd\cos\alpha.$$

Eliminating N and N' from these conditions, we find that the maximal distance d that a man can climb without making the ladder slip is

$$d = \frac{\mu}{1+\mu^2}L(\mu + \tan\alpha).$$

7-3 Consider slices of the cylinders, having unit thickness. Let m be the mass per unit length of each cylinder, and a the radius of the cylinders. Then, because of the symmetry of the configuration, the cylinders should be in equilibrium under the action of the normal forces and frictional forces (on the slices) as shown in the figure.

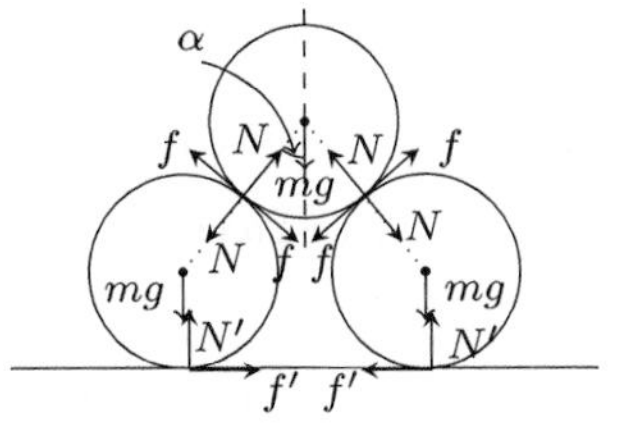

Equating the vertical component of the total force acting on the slices to zero gives

$$2N' - 3mg = 0.$$

The horizontal component of the total force is automatically zero (as the symmetry of the configuration has been used). We may then apply the equilibrium conditions to one of the lower cylinders. Now, for the force balance, we find

$$f' + f\cos\alpha - N\sin\alpha = 0,$$
$$N' - f\sin\alpha - N\cos\alpha - mg = 0,$$

while the torque balance equation (about the axis of the cylinder) gives

$$af' - af = 0.$$

After a little algebraic manipulation, the above four equations give

$$f = f' = \frac{3mg\sin\alpha}{2(1+\cos\alpha)}, \quad N' = \frac{3}{2}mg, \quad N = \frac{1}{2}mg.$$

Since $f \le \mu N$ and $f' \le \mu' N'$, it follows that

$$\frac{3\sin\alpha}{1+\cos\alpha} \le \mu, \quad \frac{\sin\alpha}{1+\cos\alpha} \le \mu'.$$

[Remark: Note that this problem has applications to the storage of cylindrical objects, for example tree trunks.]

7-4 With the coordinate origin taken at the point A, the forces acting on the rod are as follows:

(i) a force $-Mg\mathbf{k}$ at $-a\cos\alpha\mathbf{i} - a\sin\alpha\sin\theta\mathbf{j} + a\sin\alpha\cos\theta\mathbf{k}$ (: the CM position),

(ii) a normal force $\vec{N} = N\mathbf{i}$ and friction $\vec{f} = f_y\mathbf{j} + f_z\mathbf{k}$ at point B, i.e., at $-2a\cos\alpha\mathbf{i} - 2a\sin\alpha\sin\theta\mathbf{j} + 2a\sin\alpha\cos\theta\mathbf{k}$,

(iii) a reaction force $\vec{f}'$ at A.

Consider the torque balance equation *about the point* A:

$$\begin{aligned}\vec{\Gamma}_A &= (-a\cos\alpha\mathbf{i} - a\sin\alpha\sin\theta\mathbf{j}) \times (-Mg\mathbf{k}) \\ &\quad + (-2a\cos\alpha\mathbf{i} - 2a\sin\alpha\sin\theta\mathbf{j} + 2a\sin\alpha\cos\theta\mathbf{k}) \times (N\mathbf{i} + f_y\mathbf{j} + f_z\mathbf{k}) \\ &= 0\end{aligned}$$

$$\rightarrow \quad \begin{cases} a\sin\alpha(Mg\sin\theta - 2f_z\sin\theta - 2f_y\cos\theta) = 0 \\ a(-Mg\cos\alpha + 2f_z\cos\alpha + 2N\sin\alpha\cos\theta) = 0 \\ 2a(-f_y\cos\alpha + N\sin\alpha\sin\theta) = 0 \end{cases} \tag{1}$$

When the point B is on the verge of sliding, $\vec{f}$ should act along the tangent at B to the vertical circle (with center P and radius $\overline{PB} = 2a\sin\alpha$) and $|\vec{f}| = \mu N$; hence, we can identify

$$f_y = \mu N\cos\theta, \qquad f_z = \mu N\sin\theta. \tag{2}$$

Substituting these into the last equation of (1), we find

$$-\mu N\cos\alpha\cos\theta + N\sin\alpha\sin\theta = 0 \quad \rightarrow \quad \tan\theta = \mu\cot\alpha$$

and so θ is determined. [Using other equation in (1) with (2), it is possible to determine the corresponding magnitudes of f_y and f_z as well.]

7-5 The moment of inertia of the disc about O is $\frac{1}{2}Ma^2$ and the torque by the frictional force about O is $-Fa$. The equation of motion of the disc is therefore

$$\frac{1}{2}Ma^2\ddot{\theta} = -Fa.$$

Since F is constant, this can be integrated to give

$$\frac{1}{2}Ma^2\dot{\theta} = -Fat + \text{const}.$$

Now $\dot{\theta} = \omega_0$ when $t = 0$, so that the constant of integration is $\frac{1}{2}Ma^2\omega_0$. Hence

$$\frac{1}{2}Ma^2\dot{\theta} = -Fat + \frac{1}{2}Ma^2\omega_0.$$

The disc will be reduced to rest in a time T, found by putting $\dot{\theta} = 0$ and $t = T$ in the above equation. This yields

$$F = \frac{Ma\omega_0}{2T}.$$

7-6 Recall that the moment of inertia about the axis through CM is $I_{\rm CM} = \frac{1}{4}Ma^2 + \frac{1}{12}M(l_1+l_2)^2$. Then, applying the parallel axis theorem, we find the moment of inertia about the axis in the figure to be

$$\begin{aligned} I_x &= I_{\rm CM} + M\left(\frac{l_1+l_2}{2} - l_1\right)^2 \\ &= \frac{1}{4}Ma^2 + \frac{1}{12}M\Big((l_1+l_2)^2 + 3(l_2-l_1)^2\Big) \\ &= \frac{1}{4}Ma^2 + \frac{1}{3}M(l_1^2 + l_2^2 - l_1 l_2). \end{aligned}$$

Denoting the rotation angle by θ, the torque due to the gravitational force is given by $Mg\frac{l_2-l_1}{2}\sin\theta$. The equation of motion is thus given by

$$I_x\ddot{\theta} = -Mg\frac{l_2-l_1}{2}\sin\theta.$$

This is identical to the equation of motion for a simple pendulum. If the angle θ is small, this can be approximated by

$$\ddot{\theta} + \frac{Mg(l_2-l_1)}{2I_x}\theta = 0,$$

and hence the cylinder oscillates with the period

$$T = 2\pi\sqrt{\frac{2I_x}{Mg(l_2-l_1)}} = 2\pi\left(\frac{3a^2 + 4(l_1^2+l_2^2-l_1l_2)}{6g(l_2-l_1)}\right)^{1/2}.$$

7-7 The moment of inertia of the rod about the axis through the hinge is $I_m = \frac{1}{3}M(2a)^2$, and accordingly the equation of rotation is

$$\frac{4}{3}Ma^2\ddot{\theta} = -Mag\sin\theta.$$

Comparing this equation with the equation for a simple pendulum, we determine the length of the equivalent simple pendulum as $l = \frac{4}{3}a$. If the impulse I is applied at a point a distance b below the hinge, the rotational impulse is Ib and we have the equation $I_m\omega = Ib$: this determines the angular velocity $\omega = \frac{Ib}{I_m} = \frac{3}{4}\frac{Ib}{Ma^2}$. CM moves with the velocity $v_{\rm CM} = a\omega = \frac{3}{4}\frac{Ib}{Ma}$. Then we obtain the linear momentum of the rod, $Mv_{\rm CM} = \frac{3}{4}\frac{Ib}{a}$. However, if there is no impulsive reaction at the hinge, the linear momentum should be equal to the impulse I. From these considerations, we conclude that there is no impulsive reaction at the hinge only when the rod is struck at the distance

$b = \frac{4}{3}a$. When $b > \frac{4}{3}a$, there should be an impulsive reaction at the hinge of magnitude

$$I_{\text{hinge}} = \left(\frac{3b}{4a} - 1\right) I$$

and in the same direction as I. When $b < \frac{4}{3}a$, the direction of the reaction is opposite.

7-8 (i) Since there is no external force acting on the rod after collision, we can characterize the state of the rod after collision: (a) its CM moves with some constant velocity V (here, in the same direction as the incident velocity of the projectile) and (b) it rotates about its CM with some constant angular velocity ω. (See the figure.) Then, from the conservation of total linear momentum and that of total angular momentum with respect to the position of the CM of the rod before collision (the point O in the figure) and after the collision, we must have

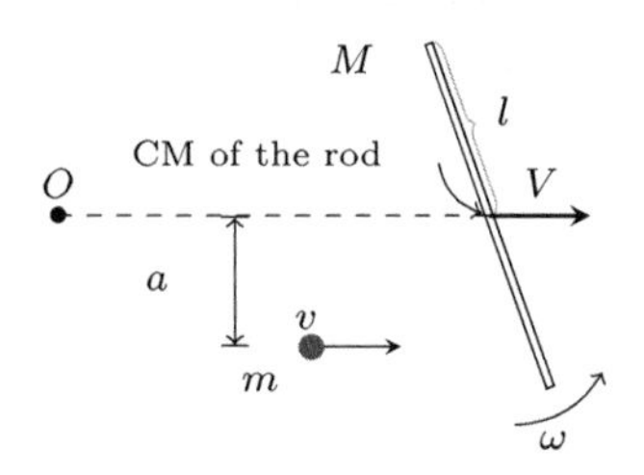

"After collision"

$$mv_0 = mv + MV, \tag{1}$$

$$mv_0 a = mva + I\omega, \tag{2}$$

where v is the speed of the projectile after collision, and I is the moment of inertia of the rod about the perpendicular axis through its CM, i.e., $I = \frac{1}{12}(2l)^2 M = \frac{1}{3}l^2 M$. Also, since the total kinetic energy must be conserved in an elastic collision, we have

$$\frac{1}{2}mv_0^2 = \frac{1}{2}mv^2 + \frac{1}{2}MV^2 + \frac{I}{2}\omega^2. \tag{3}$$

From (1) and (2) we can write

$$V = \frac{m}{M}(v_0 - v), \quad \omega = \frac{ma}{I}(v_0 - v) = \frac{m}{M}\frac{3a}{l^2}(v_0 - v). \tag{4}$$

Using these in (3), we can readily determine v:

$$v = -\left[\frac{1 - (1 + \frac{3a^2}{l^2})\frac{m}{M}}{1 + (1 + \frac{3a^2}{l^2})\frac{m}{M}}\right] v_0. \tag{5}$$

Then, from (4), we obtain

$$V = \frac{2\frac{m}{M}}{1 + (1 + \frac{3a^2}{l^2})\frac{m}{M}} v_0, \tag{6}$$

$$\omega = \frac{6\frac{m}{M}}{1 + (1 + \frac{3a^2}{l^2})\frac{m}{M}} \frac{a}{l^2} v_0. \tag{7}$$

Note that $\omega = 0$ if $a = 0$, i.e., when the rod is hit at its center.

The expression (7) as a function of the impact distance a has the following behavior:

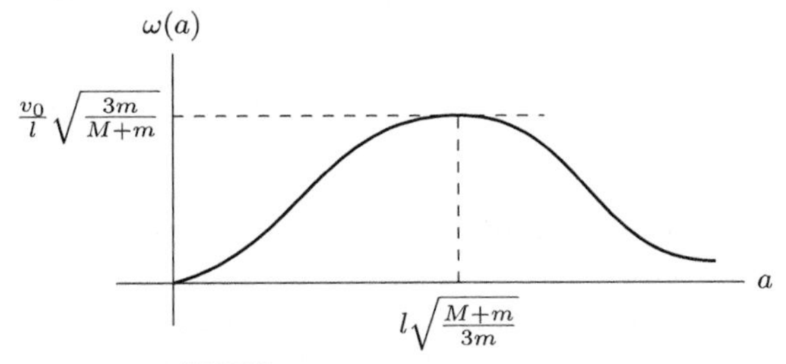

If $M \geq 2m$, then $l\sqrt{\frac{M+m}{3m}} \geq l$; in this case, $\omega(a)$ is monotonically increasing in the physical domain $0 \leq a \leq l$. So, if $M \geq 2m$, the largest angular velocity is obtained when $a = l$, i.e., when the rod is hit at the end. But, if $M < 2m$, the largest angular velocity is obtained when $a = l\sqrt{\frac{M+m}{3m}}$ $(< l)$. [If the rod is very light, i.e., for $M \ll m$, the largest angular velocity results when $a \approx l/\sqrt{3}$.]

(ii) After totally inelastic collision, suppose that the CM of the rod–projectile composite moves with constant velocity V and the composite rotates about its CM with constant angular velocity ω. Note that the CM of composite is different from the CM of the rod alone (see the figure). Then, from the conservation of total linear momentum and that of total angular momentum with respect to the position of the CM of the rod before collision, we find

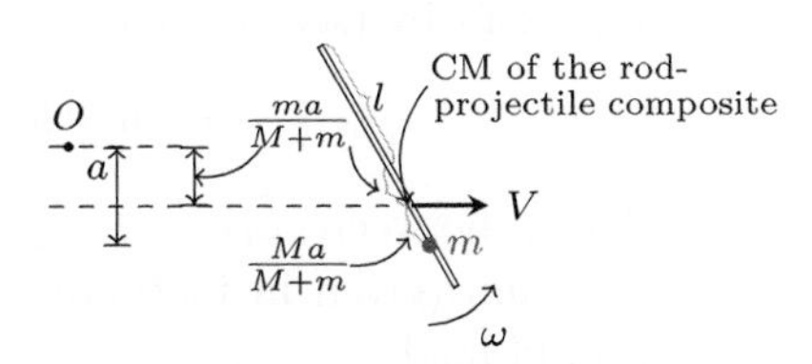

After totally inelastic collision

$$mv_0 = (M + m)V, \tag{8}$$

$$mv_0 a = (M + m)\left(\frac{m}{M + m}a\right)V + I'\omega, \tag{9}$$

where I', the moment of inertia of the composite about the perpendicular axis through its CM, is given by

$$I' = \frac{l^2}{3}M + M\left(\frac{m}{M+m}a\right)^2 + m\left(\frac{M}{M+m}a\right)^2$$
$$= \frac{l^2}{3}M + \frac{mM}{M+m}a^2.$$

From (8) and (9), we thus obtain

$$V = \frac{m}{M+m}v_0,$$
$$\omega = \frac{1}{I'}ma(v_0 - V) = \frac{3\frac{m}{M}}{1 + \left(1 + \frac{3a^2}{l^2}\right)\frac{m}{M}}\frac{a}{l^2}v_0$$

in the case of totally inelastic collision. Somewhat surprisingly, the expression obtained here for ω differs from the corresponding expression in the elastic collision case, the formula (7) above, only by an overall factor. Hence, as to the impact distance dependence of the angular velocity, we can give the same statement as made in (i).

7-9 The given pendulum together with the suspension point O defines a rigid body. Let $\vec{L}_O$ and $\vec{\Gamma}_O$ denote the angular momentum and the external torque about the suspension point O whose position we can represent by (see the figure)

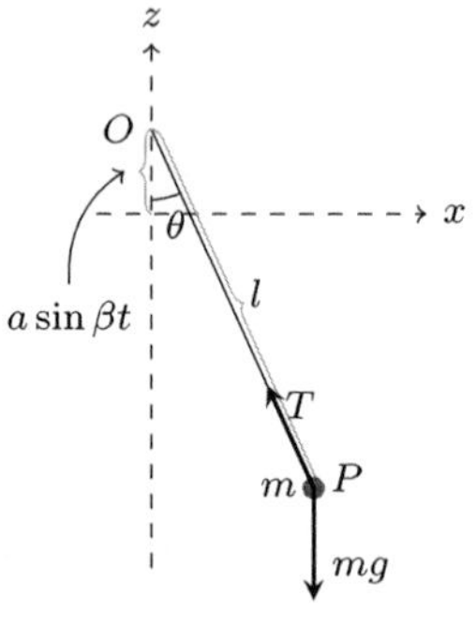

$$\vec{r}_O = a\sin\beta t\,\mathbf{k}.$$

Then, between $\vec{L}_O$ and $\vec{\Gamma}_O$, the following equation (given in footnote 1 of Chapter 7) should hold:

$$\dot{\vec{L}}_O = \vec{\Gamma}_O - m(\vec{R} - \vec{r}_O) \times \ddot{\vec{r}}_O, \qquad (1)$$

where we can identify $\vec{R} - \vec{r}_O = \overrightarrow{OP} = l\sin\theta\mathbf{i} - l\cos\theta\mathbf{k}$, since the center of mass $\vec{R}$ of the system can be taken at the position of the bob. Here

$$\dot{\vec{L}}_O = \frac{d}{dt}\{m(l\sin\theta\mathbf{i} - l\cos\theta\mathbf{k}) \times [l\dot{\theta}(\cos\theta\mathbf{i} + \sin\theta\mathbf{k})]\}$$
$$= -ml^2\ddot{\theta}\mathbf{j},$$
$$\vec{\Gamma}_O = (l\sin\theta\mathbf{i} - l\cos\theta\mathbf{k}) \times (-mg\mathbf{k}) = mgl\sin\theta\mathbf{j},$$

$$(\vec{R}-\vec{r}_O)\times\ddot{\vec{r}}_O = (l\sin\theta\mathbf{i} - l\cos\theta\mathbf{k})\times(-a\beta^2\sin\beta t\mathbf{k})$$
$$= al\beta^2\sin\beta t\sin\theta\mathbf{j},$$

and hence inserting these in (1) gives the equation of motion

$$-l^2\ddot{\theta} = gl\sin\theta - al\beta^2\sin\beta t\sin\theta$$

or

$$\ddot{\theta} + \left[\frac{g}{l} - \frac{a\beta^2}{l}\sin\beta t\right]\sin\theta = 0.$$

[Remark: The same equation of motion can be derived from Newton's laws of motion applied to the bob. Namely, if we represent the position of the bob at time t by

$$x = l\sin\theta(t),$$
$$z = a\sin\beta t - l\cos\theta(t),$$

we must have, from Newton's laws of motion,

$$m\frac{d^2x}{dt^2} = ml(\ddot{\theta}\cos\theta - \dot{\theta}^2\sin\theta) = -T\sin\theta, \tag{2}$$
$$m\frac{d^2z}{dt^2} = m\big[-a\beta^2\sin\beta t - l(-\ddot{\theta}\sin\theta - \dot{\theta}^2\cos\theta)\big] = T\cos\theta - mg, \tag{3}$$

where T denotes the tension along the string. Multiplying (2) and (3) by $\cos\theta$ and $\sin\theta$, respectively, and then adding the two equations give

$$ml\ddot{\theta} + m(-a\beta^2)\sin\beta t\sin\theta = -mg\sin\theta$$

or

$$\ddot{\theta} + \left[\frac{g}{l} - \frac{a\beta^2}{l}\sin\beta t\right]\sin\theta = 0.$$]

7-10 (i) If m is the mass of the ladder, the moment of inertia about its axis passing through the CM is $I_{\text{CM}} = \frac{1}{3}ma^2$. For the rotational motion about the CM, we thus have the equation (see the figure)

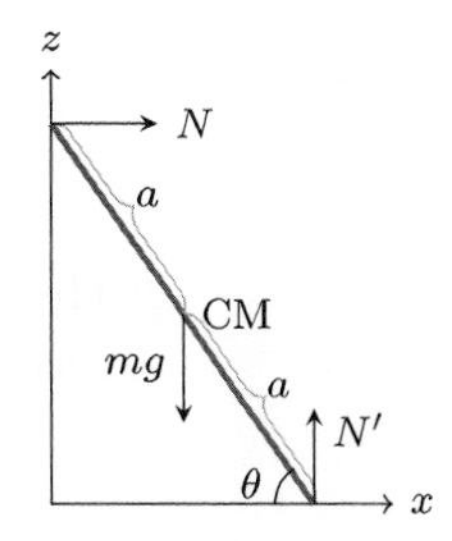

$$\frac{1}{3}ma^2\ddot{\theta} = Na\sin\theta - N'a\cos\theta, \tag{1}$$

where θ is the inclination of the ladder to the horizontal, and N and N' are the reactions at the wall and the floor, respectively. In addition, as the acceleration of the CM can be represented as

$$\begin{aligned}\vec{a}_{\rm CM} &= \frac{d^2}{dt^2}(a\cos\theta\mathbf{i} + a\sin\theta\mathbf{k})\\ &= a\frac{d}{dt}(-\dot{\theta}\sin\theta\mathbf{i} + \dot{\theta}\cos\theta\mathbf{k})\\ &= (-a\ddot{\theta}\sin\theta - a\dot{\theta}^2\cos\theta)\mathbf{i} + (a\ddot{\theta}\cos\theta - a\dot{\theta}^2\sin\theta)\mathbf{k},\end{aligned}$$

we have the following equations governing the CM motion:

$$m(-a\ddot{\theta}\sin\theta - a\dot{\theta}^2\cos\theta) = N, \tag{2}$$

$$m(a\ddot{\theta}\cos\theta - a\dot{\theta}^2\sin\theta) = N' - mg. \tag{3}$$

The motion of the ladder can be determined using (1), (2) and (3).

(ii) Eliminating the reactions N and N' between the three equations (1), (2) and (3) gives, after simplification,

$$\frac{4}{3}a\ddot{\theta} = -g\cos\theta. \tag{4}$$

Multiplying this by $\dot{\theta}$ and integrating gives

$$\frac{2}{3}a\dot{\theta}^2 = -g\sin\theta + \text{const.},$$

and then, using $\dot{\theta} = 0$ when $\theta = \alpha$, we obtain

$$\frac{2}{3}a\dot{\theta}^2 = -g\sin\theta + g\sin\alpha. \tag{5}$$

(This is essentially the energy conservation equation.) Now, eliminating $\ddot{\theta}$ and $\dot{\theta}^2$ from our equations (2) and (3) (by using (4) and (5)), we obtain the following expressions for the reactions:

$$\begin{aligned}N &= \frac{3}{4}mg(3\sin\theta\cos\theta - 2\sin\alpha\cos\theta),\\ N' &= mg + \frac{3}{4}mg(-\cos^2\theta + 2\sin^2\theta - 2\sin\theta\sin\alpha).\end{aligned}$$

The ladder will cease to touch the wall when the reaction N becomes zero. This leads to the equation

$$3\sin\theta\cos\theta = 2\sin\alpha\cos\theta.$$

The solution $\cos\theta = 0$ corresponds to the ladder standing in equilibrium in the upright position. The other solution is

$$\sin\theta = \frac{2}{3}\sin\alpha$$

so that the upper end of the ladder will indeed have fallen a distance equal to one third of its original height.

7-11 (i) Let **k** be a unit vector drawn vertically upward. Then the reaction of the table on the ball (of mass m) at the contact point P may be written as

$$\vec{f} + N\mathbf{k} \quad (N = mg),$$

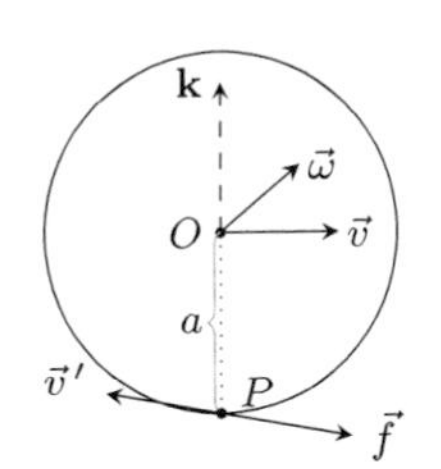

where $\vec{f}$ denotes the friction (in the horizontal direction), and $N\mathbf{k}$ the normal force. When the ball is slipping, the equation of motion for the CM is

$$m\frac{d\vec{v}}{dt} = \vec{f} = -\mu mg\frac{\vec{v}'}{|\vec{v}'|} \tag{1}$$

where $\vec{v}\,'$ — the velocity of the particle at P — is given by

$$\vec{v}\,' = \vec{v} + \vec{\omega} \times (-a\mathbf{k}). \tag{2}$$

At the same time, as the angular momentum about O can be written as $\vec{L} = I_0\vec{\omega}$ (I_0: the moment of inertia about O), the angular velocity must satisfy the equation of motion

$$I_0\frac{d\vec{\omega}}{dt} = (-a\mathbf{k}) \times (\vec{f} + mg\mathbf{k}) = -a\mathbf{k} \times \vec{f}. \tag{3}$$

(ii) For the above slipping motion, using (1) and (3) with (2) yields

$$\begin{aligned}\frac{d\vec{v}\,'}{dt} &= \frac{d\vec{v}}{dt} + \frac{d\vec{\omega}}{dt} \times (-a\mathbf{k}) \\ &= \frac{1}{m}\left(1 + \frac{ma^2}{I_0}\right)\vec{f} \quad \left(\vec{f} = -\mu mg\frac{\vec{v}\,'}{|\vec{v}\,'|}\right).\end{aligned} \tag{4}$$

Since $\frac{d\vec{v}\,'}{dt} \,//\, (-\vec{v}\,')$, the velocity of the contact point P has a *fixed direction*, say, $\vec{v}\,' = v'\mathbf{e}_0$ ($\mathbf{e}_0$: a fixed unit vector in the horizontal). Then we can also write $\vec{f} = -\mu mg\mathbf{e}_0$ and so, from (1),

$$\frac{d\vec{v}}{dt} = -\mu m g \mathbf{e}_0 \longrightarrow \left(\begin{array}{c}\text{the center, being subject to a uniform}\\ \text{acceleration, has a parabolic motion in general}\end{array}\right) \tag{5}$$

as long as slipping persists. Also, from (4),

$$\frac{dv'}{dt} = -\mu g\left(1+\frac{ma^2}{I_0}\right) \longrightarrow v'(t) = v_0' - \mu g\left(1+\frac{ma^2}{I_0}\right)t \tag{6}$$

where v_0', the initial velocity of slipping, can be determined from

$$v_0'\mathbf{e}_0 = \vec{v}_0 - a\vec{\omega}_0 \times \mathbf{k}.$$

(iii) From (6) we shall have $v' = 0$ when

$$t = \frac{v_0'}{\mu g}\frac{1}{\left(1+\frac{ma^2}{I_0}\right)}.$$

At this instant, *slipping ceases*; beyond this instant, the motion becomes a simple rolling (with vanishing friction) in a straight line with *constant values* for $\vec{v}$ and $\vec{\omega}$ (from (1) and (3) with $\vec{f} = 0$).

7-12 (i) For a uniform hemisphere, the CM is at a distance $\frac{3}{8}R$ from the center of the sphere since

$$\frac{2\pi\int_0^R dr\int_0^{\frac{\pi}{2}} d\theta\, r^2\sin\theta\, r\cos\theta}{\frac{2}{3}\pi R^3} = \frac{3}{R^3}\frac{R^4}{4}\left[-\frac{1}{4}\cos 2\theta\right]\Big|_{\theta=0}^{\theta=\frac{\pi}{2}} = \frac{3}{8}R.$$

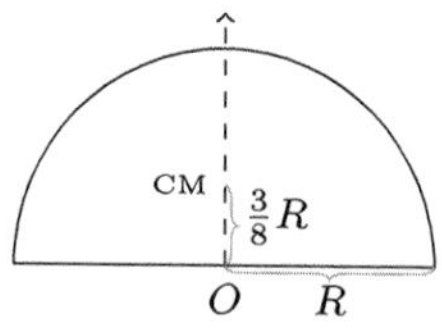

CM $\frac{3}{8}R$ O R

Accordingly, we have $z = R - \frac{3}{8}R\cos\theta$.

(ii) Here, $I_{\rm CM}$ can be obtained by using the parallel axis theorem:

$$I_O = I_{\rm CM} + \left(\frac{3}{8}R\right)^2 M \quad \left(\text{with } I_O = \frac{2}{5}MR^2\right)$$

$$\longrightarrow \quad I_{\rm CM} = \frac{2}{5}MR^2 - \frac{9}{64}MR^2 = \frac{83}{320}MR^2.$$

(iii) Since the hemisphere is subject only to the normal force and gravity, its CM can move along the vertical direction according to the equation

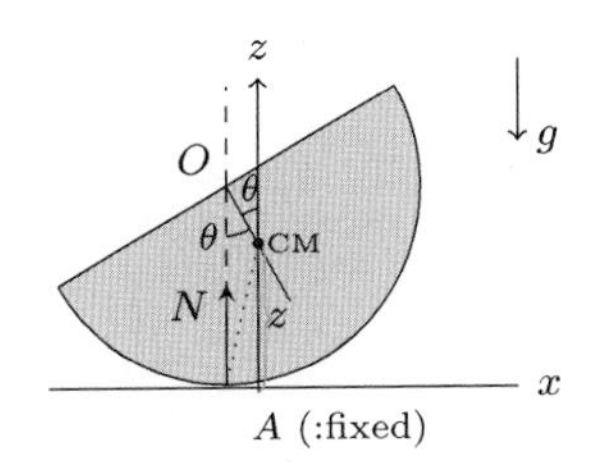

$$M\frac{d^2z}{dt^2} = N - Mg, \qquad (1)$$

where N represents the upward normal force at the contact point. On the other hand, the rotational motion about the CM is governed by

$$I_{\text{CM}}\ddot{\theta} = -\left(-\frac{3}{8}R\sin\theta\mathbf{i} - z\mathbf{k}\right) \times N\mathbf{k} = -\frac{3}{8}RN\sin\theta. \qquad (2)$$

But, since $z = R - \frac{3}{8}R\cos\theta$ (from (i)), we have from (1)

$$N = mg + \frac{3}{8}RM\frac{d}{dt}(\dot{\theta}\sin\theta) = M\left(g + \frac{3}{8}R\ddot{\theta}\sin\theta + \frac{3}{8}R\dot{\theta}^2\cos\theta\right),$$

and so inserting this into (2) gives the following equation for θ:

$$I_{\text{CM}}\ddot{\theta} = -\frac{3}{8}RM\sin\theta\left(g + \frac{3}{8}R\ddot{\theta}\sin\theta + \frac{3}{8}R\dot{\theta}^2\cos\theta\right). \qquad (3)$$

To determine the frequency of small oscillation about the equilibrium point (i.e., $\theta = 0$), note that (3) for small θ gives

$$I_{\text{CM}}\ddot{\theta} = -\frac{3}{8}RMg\theta + (\text{higher orders in } \theta).$$

Hence, for the related frequency ω, we find

$$\omega^2 = \frac{\frac{3}{8}RMg}{I_{\text{CM}}} = \frac{\frac{3}{8}RMg}{\frac{83}{320}MR^2} = \frac{120g}{83R}.$$

7-13 Obviously, by the symmetry argument, three principal axes through the center are provided by the x-, y- and z-axes. The moment of inertia about the x-axis is given by

$$I_x = \frac{M}{\frac{4}{3}\pi abc} \iiint_{\frac{x^2}{a^2}+\frac{y^2}{b^2}+\frac{z^2}{c^2}\leq 1} (y^2 + z^2) dx dy dz$$

$$= \frac{M \cancel{abc}}{\frac{4}{3}\pi \cancel{abc}} \iiint_{x'^2+y'^2+z'^2\leq 1} (b^2 y'^2 + c^2 z'^2) dx' dy' dz',$$

where we made changes of variables according to $x = ax'$, $y = by'$ and $z = cz'$. But,

$$\iiint_{\text{unit sphere}} y'^2 dx' dy' dz' = \iiint_{\text{unit sphere}} z'^2 dx' dy' dz'$$

$$= \frac{1}{2} \iiint_{\text{unit sphere}} (y'^2 + z'^2) dx' dy' dz'$$

$$= \frac{1}{2} \cdot \frac{2}{5} \left(\frac{4\pi}{3} \cdot 1^2 \right) = \frac{4\pi}{15},$$

and hence

$$I_x = \frac{3M}{4\pi} \frac{4\pi}{15} (b^2 + c^2) = \frac{M}{5} (b^2 + c^2).$$

The values for $I_y = \frac{M}{5}(c^2 + a^2)$ and $I_z = \frac{M}{5}(a^2 + b^2)$ follow in exactly the same way.

7-14 (i) Let us find the moment of inertia tensor about the center O. First we evaluate I_{xx}:

$$I_{xx} = M \frac{\int_{-L}^{L} \int_{-L}^{L} \int_{-L}^{L} dz dy dx (y^2 + z^2)}{(2L)^3}$$

$$= M \frac{(2L)\{2 \cdot 2 \cdot \frac{L^3}{3} \cdot 2L\}}{(2L)^3} = \frac{2}{3} ML^2.$$

Obviously, the x-, y- and z-axes are principal axes and so, by symmetry,

$$\mathbf{I}_O = ((I_O)_{kl}) = \begin{pmatrix} \frac{2}{3}ML^2 & 0 & 0 \\ 0 & \frac{2}{3}ML^2 & 0 \\ 0 & 0 & \frac{2}{3}ML^2 \end{pmatrix}.$$

Then the moment of inertia about an axis in the direction of $\hat{u}$ is

$$\begin{aligned} I_{O\hat{u}} &= \sum_i m_i(r_i^2 - (\vec{r}\cdot\hat{u})^2) \\ &= \hat{u}_k \left[\sum_i m_i(r_i^2\delta_{kl} - x_k x_l)\right] \hat{u}_l \\ &= \frac{2}{3}ML^2 \qquad \text{(regardless of the direction } \hat{u}\text{).} \end{aligned}$$

(ii), (iii) With the origin at the center, the corner point A is at (L, L, L). Now, we use the parallel axis theorem, i.e., $(I_A)_{kl} = (I_O)_{kl} + M(|\vec{X}|^2\delta_{kl} - X_k X_l)$ with $\vec{X} = -L\mathbf{i} - L\mathbf{j} - L\mathbf{k}$. The moment of inertia tensor about the point $A = (L, L, L)$ is given by

$$I_A = \begin{pmatrix} \frac{8}{3}ML^2 & -ML^2 & -ML^2 \\ -ML^2 & \frac{8}{3}ML^2 & -ML^2 \\ -ML^2 & -ML^2 & \frac{8}{3}ML^2 \end{pmatrix}.$$

By symmetry consideration (reflection symmetry about the plane containing O, $(L, L, \pm L)$ and $(-L, -L, \pm L)$), one principal axis can be taken to be in the direction

$$\hat{\varphi} = -\frac{1}{\sqrt{2}}\mathbf{i} + \frac{1}{\sqrt{2}}\mathbf{j}.$$

Furthermore, from (i), we know that $\hat{r} \equiv \frac{\overrightarrow{OA}}{|\overrightarrow{OA}|} = \frac{1}{\sqrt{3}}(\mathbf{i} + \mathbf{j} + \mathbf{k})$ is also a principal axis. Then the third principal axis is

$$\hat{\theta} = \hat{\varphi} \times \hat{r} = \frac{1}{\sqrt{6}}(-\mathbf{k} + \mathbf{j} - \mathbf{k} + \mathbf{i}) = \frac{1}{\sqrt{6}}(\mathbf{i} + \mathbf{j} - 2\mathbf{k}).$$

Now the corresponding principal moments are easily found as:

$$\text{about } \hat{r}: \quad I_{A\hat{r}} = \frac{2}{3}ML^2 \quad \text{(from (i))},$$

$$\text{about } \hat{\varphi}: \quad \begin{pmatrix} \frac{8}{3} & -1 & -1 \\ -1 & \frac{8}{3} & -1 \\ -1 & -1 & \frac{8}{3} \end{pmatrix}\begin{pmatrix} -1 \\ 1 \\ 0 \end{pmatrix} = \frac{11}{3}\begin{pmatrix} -1 \\ 1 \\ 0 \end{pmatrix}$$

$$\to I_{A\hat{\varphi}} = \frac{11}{3}ML^2,$$

$$\text{about } \hat{\theta}: \quad \begin{pmatrix} \frac{8}{3} & -1 & -1 \\ -1 & \frac{8}{3} & -1 \\ -1 & -1 & \frac{8}{3} \end{pmatrix} \begin{pmatrix} 1 \\ 1 \\ -2 \end{pmatrix} = \frac{11}{3} \begin{pmatrix} 1 \\ 1 \\ -2 \end{pmatrix}$$

$$\to I_{A\hat{\theta}} = \frac{11}{3} ML^2.$$

7-15 The center of mass is a point on the axis of symmetry at a distance $d = \frac{(R-r)r^3}{R^3-r^3}$ to the left of the center of the sphere [Why? We must have $(\frac{4}{3}\pi R^3 - \frac{4}{3}\pi r^3)d = \frac{4}{3}\pi r^3(R-r)$.] The body is a symmetric top. With respect to the axis of symmetry (the z-axis), the moment of inertia becomes (here, ρ denotes the mass density of the body)

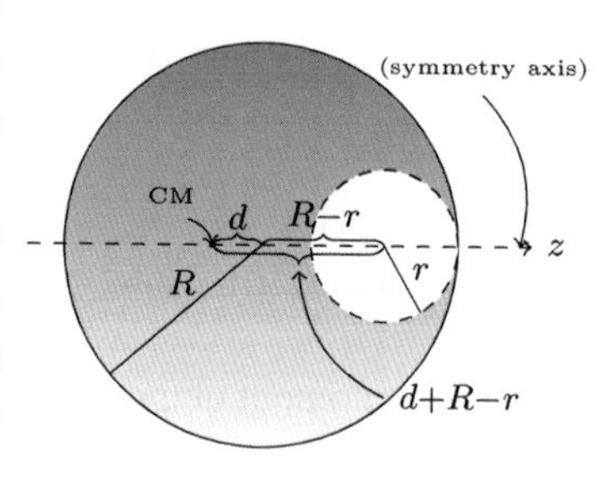

$$I_z = \frac{4}{3}\pi\rho\frac{2}{5}(R^5 - r^5).$$

On the other hand, with respect to any two perpendicular axes passing through the CM, we have the moment of inertia given by

$$\begin{aligned} I_x = I_y &= \frac{1}{2}(I_x + I_y) \\ &= \frac{1}{2}I_z + \sum_i m_i z_i'^2, \end{aligned}$$

where z' refers to the z-coordinate measured from the CM of the object. For the quantity $\sum_i m_i z_i'^2$, we may evaluate this quantity for the full sphere of radius R and that for the sphere of radius r at the location of the cavity, and then subtract the latter from the former. That is

$$\begin{aligned} \sum_i m_i z_i'^2 &= \left[\frac{4}{3}\pi\rho R^3 d^2 + \frac{1}{2}\frac{4}{3}\pi\rho\frac{2}{5}R^5\right] \\ &\quad - \left[\frac{4}{3}\pi\rho r^3(R-r+d)^2 + \frac{1}{2}\frac{4}{3}\pi\rho\frac{2}{5}r^5\right] \\ &= \frac{1}{2}I_z - \frac{4}{3}\pi\rho\frac{(R-r)^2 r^3 R^3}{R^3 - r^3}. \end{aligned}$$

Hence

$$I_x = I_y = I_z + \frac{4}{3}\pi\rho\frac{(R-r)^2}{(R^3-r^3)^2}(R^3r^6 - r^3R^6)$$
$$= \frac{4}{3}\pi\rho\left[\frac{2}{5}(R^5-r^5) - \frac{(R-r)^2r^3R^3}{R^3-r^3}\right].$$

All products of inertia vanish with the above choice of body axes.

7-16 (i) Notice that

$$(\vec{F}_{\text{tot}})_i = -\int d^3\vec{r}'\rho(r')\frac{\partial}{\partial x_i'}\mathcal{G}_{\text{ext}}(\vec{r}')$$
$$= -\int_0^a r'^2dr'\rho(r')\int d\Omega\frac{\partial}{\partial x_i'}\mathcal{G}_{\text{ext}}(\vec{r}')$$
$$= -\left(4\pi\int_0^a r'^2dr'\rho(r')\right)\left[\frac{\partial}{\partial x_i}\mathcal{G}_{\text{ext}}(\vec{r})\right]\Bigg|_{\vec{r}=0},$$

where we used the mean value theorem for the harmonic function $\frac{\partial}{\partial x_i'}\mathcal{G}_{\text{ext}}(\vec{r}')$. Since $4\pi\int_0^a r'^2dr'\rho(r') = M$, we thus find

$$\vec{F}_{\text{tot}} = -M[\vec{\nabla}\mathcal{G}_{\text{ext}}(\vec{r})]\Big|_{\vec{r}=0}.$$

(ii) Notice that

$$\vec{\Gamma}_{\text{ext}} = \int_{\text{sphere}} d^3\vec{r}'\rho(r')\vec{r}' \times [-\vec{\nabla}'\mathcal{G}_{\text{ext}}(\vec{r}')]$$
$$\longrightarrow \quad (\vec{\Gamma}_{\text{ext}})_i = -\int d^3\vec{r}'\rho(r')\epsilon_{ijk}x_j'\partial_k'\mathcal{G}_{\text{ext}}(\vec{r}')$$
$$= \int d^3\vec{r}'\epsilon_{kij}\cancel{x_j'[\partial_k'\rho(r')]\mathcal{G}_{\text{ext}}(\vec{r}')}$$
$$+ \text{(surface term)}.$$

Here the surface term is

$$-\int d^3\vec{r}'\partial_k'[\epsilon_{ijk}\rho(r')x_j'\mathcal{G}_{\text{ext}}(\vec{r}')]$$
$$= -\int a^2d\Omega'\cancel{\frac{x_k'}{a}\epsilon_{ijk}\rho(r')x_j'}\mathcal{G}_{\text{ext}}(\vec{r}')$$
$$= 0,$$

thanks to the divergence theorem. Therefore, we find $\vec{\Gamma}_{\text{ext}} = 0$.

(iii) Based on (i) and (ii), the motion of the given spherical body is governed by

$$\frac{d^2\vec{R}}{dt^2} = -\vec{\nabla}\mathcal{G}_{\text{ext}}(\vec{R}),$$

for the CM position $\vec{R}$ (= the center of the sphere)

$$\frac{d\vec{L}_{\vec{R}}}{dt} = 0 \quad (\vec{L}_{\vec{R}}: \text{ the total angular momentum about the CM}).$$

Since we can write $\vec{L}_{\vec{R}} = I_{\text{CM}}\,\vec{\omega}$ for a spherically symmetric rigid body, the second relation implies that $\vec{\omega} = \text{const.}$, i.e, the body undergoes *uniform rotation* about any given axis in space.

7-17 (i) By symmetry, body-fixed axes $(\mathbf{e}_1^*, \mathbf{e}_2^*, \mathbf{e}_3^* = \mathbf{e}_1^* \times \mathbf{e}_2^*)$ (see the figure) define the principal axes, where $\mathbf{e}_1^*$ and $\mathbf{e}_2^*$ lie in the plane of the slab while $\mathbf{e}_3^*$ is perpendicular to the slab. Corresponding principal moments are

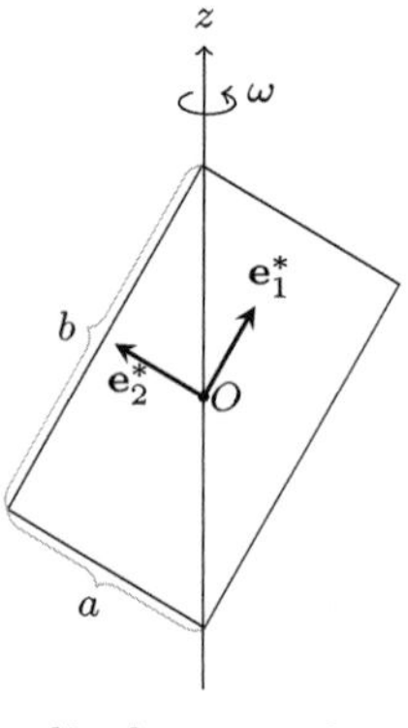

$$I_1 = \frac{m}{ab}\int_{-\frac{a}{2}}^{\frac{a}{2}} by^2 dy = \frac{m}{a}\frac{2}{3}\left(\frac{a}{2}\right)^3 = \frac{1}{12}ma^2,$$

$$I_2 = \frac{m}{ab}\int_{-\frac{b}{2}}^{\frac{b}{2}} ax^2 dx = \frac{m}{b}\frac{2}{3}\left(\frac{b}{2}\right)^3 = \frac{1}{12}mb^2,$$

$$I_3 = I_1 + I_2 = \frac{1}{12}m(a^2 + b^2).$$

In the last formula, we used the perpendicular axis theorem.

(ii) With respect to the body-fixed frame, $\vec{\omega} = \omega\hat{z}$ is expressed as

$$\vec{\omega} = \omega[(\hat{z}\cdot\mathbf{e}_1^*)\mathbf{e}_1^* + (\hat{z}\cdot\mathbf{e}_2^*)\mathbf{e}_2^*] = \frac{\omega b}{\sqrt{a^2+b^2}}\mathbf{e}_1^* + \frac{\omega a}{\sqrt{a^2+b^2}}\mathbf{e}_2^*$$

$$\longrightarrow \quad \omega_1 = \frac{b}{\sqrt{a^2+b^2}}\omega, \quad \omega_2 = \frac{a}{\sqrt{a^2+b^2}}\omega.$$

Hence, the angular momentum is given by

$$\vec{L} = I_1\omega_1\mathbf{e}_1^* + I_2\omega_2\mathbf{e}_2^* = \frac{m\omega ab}{12\sqrt{a^2+b^2}}(a\mathbf{e}_1^* + b\mathbf{e}_2^*).$$

(iii) From $\frac{d\vec{L}}{dt} = \vec{\Gamma}^{(e)}$ and $\hat{z}\cdot\vec{L} = \frac{m\omega ab}{12\sqrt{a^2+b^2}}\left(\frac{2ab}{\sqrt{a^2+b^2}}\right) = \frac{m\omega a^2b^2}{6(a^2+b^2)}$ (using the form obtained in (ii) for $\vec{L}$), the desired equation of motion, i.e., $\frac{dL_z}{dt} = \Gamma_z^{(e)}$, assumes the form

$$\frac{ma^2b^2}{6(a^2+b^2)}\frac{d\omega}{dt} = \Gamma_z^{(e)}(t).$$

(iv) If $\omega = \omega_0 = \text{const.}$ (i.e., $\vec{\omega} = \omega_0\hat{z}$),

$$\frac{d\vec{L}}{dt} = \vec{\Gamma}^{(e)} \longrightarrow \vec{\omega}\times\vec{L} = \vec{\Gamma}^{(e)},$$

since $\frac{d\mathbf{e}_i^*}{dt} = \vec{\omega}\times\mathbf{e}_i^*$. Hence the external torque that should be applied is

$$\begin{aligned}
\vec{\Gamma}^{(e)} &= \omega_0\hat{z}\times\frac{m\omega_0 ab}{12\sqrt{a^2+b^2}}(a\mathbf{e}_1^* + b\mathbf{e}_2^*)\\
&= \frac{m\omega_0^2 ab}{12(a^2+b^2)}(b\mathbf{e}_1^* + a\mathbf{e}_2^*)\times(a\mathbf{e}_1^* + b\mathbf{e}_2^*)\\
&= \frac{m\omega_0^2 ab}{12(a^2+b^2)}(b^2-a^2)\mathbf{e}_3^*.
\end{aligned}$$

7-18 Consider three particles with masses m_1, m_2 and m_3 at the location $\vec{r}_1$, $\vec{r}_2$ and $\vec{r}_3$, respectively. We can always choose the coordinate origin at their CM, so that

$$m_1\vec{r}_1 + m_2\vec{r}_2 + m_3\vec{r}_3 = 0. \tag{1}$$

The three noncollinear particles and their CM then define a plane, and then the forces will also fall within this plane. (Since gravitation is an attractive central force, this implies that the acceleration vector of any one particle must be directed toward the interior of the triangle defined by the three particles.)

Now, if the three particles move as a rigid body rotating with a uniform angular velocity $\vec{\omega}$, we can represent their velocities $\vec{v}_i \equiv \dot{\vec{r}}_i$ and accelerations $\vec{a}_i \equiv \dot{\vec{v}}_i$ by

$$\vec{v}_i = \vec{\omega}\times\vec{r}_i, \tag{2}$$

$$\vec{a}_i = \vec{\omega}\times\vec{v}_i = \omega\times(\vec{\omega}\times\vec{r}_i) = (\vec{\omega}\cdot\vec{r}_i)\vec{\omega} - \omega^2\vec{r}_i. \tag{3}$$

Since $\vec{a}_i$ is in the plane of the three particles, we should have $\vec{\omega}\cdot\vec{r}_i = 0$ for all three particles ($\vec{\omega}$ *cannot be in the plane of the three particles!*), i.e., $\vec{\omega}$ is *perpendicular to the plane*. Hence, from (3), we conclude that

$$\vec{a}_i = -\omega^2 \vec{r}_i \quad (i = 1, 2, 3). \tag{4}$$

Then use Newton's law of gravitation together with $\vec{F} = m\vec{a}$, say, to particle 1:

$$-\frac{Gm_1 m_2}{r_{12}^3}(\vec{r}_1 - \vec{r}_2) - \frac{Gm_1 m_3}{r_{13}^3}(\vec{r}_1 - \vec{r}_3) = -m_1 \omega^2 \vec{r}_1 \tag{5}$$

with the notation $r_{ij} \equiv |\vec{r}_i - \vec{r}_j|$. If we eliminate $\vec{r}_3$ in favor of $\vec{r}_1$ and $\vec{r}_2$ from this equation, the result is

$$\left(\frac{m_2}{r_{12}^3} - \frac{m_2}{r_{13}^3}\right)\vec{r}_2 = \left(\frac{m_1}{r_{13}^3} + \frac{m_2}{r_{12}^3} + \frac{m_3}{r_{13}^3} - \frac{\omega^2}{G}\right)\vec{r}_1 \,. \tag{6}$$

Since the particles are not collinear, both sides of (6) must vanish. The vanishing of the left hand side implies that $r_{12} = r_{13}$, and it is clear from symmetry that the same reasoning applied to particle 2 implies $r_{21} = r_{23}$. Hence

$$r_{12} = r_{13} = r_{23} \quad (\equiv s), \tag{7}$$

and the triangle is *equilateral.* Furthermore, the vanishing of the right hand side of (6) implies that

$$\omega^2 = G\frac{(m_1 + m_2 + m_3)}{s^3}.$$

[Remark: Because Newton's law of gravitation is not used above until the final stage, this derivation can easily be extended to treat particles that interact through any central force. In particular the equilateral triangle result (7) holds quite generally.]

8 Elements of Fluid Mechanics

8-1 (i) If a fluid of volume V expands against the given pressure p, the external force acting on an element of the boundary surface of area dA is $d\vec{F} = -(p\,dA)\hat{n}$, and accordingly the work done on the element by this force (as the element moves by $\delta\vec{r}$) becomes $d\vec{F} \cdot \delta\vec{r} = -p\,dA\hat{n} \cdot \delta r$. (See the figure.) Since $\hat{n} \cdot \delta\vec{r}$ represents the vertical distance traversed, $(dA)\hat{n} \cdot \delta\vec{r}$ equals the volume covered. Therefore, integrating over the entire surface, we find the net work

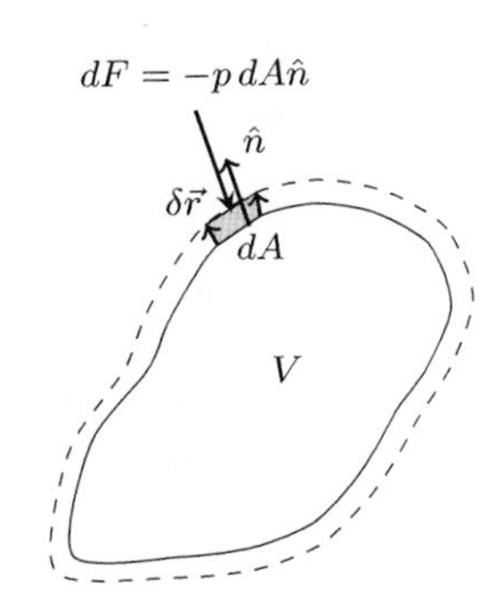

$$\delta W = -p\,dV,$$

where dV denotes the net volume change occurred in the process.

(ii) Being an adiabatic process, the total energy change is due to the external work done by pressure only; hence, we have

$$dU = -p\,dV$$

with $U = qpV$ (assuming an ideal gas). This implies that

$$q(p\,dV + V\,dp) = -p\,dV$$

or

$$\frac{dp}{p} = -\left(1 + \frac{1}{q}\right)\frac{dV}{V} = -\gamma\frac{dV}{V}$$

where $\gamma \equiv 1 + \frac{1}{q}$. Integrating the last equation yields

$$p = (\text{const.})V^{-\gamma} \qquad \text{or} \quad pV^{\gamma} = \text{const.}$$

8-2 The equation of state for an ideal gas can be written as $p = \frac{\rho}{m} k_B T$, where m is the average mass of an air molecule. Therefore, $p_0 = \frac{k_B}{m}\rho_0 T_0$ and so $\frac{k_B}{m} = \frac{p_0}{\rho_0 T_0}$. Then, we insert the formula $\rho(z) = \frac{m}{k_B}\frac{p(z)}{T(z)} = \frac{\rho_0 T_0}{p_0}\frac{p(z)}{T(z)}$ into the hydrostatic equation $\frac{dp(z)}{dz} = -\rho(z)g$, to obtain

$$\frac{dp(z)}{p(z)} = -g\frac{\rho_0 T_0}{p_0}\frac{dz}{T(z)}.$$

Integrating this yields

$$p(z) = p_0 e^{-\frac{g\rho_0 T_0}{p_0}\int_0^z \frac{dz}{T(z)}}.$$

From this formula, we find that

(i) with $T(z) = T_0\left(1 - \frac{z}{H}\right)$, we have $T_0\int_0^z \frac{dz}{T(z)} = -H\ln\left(1 - \frac{z}{H}\right)$ and so $p(z) = p_0\left(1 - \frac{z}{H}\right)^{\alpha_1}$ where $\alpha_1 = \frac{g\rho_0 H}{p_0}$, and

(ii) with $T(z) = T_0\left(1 - \frac{z^2}{H^2}\right)$, we have $T_0\int_0^z \frac{dz}{T(z)} = \frac{H}{2}\ln\left|\frac{1+z/H}{1-z/H}\right|$ and so $p(z) = p_0\left[\frac{1-z/H}{1+z/H}\right]^{\alpha_2}$ where $\alpha_2 = \frac{\alpha_1}{2} = \frac{g\rho_0 H}{2p_0}$.

8-3 (i) If we substitute the gravitational body force $\vec{f}(\vec{r}) = -\rho(r)\frac{GM(r)}{r^2}\hat{r}$ ($M(r) = 4\pi\int_0^r \rho(r')r'^2\,dr'$ is the mass interior to the radius r) into the hydrostatic equilibrium condition $\vec{\nabla}p = \vec{f}$, we find

$$\vec{\nabla}p(r) = \frac{dp(r)}{dr}\hat{r} = -\frac{GM(r)\rho(r)}{r^2}\hat{r}$$

and hence

$$\frac{dp(r)}{dr} = -\frac{GM(r)\rho(r)}{r^2}. \tag{1}$$

Note that, to counteract gravity, the pressure must decrease with increasing radius.

(ii) With constant ρ $(= M/(\frac{4}{3}\pi R^3)$, if M and R denote the mass and radius of the Earth), we have $M(r) = \frac{4}{3}\pi r^3\rho = M\left(\frac{r}{R}\right)^3$. Now, using the well-known gravitational acceleration $g = \frac{GM}{R^2} \approx 9.8\,\mathrm{m\,s^{-2}}$ at the Earth's surface. We can write (1) as

$$\frac{dp(r)}{dr} = -\frac{GM(r/R)^3\rho}{r^2} = -\rho g\left(\frac{r}{R}\right),$$

so that

$$p(r) = p_a + \frac{\rho g}{2R}(R^2 - r^2),$$

where p_a is the pressure at the Earth's surface. The maximum pressure is found at the Earth's center, with the value $p(0) = p_a + \frac{\rho g}{2}R$. We can here express the average density ρ by $\rho = gR^2/G(\frac{4}{3}\pi R^3) = \frac{3g}{4\pi GR}$, where $G \approx 6.67 \times 10^{-11}\,\mathrm{m^3\,kg^{-1}s^{-2}}$. Hence

$$p(0) = p_a + \frac{3g^2}{8\pi G} \approx 10^6\,\mathrm{atm}.$$

8-4 (i) Using the result of Problem 8-3(i), we have

$$\frac{dp(r)}{dr} = -\frac{GM(r)\rho(r)}{r^2},$$

and multiplying both sides of this equation by $4\pi r^3 dr$ and integrating from $r = 0$ to r_* yields

$$\int_0^{r_*} 4\pi r^3 \frac{dp}{dr} dr = -\int_0^{r_*} \frac{GM(r)\rho(r)}{r} 4\pi r^2 dr. \tag{1}$$

But the right hand side of this equation can be equated to the gravitational self-energy since

$$\begin{aligned}
E_{\text{gr}} &= -\frac{G}{2}\int_{r'\le r_*} d^3\vec{r}\,' \int_{r\le r_*} d^3\vec{r}\,[\theta(r'-r)+\theta(r-r')]\,\frac{\rho(r')\rho(r)}{|\vec{r}\,'-\vec{r}|} \\
&= -\frac{1}{2}\int_{r'\le r_*} d^3\vec{r}\,'\rho(r') \int_{r\le r'} d^3\vec{r}\,\frac{G\rho(r)}{|\vec{r}\,'-\vec{r}|} \\
&\quad -\frac{1}{2}\int_{r\le r_*} d^3\vec{r}\,\rho(r) \int_{r'\le r} d^3\vec{r}\,'\,\frac{G\rho(r')}{|\vec{r}\,'-\vec{r}|} \\
&= -\frac{1}{2}\int_0^{r_*} dr'\, 4\pi r'^2\,\frac{G\rho(r')M(r')}{r'} \\
&\quad -\frac{1}{2}\int_0^{r_*} dr\, 4\pi r^2\,\frac{G\rho(r)M(r)}{r} \\
&= -\int_0^{r_*} dr\, 4\pi r^2\,\frac{G\rho(r)M(r)}{r}\,.
\end{aligned} \tag{2}$$

(This in fact represents the energy that would be gained in constructing the star from the inside out, bringing from infinity shell by shell.) On the other hand, what we have on the left hand side of (1) can be written, after integrating by parts, as

$$[p(r)4\pi r^3]\big|_0^{r_*} - 3\int_0^{r_*} p(r)4\pi r^2 dr = -3\bar{p}V \tag{3}$$

(since $p(r_*) = 0$), where $\bar{p} = \frac{1}{V}\int_0^{r_*} p(r)4\pi r^2 dr$ is the volume-averaged pressure. Hence, from these results, we obtain $\bar{p} = -\frac{1}{3}\frac{1}{V}E_{\text{gr}}$.

(ii) For a star composed of a monoatomic ideal gas, we can assume that, at every point in the star, the local pressure p is related to the local thermal energy density $\mathcal{E}_{\text{th}}$ by

$$p = \frac{2}{3}\mathcal{E}_{\text{th}} \ . \tag{4}$$

[This follows from the equations $p = nk_BT$ and $\mathcal{E}_{\text{th}} = \frac{3}{2}nk_BT$ (n is the number density), valid for any monoatomic ideal gas.] Multiplying both sides of (4) by $4\pi r^2$ and integrating from $r = 0$ to r_*, we find

$$\bar{p}V = \frac{2}{3}E_{\text{th}}^{\text{tot}} \,, \tag{5}$$

where $E_{\text{th}}^{\text{tot}} = \int_0^{r_*} \mathcal{E}_{\text{th}}(r)4\pi r^2 dr$ is the total thermal energy of the star. Since $\bar{p}V = -\frac{1}{3}E_{\text{gr}}$ from (1), we thus obtain

$$E_{\text{th}}^{\text{tot}} = -\frac{1}{2}E_{\text{gr}} \ . \tag{6}$$

This equation says that when a star contracts and loses energy, i.e., its gravitational self-energy becomes more negative, then its thermal energy rises. From (6) we also see that the stars have *negative heat capacity* since their temperatures rise when they lose energy.

[Remark: The total energy of a star, both gravitational and thermal, is given by

$$E_{\text{total}} = E_{\text{th}}^{\text{tot}} + E_{\text{gr}} = \frac{1}{2}E_{\text{gr}} \,,$$

where we used (6). Thus the total energy of a star is negative, meaning that the star is bound. Since all stars constantly radiate away their energy (and hence E_{total} becomes more negative), they are doomed to collapse eventually, i.e., E_{gr} approaches negative infinity (if the star is composed of a *classical* ideal gas).]

8-5 Initially the forces acting on the ball are the gravitational force mg, the buoyant force of the water and the normal force exerted by the rim of the hole. When the buoyant force just equals the weight of the ball, the normal force becomes zero and the ball leaves the hole.

Let us calculate the buoyant force exerted on the ball when the water depth is h. We may here denote the volume of the ball immersed in water by V (see the figure). If the space under flattened part of the ball, having area πr^2, were also filled with water, the buoyant force would be

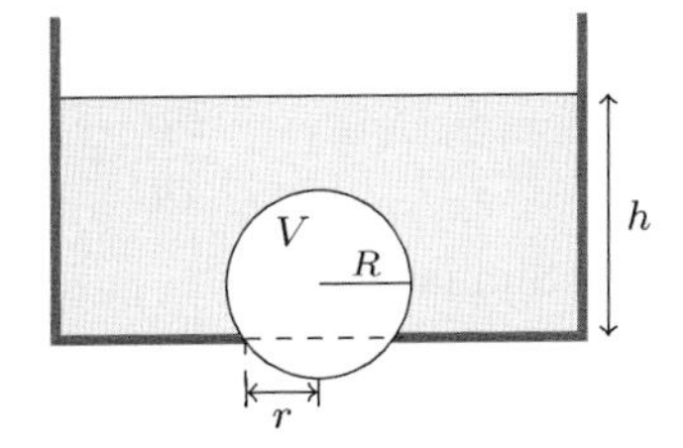

$$F' = \rho g V,$$

where ρ is the density of water. But, in our case, there is no water under the flattened part and so the actual buoyant force on the ball due to pressure, F_p is less a contribution $\rho g h \pi r^2$ from F', i.e.,

$$F_p = \rho g V - \rho g h \pi r^2.$$

Notice that, for sufficiently large h, F_p can be negative and then the "buoyant" force becomes directed downwards. Decreasing h causes F_p to increase and, provided the ball is not uncovered, it will rise when $F_p = mg$ and the corresponding water depth is h_0. Thus

$$h_0 = \frac{V}{\pi r^2} - \frac{m}{\pi r^2 \rho} \, . \tag{1}$$

For the volume of the truncated sphere, a straightforward calculation gives

$$V = \frac{\pi}{3}\left[2R^3 + (2R^2 + r^2)\sqrt{R^2 - r^2}\right].$$

Using this formula in (1), we can express h_0 as

$$h_0 = \frac{2R^3}{3r^2} + \frac{2R^2 + r^2}{3r^2}\sqrt{R^2 - r^2} - \frac{m}{\pi r^2 \rho} \, .$$

(This formula is valid if the top of the ball is still covered by water, i.e., when $h > R + \sqrt{R^2 - r^2}$.)

8-6 Note that

$$\left.\frac{\partial \rho'}{\partial t}\right|_{\vec{x}'} = \vec{u}\cdot\vec{\nabla}\rho + \frac{\partial \rho}{\partial t},$$

$$\vec{\nabla}'\cdot(\rho'\vec{v}') = \vec{\nabla}\big[\rho(\vec{v}-\vec{u})\big] = \vec{\nabla}\cdot(\rho\vec{v}) - \vec{u}\cdot\vec{\nabla}\rho,$$

and accordingly

$$\frac{\partial \rho'}{\partial t} + \vec{\nabla}'\cdot(\rho'\vec{v}') = \frac{\partial \rho}{\partial t} + \vec{\nabla}\cdot(\rho\vec{v}) = 0.$$

Similarly,

$$\left.\frac{\partial \vec{v}'}{\partial t}\right|_{\vec{x}'} = (\vec{u}\cdot\vec{\nabla})\vec{v} + \frac{\partial \vec{v}}{\partial t},$$

$$(\vec{v}'\cdot\vec{\nabla}')\vec{v}' = (\vec{v}\cdot\vec{\nabla})\vec{v} - (\vec{u}\cdot\vec{\nabla})\vec{v},$$

$$\frac{1}{\rho'}\vec{\nabla}'p' = \frac{1}{\rho}\vec{\nabla}p,$$

so that we have

$$\frac{\partial \vec{v}'}{\partial t} + (\vec{v}'\cdot\vec{\nabla}')\vec{v}' = \frac{\partial \vec{v}}{\partial t} + (\vec{v}\cdot\vec{\nabla})\vec{v} = -\frac{1}{\rho}\vec{\nabla}p = -\frac{1}{\rho'}\vec{\nabla}'p'.$$

8-7 (i) Since the fluid is incompressible, the corresponding velocity field $\vec{v}(\vec{r})$ should satisfy the equations

$$\vec{\nabla}\cdot\vec{v} = 0, \quad \vec{\nabla}\times\vec{v} = \vec{\Omega} = w(\rho)\mathbf{k}.$$

This is the same problem as that of determining the magnetic field $\vec{B}$ in space when there is an electric current $\vec{J} \propto w(\rho)\mathbf{k}$. Then, based on this magnetostatic analogy, we may set $\vec{v}(\vec{r}) = f(\rho)\hat{\varphi}$; $f(\rho)$ here can be determined using Stokes's theorem (see the figure)

$$\int_S \vec{\nabla}\times\vec{v}(\vec{r})\cdot\hat{n}\,da = \oint_{C=\partial S} \vec{v}\cdot d\vec{r}\,,$$

and this leads to

$$f(\rho) = \frac{1}{\rho}\left[\int_0^\rho w(\bar{\rho})\bar{\rho}\,d\bar{\rho}\right]. \tag{1}$$

Furthermore, with $\vec{\Omega} = w(\rho)\mathbf{k}$ and $\vec{v} = f(\rho)\hat{\varphi}$, we find $\vec{\Omega}\times\vec{v} = -w(\rho)f(\rho)\hat{\rho}$ and so $\vec{\nabla}\times[\vec{\Omega}\times\vec{v}] = 0$; hence the vorticity equation

$\frac{\partial \vec{\Omega}}{\partial t} + \vec{\nabla} \times (\vec{\Omega} \times \vec{v}) = \vec{\nabla} \times (\frac{1}{p}\vec{f})$, another equation involving $\vec{v}$ and $\vec{\Omega}$, is also fulfilled.

(ii) Substituting $w(\rho) = 2\omega_0$ (if $\rho \leq a$) and 0 (if $\rho > a$) in (1), we obtain

$$\rho \leq a: \qquad f(\rho) = \frac{1}{\rho}(2\omega_0)\frac{1}{2}\rho^2 = \omega_0 \rho,$$

$$\rho > a: \qquad f(\rho) = \frac{1}{\rho}(2\omega_0)\frac{1}{2}a^2 = \frac{a^2}{\rho}\omega_0.$$

Hence the velocity field in space is given by

$$\vec{v}(\vec{r}) = \begin{cases} -\omega_0 y\mathbf{i} + \omega_0 x\mathbf{j}, & \text{if } x^2 + y^2 \leq a^2 \\ \frac{a^2}{x^2+y^2}(-\omega_0 y\mathbf{i} + \omega_0 x\mathbf{j}), & \text{if } x^2 + y^2 \geq a^2 . \end{cases} \tag{2}$$

(iii) With $\frac{\partial \vec{v}}{\partial t} = 0$ and $\vec{f} = 0$, Euler's equation (Sec. 8.2) gives

$$-\vec{\nabla}\left(\frac{p}{\rho_0}\right) = \vec{\Omega} \times \vec{v} + \frac{1}{2}\vec{\nabla}(\vec{v}^2). \tag{3}$$

In the region $x^2 + y^2 \geq a^2$ where $\vec{\Omega} = 0$ and $\vec{v} = \frac{a^2}{\rho}\omega_0\hat{\varphi}$, we thus find

$$\frac{p}{\rho_0} = -\frac{1}{2}\frac{a^4\omega_0^2}{x^2 + y^2} + c_1\,, \tag{4}$$

where c_1 is a constant. In the region $x^2 + y^2 \leq a^2$ where $\vec{v}^2 = \omega_0^2\rho^2$ and $\vec{\Omega} \times \vec{v} = -2\omega_0^2\rho\hat{\rho} = -\omega_0^2\vec{\nabla}(\rho^2)$, we find from (3)

$$\frac{p}{\rho_0} = \frac{1}{2}\omega_0^2(x^2 + y^2) + c_2\,, \tag{5}$$

where c_2 is another constant. If the pressure on the z-axis, i.e., at $x = y = 0$ is equal to P_0, then $c_2 = P_0/\rho_0$. The pressure at $\rho = a$, obtained from formulas (4) and (5), must be equal and accordingly

$$-\frac{1}{2}a^2\omega_0^2 + c_1 = \frac{1}{2}\omega_0^2 a^2 + P_0/\rho_0.$$

From this, $c_1 = \omega_0^2 a^2 + P_0/\rho_0$. So the pressure at each point of the fluid is given by

$$p = \begin{cases} P_0 + \frac{1}{2}\omega_0^2\rho_0(x^2 + y^2), & \text{if } x^2 + y^2 \leq a^2 \\ P_0 + \omega_0^2\rho_0 a^2 - \frac{\omega_0^2\rho_0 a^4}{2(x^2+y^2)}, & \text{if } x^2 + y^2 \geq a^2. \end{cases}$$

8-8 (i) The time derivative of the energy density $\mathcal{E} = \frac{1}{2}\rho\vec{v}^2 + \rho u + \rho\mathcal{G}$ is

$$\begin{aligned}\frac{\partial\mathcal{E}}{\partial t} &= \left(\frac{1}{2}\vec{v}^2 + \mathcal{G}\right)\frac{\partial\rho}{\partial t} + \rho\vec{v}\cdot\frac{\partial\vec{v}}{\partial t} + \frac{\partial}{\partial t}(\rho u) \\ &= -\left(\frac{1}{2}\vec{v}^2 + \mathcal{G}\right)\vec{\nabla}\cdot(\rho\vec{v}) + \rho\vec{v}\cdot[-\vec{v}\cdot\vec{\nabla}\vec{v} - \vec{\nabla}h - \vec{\nabla}\mathcal{G}], \quad (1)\end{aligned}$$

where, for the second expression, we made use of the continuity equation and Euler's equation (with $\frac{1}{\rho}\vec{\nabla}p = \vec{\nabla}h$). Then, using $\rho\vec{v}\cdot[\vec{v}\cdot\vec{\nabla}\vec{v}] = \rho\vec{v}\cdot\vec{\nabla}(\frac{1}{2}\vec{v}^2)$ and $\frac{\partial}{\partial t}(\rho u) = h\frac{\partial\rho}{\partial t} = -h\vec{\nabla}\cdot(\rho\vec{v})$, it is possible to organize (1) in the form

$$\begin{aligned}\frac{\partial\mathcal{E}}{\partial t} &= -\left(\frac{1}{2}\vec{v}^2 + h + \mathcal{G}\right)\vec{\nabla}\cdot(\rho\vec{v}) - \rho\vec{v}\cdot\vec{\nabla}\left(\frac{1}{2}\vec{v}^2 + h + \mathcal{G}\right) \\ &= -\vec{\nabla}\cdot\left[\left(\frac{1}{2}\vec{v}^2 + h + \mathcal{G}\right)\rho\vec{v}\right], \quad (2)\end{aligned}$$

which is the desired result.

(ii) If we integrate (2) over an arbitrary fixed volume V and use the divergence theorem, we obtain the integral relation

$$\begin{aligned}&\frac{\partial}{\partial t}\int_V\left(\frac{1}{2}\rho\vec{v}^2 + \rho u + \rho\mathcal{G}\right)dxdydz \\ &\quad = -\int_{S=\partial V}\rho\left(\frac{1}{2}\vec{v}^2 + h + \mathcal{G}\right)\vec{v}\cdot\hat{n}\,da\,, \quad (3)\end{aligned}$$

which expresses the general energy conservation law for an ideal fluid. On the right hand side, there appears the *energy flux* across the boundary surface S; here, note that h and not the internal energy u is added to the mechanical energy $\frac{1}{2}\vec{v}^2 + \mathcal{G}$. This is because the fluid is not simply transporting the energy but also doing some work on the volume under consideration (for example, compressing it). It provides the additional energy flux $p\vec{v}$.

(iii) Using (3), one can easily obtain Bernoulli's theorem. To this end, choose for volume V the element of the small flux tube between (vertical) cross sections S_1 and S_2 and consider the case of a steady flow. Then it follows from (3) that

$$\rho_1 v_1 S_1\left(\frac{1}{2}\vec{v}_1^2 + h_1 + \mathcal{G}_1\right) = \rho_2 v_2 S_2\left(\frac{1}{2}\vec{v}_2^2 + h_2 + \mathcal{G}_2\right)$$

since $\vec{v}\cdot\hat{n} = 0$ at the lateral walls of the tube. This immediately yields Bernoulli's theorem, $\frac{1}{2}\vec{v}^2 + h + \mathcal{G} = \text{const.}$ (along the given streamline), because we have $\rho_1 v_1 S_1 = \rho_2 v_2 S_2$ from the continuity equation.

8-9 Let us consider some streamline connecting the free surface of the liquid in the vessel and the hole. We can apply Bernoulli's theorem here. At both level z_0 and level z, the pressure is atmospheric, i.e., $p = p_a$. If v_0 (≈ 0, assuming that the hole is sufficiently small) and v' denote the fluid velocities at level z_0 and level z (i.e., at the hole), Bernoulli's theorem then tells us that

$$gz + \frac{1}{2}v'^2 = gz_0.$$

Hence, $v' = \sqrt{2g(z - z_0)}$. In free fall in the gravitational field, the fluid will reach the bottom level after the time $T = \sqrt{\frac{2z}{g}}$. The horizontal distance covered during this time is $D = v'T = 2\sqrt{z(z_0 - z)}$. This attains maximum when $z = \frac{z_0}{2}$ (with the horizontal distance $D_{\max} = z_0$).

8-10 Let the height of the free surface above the orifice be z and let the radius of the free surface be x. If v_f and v_0 are the velocities of the fluid at the free surface and at the orifice respectively, then the equation of continuity yields

$$v_f \pi x^2 = v_0 \pi a^2, \tag{1}$$

where a is the radius of the orifice. The pressure at the free surface and at the orifice is atmospheric and therefore Bernoulli's theorem yields

$$\frac{1}{2}v_f^2 + gz = \frac{1}{2}v_0^2. \tag{2}$$

Eliminating v_0 from (1) and (2) gives

$$\frac{1}{2}v_f^2 + gz = \frac{1}{2}v_f^2 \frac{x^4}{a^4}$$

so that we have

$$z = \frac{v_f^2}{2ga^4}(x^4 - a^4). \tag{3}$$

Since v_f is to remain constant, the water jar should be formed by revolving the curve $z = \frac{v_f^2}{2ga^4}(x^4 - a^4)$ about z-axis. Since the orifice is small, a^4 may be neglected in comparison to x^4 and then the curve becomes

$$z = \frac{v_f^2}{2ga^4} x^4.$$

8-11 The equation of state for an ideal gas reads $p = \frac{\rho k_B T}{m}$, if m is the average molecular mass. Then, using our formula for the total internal energy $U = qNk_BT$ (see Chapter 5), the internal energy per unit mass is given by

$$u = \frac{q}{m} k_B T = q\frac{p}{\rho} = \frac{1}{\gamma - 1}\frac{p}{\rho},$$

where $\gamma = 1 + \frac{1}{q}$ denotes the adiabatic index. So the enthalpy per unit mass, h, becomes

$$h = u + \frac{p}{\rho} = \left(\frac{1}{\gamma - 1} + 1\right)\frac{p}{\rho} = \frac{\gamma}{\gamma - 1}\frac{p}{\rho},$$

and Bernoulli's theorem for an isentropic flow assumes the form

$$\frac{1}{2}\vec{v}^2 + h + \mathcal{G} = \frac{1}{2}\vec{v}^2 + \frac{\gamma}{\gamma - 1}\frac{p}{\rho} + gz = C_1 \quad (\text{: constant})$$

along each streamline.

8-12 (i) For an isentropic flow with $p = p(\rho)$ (and $p_0 = p(\rho_0)$), we have Taylor's expansion

$$p_0 + p' = p(\rho_0 + \rho') = p_0 + \left(\frac{\partial p}{\partial \rho}\right)_s\bigg|_{\rho=\rho_0} \rho' + \cdots,$$

and accordingly, if we define $c^2 \equiv \left(\frac{\partial p}{\partial \rho}\right)_s\big|_{\rho=\rho_0}$, it is possible to set

$$p'(\vec{r}, t) = c^2 \rho'(\vec{r}, t). \tag{1}$$

Then, from the equation of continuity $\frac{\partial \rho}{\partial t} + \vec{\nabla}\cdot(\rho\vec{v}) = 0$ with $\rho = \rho_0 + \rho'$ and $\vec{v} = \vec{v}'$, we find

$$\frac{1}{\rho_0}\frac{\partial \rho'}{\partial t} + \vec{\nabla}\cdot\vec{v}' = 0 \tag{2}$$

to first order in the fluctuation terms. Also, to this order, Euler's equation of motion $\frac{\partial \vec{v}}{\partial t} + (\vec{v}\cdot\vec{\nabla})\vec{v} = -\frac{1}{\rho}\vec{\nabla}p$ gives

$$\frac{\partial \vec{v}'}{\partial t} = -\frac{1}{\rho_0}\vec{\nabla}p'. \tag{3}$$

Now, if we take the divergence of (3), we find

$$\frac{\partial}{\partial t}(\vec{\nabla}\cdot\vec{v}') = -\frac{1}{\rho_0}\vec{\nabla}^2 p'.$$

Then, substituting (2) and (1) into this equation gives the desired wave equation for p':

$$\vec{\nabla}^2 p' - \frac{1}{c^2}\frac{\partial^2 p'}{\partial t^2} = 0. \tag{5}$$

The density fluctuation ρ' also satisfies the same wave equation.

(ii) When the fluid is irrotaional, i.e., $\vec{\nabla}\times\vec{v}' = 0$, we must have

$$\begin{aligned} 0 &= \vec{\nabla}\times(\vec{\nabla}\times\vec{v}') = \vec{\nabla}(\vec{\nabla}\cdot\vec{v}') - \vec{\nabla}^2\vec{v}' \\ &= -\frac{1}{\rho_0}\frac{\partial}{\partial t}(\vec{\nabla}\rho') - \vec{\nabla}^2\vec{v}', \end{aligned} \tag{6}$$

where we used (2). But, from (1) and (3), we find

$$\frac{1}{\rho_0}\frac{\partial}{\partial t}(\vec{\nabla}\rho') = \frac{1}{\rho_0}\frac{1}{c^2}\frac{\partial}{\partial t}(\vec{\nabla}p') = -\frac{1}{c^2}\frac{\partial^2\vec{v}'}{\partial t^2}. \tag{7}$$

From (6) and (7), we are led to the wave equation for $\vec{v}'$:

$$\vec{\nabla}^2\vec{v}' - \frac{1}{c^2}\frac{\partial^2\vec{v}'}{\partial t^2} = 0. \tag{8}$$

[Remark: Note that, even when $\vec{\nabla}\times\vec{v}' \neq 0$, the *time-dependent part* of $\vec{v}'$ which is present in a sound wave is irrotational due to (3). In this case, the wave equation holds only for the time-dependent part of $\vec{v}'$.]

(iii) If we assume the monochromatic wave form $p'(\vec{r},t) = A\cos(\vec{k}\cdot\vec{r} - \omega t + \alpha)$, we have

$$\begin{aligned} \frac{\partial^2}{\partial t^2}p'(\vec{r},t) &= -\omega^2 A\cos(\vec{k}\cdot\vec{r} - \omega t + \alpha) = -\omega^2 p'(\vec{r},t), \\ \vec{\nabla}^2 p'(\vec{r},t) &= \vec{\nabla}\cdot[-\vec{k}A\sin(\vec{k}\cdot\vec{r} - \omega t + \alpha)] \\ &= -\vec{k}^2 A\cos(\vec{k}\cdot\vec{r} - \omega t + \alpha) = -\vec{k}^2 p'(\vec{r},t), \end{aligned}$$

so that

$$\vec{\nabla}^2 p' - \frac{1}{c^2}\frac{\partial^2}{\partial t^2}p' = -\left(\vec{k}^2 - \frac{\omega^2}{c^2}\right)p'.$$

Hence, if the dispersion relation $\omega = c|\vec{k}|$ holds, the given monochromatic wave provides a solution to the wave equation (5). For the corresponding velocity, we use (3) to obtain

$$\vec{v}'(\vec{r},t) = \frac{1}{\rho_0 c} A \frac{\vec{k}}{|\vec{k}|} \cos(\vec{k}\cdot\vec{r} - \omega t + \alpha) = \frac{c}{\rho_0} \rho'(\vec{r},t) \frac{\vec{k}}{|\vec{k}|},$$

which shows a longitudinal wave nature (i.e., oscillation along the direction of the wave vector $\vec{k}$).

8-13 (i) In this case we have, from the equation of continuity $\frac{\partial \rho}{\partial t} + \vec{\nabla}\cdot(\rho\vec{v}) = 0$,

$$\frac{1}{\rho_0}\left(\frac{\partial}{\partial t} + \vec{v}_0\cdot\vec{\nabla}\right)\rho' + \vec{\nabla}\cdot\vec{v}' = 0. \tag{1}$$

Euler's equation of motion $\frac{\partial \vec{v}}{\partial t} + (\vec{v}\cdot\vec{\nabla})\vec{v} = -\frac{1}{\rho}\vec{\nabla}p$, on the other hand, gives rise to the following equation for the fluctuation fields:

$$\frac{\partial \vec{v}'}{\partial t} + (\vec{v}_0\cdot\vec{\nabla})\vec{v}' = -\frac{1}{\rho_0}\vec{\nabla}p'. \tag{2}$$

If we take the divergence of (2), we find

$$\frac{\partial}{\partial t}(\vec{\nabla}\cdot\vec{v}') + (\vec{v}_0\cdot\vec{\nabla})\vec{\nabla}\cdot\vec{v}' = -\frac{1}{\rho_0}\vec{\nabla}^2 p'. \tag{3}$$

Into this equation we substitute, from (1),

$$\begin{aligned}\vec{\nabla}\cdot\vec{v}' &= -\frac{1}{\rho_0}\left(\frac{\partial}{\partial t} + \vec{v}_0\cdot\vec{\nabla}\right)\rho' \\ &= -\frac{1}{\rho_0 c^2}\left(\frac{\partial}{\partial t} + \vec{v}_0\cdot\vec{\nabla}\right)p',\end{aligned}$$

where we used $p' = c^2\rho'$, with $c^2 = \left(\frac{\partial p}{\partial \rho}\right)_s\Big|_{\rho=\rho_0}$ (as defined in Problem 8-12). We then obtain the following differential equation for the pressure fluctuation:

$$\vec{\nabla}^2 p' - \frac{1}{c^2}\left(\frac{\partial}{\partial t} + \vec{v}_0\cdot\vec{\nabla}\right)^2 p' = 0. \tag{5}$$

For a solution of (5), try again a monochromatic wave form

$$p'(\vec{r}, t) = A\cos(\vec{k} \cdot \vec{r} - \omega t + \alpha) \quad (\omega > 0). \tag{6}$$

Then observe that

$$\vec{\nabla}^2[\cos(\vec{k} \cdot \vec{r} - \omega t + \alpha)] = -\vec{k}^2 \cos(\vec{k} \cdot \vec{r} - \omega t + \alpha),$$
$$\left(\frac{\partial}{\partial t} + \vec{v}_0 \cdot \vec{\nabla}\right)^2 [\cos(\vec{k} \cdot \vec{r} - \omega t + \alpha)]$$
$$= -(\omega - \vec{v}_0 \cdot \vec{k})^2 \cos(\vec{k} \cdot \vec{r} - \omega t + \alpha).$$

Hence, for the form (6) to be a solution of (5), the frequency ω should be related to the wave vector $\vec{k}$ by

$$\omega = c|\vec{k}| + \vec{v}_0 \cdot \vec{k}, \tag{7}$$

where the vector $\vec{k}$ can arbitrarily be chosen.

(ii) Imagine observing the sound wave phenomenon inside the stationary medium — the case considered in Problem 8-12 — in a new reference frame which is moving with velocity $-\vec{v}_0$ relative to this fluid medium. The coordinates in the new frame ($\vec{r}^*$) is related to the old coordinates $\vec{r}$ by $\vec{r}^* = \vec{r} + \vec{v}_0 t$. Then, according to what we saw in Problem 8-6, it should be possible to obtain the pressure fluctuation in the new frame (i.e., $p^{*\prime}(\vec{r}^*, t)$) by applying Galilean boost to the pressure fluctuation in the old frame (i.e., $p'(\vec{r}, t)$). Taking $p'(\vec{r}, t) = A\cos(\vec{k} \cdot \vec{r} - c|\vec{k}|t + \alpha)$ (from Problem 8-12), we are then led to

$$p^{*\prime}(\vec{r}^*, t) = p'(\vec{r}, t)|_{\vec{r} = \vec{r}^* - \vec{v}_0 t}$$
$$= A\cos\left[\vec{k} \cdot \vec{r}^* - (c|\vec{k}| + \vec{v}_0 \cdot \vec{k})t + \alpha\right].$$

This is entirely consistent with our finding in (i), especially the formula (7).

8-14 Since $\hat{t} = \frac{dx(l)}{dl}\mathbf{i} + \frac{dy(l)}{dl}\mathbf{j}$ describes the unit tangent to the curve $C = (x(l), y(l))$, the unit normal can be represented as

$$\hat{n} = \frac{dy(l)}{dl}\mathbf{i} - \frac{dx(l)}{dl}\mathbf{j}.$$

Now, using $\vec{v} = \frac{\partial\psi}{dy}\mathbf{i} - \frac{\partial\psi}{dx}\mathbf{j}$, we have

$$\begin{aligned}
\mathcal{N} &= \int_1^2 \vec{v}\cdot\hat{n}\,dl \\
&= \int_1^2 \left(\frac{\partial\psi}{dy}\frac{dy(l)}{dl} + \frac{\partial\psi}{dx}\frac{dx(l)}{dl}\right) dl \\
&= \int_1^2 \left(\frac{\partial\psi}{dy}dy + \frac{\partial\psi}{dx}dx\right) \\
&= \psi(x_2, y_2) - \psi(x_1, y_1).
\end{aligned}$$

8-15 (i) In two dimensions, the gradient operator is

$$\begin{aligned}
\vec{\nabla} &= \mathbf{i}\frac{\partial}{\partial x} + \mathbf{j}\frac{\partial}{\partial y} = \hat{r}\frac{\partial}{\partial r} + \hat{\theta}\frac{1}{r}\frac{\partial}{\partial\theta} \\
&= (x\mathbf{i} + y\mathbf{j})\frac{1}{r}\frac{\partial}{\partial r} + (-y\mathbf{i} + x\mathbf{j})\frac{1}{r^2}\frac{\partial}{\partial\theta}.
\end{aligned}$$

Hence, for the Laplacian of ϕ, we have

$$\begin{aligned}
\vec{\nabla}^2\phi &= \vec{\nabla}\cdot\left[(x\mathbf{i} + y\mathbf{j})\frac{1}{r}\frac{\partial\phi}{\partial r} + (-y\mathbf{i} + x\mathbf{j})\frac{1}{r^2}\frac{\partial\phi}{\partial\theta}\right] \\
&= (x\mathbf{i} + y\mathbf{j})\cdot\vec{\nabla}\left(\frac{1}{r}\frac{\partial\phi}{\partial r}\right) + 2\frac{1}{r}\frac{\partial\phi}{\partial r} \\
&\quad + (-y\mathbf{i} + x\mathbf{j})\cdot\vec{\nabla}\left(\frac{1}{r^2}\frac{\partial\phi}{\partial\theta}\right) \\
&= r\frac{\partial}{\partial r}\left(\frac{1}{r}\frac{\partial\phi}{\partial r}\right) + 2\frac{1}{r}\frac{\partial\phi}{\partial r} + r\cdot\frac{1}{r}\frac{\partial}{\partial\theta}\left(\frac{1}{r^2}\frac{\partial\phi}{\partial\theta}\right) \\
&= \frac{\partial^2\phi}{\partial r^2} + \frac{1}{r}\frac{\partial\phi}{\partial r} + \frac{1}{r^2}\frac{\partial^2\phi}{\partial\theta^2}.
\end{aligned}$$

The Laplace equation in polar coordinates thus assumes the form

$$0 = \frac{\partial^2\phi}{\partial r^2} + \frac{1}{r}\frac{\partial\phi}{\partial r} + \frac{1}{r^2}\frac{\partial^2\phi}{\partial\theta^2} = \frac{1}{r}\frac{\partial}{\partial r}\left(r\frac{\partial\phi}{\partial r}\right) + \frac{1}{r^2}\frac{\partial^2\phi}{\partial\theta^2}. \tag{1}$$

(ii) For a solution of (1), try a simple product form $\phi(r,\theta) = R(r)\Psi(\theta)$. Inserting it into (1) gives

$$\frac{1}{r}\left[\frac{d}{dr}\left(r\frac{dR}{dr}\right)\right]\Psi + \frac{1}{r^2}R\frac{d^2\Psi}{d\theta^2} = 0$$

or

$$\frac{r}{R}\frac{d}{dr}\left(r\frac{dR}{dr}\right) = -\frac{1}{\Psi}\frac{d^2\Psi}{d\theta^2}. \tag{2}$$

In (2), the left hand side is a function of r only while the right hand side involves only the variable θ: this can be consistent only when

$$\frac{r}{R}\frac{d}{dr}\left(r\frac{dR}{dr}\right) = -\frac{1}{\Psi}\frac{d^2\Psi}{d\theta^2} = C \quad (\text{: const.}). \tag{3}$$

Then, setting $C = n^2$ where $n = 1, 2, \ldots$ (so that we have $\Psi(\theta + 2\pi) = \Psi(\theta)$, except for the case $n = 0$), we readily find from (3) that $R(r)$ and $\Psi(\theta)$ may have following forms:

$$\begin{aligned}
\text{for } n \neq 0: \;\; & R_n(r) = a_n r^n + b_n r^{-n}, \\
& \Psi_n(\theta) = A_n \cos n\theta + B_n \sin n\theta, \\
\text{for } n = 0: \;\; & R_0(r) = a_0 + b_0 \ln r, \quad \Psi_0(\theta) = A_0 + B_0\theta.
\end{aligned}$$

The Laplace equation (1) is satisfied by $\phi(r,\theta) = R_n(r)\Psi_n(\theta)$ for each value of $n = 0, 1, 2, \ldots$. Since (1) is a linear equation, the general superposition of these product solutions should also correspond to a solution. Hence the function of the form

$$\begin{aligned}
\phi(r,\theta) = {} & a_0 + b_0 \ln r + (c_0 + d_0 \ln r)\theta \\
& + \sum_{n=1}^{\infty}\Big[\left(a_n r^n + b_n r^{-n}\right)\cos n\theta \\
& + \left(c_n r^n + d_n r^{-n}\right)\sin n\theta\Big]
\end{aligned} \tag{4}$$

provides a solution to (1).

(iii) Starting from the form (4), the velocity potential describing the flow past an infinite cylinder at $r = R$ can be obtained in the following way. First, from the requirement that the velocity field $\vec{v} = \vec{\nabla}\phi$ be a single valued function of $\vec{r}$ (especially, $\vec{v}(r, \theta + 2\pi) = \vec{v}(r,\theta)$), we must set $d_0 = 0$. Then, due to the asymptotic condition $\phi(r,\theta) \to -v_0 r\cos\theta$ as $r \to \infty$, we are left with (c_0 is written as k here)

$$\phi(r,\theta) = a_0 + b_0 \ln r + k\theta + \left(-v_0 r + \frac{b_1}{r}\right)\cos\theta + \sum_{n=2}^{\infty} r^{-n}(b_n \cos n\theta + d_n \sin n\theta). \qquad (5)$$

With this expression we demand the condition $\frac{\partial\phi(r,\theta)}{\partial r}|_{r=R} = 0$ for arbitrary θ; then, $b_1 = -v_0 R^2$ and $b_0 = b_2 = b_3 = \cdots = d_2 = d_3 = \cdots = 0$. We therefore obtain the velocity potential

$$\phi(r,\theta) = a_0 + k\theta - v_0\left(r + \frac{R^2}{r}\right)\cos\theta \quad (r \geq R).$$

8-16 (i) For the given velocity potential, the velocity field is

$$\begin{aligned}\vec{v} &= \frac{\partial\phi}{\partial r}\hat{r} + \frac{\partial\phi}{\partial\theta}\frac{1}{r}\hat{\theta} \\ &= -v_0\left(1 - \frac{R^2}{r^2}\right)\cos\theta\hat{r} + v_0\left(1 + \frac{R^2}{r^2}\right)\sin\theta\hat{\theta} + \frac{k}{r}\hat{\theta}. \qquad (1)\end{aligned}$$

Because of the term $\frac{k}{r}\hat{\theta}$, the circulation $\Gamma = \oint_C \vec{v}\cdot d\vec{r}$ acquires the nonvanishing value $2\pi k$ for any circle $r = $ const.

(ii) The velocity potential $\phi = a_0 + k\theta - v_0\left(r + \frac{R^2}{r}\right)\cos\theta$ corresponds to the real part of the complex potential

$$F(z) = a_0 - ik\ln z - v_0\left(z + \frac{R^2}{z}\right). \qquad (2)$$

From the imaginary part of $F(z)$, we thus obtain the stream function in the form

$$\begin{aligned}\psi &= a_0 - k\ln r - v_0\left(1 - \frac{R^2}{r^2}\right) r\sin\theta \\ &= a_0 - \frac{k}{2}\ln(x^2+y^2) - v_0\left(1 - \frac{R^2}{x^2+y^2}\right)y.\end{aligned}$$

Therefore, the streamlines are defined by the equation

$$-v_0 y\left(1 - \frac{R^2}{x^2+y^2}\right) - \frac{k}{2}\ln(x^2+y^2) = \text{const.}$$

Streamlines of the flow with $k > 0$, against that with $k = 0$, are schematically represented in the figure.

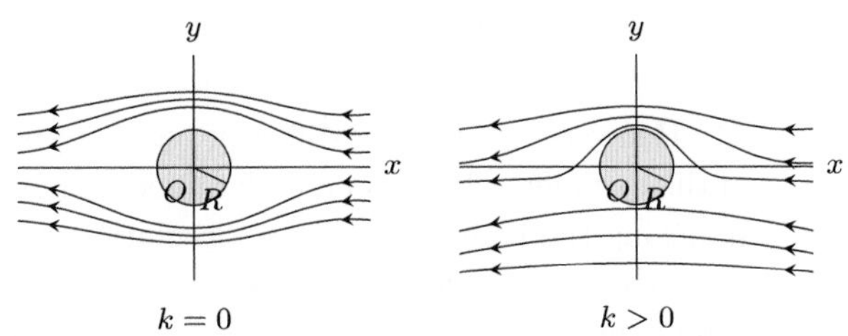

$k = 0$ $\qquad$ $k > 0$

(iii) When there is no circulation (i.e., with $k = 0$), the stagnation points are the points ($x = R$, $y = 0$) and ($x = -R$, $y = 0$) (that is, the points corresponding to $r = R$ and $\theta = 0, \pi$). When $k \neq 0$, let $\alpha \equiv \frac{k}{2v_0 R}$. If $|\alpha| < 1$, we see from (1) that the stagnation points are at

$$r = R, \qquad \theta = -\sin^{-1}\alpha,\ \pi + \sin^{-1}\alpha.$$

On the other hand, if $|\alpha| \geq 1$, there is only one stagnation point at a distance $r = R(|\alpha| + \sqrt{\alpha^2 - 1})$ from the center of the cylinder section, on the negative (positive) y-axis if α is positive (negative).

8-17 (i) From the corresponding velocity field (Problem 8-16(i)),

$$\vec{v}(r,\theta) = -v_0\left(1 - \frac{R^2}{r^2}\right)\cos\theta\hat{r} + v_0\left(1 + \frac{R^2}{r^2}\right)\sin\theta\hat{\theta} + \frac{\Gamma}{2\pi r}\hat{\theta}, \tag{1}$$

we see that, while the normal component of the velocity on the cylinder surface vanishes, there is a nonvanishing tangential component equal to

$$v_\theta = 2v_0\sin\theta + \frac{\Gamma}{2\pi R}. \tag{2}$$

Substituting this result into Bernoulli's theorem, we obtain, for the pressure distribution over the cylinder surface,

$$\begin{aligned} p &= p_0 + \frac{\rho}{2}\left[v_0^2 - \left(2v_0\sin\theta + \frac{\Gamma}{2\pi R}\right)^2\right] \\ &= p_0 + \frac{\rho}{2}\left[v_0^2(1 - 4\sin^2\theta) - \frac{2\Gamma v_0}{\pi R}\sin\theta - \left(\frac{\Gamma}{2\pi R}\right)^2\right]. \end{aligned} \tag{3}$$

(ii) The pressure distribution (3) is symmetric with respect to the midsection ($\theta = \frac{\pi}{2}$), and therefore the cylinder experiences no drag. However, the pressure is not symmetric with respect to the plane $\theta = 0$ due to the presence of the term $-\frac{\rho}{2}\left(\frac{2\Gamma v_0}{\pi R}\right)\sin\theta$ in (3). Hence the pressure is different on the upper and lower halves of the cylinder. This can be understood by noting that, by superimposing the counterclockwise circulation flow (i.e., positive Γ) upon the symmetrical flow shown in the figure below, the velocities of the flow will become greater above the cylinder compared to those below the cylinder. Then, by Bernoulli's theorem, the pressure will be greater below the cylinder, thus producing a force in the y-direction (*hydrodynamic lift*). The force per unit length of the cylinder is given by (see Sec. 8.1)

$$\begin{aligned}
F_y &= -\int_{-\pi}^{\pi} p(\sin\theta)Rd\theta \\
&= -2R\int_{\pi/2}^{\pi/2}\left[-\frac{\rho}{2}\left(\frac{2\Gamma v_0}{\pi R}\right)\sin\theta\right]\sin\theta d\theta \\
&= \rho v_0\Gamma.
\end{aligned}$$

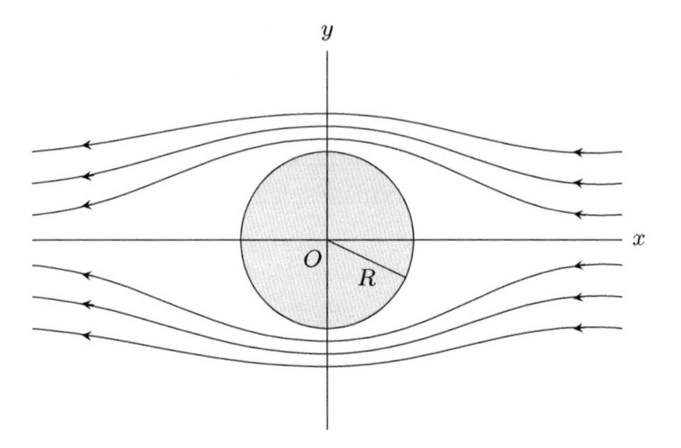

Notice that the lift is proportional to the velocity of the symmetrical flow v_0 as well as to the circulation Γ. If the circulation changes its sign, the "lift" also changes its direction.

8-18 (i) Taking the axis of the cylinder to be parallel to the z-axis, let the cylinder move with velocity $\vec{u} = u\mathbf{i}$ within the fluid. Then the corresponding velocity potential can be expressed, using polar coordinates r' and θ' (defined from the point on the axis of the cylinder), as

$$\phi = -u\frac{R^2}{r'}\cos\theta' = -R^2\frac{\vec{u}\cdot\vec{r}'}{r'^2} \quad (r' \geq R) \tag{1}$$

(see Sec. 8.3). The fluid velocity is thus given by

$$\vec{v} = \frac{R^2}{r'^4}\left[2(\vec{u}\cdot\vec{r}')\vec{r}' - \vec{u}r'^2\right] = \frac{R^2}{r'^2}\left[(2u\cos\theta')\hat{r}' - \vec{u}\right]. \tag{2}$$

The corresponding pressure can then be found using the formula $p = p_0 - \frac{\rho}{2}\vec{v}^2 - \rho\frac{\partial\phi}{\partial t}|_{\vec{r}}$ with $\frac{\partial\phi}{\partial t}|_{\vec{r}} = -\vec{u}\cdot\vec{\nabla}'\phi + \frac{\partial\phi}{\partial\vec{u}}\cdot\frac{d\vec{u}}{dt}$.

(ii) For the pressure at the surface of the cylinder, one readily obtains

$$p = p_0 + \frac{1}{2}\rho u^2(4\cos^2\theta' - 3) + \rho R\hat{r}'\cdot\frac{d\vec{u}}{dt}. \tag{3}$$

Therefore, when $\frac{\partial\vec{u}}{\partial t} = 0$, the pressure distribution is clearly reflection-symmetric about the x'- or y'-axis; hence the fluid exerts no force on the cylinder.

(iii) With $\frac{d\vec{u}}{dt} \neq 0$, we use the pressure distribution (3) to find the force per unit length of the cylinder:

$$\begin{aligned}\vec{F}_p &= -\int_0^{2\pi} R\,d\theta'\rho R\left(\hat{r}'\cdot\frac{d\vec{u}}{dt}\right)\hat{r}' \\ &= -\rho R^2\left[\int_0^{2\pi}\cos^2\theta' d\theta'\right]\frac{d\vec{u}}{dt} \\ &= -\rho\pi R^2\frac{d\vec{u}}{dt}.\end{aligned} \tag{4}$$

Hence the induced mass per unit length is $\rho\pi R^2$, i.e., equal to the mass of the displaced fluid.

8-19 (i) In a fluid the equation of motion for the ball on a spring is

$$m\ddot{x} = -kx - \tilde{m}\ddot{x}$$

where $\tilde{m} = \frac{\rho V}{2}$ (with $V = \frac{m}{\rho_0}$) is the induced mass of a sphere. Hence, when $\omega_0 = \sqrt{\frac{k}{m}}$ is the frequency of oscillation in the absence of a fluid, the frequency of oscillation in an ideal fluid is

$$\omega = \sqrt{\frac{k}{m+\tilde{m}}} = \omega_0\sqrt{\frac{1}{1+(\tilde{m}/m)}} = \omega_0\sqrt{\frac{2\rho_0}{2\rho_0+\rho}}\,.$$

(ii) Without a fluid the equation of motion for the pendulum is $ml\ddot{\theta} = -mg\theta$. In a fluid, it is

$$ml\ddot{\theta} = -mg\theta + \rho Vg\theta - \tilde{m}l\ddot{\theta},$$

where $\rho Vg\theta$ is the Archimedes buoyant force and $-\tilde{m}l\ddot{\theta}$ the force related to the induced mass. As we denote the frequency without a fluid by $\omega_0 = \sqrt{\frac{g}{l}}$, the frequency of the pendulum placed in the fluid is thus given by

$$\omega = \sqrt{\frac{mg - \rho Vg}{ml + \tilde{m}l}} = \omega_0 \sqrt{\frac{2(\rho_0 - \rho)}{2\rho_0 + \rho}}\,.$$

8-20 (i) Note that, due to the Navier–Stokes equation,

$$\begin{aligned}\frac{\partial}{\partial t}\left(\frac{1}{2}\rho\vec{v}^2\right) &= \rho v_i \frac{\partial v_i}{\partial t} \\ &= \rho v_i \left(-v_k \frac{\partial v_i}{\partial x_k} - \frac{1}{\rho}\frac{\partial p}{\partial x_i} + \frac{1}{\rho}\frac{\partial \sigma_{ij}}{\partial x_j}\right) \\ &= -\rho v_k \frac{\partial}{\partial x_k}\left(\frac{1}{2}\vec{v}^2 + \frac{p}{\rho}\right) + \frac{\partial}{\partial x_j}(v_i\sigma_{ij}) - \sigma_{ij}\frac{\partial v_i}{\partial x_j}.\end{aligned}$$

Then, using the incompressibility condition $\vec{\nabla}\cdot\vec{v} = 0$, this can be rewritten as

$$\frac{\partial}{\partial t}\left(\frac{1}{2}\rho\vec{v}^2\right) = -\vec{\nabla}\cdot\left[\rho\vec{v}\left(\frac{1}{2}\vec{v}^2 + \frac{p}{\rho}\right) - \vec{v}\cdot\vec{\sigma}\right] - \sigma_{ij}\frac{\partial v_i}{\partial x_j}. \quad (1)$$

(ii) In the expression inside the square brackets of (1), $\rho\vec{v}\left(\frac{1}{2}\vec{v}^2 + \frac{p}{\rho}\right)$ is just the energy flux density associated with the actual transfer of fluid mass (we had this term also in an ideal fluid) while the term $\vec{v}\cdot\vec{\sigma}$ represents the energy flux due to processes involving internal friction (as the presence of viscosity results in a momentum flux σ_{ij}). If we integrate (1) over the whole volume of the fluid (assuming that the fluid velocity vanishes at infinity), we find the rate of dissipation for the total kinetic energy:

$$\frac{d}{dt}\left[\int \frac{1}{2}\rho\vec{v}^2\, d^3\vec{r}\right] = -\int \sigma_{ij}\frac{\partial v_i}{\partial x_j} d^3\vec{r}$$

$$= -\eta \int \left(\frac{\partial v_j}{\partial x_i} + \frac{\partial v_i}{\partial x_j} \right) \frac{\partial v_i}{\partial x_j} d^3\vec{r}$$
$$= -\frac{\eta}{2} \int \left(\frac{\partial v_j}{\partial x_i} + \frac{\partial v_i}{\partial x_j} \right)^2 d^3\vec{r}. \qquad (2)$$

Noting that $(\vec{\nabla} \times \vec{v})^2 = \epsilon_{ijk}(\partial_j v_k)\epsilon_{ilm}\partial_l v_m = (\partial_j v_k)^2 - \partial_k(v_j \partial_j v_k)$ (by using $\epsilon_{ijk}\epsilon_{ilm} = \delta_{jl}\delta_{km} - \delta_{jm}\delta_{kl}$ and $\partial_i v_i = 0$), the above relation can also be written as

$$\frac{d}{dt}\left[\int \frac{1}{2}\rho\vec{v}^2 \, d^3\vec{r} \right] = -\eta \int (\vec{\nabla} \times \vec{v})^2 d^3\vec{r}. \qquad (3)$$

8-21 With the general solution given in (8.137), let us fix the constants a and c by imposing the boundary conditions $v = 0$ at $r = R_1$ and $r = R_2$. This gives the velocity distribution

$$v = \frac{\Delta p}{4\eta l}\left[R_2^2 - r^2 + \frac{R_2^2 - R_1^2}{\ln(R_2/R_1)} \ln \frac{r}{R_2} \right].$$

The discharge is determined as (here, $\nu \equiv \eta/\rho$)

$$Q = 2\pi\rho \int_{R_1}^{R_2} rv \, dr$$
$$= \frac{\pi \Delta p}{8\nu l}\left[R_2^4 - R_1^4 - \frac{(R_2^2 - R_1^2)^2}{\ln(R_2/R_1)} \right].$$

8-22 (i) Choosing the z-axis along the axis of the cylinders, we may invoke symmetry to write

$$v_z = v_r = 0, \quad v_\theta = v(r), \quad p = p(r) \qquad (1)$$

(r, θ denote polar coordinates in the xy-plane). Then, from the Navier–Stokes equation, we obtain

$$\frac{dp}{dr} = \frac{\rho}{r}v^2, \qquad (2a)$$
$$\frac{d^2 v}{dr^2} + \frac{1}{r}\frac{dv}{dr} - \frac{v}{r^2} = 0, \qquad (2b)$$

since $\vec{\nabla}p = \frac{dp}{dr}\hat{r}$ and

$$(\vec{v}\cdot\vec{\nabla})\vec{v} = v\frac{1}{r}\frac{\partial}{\partial\theta}(v\hat{\theta}) = -\frac{v^2}{r}\hat{r},$$
$$\vec{\nabla}^2\vec{v} = \left(\frac{1}{r}\frac{\partial}{\partial r}r\frac{\partial}{\partial r} + \frac{1}{r^2}\frac{\partial^2}{\partial\theta^2}\right)(v\hat{\theta}) = \left(\frac{d^2v}{dr^2} + \frac{1}{r}\frac{dv}{dr} - \frac{v}{r^2}\right)\hat{\theta}. \tag{3}$$

The general solution of the equation (2b) (: an Euler–Cauchy differential equation) is

$$v = ar + \frac{b}{r}. \tag{4}$$

The constants a and b are found from the boundary conditions that the fluid velocity at the inner and outer cylindrical surfaces must be equal to that of the corresponding cylinders, i.e., $v_\theta(r{=}R_1) = R_1\omega_1$ and $v_\theta(r{=}R_2) = R_2\omega_2$. As a result we find the velocity distribution

$$v_\theta = v(r) = \frac{\omega_2 R_2^2 - \omega_1 R_1^2}{R_2^2 - R_1^2}r + \frac{(\omega_1-\omega_2)R_1^2R_2^2}{R_2^2-R_1^2}\frac{1}{r}. \tag{5}$$

Then, from (2a), the pressure distribution is given as

$$\begin{aligned} p(r) &= p(R_1) + \rho\int_{R_1}^{r}\frac{1}{r}\left[\frac{\omega_2 R_2^2 - \omega_1 R_1^2}{R_2^2 - R_1^2}r + \frac{(\omega_1-\omega_2)R_1^2R_2^2}{R_2^2-R_1^2}\frac{1}{r}\right]^2 dr \\ &= p(R_1) + \frac{\rho}{R_2^2 - R_1^2}\left[2(\omega_1-\omega_2)R_1^2R_2^2(\omega_2R_2^2 - \omega_1R_1^2)\ln\frac{r}{R_1}\right. \\ &\quad + \frac{1}{2}(\omega_2R_2^2 - \omega_1R_1^2)^2(r^2 - R_1^2) \\ &\quad \left. + (\omega_1-\omega_2)^2R_1^2R_2^4\left(\frac{r^2 - R_1^2}{2r^2}\right)\right]. \end{aligned}$$

[Remark: Note that if $\omega_1 = \omega_2 = \omega$, we find $v_\theta = \omega r$ from (5), i.e., the fluid rotates rigidly with the cylinders. If the outer cylinder is absent (i.e., $R_2 = \infty$ and $\omega_2 = 0$), we have $v_\theta = \omega_1 R_1^2/r$.]

(ii) In this flow we expect viscous stresses to be the same at any plane containing the axis of the cylinders. Therefore, we may confine our attention to, say, $\sigma_{xy}\big|_{x=0,\,y=r}$. Here, since $v_x = -v(r)\frac{y}{r}$ and

$v_y = v(r)\frac{x}{r}$ (with $v(r)$ given by (5)) from (1), we have

$$\sigma_{xy} = \eta\left(\frac{\partial v_x}{\partial y} + \frac{\partial v_y}{\partial x}\right) = \eta\frac{(-y^2+x^2)}{r}\frac{d}{dr}\left(\frac{v(r)}{r}\right)$$
$$= 2\eta(\omega_1 - \omega_2)\frac{R_1^2 R_2^2}{R_2^2 - R_1^2}\frac{(y^2-x^2)}{r^4}$$

and hence

$$\sigma_{xy}\big|_{x=0,\,y=r} = 2\eta\frac{(\omega_1-\omega_2)}{R_2^2 - R_1^2}\frac{R_1^2 R_2^2}{r^2}.$$

Accordingly, the frictional force acting on unit length of the inner cylinder is $(2\pi R_1)[2\eta\frac{(\omega_1-\omega_2)}{R_2^2-R_1^2}R_2^2]$, while that on the outer cylinder is $-(2\pi R_2)[2\eta\frac{(\omega_1-\omega_2)}{R_2^2-R_1^2}R_1^2]$. (Note that the *moment* of the viscous force through the rotating fluid layer is actually independent of r, being equal to $2\pi r\sigma_{xy}|_{x=0,\,y=r}\cdot r = 4\pi\eta\frac{(\omega_1-\omega_2)}{R_2^2-R_1^2}R_1^2R_2^2$.)

8-23 A simple way to find the corresponding pressure $p(\vec{r})$ is as follows. Substituting the form (8.145) in (8.141) yields, for the gradient of p,

$$\vec{\nabla}p = \eta\vec{\nabla}^2\vec{v} = \eta\vec{\nabla}^2\left\{\vec{\nabla}(\vec{\nabla}\cdot[f\vec{u}]) - (\vec{\nabla}^2 f)\vec{u}\right\}$$
$$= \eta\vec{\nabla}^2\vec{\nabla}(\vec{\nabla}\cdot[f\vec{u}]),$$

where we used $(\vec{\nabla}^2)^2 f = 0$ (see (8.148). This implies that

$$\vec{\nabla}p = \vec{\nabla}\left\{\eta\vec{\nabla}^2(\vec{\nabla}\cdot[f\vec{u}])\right\}$$
$$= \vec{\nabla}\left\{\eta\vec{\nabla}^2(\vec{u}\cdot\vec{\nabla}f)\right\}$$
$$= \vec{\nabla}\left\{\eta\,\vec{u}\cdot\vec{\nabla}(\vec{\nabla}^2 f)\right\},$$

and accordingly

$$p(\vec{r}) = \eta\,\vec{u}\cdot\vec{\nabla}(\vec{\nabla}^2 f) + p_0,$$

where p_0 can be identified with the pressure value at infinity. Then, using $f = \frac{3R}{4}r + \frac{R^3}{4r}$ (from (8.152) and (8.155)), we are led to

$$p(\vec{r}) = p_0 + \eta\,\vec{u}\cdot\vec{\nabla}\left[\frac{1}{r^2}\frac{d}{dr}r^2\frac{d}{dr}\left(\frac{3R}{4}r\right)\right]$$
$$= p_0 - \frac{3}{2}\eta R\frac{\vec{u}\cdot\hat{r}}{r^2} \quad (r \geq R)$$

which is the desired result.

[illegible]

and hence

[illegible]

Accordingly, [illegible]

[illegible]

and so that

[illegible]

9 Motion in a Non-Inertial Reference Frame

9-1 (i) Choosing the coordinate origin at the position of the moving support, the equation of motion for the pendulum reads

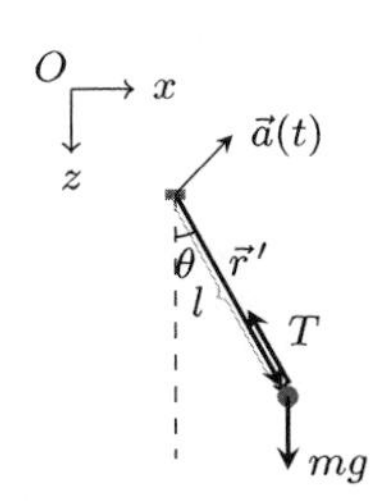

$$m\frac{d^2\vec{r}'}{dt^2} = \vec{T} + mg\mathbf{k} - m\vec{a},$$

where $\vec{T}$ $(= -T\hat{r}')$ is the tension and we included the inertial force $-m\vec{a}$ $(= -m(a_x\mathbf{i} + a_z\mathbf{k}))$ on the right hand side. This then leads to

$$\begin{aligned} m\frac{d}{dt}\left(\vec{r}' \times \frac{d\vec{r}'}{dt}\right) \\ &= \vec{r}' \times (\vec{T} + mg\mathbf{k} - m\vec{a}) \\ &= ml(\sin\theta\mathbf{i} + \cos\theta\mathbf{k}) \times [(g - a_z)\mathbf{k} - a_x\mathbf{i}] \\ &= -l(g - a_z)\sin\theta\mathbf{j} - la_x\cos\theta\mathbf{j}, \end{aligned}$$

while we have $\vec{r}' \times \frac{d\vec{r}'}{dt} = l^2\dot{\theta}\mathbf{j}$ (with $\vec{r}' = l(\sin\theta\hat{\theta} + \cos\theta\mathbf{k})$). Hence

$$\ddot{\theta} + \frac{1}{l}[g - a_z(t)]\sin\theta = -\frac{1}{l}a_x(t)\cos\theta.$$

(ii) Let $\vec{a} = \vec{a}_0 = \text{const}$. Then, since the inertial mass (i.e., m of $-m\vec{a}$) is equal to the gravitational mass (m of $mg\mathbf{k}$), we can write the equation of motion as

$$m\frac{d}{dt}\left(\vec{r}' \times \frac{d\vec{r}'}{dt}\right) = \vec{r}' \times m\vec{g}' \quad (\vec{g}' \equiv g\mathbf{k} - \vec{a}_0). \tag{1}$$

Denoting the angle between $\vec{r}'$ and $\vec{g}'$ by Θ, (1) then leads to

$$\ddot{\Theta} + \frac{1}{l} g' \sin\Theta = 0 \quad (g' \equiv |\vec{g}'|).$$

That is, with a uniformly accelerating support (or for a pendulum attached to a uniformly accelerating frame), the gravity is effectively changed from $\vec{g} = g\mathbf{k}$ to $\vec{g}' = g\mathbf{k} - \vec{a}_0$. Since the pendulum motion "knows" only $\vec{g}'$, effects due to real gravity and inertial force are not distinguishable on the basis of local experiments (here represented by the pendulum motion) alone. This also implies that effects due to real gravity $\vec{g}$ may be "canceled" by going to an accelerating frame with $\vec{a} = \vec{g}$ (i.e., a *freely falling frame*), since we then find $\vec{g}' = \vec{g} - \vec{g} = 0$. Indeed, when the moving support of the pendulum is in the free fall (i.e., $\vec{a} = g\mathbf{k}$), the equation of motion for the pendulum becomes $\ddot{\theta} = 0$, i.e., the pendulum undergoes uniform rotation (as in the case of pendulum in free space). This is the content of the equivalence principle, which played an important role when Einstein formulated his General Theory of Relativity.

9-2 We will consider the problem in two different frames of reference.

- (View in an inertial frame) If we assume that the car is travelling into the page and is turning towards the left, it must be subject to a centripetal acceleration $\frac{v^2}{a}$ to the left, provided by frictional forces acting on the tires. (A *rear* view of the system is shown in the figure.)

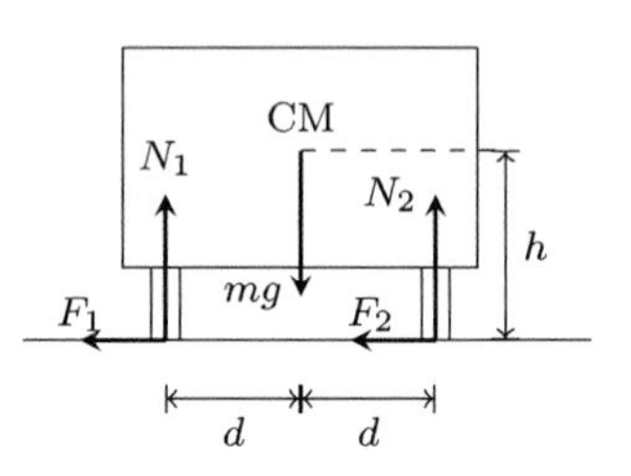

Then, considering the translational motion of the car, we infer that the following equations must hold:

$$N_1 + N_2 = mg, \tag{1}$$

$$F_1 + F_2 = m\frac{v^2}{a}, \tag{2}$$

where we denoted the normal and frictional forces by N's and F's. In addition to these, the torque balance equation about the CM

$$(F_1 + F_2)h + N_1 d = N_2 d \quad \left(\longrightarrow \quad F_1 + F_2 = (N_2 - N_1)\frac{d}{h} \right) \tag{3}$$

should be satisfied for the car not to overturn. From (2) and (3), we find that

$$N_2 - N_1 = \frac{hmv^2}{ad}. \tag{4}$$

Using (1) and (4), we can solve for N_1 and N_2 to obtain

$$N_1 = \frac{1}{2}\left(mg - \frac{hmv^2}{ad}\right), \quad N_2 = \frac{1}{2}\left(mg + \frac{hmv^2}{ad}\right).$$

The inner wheels will leave the ground, i.e., the car begins to overturn if N_1 (and so F_1 also) becomes zero, that is, when

$$mg = \frac{hmv^2}{ad}.$$

Hence the limiting speed is given by $v^2 = \frac{gad}{h}$.

• (View in the frame of reference rotating at the same rate as the car) In the reference frame rotating at an angular speed of $\frac{v}{a}$ (about the center of the circular track), we must include, among the forces acting on the car, a fictitious centrifugal force of magnitude $\frac{mv^2}{a}$ which acts through its CM and directed away from the center of the circle. At the instant when the car begins to overturn, the forces N_1 and F_1 are zero and so the only forces on the car are N_2, F_2, the weight mg, and the centrifugal force. Here, if the net force given by the vector sum of the weight and the centrifugal forces passes outside the contact point with the ground (of the outer wheels) as shown in the figure, the torque on the car will be clockwise, causing it to overturn. Clearly the critical condition, when overturning is just about to occur, happens when the net force vector passes through the contact point: this yields the condition (for overturning)

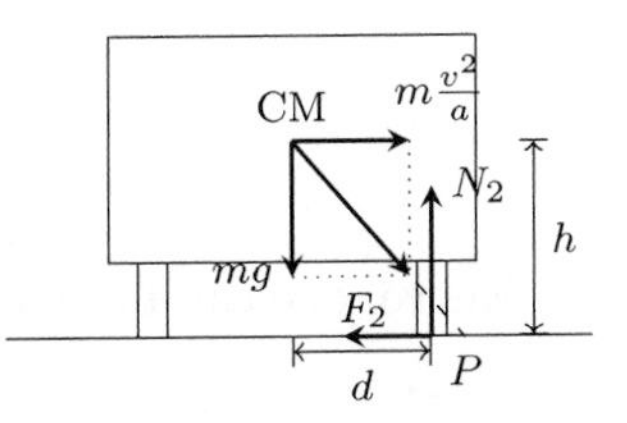

$$\frac{mv^2/a}{mg} > \frac{d}{h},$$

which gives $v^2 > \frac{gad}{h}$ as before.

9-3 From $\vec{v}_i \equiv \frac{d\vec{r}_i}{dt} = \vec{v}_i^* + \vec{\omega} \times \vec{r}_i$, it follows that $\vec{L} \equiv \sum_i m_i \vec{r}_i \times \vec{v}_i$ and $\vec{L}^* \equiv \sum_i m_i \vec{r}_i \times \vec{v}_i^*$ are related by

$$\vec{L} = \vec{L}^* + \sum_i m_i \vec{r}_i \times (\vec{\omega} \times \vec{r}_i).$$

Then, using this relation with $\frac{d\vec{L}}{dt} = \frac{d^*\vec{L}}{dt} + \vec{\omega} \times \vec{L}$, we can express the conservation of total angular momentum, $\frac{d\vec{L}}{dt} = 0$, by

$$\begin{aligned}
0 &= \frac{d^*}{dt}\left[\vec{L}^* + \sum_i m_i \vec{r}_i \times (\vec{\omega} \times \vec{r}_i)\right] + \vec{\omega} \times \left[\vec{L}^* + \sum_i m_i \vec{r}_i \times (\vec{\omega} \times \vec{r}_i)\right] \\
&= \frac{d^*\vec{L}^*}{dt} + \sum_i m_i \vec{v}_i^* \times (\vec{\omega} \times \vec{r}_i) + \sum_i m_i \vec{r}_i \times \left(\frac{d^*\vec{\omega}}{dt} \times \vec{r}_i\right) \\
&\quad + \sum_i m_i \vec{r}_i \times (\vec{\omega} \times \vec{v}_i^*) + \sum_i m_i \vec{\omega} \times (\vec{r}_i \times \vec{v}_i^*) \\
&\quad + \sum_i m_i \vec{\omega} \times \left[\vec{r}_i \times (\vec{\omega} \times \vec{r}_i)\right], \qquad (1)
\end{aligned}$$

where, as regards the derivative $\frac{d^*}{dt}$, we made use of the property mentioned in footnote 4 of this chapter. To simplify what we have on the right hand side of (1), note that

$$\begin{aligned}
&\vec{v}^* \times (\vec{\omega} \times \vec{r}_i) + \vec{r}_i \times (\vec{\omega} \times \vec{v}_i^*) + \vec{\omega} \times (\vec{r}_i \times \vec{v}_i^*) \\
&\qquad = 2(\vec{r}_i \cdot \vec{v}_i^*)\vec{\omega} - 2(\vec{r}_i \cdot \vec{\omega})\vec{v}_i^* = 2\vec{r}_i \times (\vec{\omega} \times \vec{v}_i^*), \\
&\vec{\omega} \times \left[\vec{r}_i \times (\vec{\omega} \times \vec{r}_i)\right] = \vec{r}_i \times \vec{\omega}(\vec{r}_i \cdot \vec{\omega}) = \vec{r}_i \times \left[\vec{\omega} \times (\vec{\omega} \times \vec{r}_i)\right]
\end{aligned}$$

and

$$\frac{d^*\vec{\omega}}{dt} = \frac{d\vec{\omega}}{dt}.$$

Hence, one can present (1) in the form

$$\frac{d^*\vec{L}^*}{dt} = \sum_i \vec{r}_i \times \Big[\underbrace{-2m_i(\vec{\omega} \times \vec{v}_i^*)}_{\text{Coriolis force}} \underbrace{-m_i \vec{\omega} \times (\vec{\omega} \times \vec{r}_i)}_{\text{centrifugal force}} - m_i \frac{d\vec{\omega}}{dt} \times \vec{r}_i\Big], \qquad (2)$$

i.e., in a rotating frame one must not forget "external" torques due to standard fictitious forces.

In the absence of any real external torque, the total angular momentum conservation implies that $L_l = \sum_i m_i \epsilon_{lmn} x_{m(i)} v_{n(i)} = \text{const.}$ $(l, m, n = 1, 2, 3)$ in an inertial frame. But, according to (2) (which tells us about $\frac{d^*\vec{L}^*}{dt} = \dot{L}_l^* \mathbf{e}_l^*)$, the components of the total angular momentum $\vec{L}^*$ that would be used by a rotating observer, $L_l^* = \sum_i m_i \epsilon_{lmn} x^*_{m(i)} v^*_{n(i)}$, are no longer constant but vary in time because of the torque due to fictitious forces. If there exist also torques due to real external forces $\vec{F}_i^{(e)}$, the left hand side of (1) should be taken by $\sum_i \vec{r}_i \times \vec{F}_i^{(e)}$. In such a situation, (2) should be replaced by

$$\frac{d^*\vec{L}^*}{dt} = \sum_i \vec{r}_i \times \left[\vec{F}_i^{(e)} - 2m_i(\vec{\omega} \times \vec{v}_i^*) - m_i\vec{\omega} \times (\vec{\omega} \times \vec{r}_i) - m_i \frac{d\vec{\omega}}{dt} \times \vec{r}_i\right].$$

9-4 (i) In a reference frame attached to the rotating Earth, let $r = f(\theta)$ describe the equilibrium shape of the Earth as given by the combined effect of gravitational and centrifugal forces. Here, for an object near the Earth's surface, the centrifugal potential is represented using spherical coordinates by

$$U_{\mathrm{ct}} = -\frac{1}{2}m[w^2r^2 - (\vec{w} \cdot \vec{r})^2] = -\frac{1}{2}\omega^2 r^2 \sin^2\theta$$

(ω is the rotational angular velocity of the Earth), while for the gravitational potential we are supposed to take the form

$$\mathcal{G}(r,\theta) = -G\frac{M}{r} - \frac{GQ}{4r^3}(3\cos^2\theta - 1).$$

Since the (perpendicular) force on the surface must be balanced by gravitational and centrifugal forces, the Earth's surface $r = f(\theta)$ should be an equipotential surface

$$\begin{aligned} U_{\mathrm{total}}(r,\theta) &\equiv -G\frac{M}{r} - \frac{GQ}{4r^3}(3\cos^2\theta - 1) - \frac{1}{2}\omega^2 r^2 \sin^2\theta \\ &= \text{const.}, \end{aligned} \tag{1}$$

and we will have the apparent gravitational acceleration $\vec{g}^*$ near the Earth's surface be given by $\vec{g}^*(\theta) = [-\vec{\nabla}U_{\mathrm{total}}]\big|_{r=f(\theta)}$ (which points in the direction perpendicular to the Earth's surface). Explicitly, we have

$$\begin{aligned} \vec{g}^*(\theta) = \Bigg[&\left\{-\frac{GM}{r^2} - \frac{3GQ}{4r^4}(3\cos^2\theta - 1) + \omega^2 r \sin^2\theta\right\}\hat{r} \\ &+ \left\{-\frac{3GQ}{2r^4} + \omega^2 r\right\}\sin\theta\cos\theta\,\hat{\theta}\Bigg]\Bigg|_{r=f(\theta)}. \end{aligned} \tag{2}$$

(ii) From the shape equation in (1), we should have

$$U_{\text{total}}\left(r = R, \theta = \frac{\pi}{2}\right) = U_{\text{total}}(r = R(1-\eta), \theta = 0),$$

which leads, approximately (for sufficiently small η, Q and ω), to

$$-G\frac{M}{R} + \frac{GQ}{4R^3} - \frac{1}{2}\omega^2 R^2 = -G\frac{M}{R}(1+\eta) - 2\frac{GQ}{4R^3}.$$

This gives

$$Q = \frac{4}{3}\left(-MR^2\eta + \frac{1}{2G}\omega^2 R^5\right) = -\frac{4}{3}MR^2\left(\eta - \frac{R}{2g}\omega^2\right),$$

where we used $\frac{GM}{R^2} = g \approx 9.8\,\text{m}\,\text{s}^{-2}$. Inserting the values $R \approx 6370\,\text{km}$, $\omega \approx 7.3\times10^{-5}\,\text{s}^{-1}$ and $\eta \approx 0.0034$, we thus find $Q \approx -2.2\times10^{-3}MR^2$ (which is quite close the actual value of $Q \approx -2.16\times10^{-3}MR^2$).

(iii) We use the representation of $\vec{g}^*$ given in (2). Ignoring the second and third terms in the radial component and using the approximation $r \approx R$, we find the angle α to be given by

$$\alpha \approx \frac{g_\theta^*}{g_r^*} \approx \frac{1}{g}\left(-\frac{3GQ}{2R^4} + \omega^2 R\right)\sin\theta\cos\theta.$$

Here note that $\omega^2 R \approx 34\,\text{mm}\,\text{s}^{-2}$ and $-\frac{3GQ}{2R^4} \approx \frac{3}{2}(2.2\times10^{-3})g \approx 32\,\text{mm}\,\text{s}^{-2}$ (using the value of Q obtained in (ii)), and hence the quadrupole-dependent term (related to the flattening of the Earth) must not be ignored.

9-5 Let $(\mathbf{e}_1^*, \mathbf{e}_2^*, \mathbf{k}^*)$ denote the frame fixed to the Earth's surface, with $\mathbf{k}^*$ in the vertical-up direction and the wind velocity taken as $\frac{d^*\vec{r}}{dt} = v\mathbf{e}_2^*$. At colatitude θ the Earth's rotational angular velocity is expressed as $\vec{\omega} = \vec{\omega}_\perp + \omega\cos\theta\mathbf{k}^*$, where $\vec{\omega}_\perp \cdot \mathbf{k}^* = 0$. Then a small parcel of air mass, having volume δV and mass $\delta m = \rho\delta V$, is subject to the equation of motion

$$\delta m\frac{d^{*2}\vec{r}}{dt^2} = -(\vec{\nabla}p)\delta V - 2\delta m\vec{\omega}\times\frac{d^*\vec{r}}{dt}.$$

In the horizontal direction the Coriolis force $-2\delta m\vec{\omega}\times\frac{d^*\vec{r}}{dt}$ gives rise to the term $2\delta m(\omega\cos\theta)v\mathbf{e}_1^*$. To maintain a constant wind direction, this must be balanced by the force due to the pressure gradient, i.e., $-\frac{dp}{dx}(\delta V)\mathbf{e}_1^*$. Hence,

$$-\frac{dp}{dx}(\delta V) + 2\delta m(\omega\cos\theta)v = 0$$

or, since $\delta m = \rho\,\delta V$

$$\frac{dp}{dx} = 2\rho\,\omega v\cos\theta.$$

9-6 Using the same notations as used in Sec. 9.2, we must here find, to first order in ω, the solution of

$$m\frac{d^{*2}\vec{r}}{dt^2} = m\vec{g}^* - 2m\vec{\omega}\times\frac{d^*\vec{r}}{dt} \tag{1}$$

under the initial conditions $x(t=0) = y(t=0) = z(t=0) = \dot{x}(t=0) = \dot{y}(t=0) = 0$, $\dot{z}(t{=}0) = v_0\ (>0)$. With the Coriolis force term neglected, the motion for $t > 0$ is given by

$$x(t) = y(t) = 0, \quad z(t) = v_0 t - \frac{1}{2}g^*t^2.$$

So, to first order in ω, the term $-2m\vec{\omega}\times\frac{d^*\vec{r}}{dt}$ on the right hand side of (1) may be approximated by $2m\omega(-v_0 + g^*t)\sin\theta\mathbf{i}^*$; then, the equation of motion (1) becomes

$$m\ddot{x} = 2m\omega(-v_0 + g^*t)\sin\theta, \quad m\ddot{y} = 0, \quad m\ddot{z} = -mg^*. \tag{2}$$

This has the solution in the form

$$x(t) = 2\omega\left(-\frac{1}{2}v_0t^2 + \frac{1}{6}g^*t^3\right)\sin\theta, \quad y(t) = 0, \quad z(t) = v_0t - \frac{1}{2}g^*t^2. \tag{3}$$

When the object hits the ground, i.e., after the time of $T = \frac{2v_0}{g^*}$, the deflection is

$$x(t{=}T) = 2\omega\left(-\frac{1}{2}v_0\frac{4v_0^2}{g^{*2}} + \frac{1}{6}g^*\frac{8v_0^3}{g^{*3}}\right)\sin\theta = -\frac{4}{3}\omega\sin\theta\frac{v_0^2}{g^{*2}},$$

showing that it points *toward the west*, i.e., in the direction opposite to the case when an object is dropped at rest from a certain altitude. This happened because, in the present case, the object has both upward and downward motions and the Coriolis force acts in opposite directions for these two motions. [To be more explicit, note that our object reaches the maximum height $h = \frac{v_0^2}{2g^*}$ at time $\frac{T}{2} = \frac{v_0}{g^*}$, with $x(t{=}\frac{T}{2}) = -\frac{2}{3}\omega\sin\theta\frac{v_0^2}{g^{*2}}$ and $\dot{x}(t{=}\frac{T}{2}) = -\omega\sin\theta\frac{v_0^2}{g^*}$; then, at time $T = \frac{2v_0}{g^*}$, it has $z(t{=}T) = 0$, $x(t{=}T) = -\frac{4}{3}\omega\sin\theta\frac{v_0^2}{g^{*2}}$ and $\dot{x}(t{=}T) = 0$. This informs us that the Coriolis force during the downward motion tends to *reduce* the westward velocity that the object acquired because of the Coriolis force during its upward motion.]

9-7 Without taking the Coriolis force effect into account, we have the motion

$$x(t) = 0, \quad y(t) = v_{0y}t, \quad z(t) = v_{0z}t - \frac{1}{2}gt^2. \tag{1}$$

Noting that $v_{0y} = v_{0z} = \frac{v_0}{\sqrt{2}}$, the last equation in (1) tells us that $z = 0$ at time $t = \frac{2v_{0z}}{g} = \frac{\sqrt{2}v_0}{g}$; then $y(t{=}\frac{\sqrt{2}v_0}{g}) = \frac{v_0}{\sqrt{2}}\frac{\sqrt{2}v_0}{g} = y_0$ and so $v_0 = \sqrt{gy_0}$.

Now include the Coriolis force effect due to Earth's rotation (with $g \approx g^*$); then, from (9.34) in the text,

$$\begin{aligned}\ddot{x} &= 2\omega(\dot{y}\cos\theta - \dot{z}\sin\theta)\\ &= \sqrt{2}\omega v_0(\cos\theta - \sin\theta) + 2\omega gt\sin\theta + O(\omega^2),\end{aligned}$$

where we used the approximation based on the results (1). Integrating this equation with the initial conditions $x(0) = 0$ and $\dot{x}(0) = 0$, we find (ignoring $O(\omega^2)$ effects)

$$\begin{aligned}\dot{x} &= \sqrt{2}\omega v_0(\cos\theta - \sin\theta)t + \omega gt^2\sin\theta,\\ x &= \frac{\omega}{\sqrt{2}}v_0(\cos\theta - \sin\theta)t^2 + \frac{1}{3}\omega gt^3\sin\theta,\end{aligned}$$

and so, at time $t = \frac{\sqrt{2}v_0}{g^*}$, the x-coordinate becomes

$$\begin{aligned}x &= \frac{\omega}{\sqrt{2}}v_0(\cos\theta - \sin\theta)\frac{2v_0^2}{g^{*2}} + \frac{1}{3}\omega g^*\frac{2\sqrt{2}v_0^3}{g^{*3}}\sin\theta\\ &= \omega\left(\frac{2y_0^3}{g^*}\right)^{1/2}\left(\cos\theta - \frac{1}{3}\sin\theta\right),\end{aligned}$$

where we used $v_0 = \sqrt{g^*y_0}$ to obtain the last expression. This is the result wanted.

9-8 The relevant equation of motion is

$$m\frac{d^2\vec{r}}{dt^2} = m\vec{g}^* - 2m\vec{\omega} \times \frac{d^*\vec{r}}{dt}.$$

In components (we choose our coordinate axes as in Sec. 9.2), this means that

$$\begin{aligned}\ddot{x} &= 2\omega(\dot{y}\cos\theta - \dot{z}\sin\theta),\\ \ddot{y} &= -2\omega\dot{x}\cos\theta,\\ \ddot{z} &= -g^* + 2\omega\dot{x}\sin\theta.\end{aligned} \tag{1}$$

For a particle dropped from rest at a height h from the ground, the approximate solutions to first order in ω is (see (9.37) in the text)

$$x = \frac{1}{3}\omega g^* t^3 \sin\theta + O(\omega^2),\ y = 0 + O(\omega^2),\ z = h - \frac{1}{2}g^* t^2 + O(\omega^2). \tag{2}$$

Now, according to the procedure of successive iteration, we may use this first-order solution with (1) to obtain $O(\omega^2)$ corrections as well. We then find

$$\begin{aligned}\ddot{x} &= -2\omega[-g^* t + O(\omega^2)]\sin\theta\\ \ddot{y} &= -2\omega(\omega g^* t^2 \sin\theta)\cos\theta\\ \ddot{z} &= -g^* + 2\omega(\omega g^* t^2 \sin\theta)\cos\theta.\end{aligned} \tag{3}$$

Integrating the second equation among these yields

$$\begin{aligned}\dot{y} &= -\frac{2}{3}\omega^2 g^* t^3 \sin\theta\cos\theta,\\ y &= -\frac{1}{6}\omega^2 g^* t^4 \sin\theta\cos\theta.\end{aligned}$$

But, $z = 0$ at $t = \sqrt{\frac{2h}{g^*}} + O(\omega^2)$ from the last equation of (3); the y-coordinate at this time is thus given by

$$y = -\frac{1}{6}\omega^2 g^* \left(\frac{4h^2}{g^{*2}}\right)\sin\theta\cos\theta = -\frac{2\omega^2 h^2}{3g^*}\sin\theta\cos\theta.$$

The negative sign here means deviation to the south.

9-9 We here have $\vec{g}^* = -g^*\mathbf{k}^*$ and, if θ denotes the colatitude, $\vec{\omega} = \omega\sin\theta\mathbf{j}^* + \omega\cos\theta\mathbf{k}^*$ (see Sec. 9.2). If $\vec{r}(t) = x(t)\mathbf{i}^* + y(t)\mathbf{j}^* + z(t)\mathbf{k}^*$

describes the path of the projectile, we have $\frac{d^*\vec{r}}{dt} = \dot{x}\mathbf{i}^* + \dot{y}\mathbf{j}^* + \dot{z}\mathbf{k}^*$ and so, from the given equation of motion,

$$\ddot{x} = -\beta\dot{x} - 2\omega(\dot{z}\sin\theta - \dot{y}\cos\theta), \tag{1}$$
$$\ddot{y} = -\beta\dot{y} - 2\omega\dot{x}\cos\theta, \tag{2}$$
$$\ddot{z} = -\beta\dot{z} + 2\omega\dot{x}\sin\theta - g^*, \tag{3}$$

where we defined $\beta \equiv b/m$.

The three coupled equations (1), (2) and (3) can be solved in the following way. Differentiating (1) with respect to t and then using (2) and (3) yields

$$\begin{aligned}\dddot{x} &= -\beta\ddot{x} - 2\omega(\ddot{z}\sin\theta - \ddot{y}\cos\theta)\\ &= -\beta\ddot{x} - 4\omega^2\dot{x} + 2\omega g^*\sin\theta + 2\beta\omega(\dot{z}\cos\theta - \dot{y}\sin\theta).\end{aligned} \tag{4}$$

The last term in (4) may be expressed in terms of $\ddot{x}$ and $\dot{x}$ from (1) to give

$$\dddot{x} + 2\beta\ddot{x} + (\beta^2 + 4\omega^2)\dot{x} = 2\omega g^*\sin\theta. \tag{5}$$

We can find the general solution of this equation, but it is not needed. Note that, to first order in ω, (5) can be recast into the form

$$\frac{d^2}{dt^2}\left[e^{\beta t}\frac{dx}{dt}\right] = 2e^{\beta t}\omega g^*\sin\theta. \tag{6}$$

Hence, for the initial conditions

$$x(0) = \dot{x}(0) = 0, \quad y(0) = z(0) = 0$$

and $\ddot{x}(0) = 2\omega(v_{0y}\cos\theta - v_{0z}\sin\theta)$ (from (1)), we can readily obtain the solution to (6) in the form

$$\begin{aligned}x(t) = \frac{2\omega}{\beta^2}\bigg[&-\frac{1}{\beta}(2g^*\sin\theta - \beta A)(1 - e^{-\beta t}) + g^*\sin\theta t\\ &+(g^*\sin\theta - \beta A)te^{-\beta t}\bigg],\end{aligned} \tag{7}$$

where we set $A = v_{0y}\cos\theta - v_{0z}\sin\theta$. At the same time, since we have $\dot{x} = O(\omega)$ at most, equations (2) and (3) for $y(t)$ and $z(t)$ reduce, to first order in ω, to the familiar equations without Coriolis force effects;

the solutions satisfying the given initial conditions are thus given by

$$y(t) = \frac{v_{0y}}{\beta}(1 - e^{-\beta t}), \tag{8}$$

$$z(t) = \frac{v_{0z}}{\beta}(1 - e^{-\beta t}) + \frac{g^*}{\beta^2}(1 - \beta t - e^{-\beta t}). \tag{9}$$

In the absence of the Coriolis force, the motion is entirely two-dimensional, i.e., $x(t) = 0$. But, if we include Coriolis force effects, the projectile during its motion exhibits a tendency to deviate slightly to the west (negative x value) or to the east (positive x value), depending on the signs and magnitudes of v_{0y} and v_{0z} and also on the colatitude θ. For a projectile thrown into the air, suppose that β is sufficiently small so that $\beta t_f \ll 1$, where t_f is the total time of the flight which can be determined from (9) by setting $z = 0$: $t_f = \frac{2v_{0z}}{g^*}(1 - \frac{\beta v_{0z}}{3g^*}) + O(\beta^2)$. Then, from (7) and using $A = v_{0y}\cos\theta - v_{0z}\sin\theta$, the net deviation caused by the Coriolis force becomes

$$x(t_f) = \omega\left(\frac{2v_{0z}}{g^*}\right)^2\left(v_{0y}\cos\theta - \frac{1}{3}v_{0z}\sin\theta\right)\left(1 - \beta\frac{2v_{0z}}{g^*}\right) + O(\beta^2).$$

Also, if the projectile can fall down indefinitely, we have for the terminal velocity $\vec{v}_{\text{terminal}} = -\frac{g^*}{\beta}\mathbf{k}^* + \frac{2\omega g^*}{\beta^2}\sin\theta\mathbf{i}^*$.

9-10 (i) The position of the body relative to the center of the Earth can be represented by $\vec{r}' = \vec{r}_0 + \vec{r} = r_0\mathbf{i}^* + \vec{r}$, if $\vec{r} = x\mathbf{i}^* + y\mathbf{j}^* + z\mathbf{k}^*$ denotes the position of the body relative to the orbital station. Then the acceleration of the body in the astronaut's frame is given by

$$\vec{a}^* \equiv \frac{d^{*2}\vec{r}}{dt^2} = -\frac{GM}{r'^3}\vec{r}' - \vec{\Omega}\times(\vec{\Omega}\times\vec{r}') - 2\vec{\Omega}\times\frac{d^*\vec{r}}{dt}, \tag{1}$$

where M denotes the mass of the Earth, and $\vec{\Omega} \equiv \Omega\mathbf{k}^* = \left(\frac{GM}{r_0^3}\right)^{1/2}\mathbf{k}^*$ the angular velocity of revolution of the station. (Notice that the astronauts are at the fixed position $\vec{r}_0 = r_0\mathbf{i}^*$ in the rotating frame whose origin is at the center of the Earth.)

(ii) If $r\,(\equiv \sqrt{x^2 + y^2 + z^2}) \ll r_0$, we can make the approximation

$$\frac{GM}{r'^3} \approx \frac{GM}{r_0^3}\left(1 + 2\frac{\vec{r}_0\cdot\vec{r}}{r_0^2}\right)^{-3/2} \approx \Omega^2\left(1 - 3\frac{\vec{r}_0\cdot\vec{r}}{r_0^2}\right). \tag{2}$$

Using this approximation in (1), we obtain the following expression for $\vec{a}^*$, valid up to terms linear in r/r_0:

$$\begin{aligned}\vec{a}^* &= -\Omega^2(\vec{r}_0 + \vec{r}) + 3\Omega^2 \frac{\vec{r}_0 \cdot \vec{r}}{r_0^2}\vec{r}_0 - \vec{\Omega} \times (\vec{\Omega} \times \vec{r}_0) - \vec{\Omega} \times (\vec{\Omega} \times \vec{r}) \\ &\quad - 2\vec{\Omega} \times \frac{d^*\vec{r}}{dt} \\ &= -\Omega^2\vec{r} + 3\Omega^2 \frac{\vec{r}_0 \cdot \vec{r}}{r_0^2}\vec{r}_0 - \vec{\Omega} \times (\vec{\Omega} \times \vec{r}) - 2\vec{\Omega} \times \frac{d^*\vec{r}}{dt}. \end{aligned} \tag{3}$$

For the components x, y, z giving the body's position as seen by the astronauts, this gives (with $\vec{r}_0 = r_0\mathbf{i}^*$ and $\vec{\Omega} = \Omega\mathbf{k}^*$)

$$\begin{aligned}\ddot{x} &= -\Omega^2 x + 3\Omega^2 x + \Omega^2 x + 2\Omega\dot{y} = 3\Omega^2 x + 2\Omega\dot{y}, \\ \ddot{y} &= -\Omega^2 y + \Omega^2 y - 2\Omega\dot{x} = -2\Omega\dot{x}, \\ \ddot{z} &= -\Omega^2 z. \end{aligned} \tag{4}$$

(iii) (a) The body undergoes the rectilinear oscillatory motion (with a period $T = \frac{2\pi}{\Omega}$) along the z-axis: $x(t) = y(t) = 0$, $z(t) = \frac{v_0}{\Omega}\sin\Omega t$.

(b) In this case, the second of (4) gives $\dot{y} = -2\Omega x$. Substituting this into the first of (4) gives $\ddot{x} = -\Omega^2 x$, whence $x(t) = -\frac{v_0}{\Omega}\sin\Omega t$. Then, from the equation $\dot{y} = -2\Omega x$ (with the initial condition $y(0) = 0$), we find $y(t) = -2\frac{v_0}{\Omega}[\cos\Omega t - 1]$. Since $z(t) = 0$, the motion of the body described by these equations occurs in the xy-plane, satisfying the orbit equation of an ellipse

$$\frac{x^2}{l^2} + \frac{(y-2l)^2}{(2l)^2} = 1 \quad \left(\text{with } l \equiv \frac{v_0}{\Omega}\right).$$

(c) By the same procedure as used in the case (b), one can obtain the solution

$$x(t) = \frac{2v_0}{\Omega}[1 - \cos\Omega t], \; y(t) = \frac{4v_0}{\Omega}\sin\Omega t - 3v_0 t, \; z(t) = 0.$$

The orbit is no longer closed, because $y(t)$ contains a secular term which steadily increases with time. But, in the radial direction, we still have a periodic oscillation (with the period equal to $2\pi/\Omega$).

9-11 (i) Let $(\mathbf{e}_1^*, \mathbf{e}_2^*, \mathbf{e}_3^*)$ denote a set of rotating axes, satisfying the equations $\frac{d\mathbf{e}_i^*}{dt} = \vec{\omega}_0(t) \times \mathbf{e}_i^*$ $(i = 1, 2, 3)$ for some angular velocity

$\vec{\omega}_0(t)$. If $\frac{d^*\vec{S}}{dt}$ represents the rate of change of the spin $\vec{S}$ relative to this rotating frame, we have (see Sec. 9.1)

$$\frac{d\vec{S}}{dt} = \frac{d^*\vec{S}}{dt} + \vec{\omega}_0(t) \times \vec{S}.$$

This means that the given equation of motion for the spin can be changed, in the rotating frame, to the form

$$\frac{d^*\vec{S}}{dt} + \vec{\omega}_0 \times \vec{S} = \gamma \vec{S} \times \vec{B} \quad \longrightarrow \quad \frac{d^*\vec{S}}{dt} = \gamma \vec{S} \times \left(\frac{1}{\gamma}\vec{\omega}_0 + \vec{B} \right).$$

Accordingly, if we choose $\vec{\omega}_0 = -\gamma \vec{B}$, $\frac{d^*\vec{S}}{dt} = 0$, i.e., the effect of the magnetic field can be made to vanish. If $\vec{B}$ is time-independent, this implies that the spin $\vec{S}(t)$ in an inertial frame is simply a vector of constant magnitude that rotates with angular velocity $\vec{\omega}_0 = -\gamma \vec{B}$. (This is the *Larmor precession.*) But, with a time-dependent, oscillating magnetic field $\vec{B}$, the above observation must be used with some care especially because there can be "resonance" effects.

(ii) In an inertial frame, the spin $\vec{S}(t)$ satisfies the equation of motion

$$\frac{d\vec{S}}{dt} = \gamma \vec{S} \times \left[B_0 \mathbf{k} + \vec{B}_1(t) \right] \left(\text{with } \vec{B}_1(t) = B_1(\cos\omega t \mathbf{i} + \sin\omega t \mathbf{j}) \right). \tag{1}$$

Then, with respect to the reference frame ($\mathbf{i}^* = \cos\omega t\, \mathbf{i} + \sin\omega t\, \mathbf{j}$, $\mathbf{j}^* = \mathbf{k} \times \mathbf{i}^*$, $\mathbf{k}^* = \mathbf{k}$) which rotates relative to the inertial axes with angular velocity $\omega\mathbf{k}$, the equation of motion for the spin $\vec{S} = S_x^* \mathbf{i}^* + S_y^* \mathbf{j}^* + S_z^* \mathbf{k}^*$ will be

$$\begin{aligned} \frac{d^*\vec{S}}{dt} &\equiv \dot{S}_x^* \mathbf{i}^* + \dot{S}_y^* \mathbf{j}^* + \dot{S}_z^* \mathbf{k}^* \\ &= \gamma \vec{S} \times \left[\left(\frac{1}{\gamma}\omega + B_0 \right) \mathbf{k}^* + B_1 \mathbf{i}^* \right]. \end{aligned} \tag{2}$$

This equation is much simpler to solve than (1), since the coefficients of the right hand side are now time-independent. In fact, if we define the effective field $\vec{B}^*_{\text{eff}} = \left(\frac{\omega}{\gamma} + B_0\right)\mathbf{k}^* + B_1 \mathbf{i}^*$, the solution to (2) is immediate: $\vec{S}(t)$ (in the rotating frame) is a vector of constant magnitude that rotates with angular velocity $|\vec{\omega}^*| = |\gamma \vec{B}^*_{\text{eff}}|$ about the direction $\vec{\omega}^* = -\gamma \vec{B}^*_{\text{eff}}$.

As long as the rotation frequency ω of the field $\vec{B}_1(t)$ is significantly different from that of the Larmor precession (i.e., $-\gamma B_0$), the direction of $\vec{B}^*_{\text{eff}}$ is practically along $\mathbf{k}^*$ $(= \mathbf{k})$ since $|B_1| \ll |\frac{\omega}{\gamma} + B_0|$. In this situation, the motion of the spin is always dominated by the Larmor-type precession due to the constant magnetic field $\vec{B}_0 = B_0\mathbf{k}$, the field $\vec{B}_1(t)$ causing only minute oscillations in the spin direction. [If one had the spin $\vec{S}$ which was parallel to the constant field $\vec{B}_0$ at time $t = 0$, the direction of $\vec{S}$ will not get altered appreciably as time goes on.] This is also what one would expect on the basis of (i), as the angular velocity $\vec{\omega}_0$ is approximately equal to $-\gamma\vec{B}_0$ here. But, if the rotation frequency ω of the field $\vec{B}_1(t)$ satisfies the resonance condition $\omega \sim -\gamma B_0$, the angle between the effective field $\vec{B}^*_{\text{eff}}$ and $\mathbf{k}$ becomes large. This means that it is possible to have the precession of the spin vector $\vec{S}$ about the direction very different from that of the constant magnetic field, even if the magnitude of the rotating field is quite small; at resonance, the spin $\vec{S}$ which pointed in the direction of the field $\vec{B}_0$ at time $t = 0$ can even be in a completely flipped position at later time.

9-12 (i) From

$$U(x, y, t) = \frac{1}{2}(x\cos\omega t + y\sin\omega t)^2 - \frac{1}{2}(-x\sin\omega t + y\cos\omega t)^2$$
$$= \frac{1}{2}(x^2 - y^2)\cos 2\omega t + xy\sin 2\omega t,$$

we find

$$-\frac{\partial}{\partial x}U = -x\cos 2\omega t - y\sin 2\omega t,$$
$$-\frac{\partial}{\partial y}U = y\cos 2\omega t - x\sin 2\omega t.$$

Hence the given equations of motion are equivalent to

$$\ddot{\vec{r}} = -\vec{\nabla}U \tag{1}$$

with $U(x, y, t) = U_0(x\cos\omega t + y\sin\omega t, -x\sin\omega t + y\cos\omega t)$.

(ii) If we introduce a rotating basis ($\mathbf{i}^* \equiv \mathbf{i}\cos\omega t + \mathbf{j}\sin\omega t$, $\mathbf{j}^* \equiv -\mathbf{i}\sin\omega t + \mathbf{j}\cos\omega t$), we have

$$\vec{r} = x\mathbf{i} + y\mathbf{j} = X\mathbf{i}^* + Y\mathbf{j}^*$$

with $X = x\cos\omega t + y\sin\omega t$ and $Y = -x\sin\omega t + y\cos\omega t$; also

$$\begin{aligned} -\vec{\nabla}_{(x,y)}U &\equiv -\frac{\partial}{\partial x}U(x,y,t)\mathbf{i} - \frac{\partial}{\partial y}U(x,y,t)\mathbf{j} \\ &= -\frac{\partial}{\partial X}U_0(X,Y)\mathbf{i}^* - \frac{\partial}{\partial Y}U_0(X,Y)\mathbf{j}^* \\ &(\equiv -\vec{\nabla}_{(X,Y)}U_0(X,Y)). \end{aligned}$$

Furthermore, note that $\frac{d\mathbf{i}^*}{dt} = \vec{\omega}\times\mathbf{i}^*$ and $\frac{d\mathbf{j}^*}{dt} = \vec{\omega}\times\mathbf{j}^*$ with angular velocity $\vec{\omega} = \omega\mathbf{k}$. Hence, we may go to the viewpoint of the rotating observer to recast the equation of motion in (1) by the form

$$\frac{d^{*2}\vec{r}}{dt^2} + 2\vec{\omega}\times\frac{d^*\vec{r}}{dt} + \vec{\omega}\times(\vec{\omega}\times\vec{r}) = -\vec{\nabla}_{(X,Y)}U_0(X,Y), \qquad (2)$$

where $\frac{d^*\vec{r}}{dt} = \dot{X}\mathbf{i}^* + \dot{Y}\mathbf{j}^*$ and $\frac{d^{*2}\vec{r}}{dt^2} = \ddot{X}\mathbf{i}^* + \ddot{Y}\mathbf{j}^*$. In more explicit forms, we have from (2)

$$\begin{aligned} &\ddot{X} - 2\omega\dot{Y} + (1-\omega^2)X = 0, \\ &\ddot{Y} + 2\omega\dot{X} - (1+\omega^2)Y = 0. \end{aligned} \qquad (3)$$

This has the form of coupled second-order linear ODEs with constant coefficients; hence, for the general solution, one may look for all linearly-independent solutions of the form $\binom{X}{Y} = \binom{a_1}{a_2}e^{pt}$ (a_1, a_2, p are constants, and taking the real part of the expression should be understood) and then consider their superpositions.

(iii) If we insert the form $\binom{X}{Y} = \binom{a_1}{a_2}e^{pt}$ into (3), we obtain the following equation:

$$\begin{pmatrix} p^2+1-\omega^2 & -2\omega p \\ 2\omega p & p^2-1-\omega^2 \end{pmatrix}\begin{pmatrix} a_1 \\ a_2 \end{pmatrix} = 0.$$

So, to obtain a nontrivial solution of this form, p should chosen to satisfy the equation

$$0 = \begin{vmatrix} p^2+1-\omega^2 & -2\omega p \\ 2\omega p & p^2-1-\omega^2 \end{vmatrix} = (p^2)^2 - 2\omega^2p^2 - (1-\omega^4). \qquad (4)$$

For the critical point $x = y = 0$, which is the point $X = Y = 0$, to be stable, two roots of (4) for p^2 should be *real* and *negative*. These conditions are met only when $\omega^2 > 1$. Note that the Coriolis-force-like terms in (3) (i.e., terms proportional to $\dot{X}$, $\dot{Y}$)

are responsible for this stability with $\omega^2 > 1$, just as in the case of Lagrange points L_4 and L_5 in the restricted three-body problem.

9-13 Let M_1 (M_2) denote the mass of the Earth (Moon) so that we have $\alpha \equiv -\frac{\xi_1}{\xi_2} = \frac{M_2}{M_1} \ll 1$. (The actual value for α is $\approx \frac{1}{81.3}$.) For the corresponding collinear Lagrange points, we must solve the equation

$$-\frac{\xi_2(\xi-\xi_1)}{|\xi-\xi_1|^3} + \frac{\xi_1(\xi-\xi_2)}{|\xi-\xi_2|^3} - \xi = 0, \tag{1}$$

where $\xi_1 = \frac{M_2}{M_1+M_2}$ and $\xi_2 = -\frac{M_1}{M_1+M_2}$ ($= \xi_1 - 1$). Then, writing $\rho_1 = \xi - \xi_1$, $\rho_2 = \xi - \xi_2$ ($= \rho_1 + 1$) and $\xi = \xi_1\rho_2 - \xi_2\rho_1$, we can recast (1) as the following quintic equations:

(i) (for the point L_1, with $\rho_1 < 0$ and $\rho_2 > 0$)

$$\xi_2\rho_2^2 + \xi_1\rho_1^2 = \xi_1\rho_2^3\rho_1^2 - \xi_2\rho_1^3\rho_2^2 \qquad \text{or} \qquad \frac{\rho_2^2(1+\rho_1^3)}{\rho_1^2(1-\rho_2^3)} = \alpha, \tag{2}$$

(ii) (for the point L_2, with $\rho_1 < -1$ and $\rho_2 < 0$)

$$\xi_2\rho_2^2 - \xi_1\rho_1^2 = \xi_1\rho_2^3\rho_1^2 - \xi_2\rho_1^3\rho_2^2 \qquad \text{or} \qquad \frac{\rho_2^2(1+\rho_1^3)}{\rho_1^2(1+\rho_2^3)} = -\alpha, \tag{3}$$

(iii) (for the point L_3, with $\rho_1 > 0$ and $\rho_2 > 1$)

$$-\xi_2\rho_2^2 + \xi_1\rho_1^2 = \xi_1\rho_2^3\rho_1^2 - \xi_2\rho_1^3\rho_2^2 \qquad \text{or} \qquad \frac{\rho_2^2(1-\rho_1^3)}{\rho_1^2(1-\rho_2^3)} = -\alpha. \tag{4}$$

Evidently, for small α, these equations show that the points L_1 and L_2 should be associated to the case with $\rho_2 \sim 0$ (and $\rho_1 \sim -1$), and the point L_3 to the case with $\rho_1 \sim 1$. Based on this observation, we may approximate (2) (for the point L_1) by

$$\alpha = \frac{\rho_2^2\,[1+(\rho_2-1)^3]}{(\rho_2-1)^2(1-\rho_2^3)} \sim 3\rho_2^3(1+\rho_2)$$

to find that

$$\rho_2 \sim \left(\frac{\alpha}{3}\right)^{1/3}\left[1 - \frac{1}{3}\left(\frac{\alpha}{3}\right)^{1/3}\right], \quad \text{i.e.,}$$

$$\xi = \xi_2 + \left(\frac{\alpha}{3}\right)^{1/3}\left[1 - \frac{1}{3}\left(\frac{\alpha}{3}\right)^{1/3}\right].$$

Similarly, for the point L_2, we find from (3)

$$\rho_2 \sim -\left(\frac{\alpha}{3}\right)^{1/3}\left[1+\frac{1}{3}\left(\frac{\alpha}{3}\right)^{1/3}\right], \quad \text{i.e.,}$$

$$\xi = \xi_2 - \left(\frac{\alpha}{3}\right)^{1/3}\left[1+\frac{1}{3}\left(\frac{\alpha}{3}\right)^{1/3}\right].$$

On the other hand, for the point L_3, we may solve (4) approximately by setting $\rho_1 = 1+\beta$ and $\rho_2 = 2+\beta$ (here, $|\beta| \ll 1$), viz.,

$$-\alpha = \frac{(2+\beta)^2[1-(1+\beta)^3]}{(1+\beta)^2[1-(2+\beta)^3]} \sim \frac{12}{7}\beta\left(1-\frac{12}{7}\beta\right)$$

and this gives

$$\beta = -\frac{7}{12}\alpha(1-\alpha), \quad \text{i.e.,} \quad \xi = \xi_1 + 1 - \frac{7}{12}\alpha(1-\alpha).$$

In the inertial frame the orbit of small mass m at these three collinear Lagrange points can be represented schematically in the figure (here, note that $\alpha \sim \frac{1}{81.3}$, $a \approx 3.84 \times 10^5\,\text{km}$, and $\omega = \sqrt{\frac{G(M_1+M_2)}{a^3}} \approx \frac{2\pi}{27.3\,\text{days}}$ for the Earth–Moon system).

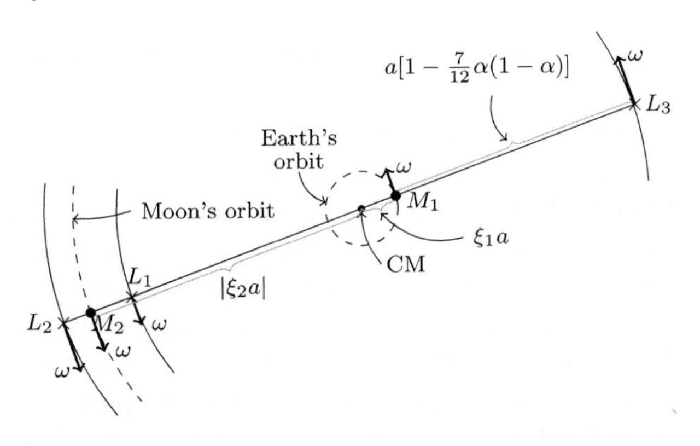

9-14 (i) First of all, from $m_1\vec{r}_1 + m_2\vec{r}_2 + m_3\vec{r}_3 = 0$ and

$$\vec{s}_1 = \vec{r}_3 - \vec{r}_2, \quad \vec{s}_2 = \vec{r}_1 - \vec{r}_3, \quad \vec{s}_3 = \vec{r}_2 - \vec{r}_1 \;(= -\vec{s}_1 - \vec{s}_2)$$

we get

$$\vec{r}_1 = \frac{1}{M}(m_3\vec{s}_2 - m_2\vec{s}_3), \quad \vec{r}_2 = \frac{1}{M}(m_1\vec{s}_3 - m_3\vec{s}_1),$$
$$\vec{r}_3 = \frac{1}{M}(m_2\vec{s}_1 - m_1\vec{s}_2), \tag{1}$$

where $M = m_1 + m_2 + m_3$. At the same time, from the given equations of motion, we find

$$\begin{aligned}
\ddot{\vec{s}}_1 &= Gm_1\frac{\vec{s}_2}{|\vec{s}_2|^3} - Gm_2\frac{\vec{s}_1}{|\vec{s}_1|^3} - Gm_3\frac{\vec{s}_1}{|\vec{s}_1|^3} + Gm_1\frac{\vec{s}_3}{|\vec{s}_3|^3} \\
&= -G(m_1+m_2+m_3)\frac{\vec{s}_1}{|\vec{s}_1|^3} + Gm_1\left(\frac{\vec{s}_1}{|\vec{s}_1|^3} + \frac{\vec{s}_2}{|\vec{s}_3|^2} + \frac{\vec{s}_3}{|\vec{s}_3|^3}\right) \\
&= -GM\frac{\vec{s}_1}{|\vec{s}_1|^3} + Gm_1\vec{S} \qquad \text{(2a)}
\end{aligned}$$

with $\vec{S} \equiv \frac{\vec{s}_1}{|\vec{s}_1|^3} + \frac{\vec{s}_2}{|\vec{s}_2|^3} + \frac{\vec{s}_3}{|\vec{s}_3|^3}$. Similarly, we also have

$$\ddot{\vec{s}}_2 = -GM\frac{\vec{s}_2}{|\vec{s}_2|^3} + Gm_2\vec{S}, \qquad \text{(2b)}$$

$$\ddot{\vec{s}}_3 = -GM\frac{\vec{s}_3}{|\vec{s}_3|^3} + Gm_3\vec{S}. \qquad \text{(2c)}$$

(ii) Clearly, if $|\vec{s}_1| = |\vec{s}_2| = |\vec{s}_3|$ at any given time, we find $\vec{S} = \frac{1}{|\vec{s}_3|^3}(\vec{s}_1 + \vec{s}_2 + \vec{s}_3) \equiv 0$ and accordingly, (2a-c) reduce to three "similar" equations

$$\ddot{\vec{s}}_1 = -GM\frac{\vec{s}_1}{|\vec{s}_1|^3}, \quad \ddot{\vec{s}}_2 = -GM\frac{\vec{s}_2}{|\vec{s}_2|^3}, \quad \ddot{\vec{s}}_3 = -GM\frac{\vec{s}_3}{|\vec{s}_3|^3}. \qquad (3)$$

With $|\vec{r}_3 - \vec{r}_2|^2 = |\vec{r}_1 - \vec{r}_3|^2 = |\vec{r}_2 - \vec{r}_1|^2$ ($= |\vec{s}_3|^2$), the three particles at any given time are located at the vertices of an equilateral triangle. At the same time we will have as the general solution to the last of (3) a general Kepler orbit in a plane, which we may represent by $\vec{s}_3(t) = R(t)(\cos\Theta(t)\mathbf{i} + \sin\Theta(t)\mathbf{j})$ (i.e., choose our xy-plane so that it contains the orbit $\vec{s}_3(t)$). Then, in the CM frame (i.e., with the center of mass $\frac{1}{M}(m_1\vec{r}_1 + m_2\vec{r}_2 + m_3\vec{r}_3)$ at the coordinate origin), we should have a picture like the one below (with $\vec{r}_1(t), \vec{r}_2(t), \vec{r}_3(t)$ all lying in the xy-plane).

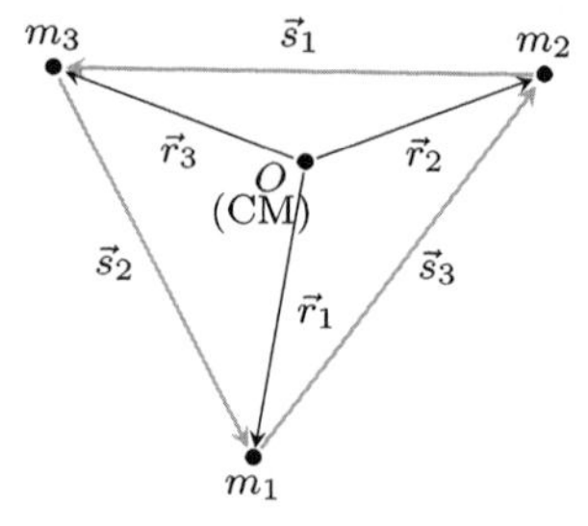

[When $|\vec{s}_1| = |\vec{s}_2| = |\vec{s}_3|$, we find from (1) and (3) that $\ddot{\vec{r}}_1 = -\frac{GM}{|\vec{s}_3|^3}\vec{r}_1$, $\ddot{\vec{r}}_2 = -\frac{GM}{|\vec{s}_3|^3}\vec{r}_2$ and $\ddot{\vec{r}}_3 = -\frac{GM}{|\vec{s}_3|^3}\vec{r}_3$, i.e., the accelerations of the three particles are all directed toward the CM.]

Now, if $\vec{s}_3(t)$ is given (in terms of $R(t)$ and $\Theta(t)$), $\vec{s}_1(t)$ and $\vec{s}_2(t)$ are also determined, viz.,

$$\begin{aligned}
\vec{s}_3(t) &= R(t)[\cos\Theta(t)\mathbf{i} + \sin\Theta(t)\mathbf{j}], \\
\vec{s}_1(t) &= R(t)\left[\cos\left(\Theta(t) + \frac{2}{3}\pi\right)\mathbf{i} + \sin\left(\Theta(t) + \frac{2}{3}\pi\right)\mathbf{j}\right], \\
\vec{s}_2(t) &= R(t)\left[\cos\left(\Theta(t) + \frac{4}{3}\pi\right)\mathbf{i} + \sin\left(\Theta(t) + \frac{4}{3}\pi\right)\mathbf{j}\right].
\end{aligned} \tag{4}$$

As the three equations in (3) are exactly similar, (4) should provide consistent solutions to all equations in (3) simultaneously. Using (1) together with (4), we can express $\vec{r}_1(t)$, $\vec{r}_2(t)$ and $\vec{r}_3(t)$ in terms of the general Kepler orbit solutions $(R(t), \Theta(t))$. For example, when $(R(t), \Theta(t))$ corresponds to an elliptical motion, the orbits of the three particles may look like the figure below.

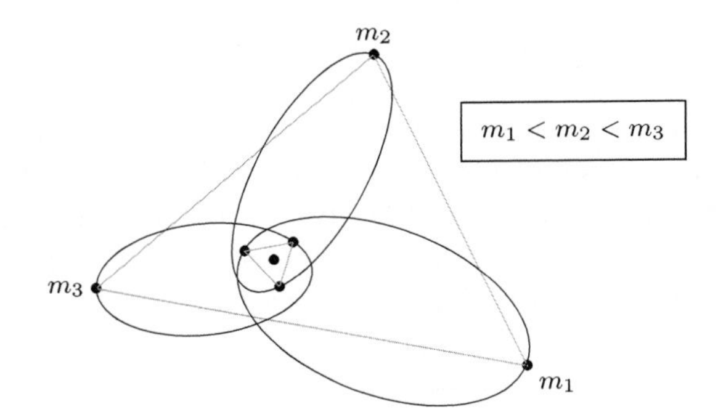

Similar orbits for hyperbolic and parabolic motions may also be constructed.

(iii) If we eliminate $\vec{S}$ from (2a) and (2b) by using (2c) (i.e., $G\vec{S} = \frac{1}{m_3}(\ddot{\vec{s}}_3 + GM\frac{\vec{s}_3}{|\vec{s}_3|^3})$), we find

$$\begin{aligned}
\ddot{\vec{s}}_1 + GM\frac{\vec{s}_1}{|\vec{s}_1|^3} &= \frac{m_1}{m_3}\left(\ddot{\vec{s}}_3 + GM\frac{\vec{s}_3}{|\vec{s}_3|^3}\right), \\
\ddot{\vec{s}}_2 + GM\frac{\vec{s}_2}{|\vec{s}_2|^3} &= \frac{m_2}{m_3}\left(\ddot{\vec{s}}_3 + GM\frac{\vec{s}_3}{|\vec{s}_3|^3}\right).
\end{aligned} \tag{5}$$

Then use the linearity conditions $\vec{s}_1 = \lambda\vec{s}_3$ and $\vec{s}_2 = -(1+\lambda)\vec{s}_3$, with λ taken to be positive (assuming that particle 2 lies between the other two particles), in (5) to obtain the two equations corresponding to a Keplerian orbit

$$(m_3\lambda - m_1)\ddot{\vec{s}}_3 = GM\left[m_1 - \frac{m_3}{\lambda^2}\right]\frac{\vec{s}_3}{|\vec{s}_3|^3},$$

$$[-m_3(1+\lambda) - m_2]\ddot{\vec{s}}_3 = GM\left[m_2 + \frac{m_3}{(1+\lambda)^2}\right]\frac{\vec{s}_3}{|\vec{s}_3|^3}.$$

Consequently, we must have

$$\frac{m_1 - (m_3/\lambda^2)}{m_3\lambda - m_1} = \frac{m_2 + m_3/(1+\lambda)^2}{-m_3(1+\lambda) - m_2},$$

and putting this in standard form gives rise to a fifth degree polynomial equation:

$$f(\lambda) \equiv (m_1+m_2)\lambda^5 + (3m_1+2m_2)\lambda^4 + (3m_1+m_2)\lambda^3 - (m_2+3m_3)\lambda^2 - (2m_2+3m_3)\lambda - (m_2+m_3) = 0. \tag{6}$$

Note that $f(0) < 0$ and $f(\lambda) \to \infty$ for $\lambda \to \infty$: therefore the polynomial equation (6) has a positive real root. By Descartes' sign rule (just one sign change in the coefficients!), it has no more than one positive real root, and so λ is a unique function of the masses. Two other solutions can be obtained by putting different particles between the other two. Thus, there are three distinct families of collinear three-body solutions. As an example, we plot a solution for the elliptic case below.

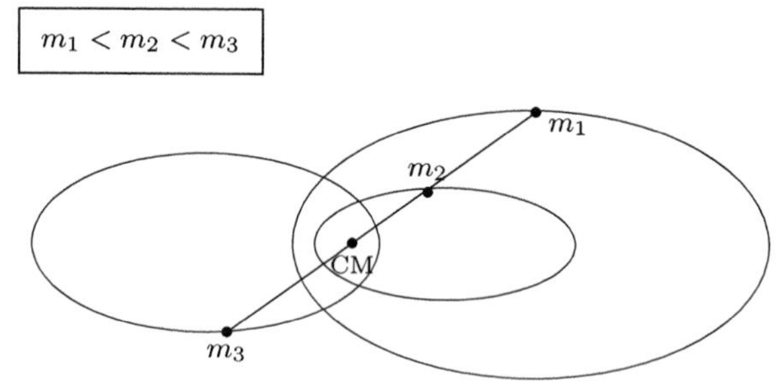

10 Lagrangian Mechanics

10-1 We will consider the problem in terms of cylindrical coordinates (ρ, θ, z). On the surface of a cylinder $\rho = \sqrt{x^2+y^2} = a$ (: const.), the length of any curve C connecting two points on the surface, (θ_0, z_0) and (θ_1, z_1), can be expressed as

$$L = \int_C ds = \int_{z_0}^{z_1} \sqrt{a^2 \left(\frac{d\theta}{dz}\right)^2 + 1}\, dz \quad \text{(: a functional of } \theta = \theta(z)\text{)}.$$

For the geodesic, we demand $\frac{\delta L}{\delta \theta(z)} = 0$, i.e., the Euler–Lagrange equation which reads

$$\frac{d}{dz}\left(\frac{a^2 \frac{d\theta}{dz}}{\sqrt{a^2\left(\frac{d\theta}{dz}\right)^2 + 1}}\right) = 0 \quad \to \quad \frac{d\theta}{dz} = \text{const}, \quad \text{i.e., } \theta = Cz + D.$$

If we use the end-point conditions, $\theta = \left(\frac{\theta_1 - \theta_0}{z_1 - z_0}\right)(z - z_0) + \theta_0$ $(z_0 \leq z \leq z_1)$; this represents a *helix*. Note that, if a flat sheet is rolled up to make a cylinder, a *straight line* on the sheet turns into a helix.

10-2 A convenient parametrization for points of the given surface is

$$x = g(u)\cos v, \quad y = g(u)\sin v, \quad z = u,$$

where $g(u) \equiv a\sqrt{1 - u^2/c^2}$. So we may represent any path lying on this surface (between two given points) by $v = v(u)$. Then the total length of the path will be

$$L = \int_1^2 \sqrt{1 + [g'(u)]^2 + [g(u)]^2 \left(\frac{dv}{du}\right)^2}\, du. \tag{1}$$

The geodesic curve, which makes this integral (as a functional of $v(u)$) extremal, should satisfy the Euler–(Lagrange) equation

$$\frac{d}{du}\left(\frac{[g(u)]^2\frac{dv}{du}}{\sqrt{1+[g'(u)]^2+[g(u)]^2\left(\frac{dv}{du}\right)^2}}\right)=0. \tag{2}$$

(Note that the integrand of (1) contains $\frac{dv}{du}$, but not v.) From (2), we find

$$\frac{[g(u)]^2\frac{dv}{du}}{\sqrt{1+[g'(u)]^2+[g(u)]^2\left(\frac{dv}{du}\right)^2}}=C_1 \quad \text{or}$$

$$\frac{dv}{du}=C_1\frac{1}{g(u)}\sqrt{\frac{1+[g'(u)]^2}{[g(u)]^2-C_1^2}}$$

(C_1 is a constant), and so v can be expressed by the form

$$v=C_1\int\frac{1}{g(u)}\sqrt{\frac{1+[g'(u)]^2}{[g(u)]^2-C_1^2}}du. \tag{3}$$

[Remark: In the case of a sphere, i.e., with $c=a$, the integral in (3) can be performed explicitly. Indeed, with $g(u)=\sqrt{a^2-u^2}$ and so $1+[g'(u)]^2=\frac{a^2}{a^2-u^2}$, we can make the substitution $u=a\cos\theta$ in (3) to write

$$\begin{aligned} v &= C_1\int\frac{d\theta}{\sin\theta\sqrt{a^2\sin^2\theta-C_1^2}} \\ &= \int\frac{\csc^2\theta d\theta}{\sqrt{[(a/C_1)^2-1]-\cot^2\theta}} = -\sin^{-1}\frac{\cot\theta}{\sqrt{(a/C_1)^2-1}}+C_2. \end{aligned}$$

Hence, it follows that

$$(\sin C_2)a\sin\theta\cos v-(\cos C_2)a\sin\theta\sin v-\frac{a\cos\theta}{\sqrt{(a/C_1)^2-1}}=0$$

and, in view of $x=a\sin\theta\cos v$, $y=a\sin\theta\sin v$ and $z=a\cos\theta$, this sphere geodesic lies on the plane

$$(\sin C_2)x-(\cos C_2)y-\frac{z}{\sqrt{(a/C_1)^2-1}}=0,$$

which passes through the center of the sphere. This is a great circle, of course.]

10-3 If $y = y(x)$ describes a light ray connecting two points (x_0, y_0) and (x_1, y_1), it must correspond to an extremal of the functional

$$T = \frac{1}{c}\int_{x_0}^{x_1} n(y)(1+y'^2)^{1/2}dx \quad \left(y' = \frac{dy}{dx}\right).$$

Since x does not appear in the integrand, the second form of the Euler equation then tells us that

$$n(y)(1+y'^2)^{1/2} - y'n(y)\frac{y'}{(1+y'^2)^{1/2}} = \frac{n(y)}{(1+y'^2)^{1/2}} = \text{const.}$$

The slope angle ψ is related to y' by $\tan\psi = y'$ and using this in the above equation gives

$$n\cos\psi = \text{const.},$$

which is the desired relation.

10-4 We note that the area of the surface of revolution can be expressed by

$$A = \int_{x_1}^{x_2} 2\pi x\sqrt{1+(y'(x))^2}\,dx \quad (\text{: a functional of } y(x)).$$

Hence, to find $y(x)$ leading to a minimum surface area, we solve $\frac{\delta A}{\delta y(x)} = 0$:

$$\frac{d}{dx}\left(\frac{xy'(x)}{\sqrt{1+(y'(x))^2}}\right) = 0 \quad \rightarrow \quad \frac{xy'(x)}{\sqrt{1+(y'(x))^2}} = C_1 \quad (\text{: const.}).$$

We can recast this as

$$x^2(y'(x))^2 = C_1^2[1+(y'(x))^2] \quad \rightarrow \quad \frac{dy}{dx} = \frac{C_1}{\sqrt{x^2 - C_1^2}}, \tag{1}$$

and so find the integral

$$y = C_1\cosh^{-1}\frac{x}{C_1} + C_2$$

(or $x = C_1\cosh\frac{y-C_2}{C_1}$), where the constants C_1 and C_2 will be determined through the conditions $y(x{=}x_1) = y_1$ and $y(x{=}x_2) = y_2$. [Remark: The area A above may also be represented by

$$A = \int_{y_1}^{y_2} 2\pi x(y)\sqrt{1+(x'(y))^2}\,dy \quad (\text{: a functional of } x(y)).$$

Then the related Euler–Lagrange equation takes the form

$$\sqrt{1+(x'(y))^2} - \frac{d}{dy}\left(\frac{x(y)x'(y)}{\sqrt{1+(x'(y))^2}}\right) = 0.$$

Or, using Euler's second form, we have

$$x\sqrt{1+(x')^2} - x'\frac{xx'}{\sqrt{1+(x')^2}} = C \quad (\text{: const.}).$$

This implies that

$$\frac{dx}{dy} = \frac{\sqrt{x^2-C^2}}{C} \quad \rightarrow \quad \frac{dy}{dx} = \frac{C}{\sqrt{x^2-C^2}}$$

which corresponds to the last formula in (1).]

10-5 The line element ds of the chain contributes to the potential energy by $-(\lambda ds)gy = -\lambda gy\sqrt{1+(y')^2}dx$. The shape of the chain $y = y(x)$ will be such that the potential energy integral

$$V = -\lambda g \int_{x_1}^{x_2} y\sqrt{1+(y')^2}\,dx \tag{1}$$

is a minimum under the constraint that the length of the chain is fixed, i.e.,

$$\int_{x_1}^{x_2} \sqrt{1+(y')^2}\,dx = L. \tag{2}$$

We may then look for the extremum of the integral

$$J = \int_{x_1}^{x_2} \left[-\lambda gy\sqrt{1+(y')^2} + \mu\sqrt{1+(y')^2}\right]dx, \tag{3}$$

where μ is a Lagrange multiplier (independent of x). Now, using the second form of the Euler equation, we are led to

$$(-\lambda gy+\mu)\sqrt{1+(y')^2} - y'\frac{\partial}{\partial y'}\left[(-\lambda gy+\mu)\sqrt{1+(y')^2}\right]$$
$$= \frac{(-\lambda gy+\mu)}{\sqrt{1+(y')^2}} = \text{const.} \tag{4}$$

or

$$\frac{dy}{\sqrt{(y-\frac{\mu}{\lambda g})^2 - C^2}} = \frac{1}{C}dx, \tag{5}$$

where C is a constant. Integrating our shape equation (5), we find

$$y = C\cosh\left(\frac{x+a}{c}\right) + \frac{\mu}{\lambda g} \tag{6}$$

where the three constants C, μ and a are to be fixed using the constraint (2) and the two end-point conditions $y(x_1) = y_1$ and $y(x_2) = y_2$. Note that (6) corresponds to the canonical form of a catenary.

10-6 The area bounded by a closed curve C can be expressed as

$$\begin{aligned} A &= \frac{1}{2}\int_C (xdy - ydx) \\ &= \frac{1}{2}\int_0^l (xy' - yx')ds \quad \left(y' \equiv \frac{dy}{ds}, x' \equiv \frac{dx}{ds}\right) \end{aligned}$$

where s denotes the arc length along the curve and the curve C is parametrized as $x = x(s)$ and $y = y(s)$. Then $x(s)$ and $y(s)$ should satisfy the *constraint equation*

$$x'^2 + y'^2 = 1.$$

Hence, to find the curve leading to the maximum enclosed area, we consider the functional

$$J[x(s), y(s)] = \int_0^l \left[\frac{1}{2}(xy' - yx') - \lambda(x'^2 + y'^2)\right] ds$$

(λ is a Lagrange multiplier), and solve the corresponding Euler–Lagrange equations:

$$\begin{aligned} &\frac{1}{2}y' - \frac{d}{ds}\left(-\frac{1}{2}y - 2\lambda x'\right) = 0 \quad \rightarrow \quad \frac{d}{ds}(y + 2\lambda x') = 0 \\ &-\frac{1}{2}x' - \frac{d}{ds}\left(\frac{1}{2}x - 2\lambda y'\right) = 0 \quad \rightarrow \quad \frac{d}{ds}(x - 2\lambda y') = 0. \end{aligned}$$

From these, we find

$$\begin{cases} y + 2\lambda x' = c_1 \\ x - 2\lambda y' = c_2 \end{cases} \longrightarrow \quad \frac{dy}{dx} = -\left(\frac{c_2 - x}{c_1 - y}\right)$$

or

$$(c_1 - y)dy + (c_2 - x)dx = 0.$$

Integrating the last equation yields

$$(x - c_2)^2 + (y - c_1)^2 = R^2,$$

which corresponds to a circle of radius $R = \frac{l}{2\pi}$.

10-7 Using spherical coordinates, the Lagrangian for a particle in an external potential is expressed as

$$L = \frac{1}{2}m(\dot{r}^2 + r^2\dot{\theta}^2 + r^2\sin^2\theta\dot{\varphi}^2) - V(r, \theta, \varphi).$$

The corresponding Lagrange equations are

$$0 = \frac{d}{dt}\left(\frac{\partial L}{\partial \dot{r}}\right) - \frac{\partial L}{\partial r} = m\ddot{r} - mr\dot{\theta}^2 - mr\sin^2\theta\dot{\varphi}^2 + \frac{\partial V}{\partial r},$$

$$0 = \frac{d}{dt}\left(\frac{\partial L}{\partial \dot{\theta}}\right) - \frac{\partial L}{\partial \theta} = \frac{d}{dt}(mr^2\dot{\theta}) - mr^2\sin\theta\cos\theta\dot{\varphi}^2 + \frac{\partial V}{\partial \theta},$$

$$0 = \frac{d}{dt}\left(\frac{\partial L}{\partial \dot{\varphi}}\right) - \frac{\partial L}{\partial \varphi} = \frac{d}{dt}(mr^2\sin^2\theta\dot{\varphi}) + \frac{\partial V}{\partial \varphi},$$

while we can write $\frac{\partial V}{\partial r} = -F_r$, $\frac{\partial V}{\partial \theta} = -rF_\theta$ and $\frac{\partial V}{\partial \varphi} = -r\sin\theta F_\varphi$ if $\vec{F} = -\vec{\nabla}V$. These agree with our earlier equations derived by rewriting $m\frac{d^2\vec{r}}{dt^2} = \vec{F}$ $(= F_r\hat{r} + F_\theta\hat{\theta} + F_\varphi\hat{\varphi})$ directly.

10-8 Using coordinates $\vec{r} = (x, y, z) = x^i\mathbf{e}_i$ attached to an inertial frame, suppose that the Lagrangian for a particle is

$$L = \frac{1}{2}m\vec{v}^2 - V(\vec{r}, t) \quad \left(\vec{v} \equiv \frac{d\vec{r}}{dt},\ \frac{d}{dt}\mathbf{e}_i = 0\right).$$

Now, let us look at this system from the viewpoint of a rotating observer whose axes $\{\mathbf{e}'_i\}$ rotate with angular velocity $\vec{\omega}$ relative to $\{\mathbf{e}_i\}$, i.e., $\frac{d}{dt}\mathbf{e}'_i = \vec{\omega} \times \mathbf{e}'_i$ $(i = 1, 2, 3)$. The rotating observer will then assign to the particle the coordinates (x', y', z') and velocities $(\dot{x}', \dot{y}', \dot{z}')$ which are related to the quantities of the inertial system by

$$\vec{r} = x^i\mathbf{e}_i = x'^i(\mathbf{x}, t)\mathbf{e}'_i,$$

$$\vec{v} = \frac{d\vec{r}}{dt} = \frac{d}{dt}(x'^i\mathbf{e}'_i(t)) = \dot{x}'^i\mathbf{e}'_i(t) + x'^i\vec{\omega} \times \mathbf{e}'_i$$

$$\equiv \vec{v}' + \vec{\omega} \times \vec{r}.$$

Accordingly, the above Lagrangian can be rewritten using $\vec{r} = x'^i \mathbf{e}'_i$ and $\vec{v}' = \dot{x}'^i \mathbf{e}'_i$ as

$$\begin{aligned} L &= \frac{1}{2}m(\vec{v}' + \vec{\omega} \times \vec{r}) \cdot (\vec{v}' + \vec{\omega} \times \vec{r}) - V \\ &= \frac{1}{2}m\vec{v}'^2 + m\vec{v}' \cdot (\vec{\omega} \times \vec{r}) + \frac{1}{2}m(\vec{\omega} \times \vec{r})^2 - V. \end{aligned}$$

Therefore, taking $x'^i = x'^i(\mathbf{x}, t)$ as our generalized coordinates, we obtain the Lagrange equations

$$\begin{aligned} &\frac{d}{dt}\left(\frac{\partial L}{\partial \dot{x}'^i}\right) = \frac{\partial L}{\partial x'^i} \\ &\longrightarrow \quad \frac{d}{dt}[m\dot{x}'^i + m(\vec{\omega} \times \vec{r})_i] = [m\vec{v}' \times \vec{\omega} - m\vec{\omega} \times (\vec{\omega} \times \vec{r})]_i - \frac{\partial V}{\partial x'^i} \\ &\longrightarrow \quad m\ddot{x}'^i = -\frac{\partial V}{\partial x'^i} + [\underbrace{2m\vec{v}' \times \vec{\omega}}_{\text{Coriolis force}} \underbrace{-m\vec{\omega} \times (\vec{\omega} \times \vec{r})}_{\text{centrifugal force}}]_i + m(\vec{r} \times \dot{\vec{\omega}})_i. \end{aligned}$$

These are the desired equations of motion in the rotating frame, with appropriate fictitious forces.

10-9 With $q_i = f_i(\mathbf{q}'; t)$, we have

$$\dot{q}_i = \frac{\partial f_i}{\partial q'_j}\dot{q}'_j + \frac{\partial f_i}{\partial t}$$

and hence $\frac{\partial \dot{q}_i}{\partial \dot{q}'_j} = \frac{\partial f_i}{\partial q'_j}$. Now observe that

$$\begin{aligned} \frac{d}{dt}\left(\frac{\partial \widetilde{L}}{\partial \dot{q}'_i}\right) &= \frac{d}{dt}\left(\frac{\partial \dot{q}_j}{\partial \dot{q}'_i}\frac{\partial L}{\partial \dot{q}_j}\right) \quad \left(\text{with } \frac{\partial \dot{q}_j}{\partial \dot{q}'_i} = \frac{\partial f_j}{\partial q'_i}\right) \\ &= \frac{\partial f_j}{\partial q'_i}\frac{d}{dt}\left(\frac{\partial L}{\partial \dot{q}_j}\right) + \left(\frac{\partial^2 f_j}{\partial q'_k \partial q'_i}\dot{q}'_k + \frac{\partial^2 f_j}{\partial t \partial q'_i}\right)\frac{\partial L}{\partial \dot{q}_j} \end{aligned} \tag{1}$$

and

$$\begin{aligned} \frac{\partial \widetilde{L}}{\partial q'_i} &= \frac{\partial f_j}{\partial q'_i}\frac{\partial L}{\partial q_j} + \frac{\partial \dot{q}_j}{\partial q'_i}\frac{\partial L}{\partial \dot{q}_j} \\ &= \frac{\partial f_j}{\partial q'_i}\frac{\partial L}{\partial q_j} + \left(\frac{\partial^2 f_j}{\partial q'_i \partial q'_k}\dot{q}'_k + \frac{\partial^2 f_j}{\partial q'_i \partial t}\right)\frac{\partial L}{\partial \dot{q}_j} \end{aligned} \tag{2}$$

where we used $\dot{q}_j = \frac{\partial f_j}{\partial q'_k}\dot{q}'_k + \frac{\partial f_j}{\partial t}$ for the last relation. From (1) and (2), we thus obtain

$$\frac{\partial \widetilde{L}}{\partial q'_i} - \frac{d}{dt}\left(\frac{\partial \widetilde{L}}{\partial \dot{q}'_i}\right) = \sum_{j=1}^{n}\left[\frac{\partial L}{\partial q_j} - \frac{d}{dt}\left(\frac{\partial L}{\partial \dot{q}_j}\right)\right]\frac{\partial f_j}{\partial q'_i}.$$

For the action functional $S = \int L\,dt$, this implies that $\frac{\delta S}{\delta q_i'(t)} = \sum_{j=1}^{n} \frac{\delta S}{\delta q_j(t)} \frac{\partial f_j}{\partial q_i'}$.

10-10 (i) If $V(\vec{r}, \dot{\vec{r}}, t) = q\phi(\vec{r}, t) - q\dot{\vec{r}} \cdot \vec{A}(\vec{r}, t)$, it follows that

$$-\frac{\partial V}{\partial x_i} + \frac{d}{dt}\left(\frac{\partial V}{\partial \dot{x}_i}\right) = -q\frac{\partial \phi}{\partial x_i} + q\dot{x}_j \frac{\partial A_j}{\partial x_i} - q\frac{d}{dt}A_i$$
$$= q\left(-\frac{\partial \phi}{\partial x_i} - \frac{\partial A_i}{\partial t}\right) + q\dot{x}_j\left(\frac{\partial A_j}{\partial x_i} - \frac{\partial A_i}{\partial x_j}\right),$$

where we used $\frac{dA_i(\vec{r},t)}{dt} = \frac{\partial A_i}{\partial t} + \frac{\partial A_i}{\partial x_j}\dot{x}_j$. Since $-\frac{\partial \phi}{\partial x_i} - \frac{\partial A_i}{\partial t} = E_i$ and $\frac{\partial A_j}{\partial x_i} - \frac{\partial A_i}{\partial x_j} = \epsilon_{ijk}\epsilon_{klm}\frac{\partial A_m}{\partial x_l} = \epsilon_{ijk}(\vec{\nabla} \times \vec{A})_k = \epsilon_{ijk}B_k$, the above relation implies that

$$-\frac{\partial V}{\partial x_i} + \frac{d}{dt}\left(\frac{\partial V}{\partial \dot{x}_i}\right) = qE_i + q\epsilon_{ijk}\dot{x}_j B_k.$$

(ii) For the vector potential describing a uniform magnetic field $\vec{B} = B_0\mathbf{k}$, one can choose either $\vec{A}_{\rm I}(\vec{r}) = -\frac{1}{2}\vec{r} \times (B_0\mathbf{k})$ or $\vec{A}_{\rm II}(\vec{r}) = -B_0 y\mathbf{i}$ since

$$[\vec{\nabla} \times \vec{A}_{\rm I}(\vec{r})]_i = \epsilon_{ijk}\frac{\partial}{\partial x_j}\left(-\frac{B_0}{2}\epsilon_{kl3}x_l\right) = -\frac{B_0}{2}\epsilon_{ijk}\epsilon_{kj3} = B_0\delta_{i3}$$

and

$$[\vec{\nabla} \times \vec{A}_{\rm II}(\vec{r})]_i = \epsilon_{ijk}\frac{\partial}{\partial x_j}\left(-B_0 y\delta_{k1}\right) = -B_0\epsilon_{i21} = B_0\delta_{i3}.$$

Also, to describe a uniform electric field $\vec{E} = E_0\mathbf{i}$, one can choose a scalar potential $\tilde{\phi}(\vec{r}) = -E_0 x$ or a time-dependent vector potential $\tilde{\vec{A}}(\vec{r}, t) = -E_0 t\mathbf{i}$ since

$$-\vec{\nabla}\tilde{\phi}(\vec{r}) = -\vec{\nabla}(-E_0 x) = E_0\mathbf{i},$$

and

$$-\frac{\partial \tilde{\vec{A}}}{\partial t} = -\frac{\partial}{\partial t}(-E_0 t\mathbf{i}) = E_0\mathbf{i}.$$

(iii) Under the gauge transformation

$$\vec{A}(\vec{r},t) \longrightarrow \vec{A}'(\vec{r},t) = \vec{A}(\vec{r},t) - \vec{\nabla}\Lambda(\vec{r},t),$$
$$\phi(\vec{r},t) \longrightarrow \phi'(\vec{r},t) = \phi(\vec{r},t) + \frac{\partial\Lambda(\vec{r},t)}{\partial t},$$

the Lagrangian $L = \frac{1}{2}m\dot{\vec{r}}^2 - q\phi + q\dot{\vec{r}}\cdot\vec{A}$ goes over to the form

$$\begin{aligned} L' &= \frac{1}{2}m\dot{\vec{r}}^2 - q\phi' + q\dot{\vec{r}}\cdot\vec{A}' \\ &= L - q\frac{\partial\Lambda}{\partial t} - q\dot{\vec{r}}\cdot\vec{\nabla}\Lambda = L - \frac{d}{dt}[q\Lambda(\vec{r},t)], \end{aligned}$$

which differs from L only by a *total time-derivative term.* Such total derivative term does not alter the content of the stationary action principle, and hence we have the same Lagrange equations. This is why only the *gauge-invariant* field strengths $\vec{E}$ and $\vec{B}$ enter the related Lagrange equations.

(iv) For the case of the uniform magnetic field,

$$\vec{A}_{\text{II}}(\vec{r}) = \vec{A}_{\text{I}}(\vec{r}) - \vec{\nabla}\left(-\frac{B_0}{2}xy\right) \quad \left(\text{i.e., } \Lambda = -\frac{B_0}{2}xy\right).$$

On the other hand, in the case of the uniform electric field, we may perform the gauge transformation with the gauge function $\Lambda = E_0xt$ on the static scalar potential $\phi(\vec{r}) = -E_0x$ to obtain

$$\begin{aligned} \phi' &= -E_0x + \frac{\partial}{\partial t}(E_0xt) = 0, \\ \vec{A}' &= 0 - \vec{\nabla}(E_0xt) = -E_0t\mathbf{i}. \end{aligned}$$

10-11 As we use t as a parameter along the world line, the proper time between two points '1' and '2' in spacetime will be given by

$$\tau_{12} = \int_1^2 \sqrt{\left\{1 + \frac{2}{c^2}\mathcal{G}(\vec{r})\right\} - \frac{1}{c^2}\left\{1 - \frac{2}{c^2}\mathcal{G}(\vec{r})\right\}\dot{\vec{r}}^2}\, dt,$$

where $\dot{\vec{r}} = (\frac{dx}{dt}, \frac{dy}{dt}, \frac{dz}{dt})$. To first order in $\frac{1}{c^2}$, we may approximate this by

$$\tau_{12} \approx \int_1^2 \left[1 - \frac{1}{c^2}\left\{\frac{1}{2}\dot{\vec{r}}^2 - \mathcal{G}(\vec{r})\right\}\right] dt.$$

From this expression, it follows that the world line that extremizes the proper time between '1' and '2' should extremize the combination

$$\int_1^2 L(\vec{r}, \dot{\vec{r}})dt \quad \left(\text{with } L(\vec{r}, \dot{\vec{r}}) = \frac{1}{2}\dot{\vec{r}}^2 - \mathcal{G}(\vec{r})\right).$$

This leads to Lagrange's equation

$$0 = \frac{d}{dt}\left(\frac{\partial L}{\partial \dot{\vec{r}}}\right) - \frac{\partial L}{\partial \vec{r}} = \ddot{\vec{r}} + \vec{\nabla}\mathcal{G}(\vec{r}),$$

which coincides with the equation of motion for a nonrelativistic particle in a gravitational potential $\mathcal{G}(\vec{r})$. In this way, Newtonian gravity can be expressed completely in geometric terms in curved spacetime.

10-12 To find T by the method of virtual work, imagine a displacement δx of the foot of the ladder to the right, thus bringing the CM of the ladder down by δy. From the geometry, we may read

$$x = L\cos\theta \quad \to \quad \delta x = -L\sin\theta\delta\theta,$$
$$y_{(\mathrm{CM})} = \frac{L}{2}\sin\theta \quad \to \quad \delta y = \frac{L}{2}\cos\theta\delta\theta.$$

The relevant explicit acting forces are T and W, and so, from the principle of virtual work, we obtain

$$-T\cos\alpha\delta x + W|\delta y| = 0$$

(α is the angle the rope makes with the horizontal, i.e., $\cos\alpha = \frac{x}{\sqrt{x^2+h^2}}$), or

$$T = W\frac{\sqrt{x^2+h^2}}{x}\left|\frac{\delta y}{\delta x}\right| = W\frac{\sqrt{x^2+h^2}}{x}\frac{\frac{L}{2}\cos\theta}{L\sin\theta}$$
$$= \frac{W}{2}\frac{\sqrt{L^2\cos^2\theta + h^2}}{L\sin\theta}.$$

10-13 Since the system has a left–right symmetry, it can be surmised that the magnitude of the forces which the walls exert on the logs are equal. Thus consider the effect of moving the right-hand wall a distance δx. It is to be expected that in that case the upper log will push the two lower logs apart and descend a distance δy, which is a function of δx. The centers of logs form a triangle (see the figure) and, calling the distance between the centers of two lower logs $2x$, we find that

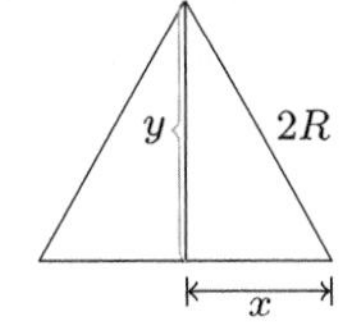

$$x^2 + y^2 = (2R)^2.$$

Hence

$$2x\delta x + 2y\delta y = 0, \quad \text{or } \delta y = -\frac{x}{y}\delta x.$$

For our example in which the logs remain in contact, we have

$$x = R, \ y = \sqrt{3}R, \quad \text{and} \quad \delta y = -\frac{1}{\sqrt{3}}\delta x.$$

The virtual work is the sum of the virtual works performed by the gravitational force and the horizontal force of the displaced wall acting on the logs. That is,

$$\delta W = Mg|\delta y| - F|\delta x|.$$

Setting the virtual work equal to zero yields the desired solution

$$F = \frac{1}{\sqrt{3}}Mg.$$

10-14 As generalized coordinates, we take the inclinations (to the downward vertical) $\theta_1, \theta_2, \ldots, \theta_n$ of the rods in order. If each rod has length a and mass m, the potential energy in a general configuration is

$$\begin{aligned}V &= -mg\frac{a}{2}\cos\theta_1 - mg\left(a\cos\theta_1 + \frac{a}{2}\cos\theta_2\right) \\ &\quad - \cdots - mg\left(a\cos\theta_1 + \cdots + a\cos\theta_{n-1} + \frac{a}{2}\cos\theta_n\right) \\ &= -mg\frac{a}{2}[(2n-1)\cos\theta_1 + (2n-3)\cos\theta_2 \\ &\quad + \cdots + 3\cos\theta_{n-1} + \cos\theta_n].\end{aligned}$$

In a small virtual displacement the work done by gravity is

$$\begin{aligned}-\delta V = -mg\frac{a}{2}[&(2n-1)\sin\theta_1\delta\theta_1 + (2n-3)\sin\theta_2\delta\theta_2 \\ &+ \cdots + 3\sin\theta_{n-1}\delta\theta_{n-1} + \sin\theta_n\delta\theta_n].\end{aligned}$$

The work done by the horizontal force $\vec{F}$ is the product of F and the horizontal displacement of A_{n+1}, i.e.,

$$F\delta(a\sin\theta_1 + a\sin\theta_2 + \cdots + a\sin\theta_n).$$

Adding these two, we find, for the total virtual work done,

$$\begin{aligned}\delta W &= [Fa\cos\theta_1 - (2n-1)mg\frac{a}{2}\sin\theta_1]\delta\theta_1 \\ &\quad + [Fa\cos\theta_2 - (2n-3)mg\frac{a}{2}\sin\theta_2]\delta\theta_2 \\ &\quad + \cdots + [Fa\cos\theta_n - mg\frac{a}{2}\sin\theta_n]\delta\theta_n.\end{aligned}$$

This must vanish for arbitrary values of $\delta\theta_1, \delta\theta_2, \ldots, \delta\theta_n$ if the displacement is from a position of equilibrium. Hence equating to zero the brackets appearing above, we find

$$\tan\theta_1 = \frac{2F}{mg(2n-1)}, \quad \tan\theta_2 = \frac{2F}{mg(2n-3)}, \quad \cdots, \quad \tan\theta_n = \frac{2F}{mg}.$$

These give the inclinations of the rods in the equilibrium configuration.

10-15 Assume that the circular wire rotates in the xy-plane about the point P and let O be the center of the wire. With the angles θ and ϕ as indicated in the figure, we have $\phi = \omega t$ and the coordinates (x, y) of this bead expressed by

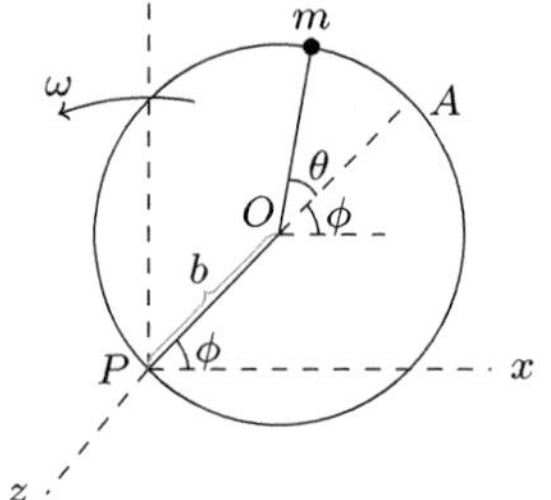

$$x = b\cos\omega t + b\cos(\omega t + \theta),$$
$$y = b\sin\omega t + b\sin(\omega t + \theta).$$

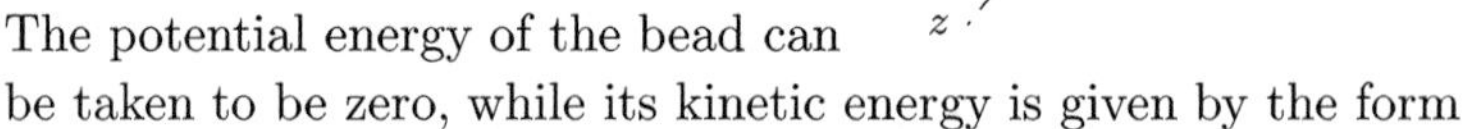

The potential energy of the bead can be taken to be zero, while its kinetic energy is given by the form

$$\begin{aligned}T &= \frac{m}{2}(\dot{x}^2 + \dot{y}^2) \\ &= \frac{m}{2}[\{-b\omega\sin\omega t - b(\omega + \dot{\theta})\sin(\omega t + \theta)\}^2 \\ &\quad + \{b\omega\cos\omega t + b(\omega + \dot{\theta})\cos(\omega t + \theta)\}^2] \\ &= \frac{mb^2}{2}[\omega^2 + (\omega + \dot{\theta})^2 + 2\omega(\omega + \dot{\theta})\cos\theta],\end{aligned}$$

which coincides with the Lagrangian of the system. The Lagrange equation thus takes the form

$$\begin{aligned}0 &= \frac{d}{dt}\left(\frac{\partial L}{\partial \dot{\theta}}\right) - \frac{\partial L}{\partial \theta} \\ &= mb^2[\ddot{\theta} - \omega\dot{\theta}\sin\theta + \omega(\omega + \dot{\theta})\sin\theta] \\ &\longrightarrow \quad \ddot{\theta} + \omega^2 \sin\theta = 0.\end{aligned}$$

We see that the bead oscillates about the line PA (see the figure) like a simple pendulum of length $l = g/\omega^2$.

10-16 Let the length of the string connecting the mass $4m$ and the lower pulley be l_1 and that of the string connecting the mass $3m$ and m be l_2 (ignoring the sizes of the pulleys). We also assign the coordinates x_1 and x_2 as in the figure. Then the potential energy of the system is

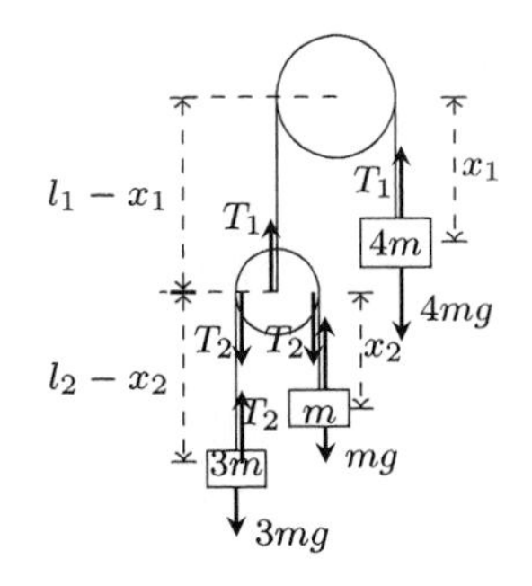

$$\begin{aligned}V =& - 4mgx_1 - mg(l_1 - x_1 + x_2) \\ &- 3mg(l_1 - x_1 + l_2 - x_2)\end{aligned}$$

and the kinetic energy is

$$\begin{aligned}T =& \frac{1}{2}(4m)\dot{x}_1^2 + \frac{1}{2}m\left\{\frac{d}{dt}[l_1 - x_1 + x_2]\right\}^2 \\ &+ \frac{1}{2}(3m)\left\{\frac{d}{dt}[l_1 - x_1 + l_2 - x_2]\right\}^2.\end{aligned}$$

Hence the Lagrangian of the system is

$$L = 2m\dot{x}_1^2 + \frac{1}{2}m(\dot{x}_1 - \dot{x}_2)^2 + \frac{3}{2}m(\dot{x}_1 + \dot{x}_2)^2 - 2mgx_2 + \text{const.}$$

Thus the equations of motion are

$$\begin{aligned}4m\ddot{x}_1 + m(\ddot{x}_1 - \ddot{x}_2) + 3m(\ddot{x}_1 + \ddot{x}_2) &= 0, \\ -m(\ddot{x}_1 - \ddot{x}_2) + 3m(\ddot{x}_1 + \ddot{x}_2) &= -2mg,\end{aligned}$$

from which we obtain

$$\ddot{x}_1 = \frac{1}{7}g, \quad \ddot{x}_2 = -\frac{4}{7}g.$$

Then the downward accelerations of the three masses $4m$, m, and $3m$ are $\ddot{x}_1 = \frac{1}{7}g$, $\frac{d^2}{dt^2}[l_1 - x_1 + x_2] = -\frac{5}{7}g$ and $\frac{d^2}{dt^2}[l_1 - x_1 + l_2 - x_2] = \frac{3}{7}g$, respectively. To find the tension T_1 use Newton's equation of motion for the mass $4m$, i.e, $4mg - T_1 = 4m\ddot{x}_1 = 4m(\frac{1}{7}g)$; we thus find $T_1 = \frac{24}{7}mg$. We also find $T_2 = \frac{12}{7}mg$, from Newton's equation of motion applied to the mass m, $mg - T_2 = m(-\frac{5}{7}g)$.

10-17 (i) The Lagrangian for this system is

$$L = T - V = \frac{1}{2}m_1\dot{x}_1^2 + \frac{1}{2}m_2\dot{x}_2^2 - \frac{1}{2}k(x_2 - x_1 - l_0)^2,$$

and hence we have equations of motion

$$m_1\ddot{x}_1 = k(x_2 - x_1 - l_0),$$
$$m_2\ddot{x}_2 = -k(x_2 - x_1 - l_0).$$

(ii) Equations of motion found in (i) are equivalent to

$$\frac{d}{dt}(m_1\dot{x}_1 + m_2\dot{x}_2) = 0 \quad \to \quad m_1\dot{x}_1 + m_2\dot{x}_2 = C \quad (\text{: const.}) \tag{1}$$

and

$$\ddot{x}_2 - \ddot{x}_1 = -\left(\frac{k}{\mu}\right)(x_2 - x_1 - l_0), \tag{2}$$

where μ denotes the reduced mass $\frac{m_1 m_2}{m_1+m_2}$. Introducing a new variable $r = x_2 - x_1 - l_0$, (2) becomes

$$\ddot{r} + \omega^2 r = 0 \quad (\omega = \sqrt{k/\mu}),$$

and this has a general solution $r(t) = A\cos(\omega t + \beta)$ with two arbitrary constants A and β.

Now use the initial conditions (we take the initial position of m_1 as the origin)

$$x_1 = 0, \quad \dot{x}_1 = \frac{I}{m_1}, \quad x_2 = l, \quad \dot{x}_2 = 0 \quad \text{at } t = 0.$$

Then, $C = I$, i.e.,

$$m_1\dot{x}_1 + m_2\dot{x}_2 = I \quad \to \quad m_1 x_1 + m_2 x_2 = It + m_2 l. \tag{3}$$

Also, from $r(t{=}0) = 0$ and $\dot{r}(t{=}0) = -\frac{I}{m_1}$, the constants A and β must have the values

$$A\cos\beta = 0, \quad A\omega\sin\beta = \frac{I}{m_1}.$$

Accordingly, $\beta = \frac{\pi}{2}$ and $A = \frac{I}{m_1\omega}$, to yield the expression

$$x_2(t) - x_1(t) = l_0 + \frac{I}{m_1\omega}\cos\left(\omega t + \frac{\pi}{2}\right). \tag{4}$$

From (3) and (4), we get

$$x_2(t) = l_0 + \frac{It}{m_1 + m_2} - \frac{I\sin\omega t}{(m_1 + m_2)\omega}$$

and thus

$$\dot{x}_2(t) = \frac{I}{m_1 + m_2} - \frac{I\cos\omega t}{m_1 + m_2}.$$

We see that m_2 comes to rest for the first time at time $\frac{2\pi}{\omega}$, and at that time m_2 has moved the distance

$$\begin{aligned}\Delta = x_2\left(t{=}\frac{2\pi}{\omega}\right) - x_2(t{=}0) &= \frac{2\pi I}{\omega(m_1 + m_2)} \\ &= 2\pi I\sqrt{\frac{m_1 m_2}{k}}(m_1 + m_2)^{-3/2}.\end{aligned}$$

10-18 The rod AB is rotating about the fixed point A and so has kinetic energy

$$\frac{1}{2}I_A\dot{\theta}^2 = \frac{1}{6}ml^2\dot{\theta}^2$$

(with the moment of inertia I_A equal to $\frac{1}{3}ml^2$). On the other hand, for the disk which has its CM at the position ($x = l\sin\theta + R\sin\varphi$, $z = l\cos\theta + R\cos\varphi$), we have

$$\begin{aligned}(\text{CM kinetic energy}) &= \frac{1}{2}M[(l\cos\theta\dot{\theta} + R\cos\varphi\dot{\varphi})^2 \\ &\quad + (-l\sin\theta\dot{\theta} - R\sin\varphi\dot{\varphi})^2] \\ &= \frac{1}{2}M[l^2\dot{\theta}^2 + R^2\dot{\varphi}^2 + 2lR\cos(\theta - \varphi)\dot{\theta}\dot{\varphi}],\end{aligned} \tag{1}$$

$$(\text{rotational energy about the CM}) = \frac{1}{2}\left(\frac{1}{2}MR^2\right)\dot{\varphi}^2, \tag{2}$$

so that the total kinetic energy of the disk is given by

$$\frac{1}{2}Ml^2\dot{\theta}^2 + \frac{3}{4}MR^2\dot{\varphi}^2 + MlR\cos(\theta - \varphi)\dot{\theta}\dot{\varphi}.$$

The potential energy is

$$V = -mg\left(\frac{l}{2}\cos\theta\right) - Mg(l\cos\theta + R\cos\varphi).$$

Therefore the Lagrangian of the system is

$$L = \frac{1}{2}\left(M + \frac{m}{3}\right)l^2\dot{\theta}^2 + \frac{3}{4}MR^2\dot{\varphi}^2 + MlR\cos(\theta - \varphi)\dot{\theta}$$
$$+ \left(\frac{1}{2}m + M\right)gl\cos\theta + MgR\cos\varphi.$$

10-19 Representing the kinetic energy of the rolling cylinder by the sum of the rotational energy about its CM and that related to its CM motion, the Lagrangian of the system can be written as

$$L = \frac{1}{2}I\omega^2 + \frac{1}{2}mv_{\mathrm{CM}}^2 - mgh. \tag{1}$$

Here, I is the moment of inertia of the cylinder about the CM (i.e., $I = \frac{1}{2}ma^2$), v_{CM} $(= (R-a)\dot{\phi})$ is the speed of its center of mass, and h is the height of the cylinder from the equilibrium point, i.e., $h = (R-a)(1-\cos\phi)$. Since the cylinder rolls without slipping, we here have

$$\omega = \frac{v_{\mathrm{CM}}}{a} = \frac{R-a}{a}\dot{\phi}.$$

Inserting this relation into (1), we obtain

$$L = \frac{3}{4}m(R-a)^2\dot{\phi}^2 - mg(R-a)(1-\cos\phi).$$

Hence the equation of motion for this system is

$$\ddot{\phi} + \frac{g}{\frac{3}{2}(R-a)}\sin\phi = 0,$$

which has the same form as that of a simple pendulum of length $\frac{3}{2}(R-a)$. For small oscillation, i.e., with $\phi \ll 1$, it becomes

$$\ddot{\phi} + \frac{g}{\frac{3}{2}(R-a)}\phi = 0.$$

The associated period is

$$T = 2\pi\sqrt{\frac{\frac{3}{2}(R-a)}{g}}.$$

10-20 (i) Let P denote the point on the sphere which happens to be in contact with the triangular block at $t = 0$, and represent its location at time t using the angle of rotation θ (see the figure). Then, as the sphere rolls down the inclined plane by the distance $\xi = \xi_0 + R\theta$, its center of mass $(\bar{x}, \bar{y})$ will acquire the values

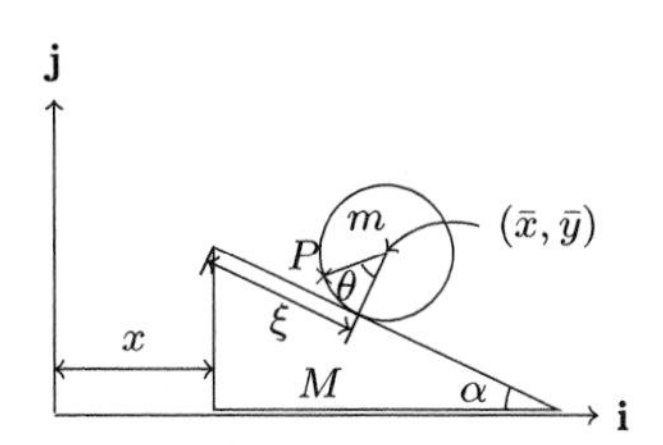

$$\bar{x} = x + (\xi_0 + R\theta)\cos\alpha, \quad \bar{y} = h - R\theta\sin\alpha,$$

assuming that, at $t = 0$, we have $x = \dot{x} = 0$, $\theta = \dot{\theta} = 0$, $\xi = \xi_0$ and $\bar{y} = h$. The Lagrangian for the system can now be written as

$$\begin{aligned} L &= \frac{1}{2}M\dot{x}^2 + \frac{1}{2}m(\dot{\bar{x}}^2 + \dot{\bar{y}}^2) + \frac{1}{2}\left(\frac{2}{5}mR^2\right)\dot{\theta}^2 - mg\bar{y} \\ &= \frac{1}{2}M\dot{x}^2 + \frac{1}{2}m(\dot{x}^2 + R^2\dot{\theta}^2 + 2R\dot{x}\dot{\theta}\cos\alpha) + \frac{1}{5}mR^2\dot{\theta}^2 \\ &\quad - mg(h - R\theta\sin\alpha). \end{aligned}$$

Then the associated Lagrange equations give

$$\begin{aligned} &(M + m)\ddot{x} + mR\ddot{\theta}\cos\alpha = 0, \\ &\ddot{x}\cos\alpha + \frac{7}{5}R\ddot{\theta} - g\sin\alpha = 0. \end{aligned} \tag{1}$$

(ii) Eliminating $\ddot{x}$ from (1) gives

$$\left(\frac{7}{5} - \frac{m\cos^2\alpha}{m + M}\right)\ddot{\theta} = \frac{g\sin\alpha}{R},$$

and hence, integrating with the initial conditions $\theta(t{=}0) = \dot{\theta}(t{=}0) = 0$,

$$\theta(t) = \frac{5(m + M)\sin\alpha}{2\left[7(m + M) - 5m\cos^2\alpha\right]}\frac{g}{R}t^2.$$

Then, from the first equation of (1), we also find

$$x(t) = -\frac{mR\cos\alpha}{m + M}\theta(t) = -\frac{5m\sin 2\alpha}{4\left[7(m + M) - 5m\cos^2\alpha\right]}gt^2.$$

Note that, as the sphere rolls down the inclined plane, the block moves to the left (as expected from the conservation of momentum).

10-21 (i) Clearly, the position of m_2 can be represented by ($x_2 = x_1 + l\sin\theta$, $y_2 = -l\cos\theta$). Hence the Lagrangian is

$$L = \frac{1}{2}m_1\dot{x}_1^2 + \frac{1}{2}m_2[(\dot{x}_1 + l\dot{\theta}\cos\theta)^2 + l^2\sin^2\theta\,\dot{\theta}^2] + m_2 gl\cos\theta.$$

(ii) As the coordinate x_1 is cyclic, we have

$$m_1\dot{x}_1 + m_2(\dot{x}_1 + l\dot{\theta}\cos\theta) = C \ (:\ \text{const.}). \tag{1}$$

[Note that the constant C here coincides with the total momentum in the x-direction.] Using this relation with the Lagrange equation related to the θ-variation

$$\begin{aligned} 0 &= \frac{d}{dt}\left[\cos\theta(\dot{x}_1 + l\dot{\theta}\cos\theta) + l\sin^2\theta\,\dot{\theta}\right] + (\dot{x}_1 + l\dot{\theta}\cos\theta)\dot{\theta}\sin\theta \\ &\quad - l\cos\theta\sin\theta\,\dot{\theta}^2 + g\sin\theta \\ &= -\dot{\theta}\sin\theta(\dot{x}_1 + l\dot{\theta}\cos\theta) + \cos\theta(\ddot{x}_1 + l\ddot{\theta}\cos\theta - l\dot{\theta}^2\sin\theta) \\ &\quad + 2l\sin\theta\cos\theta\,\dot{\theta}^2 + l\sin^2\theta\,\ddot{\theta} + (\dot{x}_1 + l\dot{\theta}\cos\theta)\dot{\theta}\sin\theta \\ &\quad - l\cos\theta\sin\theta\,\dot{\theta}^2 + g\sin\theta, \end{aligned}$$

we obtain

$$\ddot{x}_1\cos\theta + l\ddot{\theta} + g\sin\theta = 0. \tag{2}$$

If we eliminate x_1 from (1) and (2), we obtain the following equation for the angle $\theta(t)$:

$$\begin{aligned} &\ddot{\theta} - \cos^2\theta\left(\frac{m_2}{m_1 + m_2}\right)\ddot{\theta} + \cos\theta\sin\theta\left(\frac{m_2}{m_1 + m_2}\right)\dot{\theta}^2 \\ &\quad + \frac{g}{l}\sin\theta = 0. \end{aligned}$$

For small θ, this reduces to $\ddot{\theta}\frac{m_1}{m_1+m_2} + \frac{g}{l}\theta = 0$; i.e., a simple harmonic motion with angular frequency $\omega = \sqrt{\frac{m_1+m_2}{m_1}\frac{g}{l}}$ results.

10-22 Let $r(t)$ denote the radial distance (from the axis of rotation for the plane) assumed by the given particle at time t. We can then represent the particle's position by (see the figure)

$$x(t) = r(t)\cos\alpha t, \quad z(t) = r(t)\sin\alpha t.$$

Hence we have the Lagrangian

$$\begin{aligned} L &= \frac{1}{2}m(\dot{x}^2 + \dot{z}^2) - mgz \\ &= \frac{1}{2}m(\dot{r}^2 + r^2\alpha^2) - mgr\sin\alpha t, \end{aligned}$$

and accordingly the Lagrange equation

$$m\ddot{r} - m\alpha^2 r = -mg\sin\alpha t.$$

This has the general solution

$$r(t) = Ae^{\alpha t} + Be^{-\alpha t} + \frac{g}{2\alpha^2}\sin\alpha t \quad (A, B\text{: constants}).$$

Now, using the initial conditions $r(0) = r_0$ and $\dot{r}(0) = 0$, we obtain the motion described by

$$r(t) = r_0\cosh\alpha t + \frac{g}{2\alpha^2}(\sin\alpha t - \sinh\alpha t).$$

10-23 Let us assign the coordinates x_1, x_2, y_1 and y_2 as in the figure. Then the potential energy of the system is

$$V = -4mgx_1 - mg(x_2 + y_1) - 3mg(x_2 + y_2)$$

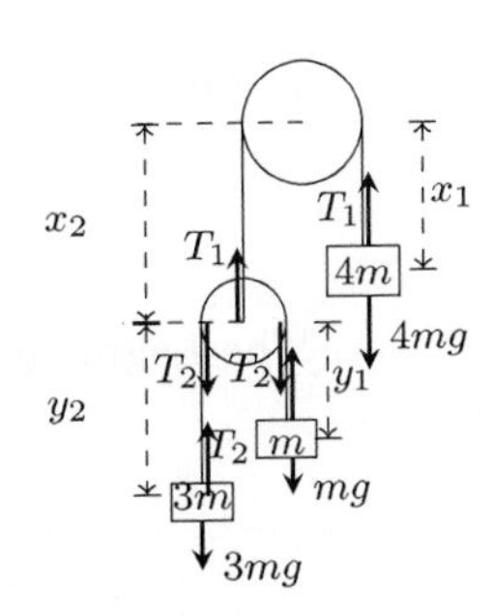

and the kinetic energy is

$$\begin{aligned} T = &\frac{1}{2}(4m)\dot{x}_1^2 + \frac{1}{2}m(\dot{x}_2 + \dot{y}_1)^2 \\ &+ \frac{1}{2}(3m)(\dot{x}_2 + \dot{y}_2)^2. \end{aligned}$$

Let the length of the string connecting the mass $4m$ and the lower pulley be l_1, and that of the string connecting the mass $3m$ and m be l_2 (ignoring the sizes of the pulleys). Then there are two constraints

$$h_1 = x_1 + x_2 - l_1 = 0, \qquad h_2 = y_1 + y_2 - l_2 = 0. \tag{1}$$

Introducing two Lagrange multipliers λ_1 and λ_2, we consider the modified Lagrangian

$$\tilde{L} = T + V + \lambda_1(x_1 + x_2 - l_1) + \lambda_2(y_1 + y_2 - l_2). \tag{2}$$

From (2), we obtain a set of Euler–Lagrange equations

$$\begin{aligned} 4m\ddot{x}_1 &= 4mg + \lambda_1, \\ m(\ddot{x}_2 + \ddot{y}_1) + 3m(\ddot{x}_2 + \ddot{y}_2) &= mg + 3mg + \lambda_1, \\ m(\ddot{x}_2 + \ddot{y}_1) &= mg + \lambda_2, \\ 3m(\ddot{x}_2 + \ddot{y}_2) &= 3mg + \lambda_2. \end{aligned}$$

Using $\ddot{x}_2 = -\ddot{x}_1$ and $\ddot{y}_2 = -\ddot{y}_1$ from the constraints (1), the above equations become

$$\begin{aligned} 4m\ddot{x}_1 &= 4mg + \lambda_1, \\ -4m\ddot{x}_1 - 2m\ddot{y}_1 &= 4mg + \lambda_1, \\ -m\ddot{x}_1 + m\ddot{y}_1 &= mg + \lambda_2, \\ -3m\ddot{x}_1 - 3m\ddot{y}_1 &= 3mg + \lambda_2. \end{aligned}$$

After eliminating λ_1 and λ_2, we obtain

$$4\ddot{x}_1 + \ddot{y}_1 = 0, \qquad 2\ddot{x}_1 + 4\ddot{y}_1 = -2g,$$

to find that

$$\ddot{x}_1 = \frac{1}{7}g = -\ddot{x}_2, \quad \ddot{y}_1 = -\frac{4}{7}g = -\ddot{y}_2.$$

The Lagrange multipliers are determined as

$$\lambda_1 = -\frac{24}{7}mg, \quad \lambda_2 = -\frac{12}{7}mg,$$

and they correspond to the constraint forces (or the tensions with the opposite sign in this case).

10-24 Incorporate the constraint by considering the Lagrangian

$$\tilde{L} = \frac{1}{2}m(\dot{x}^2 + \dot{y}^2 + \dot{z}^2) + \lambda(x^2 + y^2 + z^2 - a^2)$$

and this leads to Lagrange equations

$$m\ddot{x} = \lambda x, \quad m\ddot{y} = \lambda y, \quad m\ddot{z} = \lambda z. \tag{1}$$

The constraint $x^2 + y^2 + z^2 = a^2$ implies that

$$x\dot{x} + y\dot{y} + z\dot{z} = 0, \quad x\ddot{x} + y\ddot{y} + z\ddot{z} = -(\dot{x}^2 + \dot{y}^2 + \dot{z}^2). \tag{2}$$

Then, from (1), we have

$$m(\dot{x}\ddot{x} + \dot{y}\ddot{y} + \dot{z}\ddot{z}) = \lambda(x\dot{x} + y\dot{y} + z\dot{z}) = 0.$$

Since $\dot{x}\ddot{x}+\dot{y}\ddot{y}+\dot{z}\ddot{z} = \frac{1}{2}\frac{d}{dt}(\dot{x}^2+\dot{y}^2+\dot{z}^2)$, we may now set $\dot{x}^2+\dot{y}^2+\dot{z}^2 = v^2$ (: constant). We also have

$$m(x\ddot{x} + y\ddot{y} + z\ddot{z}) = \lambda(x^2 + y^2 + z^2) \quad \longrightarrow \quad \lambda = -m\frac{v^2}{a^2},$$

where we made use of the second equation of (2). Now, setting $\frac{v^2}{a^2} = \omega^2$, the equations in (1) tell us that

$$\ddot{x} + \omega^2 x = 0, \quad \ddot{y} + \omega^2 y = 0, \quad \ddot{z} + \omega^2 z = 0,$$

and these lead to the solution of the form

$$\vec{r}(t) = \vec{A}\cos\omega t + \vec{B}\sin\omega t \quad (\vec{A}, \vec{B} : \text{constant vectors}).$$

But, from $\vec{r}\cdot\vec{r} = a^2$, we can conclude that $|\vec{A}|^2 = |\vec{B}|^2 = a^2$ and $\vec{A}\cdot\vec{B} = 0$. So we may set $\vec{A} = a\mathbf{e}_1$ and $\vec{B} = a\mathbf{e}_2$, $\mathbf{e}_1$ and $\mathbf{e}_2$ being orthogonal, to write $\vec{r} = a\cos\omega t\mathbf{e}_1 + a\sin\omega t\mathbf{e}_2$: this represents a *uniform rotation along a circle of radius a (i.e. a great circle)* on the surface of the given sphere.

10-25 (i) We here have a system whose motion is constrained by the conditions of the general form

$$\sum_{j=1}^{n} a_{lj}(\mathbf{q}, t)dq_j + b_l(\mathbf{q}, t)dt = 0 \quad (l = 1, \ldots, k). \tag{1}$$

[With $a_{lj} = \frac{\partial h_l}{\partial q_j}$ and $b_l = \frac{\partial h_l}{\partial t}$ for certain functions $h_l(\mathbf{q}, t)$ $(l = 1, \ldots, k)$, these reduce to (in general time-dependent) holonomic constraints.] Then, as in the case of holonomic constraints, we may assume that the virtual work done by the related constraint forces vanishes as long as the considered virtual displacements δq_i (performed at a given time, i.e., with $dt = 0$) are consistent with (1), i.e., satisfy

$$\sum_{j=1}^{n} a_{lj}(\mathbf{q}, t)\delta q_i = 0 \quad (l = 1, \ldots, k). \tag{2}$$

Then we are immediately led to conclude that, if $\mathbf{q} = \mathbf{q}^*(t)$ corresponds to physical trajectories and $\{\delta q_i\}$ satisfy the conditions (2), we have the variational formula (see Sec. 10.2)

$$\delta \left[\int_{t_0}^{t_1} T\, dt \right] \Bigg|_{\mathbf{q}=\mathbf{q}^*(t)} = - \int_{t_0}^{t_1} \sum_i Q_i(\mathbf{q}^*(t))\, \delta q_i(t)\, dt \tag{3}$$

with explicitly given forces Q_i. If the Q's can be derived from a potential, this turns into a relation for the Lagrangian, i.e.,

$$\delta \left[\int_{t_0}^{t_1} L\, dt \right] \Bigg|_{\mathbf{q}=\mathbf{q}^*(t)} \equiv \int_{t_0}^{t_1} \sum_i \left[\frac{\partial L}{\partial q_i} - \frac{d}{dt}\frac{\partial L}{\partial \dot{q}_i} \right] \Bigg|_{\mathbf{q}=\mathbf{q}^*(t)} \delta q_i dt = 0 \tag{4}$$

for arbitrary virtual displacements satisfying the conditions (2). For *unrestricted* displacements δq_i (other than the usual end-point conditions), (4) will be the case if and only if

$$\sum_i \left[\frac{\partial L}{\partial q_i} - \frac{d}{dt}\frac{\partial L}{\partial \dot{q}_i} \right] \Bigg|_{\mathbf{q}=\mathbf{q}^*(t)} \delta q_i = - \sum_{l=1}^{k} \lambda^l(t) \left(\sum_{i=1}^{n} a_{li}(\mathbf{q}^*, t)\delta q_i(t) \right), \tag{5}$$

for some $\lambda^l(t)$ $(l = 1, \ldots, k)$. From (5) we thus find the n equations

$$\frac{d}{dt}\left(\frac{\partial L}{\partial \dot{q}_i}\right) - \frac{\partial L}{\partial q_i} = \sum_{l=1}^{k} \lambda^l a_{li} \quad (i = 1, \ldots, n), \tag{6}$$

which may be considered together with the original constraints

$$\sum_{j=1}^{n} a_{lj}\dot{q}_j + b_l = 0 \quad (l = 1, \ldots, k) \tag{7}$$

as the equations of motion for $q_1, \ldots, q_n, \lambda^1, \ldots, \lambda^k$.

[Remark: Note that, for given nonholonomic constraints, one must *not* take the variational equation in the form

$$\delta\left[\int_{t_0}^{t_1}\left(L+\sum_{l=1}^{k}\lambda^l[a_{lj}\dot{q}_j+b_l]\right)dt\right]=0,$$

as this yields the dynamical equations that disagree with those obtained by applying Newton's second law directly.]

(ii) If $(X, Y, 0)$ denotes the contact point, the center of mass of the disk is at $\vec{R} = X\mathbf{i} + Y\mathbf{j} + a\mathbf{k}$. Also any given point P of the disk, at distance r' from the center, can be described by the vector sum

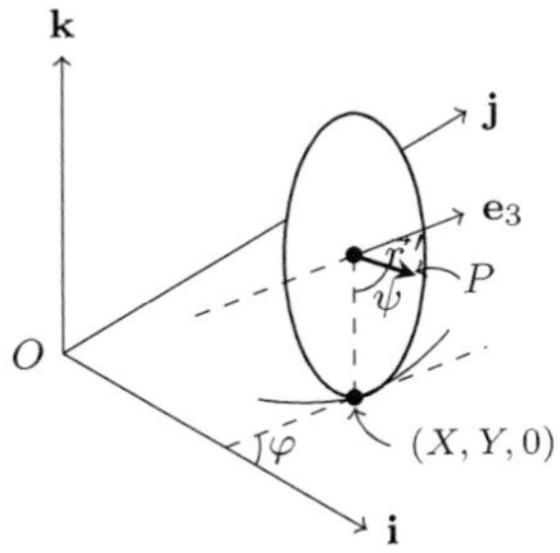

$$\vec{r}(t) = \vec{R}(t) + \vec{r}\,'(t).$$

Here $\vec{r}\,'(t)$, the relative position of P from the center, is given using angles φ and ψ (see the figure) by

$$\vec{r}\,'(t) = r'\sin\psi(t)[-\sin\varphi(t)\mathbf{i} + \cos\varphi(t)\mathbf{j}] - r'\cos\psi(t)\mathbf{k}, \quad (8)$$

when the symmetry axis of the disk is denoted as $\mathbf{e}_3 = \cos\varphi\mathbf{i} + \sin\varphi\mathbf{j}$. Due to the constraint that the disk rolls without slipping, the contact point of the disk (i.e., $\vec{r}\,'$ with $r' = a$ and $\psi = 0$) should have zero instantaneous velocity, i.e.,

$$\begin{aligned} 0 &= \left.\left[\frac{d\vec{R}(t)}{dt} + \frac{d\vec{r}\,'(t)}{dt}\right]\right|_{r'=a,\,\psi=0} \\ &= (\dot{X} - a\dot{\psi}\sin\varphi)\mathbf{i} + (\dot{Y} + a\dot{\psi}\cos\varphi)\mathbf{j}, \end{aligned} \quad (9)$$

where we used the result (from (8))

$$\begin{aligned} \frac{d\vec{r}\,'}{dt} &= \dot{\psi}\{r'\cos\psi[-\sin\varphi\mathbf{i} + \cos\varphi\mathbf{j}] + r'\sin\psi\mathbf{k}\} \\ &\quad + \dot{\varphi}(-r'\sin\psi)[\cos\varphi\mathbf{i} + \sin\varphi\mathbf{j}]. \end{aligned} \quad (10)$$

Based on (10) and the formula $\frac{d\vec{r}\,'}{dt} = \vec{\omega} \times \vec{r}\,'$, one can readily find the angular velocity vector for the disk to be given by

$$\vec{\omega} = \dot{\psi}\mathbf{e}_3 + \dot{\varphi}\mathbf{k}. \quad (11)$$

Then, noting that we can express the angular momentum with respect to the center of mass of the disk by $\vec{L} = I_3\dot{\psi}\mathbf{e}_3 + I_1\dot{\varphi}\mathbf{k}$ ($I_3 = \frac{1}{2}ma^2$, $I_1 = \frac{1}{4}ma^2$), the rotational kinetic energy of the disk can be taken as $\frac{1}{2}\vec{\omega}\cdot\vec{L} = \frac{1}{2}I_3\dot{\psi}^2 + \frac{1}{2}I_1\dot{\varphi}^2$. Hence our system has the Lagrangian

$$L = \frac{1}{2}\dot{X}^2 + \frac{1}{2}\dot{Y}^2 + \frac{1}{2}I_3\dot{\psi}^2 + \frac{1}{2}I_1\dot{\varphi}^2$$

with nonholonomic constraints from (9)

$$\dot{X} = a\dot{\psi}\sin\varphi, \quad \dot{Y} = -a\dot{\psi}\cos\varphi, \tag{13}$$

which imply for virtual displacements the conditions

$$\delta X - a\sin\varphi\delta\psi = 0, \tag{14a}$$

$$\delta Y + a\cos\varphi\delta\psi = 0. \tag{14b}$$

Then, from (i), we are led to the Lagrange equations (λ_1, λ_2 are Lagrange multipliers)

$$m\ddot{X} = \lambda_1, \tag{15a}$$

$$m\ddot{Y} = \lambda_2, \tag{15b}$$

$$I_1\ddot{\varphi} = 0, \tag{15c}$$

$$I_3\ddot{\psi} = -\lambda_1 a\sin\varphi + \lambda_2 a\cos\varphi. \tag{15d}$$

From the third of these, we have $\varphi(t) = \alpha t + \varphi_0$, i.e., the axis of the disk rotates with constant angular velocity. By inserting this into (13), we can calculate $\ddot{X}$ and $\ddot{Y}$, which in turn fixes λ_1 and λ_2 through (15a,b):

$$\lambda_1 = m\ddot{X} = m\{a\ddot{\psi}\sin(\alpha t + \varphi_0) + a\alpha\dot{\psi}\cos(\alpha t + \varphi_0)\}, \tag{16a}$$

$$\lambda_2 = m\ddot{Y} = -m\{a\ddot{\psi}\cos(\alpha t + \varphi_0) - a\alpha\dot{\psi}\sin(\alpha t + \varphi_0)\}. \tag{16b}$$

Using these in (15d), we then find, after some calculations,

$$I_3\ddot{\psi} = -ma^2\ddot{\psi} \quad (\text{i.e., } \ddot{\psi} = 0). \tag{17}$$

Accordingly, $\dot{\psi} = \beta$ (: const.) and so, from (13), we find, if $\alpha \neq 0$,

$$X(t) = -\frac{a\beta}{\alpha}\cos(\alpha t + \varphi_0) + X_0,$$
$$Y(t) = -\frac{a\beta}{\alpha}\sin(\alpha t + \varphi_0) + Y_0, \tag{18}$$

i.e., $(X(t), Y(t))$ describes a circle. [The case with $\alpha = 0$ describes a disk rolling along a straight line: in this case, from (16a,b), we find $\lambda_1 = \lambda_2 = 0$, i.e., the constraint forces vanish.]

10-26 The energy function is given by

$$\begin{aligned} E &= \sum_i \dot{x}_i \frac{\partial L}{\partial \dot{x}_i} - L \\ &= \sum_i \dot{x}_i(m\dot{x}_i + qA_i) - \frac{1}{2}m\dot{\vec{r}}^2 + q\phi - q\dot{x}_i A_i \\ &= \frac{1}{2}m\dot{\vec{r}}^2 + q\phi(\vec{r}), \end{aligned}$$

and this is a conserved quantity since $\frac{\partial L}{\partial t} = 0$.

10-27 (i) From the condition $\delta L = \epsilon \frac{d}{dt}\Lambda(\mathbf{q}, t)$ for $\delta q_i = \epsilon f_i(\mathbf{q}, t)$, we can write

$$\frac{\partial L}{\partial q_i} f_i(\mathbf{q}, t) + \frac{\partial L}{\partial \dot{q}_i}\frac{df_i(\mathbf{q}, t)}{dt} = \frac{d}{dt}\Lambda(\mathbf{q}, t). \tag{1}$$

(Note that we have $\delta \dot{q}_i = \epsilon \frac{df_i(\mathbf{q},t)}{dt} = \epsilon\left[\frac{\partial f_i}{\partial q_j}\dot{q}_j + \frac{\partial f_i}{\partial t}\right]$.) Then, when $q_i(t)$ represent physical trajectories (i.e., satisfy the Lagrange equations), the quantity $I \equiv \frac{\partial L}{\partial \dot{q}_i} f_i - \Lambda$ is necessarily a constant of motion since

$$\begin{aligned} \frac{dI}{dt} &= \left[\frac{d}{dt}\left(\frac{\partial L}{\partial \dot{q}_i}\right)\right] f_i + \frac{\partial L}{\partial \dot{q}_i}\frac{df_i}{dt} - \frac{d\Lambda}{dt} \\ &= \frac{\partial L}{\partial q_i} f_i + \frac{\partial L}{\partial \dot{q}_i}\frac{df_i}{dt} - \frac{d\Lambda}{dt} \\ &= 0, \end{aligned} \tag{2}$$

where in the last step we used the relation (1).

(ii) Under Galilean transformations $\delta\vec{r}_i = -\epsilon\vec{u}t$ ($\vec{u}$: an arbitrary constant vector) with the Lagrangian $L = \frac{1}{2}\sum_i m_i \dot{\vec{r}}_i^2$, we have

$$\delta L = \sum_i \frac{\partial L}{\partial \dot{\vec{r}}_i}\cdot(-\epsilon\vec{u}) = -\epsilon\vec{u}\cdot\sum_i m_i\dot{\vec{r}}_i = \epsilon\frac{d}{dt}(-M\vec{u}\cdot\vec{R})$$

($\vec{R}$ is the position of the center of mass, and $M = \sum_i m_i$) and hence we can identify $\Lambda = -M\vec{u}\cdot\vec{R}$. Now the related constant of motion is

$$\begin{aligned} I &= -\left(\sum_i \frac{\partial L}{\partial \dot{\vec{r}}_i}\right)\cdot\vec{u}\,t + M\vec{u}\cdot\vec{R} \\ &= M\vec{u}\cdot(\vec{R} - \dot{\vec{R}}t). \end{aligned}$$

If $\vec{R}_0$ is the position of the center of mass at $t = 0$, one may identify this conserved quantity as $I = M\vec{u}\cdot\vec{R}_0$. (Also, if one solves the differential equation $\vec{R} - \dot{\vec{R}}t = \vec{R}_0$, one finds that $\vec{R}(t) = \vec{R}_0 + \vec{V}_0\,t$, i.e., the CM moves with some uniform velocity $\vec{V}_0$.)

11 Application of the Lagrangian Method: Small Oscillations

11-1 (i) Clearly, $V = Mg(R - d\cos\theta)$. Before we find the expression for the kinetic energy T, note that the center of mass has the position

$$\vec{r}_{\rm CM} = (x + d\sin\theta)\mathbf{i} + (R - d\cos\theta)\mathbf{k},$$

while the position of any given point Q on the rim of the body is

$$\vec{r}_Q = (x + R\sin\varphi)\mathbf{i} + (R - R\cos\varphi)\mathbf{k}$$

and its velocity is

$$\vec{v}_Q = (\dot{x} + R\dot{\varphi}\cos\varphi)\mathbf{i} + R\dot{\varphi}\sin\varphi\mathbf{k}.$$

If the body is rolling without slipping, this velocity has a zero value at the contact point $\varphi = 0$, i.e., $\dot{x} + R\dot{\varphi} = 0$, and then

$$\dot{x} = -R\dot{\varphi} = -R\dot{\theta}.$$

Hence we have

$$\begin{aligned} T &= \frac{1}{2}M\dot{\vec{r}}_{\rm CM}^{2} + \frac{1}{2}I_{\rm CM}\dot{\theta}^2 \\ &= \frac{1}{2}M[(\dot{x} + d\dot{\theta}\cos\theta)^2 + (d\dot{\theta}\sin\theta)^2] + \frac{1}{2}I_{\rm CM}\dot{\theta}^2 \\ &= \frac{1}{2}M\dot{\theta}^2[R^2 - 2Rd\cos\theta + d^2] + \frac{1}{2}I_{\rm CM}\dot{\theta}^2. \end{aligned}$$

Accordingly, the Lagrangian of this system is given as

$$L = T - V = \frac{1}{2}M\dot{\theta}^2\left[R^2 + d^2 + \frac{I_{\rm CM}}{M} - 2Rd\cos\theta\right] + Mgd\cos\theta,$$

removing an irrelevant constant term in the potential.

(ii) As L does not contain any t-dependent piece explicitly, we have the energy conservation law

$$\frac{1}{2}M\dot{\theta}^2\left[R^2+d^2+\frac{I_{\rm CM}}{M}-2Rd\cos\theta\right]-Mgd\cos\theta=E\quad(\text{: const.})$$

and so the motion can be determined from

$$\dot{\theta}=\pm\left[\frac{2(E+Mgd\cos\theta)}{M(R^2+d^2+\frac{I_{\rm CM}}{M}-2Rd\cos\theta)}\right]^{1/2}.$$

For a given value of E, the direction of the line $\overline{OG}$ oscillates from the vertical within the angular range $(\theta_{\rm min},\theta_{\rm max})$ ($\theta_{\rm min}$, $\theta_{\rm max}$: two roots of $-Mgd\cos\theta=E$). At the same time the coordinate x, which determines the positions of O and the contact point, oscillates according to $x+R\theta=$const.

(iii) Clearly, $\theta=0$ corresponds to the stable equilibrium point, and for small enough θ the above energy conservation law can be approximated as

$$\frac{1}{2}\left[M(R-d)^2+I_{\rm CM}\right]\dot{\theta}^2+\frac{1}{2}Mgd\theta^2=E+Mgd\,,$$

from which one can easily read off the angular frequency of small oscillation

$$\omega=\sqrt{\frac{Mgd}{M(R-d)^2+I_{\rm CM}}}.$$

11-2 (i) The Lagrangian for this system is

$$L=\frac{1}{2}m(\dot{r}^2+r^2\dot{\theta}^2)+\frac{1}{2}M\dot{r}^2+Mg(l-r)$$

(ii) The Lagrange equations give

$$mr^2\dot{\theta}=C\quad(\text{: const.}),\tag{1}$$

$$(m+M)\ddot{r}-mr\dot{\theta}^2+Mg=0.\tag{2}$$

With $r=R$ (: const.), we find, from (2), $mR\dot{\theta}^2=Mg$, i.e., $\dot{\theta}=\sqrt{\frac{M}{m}\frac{g}{R}}$ and thus the period of *revolution* is

$$T_0=\frac{2\pi}{\sqrt{\frac{M}{m}\frac{g}{R}}}=2\pi\sqrt{\frac{m}{M}\frac{R}{g}}.$$

For this circular orbit, the constant C has the value $mR^2\sqrt{\frac{M}{m}\frac{g}{R}}$ (or $C^2 = mMR^3g$).

(iii) As the energy of this system is conserved, we have

$$\frac{1}{2}(m+M)\dot{r}^2 + \frac{1}{2}mr^2\dot{\theta}^2 - Mg(l-r) = E \quad (\text{: const.})$$

or, using (1),

$$\frac{1}{2}(m+M)\dot{r}^2 + V_{\text{eff}}(r) = E, \tag{3}$$

with $V_{\text{eff}}(r) \equiv \frac{1}{2}\frac{C^2}{mr^2} - Mg(l-r)$. If $V_{\text{eff}}(r)$ has a minimum at $r = R$, then

$$V'_{\text{eff}}(r{=}R) = \left[-\frac{C^2}{mr^3} + Mg\right]\Bigg|_{r=R} = 0,$$

so that we have $C^2 = mMR^3g$ as in (i). We are here interested in the case where E is slightly above the value $E_0 \equiv V_{\text{eff}}(r{=}R)$; in this case, (3) takes the approximate form

$$\begin{aligned}\frac{1}{2}(m+M)\dot{r}^2 &+ \frac{1}{2}V''_{\text{eff}}(r{=}R)\,(r-R)^2 \\ &= E - E_0 \quad \left(V''_{\text{eff}}(r=R) = \frac{3Mg}{R}\right).\end{aligned}$$

Hence the period of small radial oscillation is

$$\begin{aligned}T &= \frac{2\pi}{\sqrt{V''_{\text{eff}}(R)/(m+M)}} = 2\pi\sqrt{\frac{(m+M)R}{3Mg}} \\ &= \sqrt{\frac{m+M}{3m}}T_0.\end{aligned}$$

11-3 (i) The Lagrangian $L = \frac{1}{2}ml^2(\dot{\theta}^2 + \sin^2\theta\dot{\varphi}^2)$ is invariant under the three-parameter transformations

$$\begin{aligned}\delta\theta &= -\epsilon_1\sin\varphi + \epsilon_2\cos\varphi, \\ \delta\varphi &= -\epsilon_1\cot\theta\cos\varphi - \epsilon_2\cot\theta\sin\varphi + \epsilon_3,\end{aligned} \tag{1}$$

since

$$\begin{aligned}
&\left(\dot{\theta}+\frac{d}{dt}\delta\theta\right)^2+\sin^2(\theta+\delta\theta)\left[\dot{\varphi}+\frac{d}{dt}\delta\varphi\right]^2-(\dot{\theta}^2+\sin^2\theta\dot{\varphi}^2)\\
&=2\dot{\theta}[-\epsilon_1\cos\varphi-\epsilon_2\sin\varphi]\dot{\varphi}+2\dot{\varphi}\sin^2\theta\big[\epsilon_1\dot{\theta}\csc^2\theta\cos\varphi\\
&\qquad+\epsilon_1\dot{\varphi}\cot\theta\sin\varphi+\epsilon_2\dot{\theta}\csc^2\theta\sin\varphi-\epsilon_2\dot{\varphi}\cot\theta\cos\varphi\big]\\
&\quad+2\dot{\varphi}^2\sin\theta\cos\theta(-\epsilon_1\sin\varphi+\epsilon_2\cos\varphi)+O(\epsilon^3)\\
&=0.
\end{aligned}$$

The above transformations actually describe the rotational symmetry

$$\vec{r}(\theta+\delta\theta,\varphi+\delta\varphi)=\vec{r}(\theta,\varphi)+\vec{\epsilon}\times\vec{r}(\theta,\varphi) \tag{2}$$

with

$$\begin{aligned}
\vec{r}(\theta+\delta\theta,\varphi+\delta\varphi)&=r\big[\sin(\theta+\delta\theta)\cos(\varphi+\delta\varphi)\mathbf{i}\\
&\qquad+\sin(\theta+\delta\theta)\sin(\varphi+\delta\varphi)\mathbf{j}+\cos(\theta+\delta\theta)\mathbf{k}\big]\\
&=\vec{r}(\theta,\varphi)+r\big[(\cos\theta\cos\varphi\mathbf{i}+\cos\theta\sin\varphi\mathbf{j}-\sin\theta\mathbf{k})\delta\theta\\
&\qquad+(-\sin\theta\sin\varphi\mathbf{i}+\sin\theta\cos\varphi\mathbf{j})\delta\varphi\big].
\end{aligned}$$

Indeed, (2) corresponds to

$$\begin{aligned}
r\cos\theta\cos\varphi\,\delta\theta-r\sin\theta\sin\varphi\,\delta\varphi&=\epsilon_2 r\cos\theta-\epsilon_3 r\sin\theta\sin\varphi,\\
r\cos\theta\sin\varphi\,\delta\theta+r\sin\theta\cos\varphi\,\delta\varphi&=\epsilon_3 r\sin\theta\cos\varphi-\epsilon_1 r\cos\theta,\\
-r\sin\theta\,\delta\theta&=\epsilon_1 r\sin\theta\sin\varphi-\epsilon_2 r\sin\theta\cos\varphi,
\end{aligned}$$

from which we can make the identifications

$$\begin{aligned}
\delta\theta&=-\epsilon_1\sin\varphi+\epsilon_2\cos\varphi,\\
\delta\varphi&=-\epsilon_1\cot\theta\cos\varphi-\epsilon_2\cot\theta\sin\varphi+\epsilon_3.
\end{aligned}$$

(ii) From Noether's theorem, constants of motion related to the above symmetry transformations, I_1, I_2, I_3, are obtained such as

$$\begin{aligned}
\epsilon_i I_i &= \frac{\partial L}{\partial \dot{\theta}}\delta\theta + \frac{\partial L}{\partial \dot{\varphi}}\delta\varphi \\
&= ml^2\dot{\theta}(-\epsilon_1 \sin\varphi + \epsilon_2 \cos\varphi) \\
&\quad + ml^2 \sin^2\theta\dot{\varphi}(-\epsilon_1 \cot\theta \cos\varphi - \epsilon_2 \cot\theta \sin\varphi + \epsilon_3) \\
&= \epsilon_1 ml^2(-\sin\varphi\,\dot{\theta} - \sin\theta\cos\theta\cos\varphi\,\dot{\varphi}) \\
&\quad + \epsilon_2 ml^2(\cos\varphi\,\dot{\theta} - \sin\theta\cos\theta\sin\varphi\,\dot{\varphi}) + \epsilon_3 ml^2 \sin^2\theta\,\dot{\varphi},
\end{aligned}$$

viz.,

$$\begin{aligned}
I_1 &= ml^2(-\sin\varphi\,\dot{\theta} - \sin\theta\cos\theta\cos\varphi\,\dot{\varphi}) \quad \leftarrow \quad L_x \\
I_2 &= ml^2(\cos\varphi\,\dot{\theta} - \sin\theta\cos\theta\sin\varphi\,\dot{\varphi}) \quad \leftarrow \quad L_y \\
I_3 &= ml^2 \sin^2\theta\,\dot{\varphi} \quad \leftarrow \quad L_z.
\end{aligned} \tag{3}$$

[Remark: These coincide with the orbital angular momentum (L_x, L_y, L_z) in spherical coordinates, as can be seen from

$$\begin{aligned}
\vec{L} &= m\vec{r} \times \vec{v} = ml^2\hat{r} \times (\dot{\theta}\hat{\theta} + \sin\theta\,\dot{\varphi}\hat{\varphi}) \\
&= ml^2\hat{\varphi} - ml^2\sin\theta\,\dot{\varphi}\hat{\theta} \\
&= ml^2(-\sin\varphi\,\dot{\theta} - \sin\theta\cos\theta\cos\varphi\,\dot{\varphi})\mathbf{i} \\
&\quad + ml^2(\cos\varphi\,\dot{\theta} - \sin\theta\cos\theta\sin\varphi\,\dot{\varphi})\mathbf{j} + ml^2\sin^2\theta\,\dot{\varphi}\mathbf{k}.
\end{aligned}$$

We also note that $I_1^2 + I_2^2 + I_3^2 = m^2l^4(\dot{\theta}^2 + \sin^2\theta\,\dot{\varphi}^2)$; from this, the *energy* E of the system — another constant of motion — is equal to $\frac{1}{2(ml^2)}\vec{I}^2$. Furthermore, we note that

$$\begin{aligned}
\vec{I}\cdot\hat{r} &= I_1 \sin\theta\cos\varphi + I_2 \sin\theta\sin\varphi + I_3\cos\theta \\
&= ml^2(-\sin\theta\cos\varphi\sin\varphi\,\dot{\theta} - \sin^2\theta\cos\theta\cos^2\varphi\dot{\varphi}) \\
&\quad + ml^2(\sin\theta\cos\varphi\sin\varphi\,\dot{\theta} - \sin^2\theta\cos\theta\sin^2\varphi\dot{\varphi}) \\
&\quad + ml^2\cos\theta\sin^2\theta\dot{\varphi} \\
&= 0,
\end{aligned}$$

which implies that $I_1x + I_2y + I_3z = 0$, i.e., the trajectory lies on a plane perpendicular to $\vec{I}$ while passing through the origin. The particle moves along a great circle with a uniform velocity (since $E = \frac{1}{2}m\vec{v}^2 = \text{const.}$).]

11-4 (i) Using spherical coordinates $(r, \theta = \alpha, \varphi)$, the Lagrangian is

$$L = \frac{1}{2}m(\dot{r}^2 + r^2 \sin^2 \alpha\, \dot{\varphi}^2) - mgr \cos \alpha.$$

(ii) Since the Lagrangian has no explicit t-dependence, the total energy is conserved, i.e.,

$$\frac{1}{2}m(\dot{r}^2 + r^2 \sin^2 \alpha\, \dot{\varphi}^2) + mgr \cos \alpha = E \quad (\text{: const.}). \tag{1}$$

Also, since φ is cyclic, the corresponding conjugate momentum is conserved:

$$p_\varphi = \frac{\partial L}{\partial \dot{\varphi}} = mr^2 \sin^2 \alpha \dot{\varphi} = L \quad (\text{: const.}). \tag{2}$$

Using $\dot{\varphi} = \frac{L}{mr^2 \sin^2 \alpha}$ in the energy equation (1) gives rise to

$$\frac{1}{2}m\dot{r}^2 + \frac{L^2}{2mr^2 \sin^2 \alpha} + mgr \cos \alpha = E. \tag{3}$$

From (3), after introducing $V_{\text{eff}}(r) = \frac{L^2}{2mr^2 \sin^2 \alpha} + mgr \cos \alpha$, we obtain

$$\frac{dr}{dt} = \pm \sqrt{\frac{2}{m}\Big(E - V_{\text{eff}}(r)\Big)} \tag{4}$$

Integrating (4) yields

$$\int^r \frac{dr}{\sqrt{\frac{2}{m}(E - V_{\text{eff}}(r))}} = \pm t + \text{const.}$$

The function $\varphi = \varphi(t)$ can then be obtained using $\dot{\varphi} = \frac{L}{mr^2 \sin^2 \alpha}$ also.

(iii) For given L, we find a circular orbit if E is *equal to* $\min[V_{\text{eff}}]$. If the minimum of $V_{\text{eff}}(r)$ is at $r = R$, we should have

$$V'_{\text{eff}}(R) = -\frac{L^2}{mR^3 \sin^2 \alpha} + mg \cos \alpha = 0$$
$$\rightarrow L^2 = m^2 g R^3 \sin^2\alpha \cos \alpha. \tag{5}$$

Hence, for a circular orbit at $r = R$, we must have from (2) and (5)

$$(mR^2 \sin^2\alpha\, \dot{\varphi})^2 = m^2 g R^3 \sin^2 \alpha \cos \alpha \quad \rightarrow \quad \dot{\varphi}^2 = \frac{g \cos \alpha}{R \sin^2 \alpha}.$$

11-5 (i) The ρ and z-coordinates of the point mass are given by

$$\rho = a + b\cos\psi, \quad z = -b\sin\psi.$$

Hence the kinetic energy can be expressed as

$$\begin{aligned} T &= \frac{1}{2}m(\dot{\rho}^2 + \rho^2\dot{\varphi}^2 + \dot{z}^2) \\ &= \frac{1}{2}mb^2\dot{\psi}^2 + \frac{1}{2}m(a + b\cos\psi)^2\dot{\varphi}^2, \end{aligned}$$

while we have the potential energy

$$V = mgz = -mgb\sin\psi.$$

Accordingly, the Lagrangian is

$$L = T - V = \frac{1}{2}mb^2\dot{\psi}^2 + \frac{1}{2}m(a + b\cos\psi)^2\dot{\varphi}^2 + mgb\sin\psi.$$

(ii) As our L has no explicit φ-dependence, the corresponding generalized momentum

$$L_z = \frac{\partial L}{\partial \dot{\varphi}} = m(a + b\cos\psi)^2\dot{\varphi} \tag{1}$$

is a constant of motion. We also have the energy conservation

$$\begin{aligned} E &= \frac{1}{2}mb^2\dot{\psi}^2 + \frac{1}{2}m(a + b\cos\psi)^2\dot{\varphi}^2 - mgb\sin\psi \\ &= \frac{1}{2}mb^2\dot{\psi}^2 + \frac{L_z^2}{2m}\frac{1}{(a + b\cos\psi)^2} - mgb\sin\psi, \end{aligned}$$

where we used (1) for the second expression, and hence

$$\frac{1}{2}mb^2\dot{\psi}^2 + V_{\text{eff}} = E \quad (\text{: const.})$$

with $V_{\text{eff}}(\psi) = \frac{L_z^2}{2m(a+b\cos\psi)^2} - mgb\sin\psi$.
See the plot of the effective potential V_{eff}. If V_{eff} has a minimum at $\psi = \psi_0$, then

$$\begin{aligned} 0 &= V'_{\text{eff}}(\psi_0) \\ &= \frac{L_z^2 b\sin\psi_0}{m(a + b\cos\psi_0)^3} - mgb\cos\psi_0, \end{aligned}$$

that is,

$$\tan\psi_0 = \frac{m^2 g(a + b\cos\psi_0)^3}{L_z^2}. \tag{2}$$

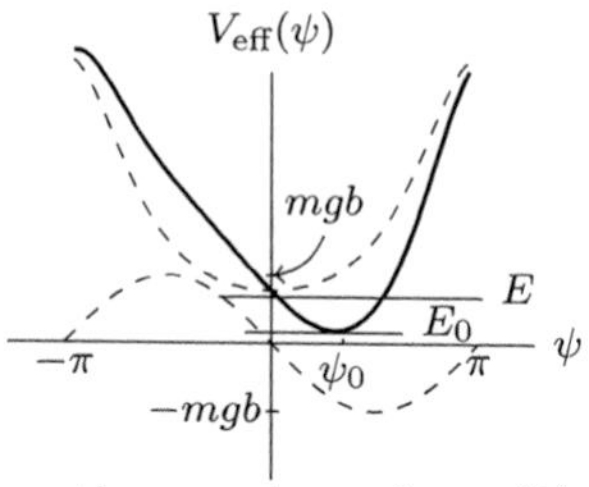

On the other hand, for this circular motion, we have from (1) $\dot{\varphi} \equiv \Omega = \frac{L_z}{m(a+b\cos\psi_0)^2}$ and using this with (2) yields

$$\tan\psi_0 = \frac{m^2 g(a + b\cos\psi_0)^3}{m^2\Omega^2(a + b\cos\psi_0)^4}$$

or

$$\Omega^2 = \frac{g}{\tan\psi_0\,(a + b\cos\psi_0)} = \frac{g}{a\tan\psi_0 + b\sin\psi_0}.$$

(iii) For E slightly larger than $E_0 = \Big[V_{\text{eff}}(\psi)\Big]_{\psi=\psi_0}$, we have approximately

$$\frac{1}{2}mb^2\dot{\psi}^2 + \frac{1}{2}V''_{\text{eff}}(\psi_0)(\psi - \psi_0)^2 = E - E_0,$$

and so the desired angular frequency of small oscillation is $\omega = \sqrt{\frac{1}{mb^2}V''_{\text{eff}}(\psi_0)}$. But

$$\begin{aligned}
V''_{\text{eff}}(\psi_0) &= \frac{L_z^2\, b\cos\psi_0}{m(a + b\cos\psi_0)^3} + \frac{3L_z^2\, b^2\sin^2\psi_0}{m(a + b\cos\psi_0)^4} + mgb\sin\psi_0 \\
&= m\Omega^2 b\cos\psi_0(a + b\cos\psi_0) + 3m\Omega^2 b^2\sin^2\psi_0 + mgb\sin\psi_0 \\
&= 3m\Omega^2 b^2\sin^2\psi_0 + \frac{mgb}{\sin\psi_0}
\end{aligned}$$

(we have here used (2) and $L_z = m(a + b\cos\psi_0)^2\Omega$), and hence

$$\omega = \sqrt{3\Omega^2\sin^2\psi_0 + \frac{g}{b\sin\psi_0}}.$$

[Remark: As the angle ψ undergoes small oscillation, i.e., $\psi = \psi_0 + \epsilon\sin(\omega t + \alpha)$, $\dot{\varphi}$ also oscillates according to

$$\dot{\varphi} = \frac{L_z}{m(a + b\cos\psi)^2}$$

$$= \frac{L_z}{m}\{a + b\cos[\psi_0 + \epsilon\sin(\omega t + \alpha)]\}^{-2}$$

$$\approx \Omega\left\{1 + \epsilon\left(\frac{2b\sin\psi_0}{a + b\cos\psi_0}\right)\sin(\omega t + \alpha)\right\}.]$$

11-6 (i) In the case of the vertical oscillation, we can represent the position of m by ($x = l\sin\theta$, $z = -l\cos\theta + d\sin\beta t$); hence, we have the Lagrangian

$$L = \frac{1}{2}m(\dot{x}^2 + \dot{z}^2) - mgz$$
$$= \frac{1}{2}ml^2\dot{\theta}^2 + ml\beta d\dot{\theta}\cos\beta t\,\sin\theta + mgl\cos\theta,$$

where we omitted terms depending only on time (as such terms do not affect the equation of motion). This gives the equation of motion

$$ml^2\ddot{\theta} + ml\beta d\frac{d}{dt}(\cos\beta t\sin\theta) - ml\beta d\,\dot{\theta}\cos\beta t\cos\theta$$
$$+ mgl\sin\theta = 0$$

or

$$\ddot{\theta} + \frac{1}{l}(g - \beta^2 d\sin\beta t)\sin\theta = 0.$$

On the other hand, when the support oscillates in the horizontal direction, we have the coordinates of m given by ($x = l\sin\theta + d\sin\beta t$, $z = -l\cos\theta$). This leads to the Lagrangian

$$L = \frac{1}{2}ml^2\dot{\theta}^2 + ml\beta d\dot{\theta}\cos\beta t\cos\theta + mgl\cos\theta$$

with the corresponding equation of motion

$$\ddot{\theta} + \frac{g}{l}\sin\theta = \frac{1}{l}\beta^2 d\sin\beta t\cos\theta.$$

These equations of motion are consistent with the result of Problem 9-1. [Note that the z-direction chosen here is opposite to that of Problem 9-1.]

(ii) The equations of motion we have are of the form

$$\ddot{\theta} + \frac{g}{l}\sin\theta = f(\theta)\sin\beta t \tag{1}$$

with

$$f(\theta) = \begin{cases} \frac{d\beta^2}{l} \sin\theta & \text{(vertical oscillation)} \\ \frac{d\beta^2}{l} \cos\theta & \text{(horizontal oscillation)}\,. \end{cases} \tag{2}$$

Here, d is supposed to be small compared to l. It does not imply that $f(\theta)$ is also small if the frequency β is sufficiently high, and this makes a rigorous mathematical treatment of (1) very difficult. But, for very large β, we expect on physical grounds that the motion of the mass be such that it traverses a smooth path and at the same time executes small-amplitude oscillations of frequency β about the very smooth path. We may thus represent our solution $\theta(t)$ by a sum

$$\theta(t) = \Theta(t) + \xi(t), \tag{3}$$

where $\xi(t)$ corresponds to the small oscillation term (but of high frequency β) and $\Theta(t)$ describes the smooth position of the mass (which changes only slightly over the time scale $\frac{2\pi}{\beta}$). To be more precise, we set

$$\Theta(t) = \overline{\theta(t)}$$

($\overline{\theta(t)}$ denotes the time average of $\theta(t)$ over the period $\frac{2\pi}{\beta}$), and then $\overline{\xi(t)} = 0$. Over the time scale $\frac{2\pi}{\beta}$, the "averaged angular position" $\Theta(t)$ behaves like a constant.

Substituting (3) into (1) gives, to first order in ξ,

$$\ddot{\Theta} + \frac{g}{l}\sin\Theta + \ddot{\xi} + \frac{g}{l}\xi\cos\Theta = f(\Theta)\sin\beta t + \xi f'(\Theta)\sin\beta t. \tag{4}$$

We may first consider the implication of this equation on the terms which oscillate very rapidly — from (4), we are then led to

$$\ddot{\xi} = f(\Theta)\sin\beta t, \tag{5}$$

ignoring terms containing the small factor ξ (but not the derivative $\ddot{\xi}$ as it will be proportional to the large quantity β^2). Integrating (5), we may then write

$$\xi(t) = -\frac{f(\Theta)}{\beta^2}\sin\beta t, \tag{6}$$

which has indeed a small amplitude if d is small. Next, we look at the time-averaged form (over the period $\frac{2\pi}{\beta}$) of (4) to conclude that

$$\ddot{\Theta} + \frac{g}{l}\sin\Theta = \overline{(\xi(t) f'(\Theta)\sin\beta t)} \tag{7}$$

since $\bar{\xi} = \overline{(\sin\beta t)} = 0$. Using (6) in (7), we can simplify this to

$$\begin{aligned}\ddot{\Theta} + \frac{g}{l}\sin\Theta &= -\frac{1}{2\beta^2}\left[\frac{d}{d\Theta}f^2(\Theta)\right]\overline{(\sin^2\beta t)}\\ &= -\frac{1}{4\beta^2}\frac{d}{d\Theta}f^2(\Theta).\end{aligned}$$

From this, we obtain the following equation of motion for the averaged position $\Theta(t)$

$$\ddot{\Theta} = -\frac{d}{d\Theta}U_{\text{eff}}(\Theta), \tag{8}$$

where $U_{\text{eff}}(\Theta)$ denotes the effective potential

$$U_{\text{eff}}(\Theta) = -\frac{g}{l}\cos\Theta + \frac{1}{4\beta^2}f^2(\Theta).$$

Based on (8), the motion of our pendulum can be addressed. Using $f(\theta)$ in (2), we obtain the following effective potentials

$$U_{\text{eff}}(\Theta) = \begin{cases} \frac{g}{l}\left(-\cos\Theta + \frac{d^2\beta^2}{4gl}\sin^2\Theta\right) & \text{(vertical oscillation)} \\ \frac{g}{l}\left(-\cos\Theta + \frac{d^2\beta^2}{4gl}\cos^2\Theta\right) & \text{(horizontal oscillation).} \end{cases} \tag{10}$$

We see from (10) that, in the case of the vertical oscillation, the downward vertical position ($\Theta = 0$) is always stable; but, if $d^2\beta^2 > 2gl$, the vertical upward position ($\Theta = \pi$) becomes also stable. In the case of the horizontal oscillation, on the other hand, the effective potential is such that the position $\Theta = 0$ is stable if $d^2\beta^2 < 2gl$; but, if $d^2\beta^2 > 2gl$, the stable equilibrium position is changed to the value of Θ satisfying the condition $\cos\Theta = \frac{2gl}{d^2\beta^2}$.

11-7 (i) As we choose our generalized coordinates as indicated in the problem, the CM of the smaller disk has coordinates

$$x + b\cos\theta, \quad y + b\sin\theta$$

and velocity components

$$\dot{x} - b\dot{\theta}\sin\theta, \quad \dot{y} + b\dot{\theta}\cos\theta.$$

Hence the total kinetic energy of the system of the two disks is

$$\begin{aligned} T &= \frac{1}{2}M(\dot{x}^2 + \dot{y}^2) + \frac{1}{4}MR^2\dot{\theta}^2 \\ &\quad + \frac{1}{2}m[(\dot{x} - b\dot{\theta}\sin\theta)^2 + (\dot{y} + b\dot{\theta}\cos\theta)^2] + \frac{1}{4}mr^2\dot{\varphi}^2 \end{aligned}$$

and the Lagrangian is

$$\begin{aligned} L &= T \\ &= \frac{1}{2}M(\dot{x}^2 + \dot{y}^2) + \frac{1}{4}MR^2\dot{\theta}^2 \\ &\quad + \frac{1}{2}m(\dot{x}^2 + \dot{y}^2 + b^2\dot{\theta}^2 - 2b\dot{x}\dot{\theta}\sin\theta + 2b\dot{y}\dot{\theta}\cos\theta) + \frac{1}{4}mr^2\dot{\varphi}^2. \end{aligned}$$

(ii) Since this Lagrangian does not involve x, y and φ, the related conjugate momenta are conserved:

$$p_x = \frac{\partial L}{\partial \dot{x}} = (M+m)\dot{x} - mb\dot{\theta}\sin\theta = \text{const.}, \tag{1}$$

$$p_y = \frac{\partial L}{\partial \dot{y}} = (M+m)\dot{y} + mb\dot{\theta}\cos\theta = \text{const.}, \tag{2}$$

$$l_\varphi = \frac{\partial L}{\partial \dot{\varphi}} = \frac{1}{2}mr^2\dot{\varphi} = \text{const.} \tag{3}$$

The Euler–Lagrange equation for the variable θ is

$$\begin{aligned} 0 &= \frac{d}{dt}\left(\frac{1}{2}MR^2\dot{\theta} + mb^2\dot{\theta} - mb\dot{x}\sin\theta + mb\dot{y}\cos\theta\right) \\ &\quad - (-mb\dot{x}\dot{\theta}\cos\theta - mb\dot{y}\dot{\theta}\sin\theta). \end{aligned} \tag{4}$$

Removing $\dot{x}$ and $\dot{y}$ in (4) with the help of (1) and (2), we find that $\ddot{\theta} = 0$, i.e., $\dot{\theta}$ is a constant. Hence there are four constants of motion: $p_x, p_y, \dot{\varphi}$ and $\dot{\theta}$. Also the total kinetic energy is constant.

11-8 It is not difficult to guess the Lagrangian:

$$\begin{aligned}
L &= T - V \\
&= \frac{1}{2}m\dot{\vec{r}}_1^2 + \frac{1}{2}m\dot{\vec{r}}_2^2 + \frac{1}{2}M\dot{\vec{r}}_3^2 \\
&\quad - \frac{k}{2}|\vec{r}_1 - \vec{r}_2|^2 - \frac{K}{2}|\vec{r}_2 - \vec{r}_3|^2 - \frac{K}{2}|\vec{r}_3 - \vec{r}_1|^2.
\end{aligned}$$

We will recast the Lagrangian into the sum of three decoupled ones (by noticing the *spatial translational invariance* and the symmetry under the interchange $\vec{r}_1 \leftrightarrow \vec{r}_2$). First, using $\vec{R} = \frac{1}{2m+M}[m(\vec{r}_1 + \vec{r}_2) + M\vec{r}_3]$, $\vec{r}_1' = \vec{r}_1 - \vec{R}$ and $\vec{r}_2' = \vec{r}_2 - \vec{R}$ as generalized coordinates, we can write (here, $\vec{r}_3' \equiv \vec{r}_3 - \vec{R} = -\frac{m}{M}(\vec{r}_1' + \vec{r}_2')$)

$$\begin{aligned}
T &= \frac{1}{2}(2m+M)\dot{\vec{R}}^2 + \frac{1}{2}m(\dot{\vec{r}}_1'^2 + \dot{\vec{r}}_2'^2) + \frac{1}{2}M\dot{\vec{r}}_3'^2 \\
&= \frac{1}{2}(2m+M)\dot{\vec{R}}^2 + \frac{1}{4}m[(\dot{\vec{r}}_1' + \dot{\vec{r}}_2')^2 + (\dot{\vec{r}}_1' - \dot{\vec{r}}_2')^2] \\
&\quad + \frac{1}{2}\frac{m^2}{M}(\dot{\vec{r}}_1' + \dot{\vec{r}}_2')^2, \\
V &= \frac{k}{2}|\vec{r}_1' - \vec{r}_2'|^2 + \frac{K}{2M^2}|m\vec{r}_1' + (M+m)\vec{r}_2'|^2 \\
&\quad + \frac{K}{2M^2}|(M+m)\vec{r}_1' + m\vec{r}_2'|^2 \\
&= \frac{k}{2}|\vec{r}_1' - \vec{r}_2'|^2 + \frac{K}{4M^2}[m^2 + (M+m)^2][(\vec{r}_1' + \vec{r}_2')^2 + (\vec{r}_1' - \vec{r}_2')^2] \\
&\quad + \frac{K}{2M^2}m(M+m)[(\vec{r}_1' + \vec{r}_2')^2 - (\vec{r}_1' - \vec{r}_2')^2].
\end{aligned}$$

So, if we choose $\vec{R}$, $\vec{q}_1 = \frac{\vec{r}_1' + \vec{r}_2'}{\sqrt{2}}$ and $\vec{q}_2 = \frac{\vec{r}_1' - \vec{r}_2'}{\sqrt{2}}$ as generalized coordinates, we obtain the Lagrangian

$$\begin{aligned}
L &= \frac{1}{2}(2m+M)\dot{\vec{R}}^2 + \frac{1}{2}m(\dot{\vec{q}}_1^2 + \dot{\vec{q}}_2^2) + \frac{m^2}{M}\dot{\vec{q}}_1^2 - k\vec{q}_2^2 \\
&\quad - \frac{K}{2M^2}[m^2 + (M+m)^2](\vec{q}_1^2 + \vec{q}_2^2) - \frac{K}{M^2}m(M+m)(\vec{q}_1^2 - \vec{q}_2^2) \\
&= \frac{1}{2}(2m+M)\dot{\vec{R}}^2 + \frac{2m+M}{M}\left\{\frac{1}{2}m\dot{\vec{q}}_1^2 - \frac{K}{2}\left(\frac{2m+M}{M}\right)\vec{q}_1^2\right\} \\
&\quad + \left\{\frac{1}{2}m\dot{\vec{q}}_2^2 - \frac{1}{2}(2k+K)\vec{q}_2^2\right\}.
\end{aligned}$$

Hence there are three independent motions, i.e., a free motion described by $\vec{R}$, a simple harmonic motion with frequency

$\omega_1 = \sqrt{\frac{K}{m}\frac{2m+M}{M}}$ described by $\vec{q}_1$, and a simple harmonic motion with frequency $\omega_2 = \sqrt{\frac{2k+K}{m}}$ described by $\vec{q}_2$.

11-9 Using the Cartesian coordinates (x, y) for the particle's position, the electrostatic potential energy is given by

$$V(x,y) = \frac{q^2}{4\pi\epsilon_0}\left[\frac{1}{\sqrt{x^2+(y-a)^2}} + \frac{1}{\sqrt{x^2+(y+a)^2}} + \frac{6}{\sqrt{(x-2a)^2+y^2}} + \frac{6}{\sqrt{(x+2a)^2+y^2}}\right].$$

For the particle located near the coordinate origin, it feels the potential energy (use the approximation $\frac{1}{\sqrt{1+r}} = 1 - \frac{1}{2}r + \frac{3}{8}r^2 + \cdots$ when $|r| \ll 1$)

$$\begin{aligned} V(x,y) &= \frac{q^2}{4\pi\epsilon_0}\left[\frac{1}{a\sqrt{1-\frac{2y}{a}+\frac{x^2+y^2}{a^2}}} + \frac{1}{a\sqrt{1+\frac{2y}{a}+\frac{x^2+y^2}{a^2}}} + \frac{6}{2a\sqrt{1-\frac{x}{a}+\frac{x^2+y^2}{4a^2}}} + \frac{6}{2a\sqrt{1+\frac{x}{a}+\frac{x^2+y^2}{4a^2}}}\right] \\ &= V_0 + \frac{q^2}{4\pi\epsilon_0}\frac{1}{a^3}\left[\frac{1}{2}x^2 + \frac{5}{4}y^2\right] + \cdots, \end{aligned}$$

where $V_0 = V(x{=}0, y{=}0)$. Since the coefficients of x^2 and y^2 are positive, the origin corresponds to a *stable* equilibrium point. The Lagrangian for small oscillations is

$$L = \frac{1}{2}m\dot{x}^2 - \frac{1}{2}\frac{q^2}{4\pi\epsilon_0 a^3}x^2 + \frac{1}{2}m\dot{y}^2 - \frac{1}{2}\frac{5q^2}{8\pi\epsilon_0 a^3}y^2,$$

and the corresponding normal modes are provided by

$$\vec{r}_{(1)} = C_1 \cos\left(\sqrt{\frac{1}{m}\frac{q^2}{4\pi\epsilon_0 a^3}}\,t + \alpha_1\right)\mathbf{i},$$

$$\vec{r}_{(2)} = C_2 \cos\left(\sqrt{\frac{1}{m}\frac{5q^2}{8\pi\epsilon_0 a^3}}\,t + \alpha_2\right)\mathbf{j}.$$

[Remark: We have this stable equilibrium point because the motion has been *restricted to the xy-plane.* If we allow the z-directional motion also, it is unstable (in agreement with the *general theorem* that any electrostatic potential satisfying the Laplace equation admits no stable equilibrium point).]

11-10 (i) Since we have

$$\frac{\partial V}{\partial x} = -kx - ky + \lambda x^3, \tag{1}$$

$$\frac{\partial V}{\partial y} = -kx + ky, \tag{2}$$

the conditions $y_0 = x_0$ and $2kx_0 = \lambda x_0^3$ must be satisfied at an equilibrium point (x_0, y_0). Thus there are two stable equilibrium points $(\sqrt{\frac{2k}{\lambda}}, \sqrt{\frac{2k}{\lambda}})$ and $(-\sqrt{\frac{2k}{\lambda}}, -\sqrt{\frac{2k}{\lambda}})$ [see (ii) below for the reason], and one unstable point $(0, 0)$. [The unstable nature of the point $(0, 0)$ is obvious from the given potential form.]

(ii) Around the stable equilibrium points, write $x = x_0 + \bar{x}$, $y = y_0 + \bar{y}$ (with $x_0 = y_0 = \pm\sqrt{\frac{2k}{\lambda}}$); then, the potential energy V becomes

$$\begin{aligned} V(x, y) &= -\frac{1}{2}k(x_0 + \bar{x})^2 - k(x_0 + \bar{x})(y_0 + \bar{y}) + \frac{1}{2}k(y_0 + \bar{y})^2 \\ &\quad + \frac{1}{4}\lambda(x_0 + \bar{x})^4 \\ &= (\text{const.}) - \frac{1}{2}k\bar{x}^2 - k\bar{x}\bar{y} + \frac{1}{2}k\bar{y}^2 + \frac{3}{2}\lambda x_0^2\bar{x}^2 \\ &\quad + \text{higher order terms.} \end{aligned}$$

The quadratic part of the potential (with $x_0^2 = \frac{2k}{\lambda}$),

$$\begin{aligned} V^{(2)}(\bar{x}, \bar{y}) &= \frac{5}{2}k\bar{x}^2 - k\bar{x}\bar{y} + \frac{1}{2}k\bar{y}^2 \\ &= 2k\bar{x}^2 + \frac{1}{2}k(\bar{x} - \bar{y})^2, \end{aligned}$$

is positive definite, and so $x_0 = y_0 = \pm\sqrt{\frac{2k}{\lambda}}$ correspond to stable equilibrium points. The Lagrangian for corresponding small oscillations is given by

$$\bar{L}(\bar{x}, \bar{y}) = \frac{1}{2}m\dot{\bar{x}}^2 + \frac{1}{2}m\dot{\bar{y}}^2 - \frac{5}{2}k\bar{x}^2 + k\bar{x}\bar{y} - \frac{1}{2}k\bar{y}^2.$$

(iii) The Lagrange equations for $\bar{x}$ and $\bar{y}$ read

$$m\ddot{\bar{x}} + 5k\bar{x} - k\bar{y} = 0,$$
$$m\ddot{\bar{y}} - k\bar{x} + k\bar{y} = 0.$$

Try normal modes of the form $\bar{x}_i = A_i e^{i\omega t}$ $(i = 1, 2)$; then,

$$\begin{pmatrix} -m\omega^2 + 5k & -k \\ -k & -m\omega^2 + k \end{pmatrix} \begin{pmatrix} A_1 \\ A_2 \end{pmatrix} = 0,$$

and ω^2 should be a root of the secular equation

$$\begin{vmatrix} -m\omega^2 + 5k & -k \\ -k & -m\omega^2 + k \end{vmatrix} = 0.$$

This leads to the equation

$$(m\omega^2)^2 - 6k(m\omega^2) + 4k^2 = 0,$$

which has solutions $m\omega^2 = (3 \pm \sqrt{5})k$. Hence there are two normal modes with frequencies $\omega_1 = (3+\sqrt{5})^{1/2}\sqrt{\frac{k}{m}}$ and $\omega_2 = (3 - \sqrt{5})^{1/2}\sqrt{\frac{k}{m}}$.

11-11 (i) For this system, the kinetic energy can be expressed as

$$\begin{aligned} T &= T_{(\text{hoop})} + T_{(\text{bead})} \\ &= \frac{1}{2}(2Ma^2)\dot{\theta}^2 \\ &\quad + \frac{1}{2}m\left\{a^2\left[\frac{d}{dt}(\sin\theta + \sin\varphi)\right]^2 \right. \\ &\quad \left. + a^2\left[\frac{d}{dt}(-\cos\theta - \cos\varphi)\right]^2\right\} \end{aligned}$$

where we used the fact that the moment of inertia of the hoop is $I = Ma^2 + Ma^2 = 2Ma^2$ (by the parallel axis theorem). Simplifying this, we find

$$T = \frac{1}{2}a^2\left[(2M + m)\dot{\theta}^2 + m\dot{\varphi}^2 + 2m\dot{\theta}\dot{\varphi}\cos(\varphi - \theta)\right].$$

On the other hand, the potential energy is

$$V = -Mga\cos\theta - mga(\cos\theta + \cos\varphi) + \text{const.}$$

Hence we have the Lagrangian

$$\begin{aligned} L &= T - V \\ &= \frac{1}{2}a^2\left[(2M+m)\dot{\theta}^2 + m\dot{\varphi}^2 + 2m\dot{\theta}\dot{\varphi}\cos(\varphi-\theta)\right] \\ &\quad + Mga\cos\theta + mga(\cos\theta + \cos\varphi). \end{aligned}$$

(ii) For small oscillations about the equilibrium point $\theta = \varphi = 0$, the above Lagrangian reduces to

$$\begin{aligned} L = \frac{1}{2}a^2\left[(2M+m)\dot{\theta}^2 + m\dot{\varphi}^2 + 2m\dot{\theta}\dot{\varphi}\right] \\ - \frac{1}{2}(M+m)ga\theta^2 - \frac{1}{2}mga\varphi^2, \end{aligned}$$

ignoring an irrelevant constant. So we have the equations of motion

$$\begin{aligned} (2M+m)\ddot{\theta} + m\ddot{\varphi} + (M+m)\frac{g}{a}\theta = 0, \\ \ddot{\varphi} + \ddot{\theta} + \frac{g}{a}\varphi = 0. \end{aligned}$$

Try $\theta(t) = \bar{\theta}e^{i\omega t}$, $\varphi(t) = \bar{\varphi}e^{i\omega t}$, and then

$$\begin{aligned} \left[-(2M+m)\omega^2 + (M+m)\frac{g}{a}\right]\bar{\theta} - m\omega^2\bar{\varphi} = 0, \\ -\omega^2\bar{\theta} + \left[-\omega^2 + \frac{g}{a}\right]\bar{\varphi} = 0. \end{aligned}$$

For nontrivial solutions of $\bar{\theta}$ and $\bar{\varphi}$, ω^2 should satisfy the characteristic equation

$$\begin{vmatrix} (M+m)\frac{g}{a} - (2M+m)\omega^2 & -m\omega^2 \\ -\omega^2 & \frac{g}{a} - \omega^2 \end{vmatrix} = 0$$

or

$$\begin{aligned} &\left[(2M+m)\omega^2 - (M+m)\frac{g}{a}\right]\left[\omega^2 - \frac{g}{a}\right] - m\omega^4 = 0 \\ \longrightarrow \quad &\omega^4 - \frac{g}{a}\frac{3M+2m}{2M}\omega^2 + \frac{g^2}{2a^2}\frac{M+m}{M} = 0. \end{aligned}$$

Hence we obtain two normal modes with frequencies $\omega_1 = \sqrt{\frac{g}{2a}}$ and $\omega_2 = \sqrt{\frac{m+M}{M}\frac{g}{a}}$. Corresponding eigenvectors are easily found: for $\omega = \omega_1$,

$$(-\omega_1^2)\bar{\theta} + \left(\frac{g}{a} - \omega_1^2\right)\bar{\varphi} = -\frac{g}{2a}\bar{\theta} + \frac{g}{2a}\bar{\varphi} = 0 \quad \rightarrow \quad \mathbf{q}_{(1)} = \begin{pmatrix} 1 \\ 1 \end{pmatrix},$$

and, for $\omega = \omega_2$,

$$(-\omega_2^2)\bar{\theta} + \left(\frac{g}{a} - \omega_2^2\right)\bar{\varphi} = -\frac{1}{M}(m+M)\frac{g}{a}\bar{\theta} - \frac{m}{M}\frac{g}{a}\bar{\varphi} = 0$$
$$\rightarrow \quad \mathbf{q}_{(2)} = \frac{1}{M}\begin{pmatrix} -m \\ m+M \end{pmatrix}.$$

Based on this, the general solution is

$$\begin{pmatrix} \theta(t) \\ \varphi(t) \end{pmatrix} = \begin{pmatrix} 1 \\ 1 \end{pmatrix} C_1 \cos\left(\sqrt{\frac{g}{2a}}\, t + \alpha_1\right)$$
$$+ \begin{pmatrix} -\frac{m}{M} \\ 1 + \frac{m}{M} \end{pmatrix} C_2 \cos\left(\sqrt{\left(1 + \frac{m}{M}\right)\frac{g}{a}}\, t + \alpha_2\right).$$

The first term describes a motion in which the bead and the hoop move together in phase, while the second describes a motion in which the bead and the hoop move in opposite directions.

11-12 Let the relaxed length of the spring be a. Then we have

$$L = \frac{1}{2}m\dot{x}_1^2 + \frac{1}{2}m\dot{x}_2^2 - \frac{1}{2}k(x_1 - a)^2$$
$$- \frac{1}{2}k(x_2 - x_1 - a)^2 + mgx_1 + mgx_2.$$

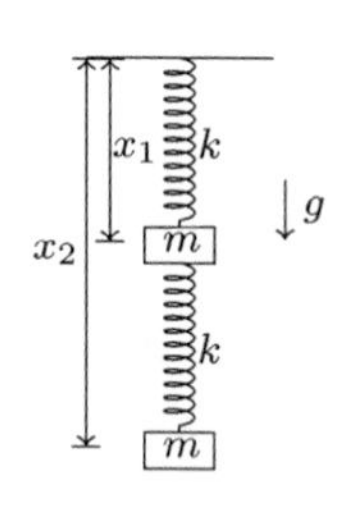

Since at equilibrium we have $x_1 = a + \frac{2mg}{k}$, $x_2 = 2a + \frac{3mg}{k}$, let us introduce new coordinates $\bar{x}_1$, $\bar{x}_2$ by writing $x_1 = a + \frac{2mg}{k} + \bar{x}_1$ and $x_2 = 2a + \frac{3mg}{k} + \bar{x}_2$. In terms of these new generalized coordinates the Lagrangian becomes

$$L = \frac{1}{2}m\dot{\bar{x}}_1^2 + \frac{1}{2}m\dot{\bar{x}}_2^2 - \frac{1}{2}k\bar{x}^2 - \frac{1}{2}k(\bar{x}_2 - \bar{x}_1)^2.$$

Hence we obtain the equations of motion

$$m\ddot{\bar{x}}_1 + k\bar{x}_1 - k(\bar{x}_2 - \bar{x}_1) = 0,$$
$$m\ddot{\bar{x}}_2 + k(\bar{x}_2 - \bar{x}_1) = 0.$$

Now, after setting $x_i(t) = A_i e^{i\omega t}$ $(i = 1, 2)$, we find

$$-m\omega^2 A_1 + 2kA_1 - kA_2 = 0,$$
$$-m\omega^2 A_2 + kA_2 - kA_1 = 0.$$

Hence, ω^2 should satisfy the characteristic equation

$$\begin{vmatrix} 2k - m\omega^2 & -k \\ -k & k - m\omega^2 \end{vmatrix} = 0 \quad \rightarrow \quad (m\omega^2)^2 - 3k(m\omega^2) + k^2 = 0. \tag{1}$$

This equation has two roots $m\omega^2 = \frac{3\pm\sqrt{5}}{2}k$. For the associated normal mode amplitudes, we should have

$$\frac{A_1}{A_2} = \frac{k}{k - m\omega^2} \quad \rightarrow \quad \frac{A_1}{A_2} = -\frac{2}{\sqrt{5}+1} \text{ or } \frac{2}{\sqrt{5}-1}.$$

11-13 (i) When the block is in equilibrium, the sum of forces parallel to the inclined surface is zero:

$$mg\sin\alpha - k(s_0 - d) = 0,$$

yielding

$$s_0 = \frac{mg\sin\alpha}{k} + d.$$

(ii) Let the height of the wedge be h and the horizontal coordinate of the left side of the wedge be x. Then the mass m will have coordinates $(x + s\cos\alpha, h - s\sin\alpha)$. The Lagrangian of the system is then

$$\begin{aligned} L &= \frac{1}{2}M\dot{x}^2 + \frac{1}{2}m\left[(\dot{x} + \dot{s}\cos\alpha)^2 + (\dot{s}\sin\alpha)^2\right] \\ &\quad - \frac{1}{2}k(s-d)^2 - mg(h - s\sin\alpha) \\ &= \frac{1}{2}(m+M)\dot{x}^2 + \frac{1}{2}m\dot{s}^2 + m\dot{x}\dot{s}\cos\alpha - \frac{1}{2}k(s-d)^2 \\ &\quad - mg(h - s\sin\alpha). \end{aligned}$$

Then the Lagrange equations give the equations of motion

$$(m+M)\ddot{x} + m\ddot{s}\cos\alpha = 0,$$
$$m\ddot{x}\cos\alpha + m\ddot{s} + ks - (kd + mg\sin\alpha) = 0.$$

(iii) By introducing a change of variable $s = s' + \frac{kd+mg\sin\alpha}{k}$, we can write the above equations as

$$(m+M)\ddot{x} + m\ddot{s}'\cos\alpha = 0,$$
$$m\ddot{x}\cos\alpha + m\ddot{s}' + ks' = 0.$$

Considering a solution of the form $x = Ae^{i\omega t}$, $s' = Be^{i\omega t}$, we are then led to the secular equation

$$\begin{vmatrix} -(m+M)\omega^2 & -m\omega^2\cos\alpha \\ -m\omega^2\cos\alpha & k - m\omega^2 \end{vmatrix} = 0,$$

yielding

$$\omega_1 = 0, \quad \omega_2 = \sqrt{\frac{k(1+\frac{m}{M})}{m(1+\frac{m}{M}\sin^2\alpha)}}.$$

As the motion related to ω_1 is not oscillatory but translational as a whole along the x-axis, there is only one natural frequency of vibration, ω_2.

11-14 (i) The center of mass of the rod has coordinates $(l\sin\varphi + \frac{3}{4}l\sin\theta, -l\cos\varphi - \frac{3}{4}l\cos\theta)$ and velocity $(l\dot{\varphi}\cos\varphi + \frac{3}{4}l\dot{\theta}\cos\theta, l\dot{\varphi}\sin\varphi + \frac{3}{4}l\dot{\theta}\sin\theta)$. The moment of inertia of the rod is $\frac{m}{12}(\frac{3l}{2})^2 = \frac{3}{16}ml^2$. Hence its Lagrangian has the form

$$L = \frac{1}{2}ml^2\left[\dot{\varphi}^2 + \frac{9}{16}\dot{\theta}^2 + \frac{3}{2}\dot{\theta}\dot{\varphi}\cos(\theta-\varphi)\right] + \frac{3}{32}ml^2\dot{\theta}^2$$
$$+ mgl\left(\cos\varphi + \frac{3}{4}\cos\theta\right).$$

(ii) For small oscillations, we expand the coordinates θ, φ around the equilibrium point $\theta = 0, \varphi = 0$. Then the above Lagrangian reduces to

$$L \approx \frac{1}{2}ml^2\left[\dot{\varphi}^2 + \frac{3}{4}\dot{\theta}^2 + \frac{3}{2}\dot{\theta}\dot{\varphi}\right] + \frac{7}{4}mgl - \frac{1}{2}mgl\left(\varphi^2 + \frac{3}{4}\theta^2\right),$$

retaining only terms of up to the second order of the small quantities $\theta, \varphi, \dot{\theta}, \dot{\varphi}$. From this Lagrangian, we obtain the equations of motion

$$\frac{3}{4}l\ddot{\theta} + l\ddot{\varphi} + g\varphi = 0,$$
$$l\ddot{\theta} + l\ddot{\varphi} + g\theta = 0.$$

With a solution of the form $\theta(t) = Ae^{i\omega t}$ and $\varphi(t) = Be^{i\omega t}$, the above equations give

$$\begin{pmatrix} -\frac{3}{4}l\omega^2 & g - l\omega^2 \\ g - l\omega^2 & -l\omega^2 \end{pmatrix} \begin{pmatrix} A \\ B \end{pmatrix} = 0.$$

The secular equation

$$\begin{vmatrix} -\frac{3}{4}l\omega^2 & g - l\omega^2 \\ g - l\omega^2 & -l\omega^2 \end{vmatrix} = 0 \quad \to \quad l^2\omega^4 - 8lg\omega^2 + 4g^2 = 0,$$

has solutions $\omega^2 = (\sqrt{3} \pm 1)^2 \frac{g}{l}$. Hence the normal-mode angular frequencies are

$$\omega_+ = (\sqrt{3} + 1)\sqrt{\frac{g}{l}}, \qquad \omega_- = (\sqrt{3} - 1)\sqrt{\frac{g}{l}}.$$

The ratio of amplitudes is

$$\frac{B}{A} = \frac{g - l\omega^2}{l\omega^2} = \begin{cases} -\frac{\sqrt{3}}{2} & \text{for } \omega = \omega_+ \\ \frac{\sqrt{3}}{2} & \text{for } \omega = \omega_-. \end{cases}$$

Thus in the normal mode given by ω_+, θ and φ are opposite in phase; while in that given by ω_-, they are in phase. In both cases the ratio of the amplitude of φ to that of θ is $\sqrt{3} : 2$.

11-15 (i) The positions of M and m are represented by $(x, 0)$ and $(x + l\sin\theta, -l\cos\theta)$ respectively, and their velocities by $(\dot{x}, 0)$ and $(\dot{x} + l\dot{\theta}\cos\theta, l\dot{\theta}\sin\theta)$. The Lagrangian of the system is

$$\begin{aligned} L = &\frac{1}{2}M\dot{x}^2 + \frac{1}{2}m(\dot{x}^2 + l^2\dot{\theta}^2 + 2l\dot{x}\dot{\theta}\cos\theta) \\ &- \frac{1}{2}k(x - x_0)^2 + mgl\cos\theta. \end{aligned}$$

The Lagrange equations give, for $\bar{x} \equiv x - x_0$ and θ,

$$\begin{aligned} (M + m)\ddot{\bar{x}} - ml\dot{\theta}^2\sin\theta + ml\ddot{\theta}\cos\theta + k\bar{x} = 0 \\ l\ddot{\theta} + \ddot{\bar{x}}\cos\theta + g\sin\theta = 0. \end{aligned} \qquad (1)$$

(ii) For small oscillations, neglect all terms of orders higher than two in $\bar{x}$, θ, $\dot{\bar{x}}$, $\dot{\theta}$ from (1):

$$\begin{aligned} (M + m)\ddot{\bar{x}} + ml\ddot{\theta} + k\bar{x} = 0 \\ l\ddot{\theta} + \ddot{\bar{x}} + g\theta = 0. \end{aligned} \qquad (2)$$

Setting $\bar{x} = Ae^{i\omega t}$ and $\theta = Be^{i\omega t}$, (2) leads to

$$\begin{pmatrix} k-(M+m)\omega^2 & -ml\omega^2 \\ -\omega^2 & g-l\omega^2 \end{pmatrix}\begin{pmatrix} A \\ B \end{pmatrix} = 0.$$

The secular equation

$$\begin{vmatrix} k-(M+m)\omega^2 & -ml\omega^2 \\ -\omega^2 & g-l\omega^2 \end{vmatrix} = Ml\omega^4 - \left[g(M+m)+kl\right]\omega^2 + gk = 0$$

has two positive roots

$$\omega_1 = \frac{1}{\sqrt{2Ml}}\left[g(M+m)+kl + \sqrt{[g(M+m)+kl]^2 - 4Mglk}\right]^{1/2},$$

$$\omega_2 = \frac{1}{\sqrt{2Ml}}\left[g(M+m)+kl - \sqrt{[g(M+m)+kl]^2 - 4Mglk}\right]^{1/2},$$

which are the normal-mode angular frequencies of the system.

11-16 (i) Let us introduce a new variable θ describing the inclination of the rod and X the position of center of mass. They are related to x_1 and x_2 through the relation

$$\tan\theta = \frac{x_1 - x_2}{l} \sim \theta \quad (\text{for } \theta \ll 1), \qquad X = \frac{x_1 + x_2}{2}.$$

The rotational kinetic energy of the bar is

$$\frac{1}{2}I\dot{\theta}^2 = \frac{1}{2}\left(\frac{1}{12}ml^2\right)\left(\frac{\dot{x}_1 - \dot{x}_2}{l}\right)^2 = \frac{1}{24}m(\dot{x}_1 - \dot{x}_2)^2 = \frac{1}{24}m\dot{\bar{x}}^2,$$

where $\bar{x} \equiv x_1 - x_2$, and the translational kinetic energy is $\frac{1}{2}m\dot{X}^2$. Hence, the Lagrangian of the system is

$$L = \frac{1}{2}m\dot{X}^2 + \frac{1}{24}m\dot{\bar{x}}^2 - (-mgX) - \frac{1}{2}k(x_1^2 + x_2^2). \qquad (1)$$

(ii) The last term in (1) can be written as $-k(X^2 + \frac{1}{4}\bar{x}^2)$ using X and $\bar{x}$. Then the Lagrangian becomes

$$\begin{aligned} L &= \frac{1}{2}m\dot{X}^2 + mgX - kX^2 + \frac{1}{24}m\dot{\bar{x}}^2 - \frac{1}{4}k\bar{x}^2 \\ &= \frac{1}{2}m\dot{\widetilde{X}}^2 - k\widetilde{X}^2 + \frac{1}{24}m\dot{\bar{x}}^2 - \frac{1}{4}k\bar{x}^2 + \text{const.}, \end{aligned}$$

where $\widetilde{X} \equiv X - \frac{mg}{2k}$. This Lagrangian is a sum of two independent oscillator Lagrangians: one with angular frequency $\omega_1 = \sqrt{\frac{2k}{m}}$ and the other with $\omega_2 = \sqrt{\frac{6k}{m}}$. Using the rescaled variables $y_1 = \sqrt{m}\widetilde{X}$ and $y_2 = \sqrt{\frac{m}{12}}\bar{x}$, the above Lagrangian can be written as

$$L = \frac{1}{2}\dot{y}_1^2 - \frac{1}{2}\omega_1^2 y_1^2 + \frac{1}{2}\dot{y}_2^2 - \frac{1}{2}\omega_2^2 y_2^2 .$$

Obviously, y_1 and y_2 are normal coordinates.

(iii) The mode described by $\widetilde{X}$ (or y_1 after scaling) corresponds to a motion where the two springs are moving in phase, while the other described by $\bar{x}$ (or y_2) corresponds to a motion where the two springs are moving in opposite ways.

11-17 (i) The Lagrangian of the system is

$$\begin{aligned} L = \frac{1}{2}m(\dot{x}_1^2 + \dot{x}_2^2 + \dot{x}_3^2) - \frac{1}{2}k(x_1 - x_2)^2 \\ - \frac{1}{2}k(x_2 - x_3)^2 + x_1 F(t). \end{aligned} \tag{1}$$

This can be written in the form

$$L = \frac{1}{2}T_{ij}\dot{x}_i\dot{x}_j - \frac{1}{2}V_{ij}x_i x_j + x_1 F(t) \quad (i, j = 1, 2, 3)$$

if we employ matrix notations

$$(T_{ij}) = \begin{pmatrix} m & 0 & 0 \\ 0 & m & 0 \\ 0 & 0 & m \end{pmatrix}, \qquad (V_{ij}) = \begin{pmatrix} k & -k & 0 \\ -k & 2k & -k \\ 0 & -k & k \end{pmatrix}.$$

(ii) First, without the external force, i.e., $F(t) = 0$, the equations of motion are

$$T_{ij}\ddot{x}_j + V_{ij}x_j = 0. \tag{2}$$

We try the solutions of the form

$$x_i(t) = A_i e^{i\omega t} \quad (i = 1, 2, 3)$$

with (2), to obtain

$$\left(-\omega^2 T_{ij} + V_{ij}\right) A_j = 0.$$

For this system of equations to admit a nontrivial solution, ω should be a root of the characteristic equation

$$\det(V - \omega^2 T) = 0.$$

This gives the values

$$\omega_1 = 0, \quad \omega_2 = \sqrt{\frac{k}{m}}, \quad \omega_3 = \sqrt{\frac{3k}{m}},$$

and the associated eigenvectors are

$$\vec{A}^{(1)} = \frac{1}{\sqrt{3m}} \begin{pmatrix} 1 \\ 1 \\ 1 \end{pmatrix}, \quad \vec{A}^{(2)} = \frac{1}{\sqrt{2m}} \begin{pmatrix} 1 \\ 0 \\ 1 \end{pmatrix},$$

$$\vec{A}^{(3)} = \frac{1}{\sqrt{6m}} \begin{pmatrix} 1 \\ -2 \\ 1 \end{pmatrix}.$$

We may use these eigenvectors as a basis to write

$$\vec{x}(t) \equiv \begin{pmatrix} x_1(t) \\ x_2(t) \\ x_3(t) \end{pmatrix} = \xi^{(1)}(t)\vec{A}^{(1)} + \xi^{(2)}(t)\vec{A}^{(2)} + \xi^{(3)}(t)\vec{A}^{(3)},$$

and then the Lagrangian (1), now turning on $F(t)$, can be written using normal coordinates $\xi^{(1)}$, $\xi^{(2)}$ and $\xi^{(3)}$ as

$$L = \frac{1}{2}\left[(\dot{\xi}^{(1)})^2 + (\dot{\xi}^{(2)})^2 + (\dot{\xi}^{(3)})^2 - \frac{k}{m}(\xi^{(2)})^2 - \frac{3k}{m}(\xi^{(3)})^2\right]$$
$$+ \left[\frac{1}{\sqrt{3m}}\xi^{(1)} + \frac{1}{\sqrt{2m}}\xi^{(2)} + \frac{1}{\sqrt{6m}}\xi^{(3)}\right] F(t).$$

When the external force has the form $F(t) = f\cos\omega t$, the normal coordinates satisfy the equations of motion

$$\ddot{\xi}^{(1)} = \frac{1}{\sqrt{3m}} f \cos\omega t,$$

$$\ddot{\xi}^{(2)} + \frac{k}{m}\xi^{(2)} = \frac{1}{\sqrt{2m}} f \cos\omega t,$$

$$\ddot{\xi}^{(3)} + \frac{3k}{m}\xi^{(3)} = \frac{1}{\sqrt{6m}} f \cos\omega t,$$

with initial conditions $\xi^{(i)} = \dot{\xi}^{(i)} = 0$ $(i = 1, 2, 3)$. The corresponding solutions are easily found:

$$\xi^{(1)}(t) = \frac{f}{\sqrt{3m}}\frac{1}{\omega^2}(1 - \cos\omega t),$$

$$\xi^{(2)}(t) = \frac{f}{\sqrt{2m}}\frac{1}{\left(\frac{k}{m} - \omega^2\right)}\left(\cos\omega t - \cos\sqrt{\frac{k}{m}}t\right),$$

$$\xi^{(3)}(t) = \frac{f}{\sqrt{6m}}\frac{1}{\left(\frac{3k}{m} - \omega^2\right)}\left(\cos\omega t - \cos\sqrt{\frac{3k}{m}}t\right).$$

Based on these results, we find the following expression for $x_3(t)$:

$$\begin{aligned} x_3(t) &= \frac{1}{\sqrt{3m}}\xi^{(1)}(t) - \frac{1}{\sqrt{2m}}\xi^{(2)}(t) + \frac{1}{\sqrt{6m}}\xi^{(3)}(t) \\ &= \frac{f}{m}\left[\frac{1}{3\omega^2}(1 - \cos\omega t) - \frac{1}{2\left(\frac{k}{m} - \omega^2\right)}\right. \\ &\quad \times \left(\cos\omega t - \cos\sqrt{\frac{k}{m}}t\right) \\ &\quad \left. + \frac{1}{6(\frac{3k}{m} - \omega^2)}\left(\cos\omega t - \cos\sqrt{\frac{3k}{m}}t\right)\right]. \end{aligned}$$

11-18 Let θ_1, θ_2, θ_3 be respectively the angular displacements of m, M, m from their equilibrium positions. The Lagrangian of the system can then be expressed as

$$\begin{aligned} L = \frac{1}{2}mr^2(\dot{\theta}_1^2 + \dot{\theta}_3^2) + \frac{1}{2}Mr^2\dot{\theta}_2^2 - \frac{1}{2}kr^2[(\theta_2 - \theta_1)^2 \\ + (\theta_3 - \theta_2)^2 + (\theta_1 - \theta_3)^2]. \end{aligned}$$

The Lagrange equations give

$$m\ddot{\theta}_1 + k(2\theta_1 - \theta_2 - \theta_3) = 0, \quad (1)$$

$$M\ddot{\theta}_2 + k(2\theta_2 - \theta_3 - \theta_1) = 0, \quad (2)$$

$$m\ddot{\theta}_3 + k(2\theta_3 - \theta_1 - \theta_2) = 0. \quad (3)$$

The above sum up to

$$m\ddot{\theta}_1 + M\ddot{\theta}_2 + m\ddot{\theta}_3 = 0, \tag{4}$$

while we find, from (1) and (3),

$$m(\ddot{\theta}_1 - \ddot{\theta}_3) + 3k(\theta_1 - \theta_3) = 0, \tag{5}$$

$$m(\ddot{\theta}_1 + \ddot{\theta}_3) + k(\theta_1 + \theta_3 - 2\theta_2) = 0. \tag{6}$$

From (4) and (5), we immediately see that

$$\xi = \theta_1 + \frac{M}{m}\theta_2 + \theta_3, \quad \eta = \theta_1 - \theta_3 \tag{7}$$

are normal coordinates of the system (up to normalization) with related frequencies $\omega_\xi = 0$ and $\omega_\eta = \sqrt{\frac{3k}{m}}$. [The coordinate ξ describes rotation of the system as a whole.] The remaining normal mode of oscillation is that involving the coordinate

$$\zeta = \theta_1 + \theta_3 - 2\theta_2 \tag{8}$$

with related frequency $\omega_\zeta = \sqrt{\frac{2m+M}{mM}k}$, since (6) may be combined with (2) to give

$$m(\ddot{\theta}_1 + \ddot{\theta}_3 - 2\ddot{\theta}_2) + \frac{2m+M}{M}k(\theta_1 + \theta_3 - 2\theta_2) = 0. \tag{9}$$

Indeed, using ξ, η and ζ, the Lagrangian (1) can be cast as

$$L = \frac{1}{2}\frac{m^2r^2}{2m+M}\dot{\xi}^2 + \frac{1}{4}mr^2\dot{\eta}^2 + \frac{1}{4}\frac{mMr^2}{2m+M}\dot{\zeta}^2 - \frac{3}{4}kr^2\eta^2 - \frac{1}{4}kr^2\zeta^2. \tag{10}$$

11-19 (i) For the setup 1 the Lagrangian can be written as

$$L = \frac{1}{2}m\sum_{j=1}^{N}\dot{y}_j^2 - \frac{1}{2}k\left[y_1^2 + \sum_{j=1}^{N-1}(y_{j+1} - y_j)^2 + y_N^2\right],$$

where y_j is the displacement of the jth mass from the equilibrium position. This becomes identical to our Lagrangian in Sec. 11.3 if we take $k = \frac{T}{l}$. Hence, for general N, the normal modes and related frequencies are provided by the results in Sec.11.3: there are N normal modes $\{y_j^{(n)}(t) =$

$A_j^n \cos(\omega_n t + \alpha_n)$, $j = 1, \ldots, N\}$ $(n = 1, \ldots, N)$, with the nth normal mode specified by

$$A_j^n = 2 \sin \frac{jn\pi}{N+1}, \quad \omega_n^2 = 2\frac{k}{m}\left[1 - \cos \frac{n\pi}{N+1}\right].$$

[Remark: Note that ω_n can also be written as $\omega_n = 2\sqrt{\frac{k}{m}} \sin \frac{n\pi}{2(N+1)}$ and if we introduce normal coordinates $q_1, \ldots, q_N$ according to

$$y_j(t) = \sum_{n=1}^{N} \sqrt{\frac{2}{N+1}} \left(\sin \frac{jn\pi}{N+1}\right) q_n(t) \quad (j = 1, \ldots, N),$$

the given Lagrangian can be reduced to a sum of squares corresponding to a set of N independent oscillators, i.e.,

$$L = \sum_{n=1}^{N} \frac{1}{2} m(\dot{q}_n^2 - \omega_n^2 q_n^2). \,]$$

(ii) For the setup 2, we have the Lagrangian

$$L = \frac{1}{2} m \sum_{j=1}^{N} y_j^2 - \frac{1}{2} k \left[y_1^2 + \sum_{j=1}^{N-1} (y_{j+1} - y_j)^2 \right],$$

and the corresponding equations of motion can be written as

$$\ddot{y}_j = \frac{k}{m}(y_{j-1} - 2y_j + y_{j+1}), \quad (j = 1, \ldots, N) \tag{1}$$

with the additional conditions

$$y_0 = 0, \qquad y_{N+1} = y_N. \tag{2}$$

The equations of motion given in (1) are the same as those for the setup 1; the difference lies only in the boundary conditions (2) (compared to those in Sec. 11.3). This means that, for normal modes, only the conditions for the linear combination $A_j = aA_j^{(\alpha)} + bA_j^{(\alpha^{-1})}$ (here, $A_j^{(\alpha)} = \alpha^j$ and $\alpha \neq \pm 1$) need to be reconsidered. In the present case, the conditions should read

$$\text{at } j = 0: \qquad aA_0^{(\alpha)} + bA_0^{(\alpha^{-1})} = a + b = 0,$$

$$\text{at } j = N+1: \quad aA_{N+1}^{(\alpha)} + bA_{N+1}^{(\alpha^{-1})} = aA_N^{(\alpha)} + bA_N^{(\alpha^{-1})}$$
$$\text{or} \quad a\alpha^{N+1} + b\alpha^{-(N+1)} = a\alpha^N + b\alpha^{-N}.$$

This yields the condition $\alpha^{2N+1} = -1$, which has the general solution

$$\alpha = e^{i\theta_n}, \quad \text{with } \theta_n = \frac{(2n-1)\pi}{2N+1} \quad (n: \text{ integer}).$$

This in turn leads to the following set of normal mode frequencies and related normal modes:

$$\omega_n^2 = \frac{k}{m}\left[2 - e^{i\theta_n} - e^{-i\theta_n}\right] = 2\frac{k}{m}\left[1 - \cos\frac{(2n-1)\pi}{2N+1}\right],$$
$$y_j^{(n)}(t) = (\text{const.}) \sin\frac{j(2n-1)\pi}{2N+1}\cos(\omega_n t + \alpha_n) \quad (j = 1, \ldots, N),$$

where we may now restrict n to $1, 2, \ldots, N$.

(iii) For a circular array of N masses we have the Lagrangian

$$L = \frac{1}{2}m\sum_{j=1}^{N}\dot{y}_j^2 - \frac{1}{2}k\left[(y_1 - y_N)^2 + \sum_{j=1}^{N-1}(y_{j+1} - y_j)^2\right].$$

The equations of motion can again be taken as (1) if we change the boundary conditions to

$$y_0 = y_N, \qquad y_{N+1} = y_1.$$

(This corresponds to a *periodic boundary condition.*) To obtain normal modes based on the linear combination $A_j = aA_j^{(\alpha)} + bA_j^{(\alpha^{-1})}$ where $A_j^{(\alpha)} = \alpha^j$, we must now impose the following conditions:

$$\text{at } j = 0: \quad aA_0^{(\alpha)} + bA_0^{(\alpha^{-1})} = aA_N^{(\alpha)} + bA_N^{(\alpha^{-1})}$$
$$\text{or} \quad a + b = a\alpha^N + b\alpha^{-N}$$
$$\text{at } j = N+1: \quad aA_{N+1}^{(\alpha)} + bA_{N+1}^{(\alpha^{-1})} = aA_1^{(\alpha)} + bA_1^{(\alpha^{-1})}$$
$$\text{or} \quad a\alpha^{N+1} + b\alpha^{-(N+1)} = a\alpha + b\alpha^{-1}.$$

This gives the condition $\alpha^N = 1$, which has the following solution

$$\alpha = e^{i\theta_n}, \quad \text{with } \theta_n = \frac{2n\pi}{N} \quad (n = 0, 1, \ldots, N-1).$$

The normal mode frequencies are thus given by

$$\omega_n^2 = \frac{k}{m}[2 - e^{i\theta_n} - e^{-i\theta_n}] = 2\frac{k}{m}\left[1 - \cos\frac{2n\pi}{N}\right],$$

i.e., $\omega_n = 2\sqrt{\frac{k}{m}}\,|\sin\frac{n\pi}{N}|$. Here note that the frequencies ω_n and ω_{N-n} are the same, and the frequency $\omega_0 = 0$ corresponds to the motion of all masses moving along the circle with a constant velocity. The two-fold degeneracy of the frequencies (i.e., ω_n and ω_{N-n}) corresponds to the possibility of having waves moving in the opposite directions (i.e., $y_j^{(n)}(t) = \mathrm{Re}[a\, e^{i(\omega_n t + j\frac{2n\pi}{N})}]$ with $A_j = (e^{i\theta_n})^j = e^{ij\frac{2n\pi}{N}}$ and $y_j^{(N-n)}(t) = \mathrm{Re}[a'\, e^{i(\omega_n t - j\frac{2n\pi}{N})}]$ with $A_j = (e^{i\theta_{N-n}})^j = e^{-ij\frac{2n\pi}{N}}$). By considering their superposition with equal amplitudes, we can also present the normal modes in the form of standing waves, i.e.,

$$y_j^{(n)}(t) = \begin{cases} (\text{const.})\ \cos j\left(\frac{2n\pi}{N}\right)\cos(\omega_n t + \alpha_n) \\ (\text{const.})\ \sin j\left(\frac{2n\pi}{N}\right)\cos(\omega_n t + \alpha_n). \end{cases}$$

11-20 (i) For N rows of pendulums, the small-oscillation Lagrangian takes the form

$$L = \frac{1}{2}ml^2\sum_{j=1}^{N}\dot{\theta}_j^2 - \frac{1}{2}mgl\sum_{j=1}^{N}\theta_j^2 - \frac{1}{2}kl^2\sum_{j=1}^{N+1}(\theta_j - \theta_{j-1})^2$$

with $\theta_0 \equiv \theta_1$ and $\theta_{N+1} \equiv \theta_N$.

(ii) Lagrange's equations of motion read

$$ml^2\ddot{\theta}_j + mgl\theta_j + kl^2\left[(\theta_j - \theta_{j-1}) - (\theta_{j+1} - \theta_j)\right] = 0 \qquad (j = 1, \ldots, N)$$

(with $\theta_0 \equiv \theta_1$ and $\theta_{N+1} \equiv \theta_N$), and these can be written as

$$\ddot{\theta}_j + \omega_0^2\theta_j + \omega_s^2(2\theta_j - \theta_{j-1} - \theta_{j+1}) = 0 \quad (j = 1, \ldots, N), \quad (2)$$

where we defined $\omega_0 = \sqrt{g/l}$ and $\omega_s = \sqrt{k/m}$.

Now let us consider the extension of the system (2) so that the index j can take any integer value, i.e., $j = 0, \pm 1, \pm 2, \ldots$. Then, if we write the corresponding normal modes as $\theta_j(t) = \mathrm{Re}[A_j e^{i\omega t}]$ $(j = 0, \pm 1, \pm 2, \ldots!)$, we can exploit the symmetry in the system, $\theta_j \to \theta_{j+1}$ for all j simultaneously, to demand $\mathbf{A}$ to be an *eigenvector of this symmetry operation*, i.e., we set

$$A_{j+1} = \alpha A_j \quad (\alpha \neq 0) \longrightarrow A_j = \alpha^j \quad (\text{setting } A_0 = 1).$$

Accordingly, the normal modes assume the form $\theta_j^{(\alpha)}(t) = \text{Re}[\alpha^j e^{i\omega t}]$. Inserting these expressions into (2) (for arbitrary integer j), we then see that α can be related to the normal-mode frequency ω according to

$$-\omega^2 + \omega_0^2 + \omega_s^2 \left(2 - \alpha - \frac{1}{\alpha}\right) = 0. \tag{3}$$

This shows that, as long as $\alpha \neq 1$, for every given ω^2 there are two (in general complex) normal modes, $A_j^{(\alpha)}$ and $A_j^{(\alpha^{-1})}$.

Let us now impose our boundary conditions, $A_0 = A_1$ and $A_{N+1} = A_N$, with the linear combination $A_j = aA_j^{(\alpha)} + bA_j^{(\alpha^{-1})}$:

$$\begin{aligned}
&a + b = a\alpha + b\frac{1}{\alpha} \quad \to \quad b = a\alpha \\
&a\alpha^{N+1} + b\alpha^{-(N+1)} = a\alpha^N + b\alpha^{-N} \\
&\qquad \to \quad a[\alpha^{N+1} + \alpha^{-N}] = a[\alpha^N + \alpha^{-(N-1)}] \\
&\qquad \to \quad \alpha^{2N} = 1
\end{aligned}$$

So $\alpha = e^{i\theta_n}$ with $\theta_n = \frac{n\pi}{N}$ (n: integer) and then, from (3), we find the following formula for the normal-mode frequency ω_n:

$$\begin{aligned}
\omega_n^2 &= \omega_0^2 + \omega_s^2(2 - e^{i\theta_n} - e^{-i\theta_n}) \\
&= \omega_0^2 + 2\omega_s^2 \left(1 - \cos\frac{n\pi}{N}\right),
\end{aligned} \tag{4}$$

where n can be restricted to values $n = 0, 1, 2, \ldots, N-1$. The corresponding normal modes have the form

$$\begin{aligned}
\theta_j^{(n)}(t) &= \text{Re}\left[a\left(e^{i\frac{n\pi}{N}j} + e^{i\frac{n\pi}{N}} e^{-i\frac{n\pi}{N}j}\right)e^{i\omega_n t}\right] \\
&= \text{Re}\left[\bar{a}\cos\left(\frac{n\pi}{N}[j - 1/2]\right)e^{i\omega_n t}\right],
\end{aligned}$$

where we set $\bar{a} = 2ae^{i\frac{n\pi}{2N}}$.

(iii) For very large N, we may represent the position of the jth mass by $x = ja$ and introduce the wave number $k = \frac{1}{a}\left(\frac{n\pi}{N}\right)$ so that we have $e^{i\frac{n\pi}{N}j} \sim e^{ikx}$. Now, from (4), we find the dispersion relation

$$\omega^2 = \omega_0^2 + 2\omega_s^2(1 - \cos ka).$$

This shows that, for a sufficiently small wave number k, we have $\omega^2 \approx \omega_0^2 + (\omega_s^2 a^2)k^2$; here, unlike the case in Sec. 11.3, ω does not vanish as the wave number k approaches zero.

11-21 (i) The initial conditions $y(x, 0) = f(x)$ and $\frac{\partial}{\partial t} y(x, t)|_{t=0} = 0$ require that

$$y(x, 0) = \sum_{n=1}^{\infty} a_n \sin\left(\frac{n\pi x}{L}\right) = f(x), \tag{1}$$

$$\left.\frac{\partial y(x, t)}{\partial t}\right|_{t=0} = \sum_{n=1}^{\infty} b_n \frac{n\pi v}{L} \sin\left(\frac{n\pi x}{L}\right) = 0. \tag{2}$$

From (2) we conclude that $b_n = 0$ for all n. The coefficients a_n in (1) are determined as

$$\begin{aligned} a_n &= \frac{2}{L}\int_0^L f(x)\sin\left(\frac{n\pi x}{L}\right)dx \\ &= \frac{2}{L}\int_0^{\frac{L}{2}} \frac{2h}{L} x \sin\left(\frac{n\pi x}{L}\right)dx \\ &\quad + \frac{2}{L}\int_{\frac{L}{2}}^{L} \frac{2h}{L}(L - x)\sin\left(\frac{n\pi x}{L}\right)dx \\ &= \frac{8h}{n^2\pi^2}\sin\frac{n\pi}{2}. \end{aligned}$$

Hence the desired solution for $t > 0$ is

$$\begin{aligned} y(x, t) &= \frac{8h}{\pi^2}\left[\sin\frac{\pi x}{L}\cos\frac{\pi v t}{L} - \frac{1}{3^2}\sin\frac{3\pi x}{L}\cos\frac{3\pi v t}{L}\right. \\ &\quad \left. + \frac{1}{5^2}\sin\frac{5\pi x}{L}\cos\frac{5\pi v t}{L} + \cdots\right] \\ &= \frac{8h}{\pi^2}\sum_{n=1}^{\infty}\frac{(-1)^n}{(2n - 1)^2}\sin\frac{(2n - 1)\pi x}{L}\cos\frac{(2n - 1)\pi v t}{L}. \end{aligned}$$

(ii) Using the identity $\sin A \cos B = \frac{1}{2}[\sin(A - B) + \sin(A + B)]$, the above series solution can be presented as

$$y(x,t) = \frac{1}{2}\left[f^*(x - vt) + f^*(x + vt)\right]$$

where $f^*(x)$ denotes a function defined by the series

$$f^*(x) = \frac{8h}{\pi^2}\sum_{l=1}^{\infty}\frac{(-1)^{l-1}}{(2l-1)^2}\sin\left[\frac{(2l-1)\pi}{L}x\right].$$

This is in fact the Fourier sine series representation of the given function $f(x)$ when $0 < x < L$.

(iii) Here are the actual string shapes at times $t = 0, \frac{L}{4v}, \frac{L}{2v}, \frac{3L}{4v}, \frac{L}{v}$:

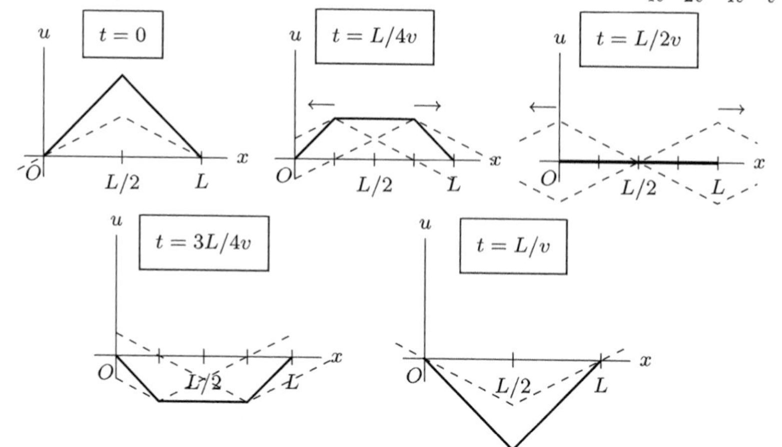

11-22 For the energy given by

$$E(t) = \int_0^L \left\{\frac{\mu}{2}[\dot{y}(x,t)]^2 + \frac{T}{2}[y'(x,t)]^2\right\} dx,$$

we have

$$\begin{aligned}\frac{d}{dt}E(t) &= \int_0^L \left\{\mu\dot{y}\ddot{y} + Ty'\left[\frac{\partial}{\partial t}y'\right]\right\} dx,\\ &= \int_0^L \left\{\mu\dot{y}\left(\ddot{y} - \frac{T}{\mu}y''\right)\right\} dx\\ &= 0,\end{aligned}$$

where for the second line we used integration by parts, and for the last we used the equation of motion. So the energy is a constant of motion. Now let us substitute

$$y(x,t) = \sum_{n=1}^{\infty}(a_n \cos\omega_n t + b_n \sin\omega_n t)\sin\frac{\omega_n x}{v} \qquad \left(\omega_n \equiv \frac{n\pi v}{L}\right)$$

into the above energy expression. Since

$$\begin{aligned}\frac{\mu}{2}\int_0^L dx\,\dot{y}^2 &= \frac{\mu}{2}\sum_{n=1}^{\infty}\sum_{m=1}^{\infty}\omega_n\omega_m(-a_n\sin\omega_n t + b_n\cos\omega_n t)\\ &\quad\times(-a_m\sin\omega_m t + b_m\cos\omega_m t)\\ &\quad\times\int_0^L dx\sin\frac{\omega_n x}{v}\sin\frac{\omega_m x}{v}\\ &= \frac{\mu L}{4}\sum_{n=1}^{\infty}\omega_n^2(-a_n\sin\omega_n t + b_n\cos\omega_n t)^2,\end{aligned}$$

$$\begin{aligned}\frac{T}{2}\int_0^L dx\,y'^2 &= \frac{T}{2}\sum_{n=1}^{\infty}\sum_{m=1}^{\infty}(a_n\cos\omega_n t + b_n\sin\omega_n t)\frac{\omega_n\omega_m}{v^2}\\ &\quad\times(a_m\cos\omega_m t + b_m\sin\omega_m t)\int_0^L dx\cos\frac{\omega_n x}{v}\cos\frac{\omega_m x}{v}\\ &= \frac{TL}{4v^2}\sum_{n=1}^{\infty}\omega_n^2(a_n\cos\omega_n t + b_n\sin\omega_n t)^2,\end{aligned}$$

we obtain (note that $v^2 = T/\mu$)

$$\begin{aligned}E &= \frac{\mu L}{4}\sum_{n=1}^{\infty}\omega_n^2\{(-a_n\sin\omega_n t + b_n\cos\omega_n t)^2 + (a_n\cos\omega_n t\\ &\quad + b_n\sin\omega_n t)^2\}\\ &= \frac{M}{4}\sum_{n=1}^{\infty}\omega_n^2(a_n^2 + b_n^2),\end{aligned}$$

where $M = \mu L$ is the total mass. In this expression, $\frac{M}{4}\omega_n^2(a_n^2 + b_n^2)$ (with $\omega_n = \frac{n\pi v}{L}$) represents the energy in the nth normal mode; clearly, higher normal modes, i.e., modes with more nodes carry higher energy. Also note that we must have $a_n, b_n \to 0$ as $n \to \infty$ to have a finite total energy.

12 Rigid Body Dynamics II

12-1 (i) With $I_1 = I_2$, Euler's equations read

$$I_1\dot{\omega}_1 = (I_1 - I_3)\omega_2\omega_3, \tag{1a}$$

$$I_2\dot{\omega}_2 = -(I_1 - I_3)\omega_1\omega_3, \tag{1b}$$

$$I_3\dot{\omega}_3 = 0. \tag{1c}$$

From (1c), we find $\omega_3 =$ constant. Then, differentiation of (1a) yields

$$I_1\ddot{\omega}_1 = (I_1 - I_3)\omega_3\dot{\omega}_2 = -\frac{(I_1 - I_3)^2}{I_1}\omega_3^2\omega_1$$

or

$$\ddot{\omega}_1 = -\Omega^2\omega_1, \quad \text{with } \Omega = \frac{(I_1 - I_3)\omega_3}{I_1}.$$

Assuming that $I_1 \neq I_3$ so that $\Omega \neq 0$, we thus obtain the solution

$$\omega_1(t) = A\cos(\Omega t + \alpha) \quad (A, \alpha : \text{ constants}),$$

and this in turn gives, using (1a),

$$\omega_2(t) = \frac{I_1\dot{\omega}_1}{(I_1 - I_3)\omega_3} = -A\sin(\Omega t + \alpha).$$

(ii) Based on (i), we can express the angular velocity and total angular momentum vectors by

$$\vec{\omega} = \vec{\omega}_\perp + \omega_3\mathbf{e}_3,$$

$$\vec{L} = I_1\vec{\omega}_\perp + I_3\omega_3\mathbf{e}_3 = I_1\vec{\omega} + (I_3 - I_1)\omega_3\mathbf{e}_3, \tag{2}$$

with

$$\vec{\omega}_\perp \equiv \omega_1\mathbf{e}_1 + \omega_2\mathbf{e}_2 = A\{\cos(\Omega t + \alpha)\mathbf{e}_1 - \sin(\Omega t + \alpha)\mathbf{e}_2\}.$$

Hence the three vectors $\vec{\omega}$, $\vec{L}$ and $\mathbf{e}_3$ are coplanar (see the figure below), and the angles $\beta = \angle(\vec{\omega}, \mathbf{e}_3)$ and $\theta = \angle(\mathbf{e}_3, \vec{L})$ are fixed in time, with

$$\tan\beta = \left|\frac{\vec{\omega}_\perp}{\omega_3}\right| = \frac{|A|}{|\omega_3|},$$
$$\tan\theta = \frac{I_1|\vec{\omega}_\perp|}{I_3|\omega_3|} = \frac{I_1}{I_3}\tan\beta.$$

The angular momentum vector $\vec{L}$ is a conserved quantity (so that we may write $\vec{L} = L\mathbf{k}$, with $|L| = \sqrt{I_1^2 A^2 + I_3\omega_3^2}$); this can be checked using our solution as

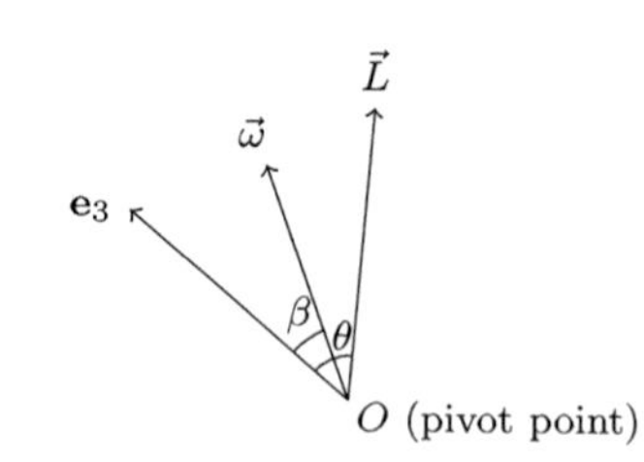

$$\begin{aligned}\frac{d\vec{L}}{dt} &= \frac{d^*\vec{L}}{dt} + \vec{\omega}\times\vec{L} \\ &= \frac{d^*}{dt}(I_1\vec{\omega}_\perp) + (I_3 - I_1)\omega_3\vec{\omega}\times\mathbf{e}_3 \\ &= I_1 A\Omega\{-\sin(\Omega t+\alpha)\mathbf{e}_1 - \cos(\Omega t+\alpha)\mathbf{e}_2\} \\ &\quad + (I_3 - I_1)\omega_3\{-\cos(\Omega t+\alpha)\mathbf{e}_2 - \sin(\Omega t+\alpha)\mathbf{e}_1\} \\ &= 0,\end{aligned}$$

where we used the expression (2) and the fact that $I_1\Omega = (I_1 - I_3)\omega_3$. Also, representing the symmetry axis $\mathbf{e}_3$ by

$$\mathbf{e}_3 = \sin\theta\cos\varphi\mathbf{i} + \sin\theta\sin\varphi\mathbf{j} + \cos\theta\mathbf{k} \tag{3}$$

($\mathbf{i}, \mathbf{j}, \mathbf{k}$ refer to space-fixed axes), see what follows from the relation $\frac{d\vec{\omega}}{dt} = \frac{d^*\vec{\omega}}{dt}$ with $\vec{\omega} = \frac{1}{I_1}\vec{L} + \frac{(I_1-I_3)\omega_3}{I_1}\mathbf{e}_3 = \frac{L}{I_1}\mathbf{k} + \Omega\mathbf{e}_3$ (from (2)). Then, from $\frac{d\vec{L}}{dt} = 0$ and $\frac{d^*\vec{\omega}}{dt} = \frac{d^*\vec{\omega}_\perp}{dt}$ since $\dot{\omega}_3 = 0$, we find that

$$\frac{d\mathbf{e}_3}{dt} = \frac{1}{\Omega}\frac{d^*\vec{\omega}_\perp}{dt} = \frac{1}{\Omega}\Omega\vec{\omega}\times\mathbf{e}_3 = \left(\frac{L}{I_1}\mathbf{k}\right)\times\mathbf{e}_3,$$

while (from (3) and $\dot{\theta} = 0$) $\frac{d\mathbf{e}_3}{dt} = \dot{\varphi}\mathbf{k}\times\mathbf{e}_3$. Hence the symmetry axis $\mathbf{e}_3$ rotates around the direction of $\vec{L}$ ($= L\mathbf{k}$) at a constant rate $\dot{\varphi} = \frac{L}{I_1}$. All these findings are fully consistent with the discussions presented in Sec. 12.1.

12-2 Take the principal axes $\mathbf{e}_1, \mathbf{e}_2, \mathbf{e}_3$ as shown in the figure ($\mathbf{e}_1$ and $\mathbf{e}_2$ are chosen on the board). Then, if ω_1, ω_2 and ω_3 denote the components of the angular velocity vector, they should satisfy Euler's equations

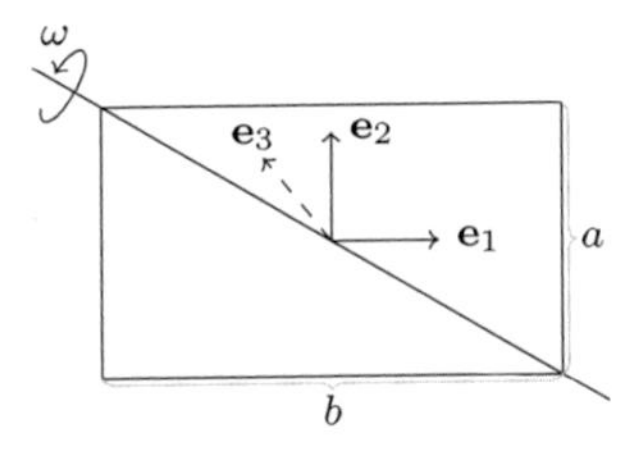

$$\begin{aligned} I_1\dot{\omega}_1 + (I_3 - I_2)\omega_2\omega_3 &= \Gamma_1^{(e)}, \\ I_2\dot{\omega}_2 + (I_1 - I_3)\omega_3\omega_1 &= \Gamma_2^{(e)}, \\ I_3\dot{\omega}_3 + (I_2 - I_1)\omega_1\omega_2 &= \Gamma_3^{(e)}, \end{aligned} \tag{1}$$

where

$$I_1 = \frac{1}{12}ma^2, \quad I_2 = \frac{1}{12}mb^2$$

and

$$I_3 = I_1 + I_2 = \frac{1}{12}m(a^2 + b^2).$$

In our problem, $\omega_3 = 0$ and $\vec{\omega} = \omega(\frac{b}{\sqrt{a^2+b^2}}\mathbf{e}_1 + \frac{a}{\sqrt{a^2+b^2}}\mathbf{e}_2)$, i.e., ω_1 and ω_2 are constants as given by

$$\omega_1 = \frac{\omega b}{\sqrt{a^2 + b^2}}, \quad \omega_2 = \frac{\omega a}{\sqrt{a^2 + b^2}}. \tag{2}$$

Substituting (2) (and $\omega_3 = 0$) into (1) then gives

$$\Gamma_1^{(e)} = \Gamma_2^{(e)} = 0$$

and

$$|\vec{\Gamma}^{(e)}| = |\Gamma_3^{(e)}| = \frac{m}{12}(b^2 - a^2)\frac{\omega^2 ab}{a^2 + b^2}.$$

12-3 (i) In terms of Euler's angles, the external torque due to gravity $\vec{g} = -g\mathbf{k}$ relative to the principle axes of the top has components (see footnote 5 in the text)

$$\Gamma_1^{(e)} = MgR\sin\theta\sin\psi, \quad \Gamma_2^{(e)} = MgR\sin\theta\cos\psi, \quad \Gamma_3^{(e)} = 0.$$

Hence, Euler's equations become

$$\begin{aligned} I_1\dot{\omega}_1 + (I_3 - I_1)\omega_2\omega_3 &= MgR\sin\theta\sin\psi, \\ I_1\dot{\omega}_2 + (I_1 - I_3)\omega_3\omega_1 &= MgR\sin\theta\cos\psi, \\ I_3\dot{\omega}_3 &= 0. \end{aligned} \tag{1}$$

on the other hand, for the top precessing steadily at an angle θ_0, we have for the components of the angular velocity vector

$$\omega_1 = -\Omega \sin\theta_0 \cos\psi, \quad \omega_2 = \Omega \sin\theta_0 \sin\psi, \quad \omega_3 = \Omega\cos\theta_0 + s, \tag{2}$$

since $\dot{\theta} = 0$. Using the form (2) in the first (or second) equation of (1) gives

$$\Omega^2(I_1 - I_3)\cos\theta_0 - \Omega s I_3 + MgR = 0. \tag{3}$$

Hence the following relation should be satisfied:

$$s\,(\equiv \dot{\psi}) = \frac{(I_1 - I_3)\Omega^2\cos\theta_0 + MgR}{I_3\Omega}.$$

(ii) For given values of s and the inclination angle θ_0, we can have a steady precession only when the quadratic equation (3) admits real roots for Ω. Hence it is necessary to have

$$I_3^2 s^2 - 4(I_1 - I_3)MgR\cos\theta_0 \geq 0$$

or

$$|s| \geq \frac{1}{I_3}\sqrt{4(I_1 - I_3)MgR\cos\theta_0},$$

giving the minimum spin velocity $\frac{1}{I_3}\sqrt{4(I_1 - I_3)MgR\cos\theta_0}$.

12-4 To discuss the rotational motion about the CM of the disk, it will be convenient to introduce an orthonomal triad $(\mathbf{e}_1', \mathbf{e}_2', \mathbf{e}_3' \equiv \mathbf{e}_3)$ where $\mathbf{e}_3'$ lies along the symmetry axis of the disk and $\mathbf{e}_1'$ is directed from the center of the disk to the point of contact with the horizontal plane. We can identify this triad with the intermediate set $(\mathbf{e}_1', \mathbf{e}_2', \mathbf{e}_3)$ defined in Sec. 12.1, and represent this triad using Euler's angles (θ, φ, ψ) as

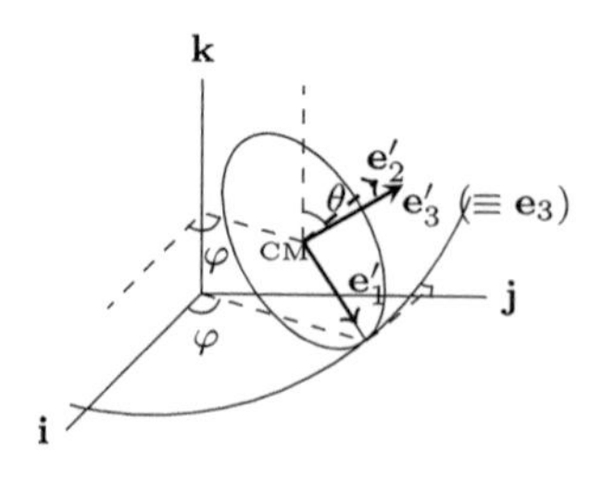

$$\mathbf{e}_3' = \sin\theta(\cos\varphi\mathbf{i} + \sin\varphi\mathbf{j}) + \cos\theta\mathbf{k},$$

$$\mathbf{e}'_1 = \cos\theta(\cos\varphi\mathbf{i} + \sin\varphi\mathbf{j}) - \sin\theta\mathbf{k},$$
$$\mathbf{e}'_2 = \mathbf{e}'_3 \times \mathbf{e}'_1 = -\sin\varphi\mathbf{i} + \cos\varphi\mathbf{j}.$$

The angular velocity vector of the disk is (see Sec. 12.1)

$$\begin{aligned}\vec{\omega} &= -\dot{\varphi}\sin\theta\mathbf{e}'_1 + \dot{\theta}\mathbf{e}'_2 + (\dot{\psi} + \dot{\varphi}\cos\theta)\mathbf{e}'_3 \quad (\equiv \omega'_i\mathbf{e}'_i)\\ &= \tilde{\vec{\omega}} + \dot{\psi}\mathbf{e}'_3 \end{aligned} \tag{1}$$

where $\tilde{\vec{\omega}} = -\dot{\varphi}\sin\theta\mathbf{e}'_1 + \dot{\theta}\mathbf{e}'_2 + \dot{\varphi}\cos\theta\mathbf{e}'_3$ represents the angular velocity associated with the rotation of our triad $(\mathbf{e}'_1, \mathbf{e}'_2, \mathbf{e}'_3)$, i.e., $\frac{d\mathbf{e}'_i}{dt} = \tilde{\vec{\omega}} \times \mathbf{e}'_i$ for each $i = 1, 2, 3$.

The equation of motion for the CM of the disk reads

$$m\frac{d^2\vec{r}_{\text{CM}}}{dt^2} = \vec{f} - mg\mathbf{k},$$

when $\vec{f}$ represents the force given to the disk at the point of contact by the plane. Due to the constraint that the disk rolls without slipping, the instantaneous velocity of the contact point of the disk should vanish, i.e.,

$$\vec{v}_{\text{contact}} = \frac{d\vec{r}_{\text{CM}}}{dt} + \omega \times (a\mathbf{e}'_1) = 0,$$

and therefore the instantaneous velocity of the CM is given by

$$\frac{d\vec{r}_{\text{CM}}}{dt} = -a\vec{\omega} \times \mathbf{e}'_1 = a\dot{\theta}\mathbf{e}'_3 - a(\dot{\psi} + \dot{\varphi}\cos\theta)\mathbf{e}'_2, \tag{2}$$

where we used (1). When the disk is in the steady motion as prescribed, we have

$$\dot{\theta} = 0, \quad \dot{\varphi} = \Omega, \quad \frac{d\vec{r}_{\text{CM}}}{dt} = \Omega b\mathbf{e}'_2, \tag{3}$$

and

$$\frac{d^2\vec{r}_{\text{CM}}}{dt^2} = \Omega b\frac{d\mathbf{e}'_2}{dt} = \Omega b\tilde{\vec{\omega}} \times \mathbf{e}'_2 = -\Omega^2 b(\sin\theta_0\mathbf{e}'_3 + \cos\theta_0\mathbf{e}'_1).$$

Using the values (3) with (2), we are led to the relation

$$\Omega b\mathbf{e}'_2 = -a\omega'_3\mathbf{e}'_2 \quad (\text{with } \omega'_3 = \dot{\psi} + \dot{\varphi}\cos\theta_0)$$

and hence $\omega'_3 = -\frac{b}{a}\Omega$ if the disk is in steady motion.

We now consider the equation governing the rotational motion about the CM. The angular momentum of the disk with respect to its CM is

$$\begin{aligned}\vec{L} &= I_1(\omega_1'\mathbf{e}_1' + \omega_2'\mathbf{e}_2') + I_3\omega_3'\mathbf{e}_3' \quad \left(I_1 = \frac{1}{4}ma^2,\ I_3 = \frac{1}{2}ma^2\right)\\ &= I_1\widetilde{\vec{\omega}} + (I_3\omega_3' - I_1\dot{\varphi}\cos\theta)\mathbf{e}_3',\end{aligned}$$

where $\omega_1' = -\dot{\varphi}\sin\theta$, $\omega_2' = \dot{\theta}$ and $\omega_3' = \dot{\psi} + \dot{\varphi}\cos\theta$. It should satisfy the equation

$$\begin{aligned}\frac{d\vec{L}}{dt} &= (a\mathbf{e}_1') \times \vec{f} \qquad (4)\\ &= a\mathbf{e}_1' \times \left(m\frac{d^2\vec{r}_{\text{CM}}}{dt^2} + mg\mathbf{k}\right).\end{aligned}$$

For the steady motion we are interested in, we have

$$\begin{aligned}&\theta = \theta_0, \quad \omega_1' = -\Omega\sin\theta_0, \quad \omega_2' = 0, \quad \omega_3' = -\frac{b}{a}\Omega,\\ &\dot{\omega}_1' = \dot{\omega}_2' = \dot{\omega}_3' = 0\end{aligned}$$

and accordingly

$$\begin{aligned}\frac{d\vec{L}}{dt} &= \widetilde{\vec{\omega}} \times \vec{L}\\ &= (-\Omega\sin\theta_0\mathbf{e}_1' + \Omega\cos\theta_0\mathbf{e}_3') \times (-I_1\Omega\sin\theta_0\mathbf{e}_1' + I_3\omega_3'\mathbf{e}_3')\\ &= (\Omega I_3\omega_3'\sin\theta_0 - \Omega^2 I_1\cos\theta_0\sin\theta_0)\mathbf{e}_2',\end{aligned}$$

$$\begin{aligned}&a\mathbf{e}_1' \times \left(m\frac{d^2\vec{r}_{\text{CM}}}{dt^2} + mg\mathbf{k}\right)\\ &\quad = a\mathbf{e}_1' \times \left[-m\Omega^2 b(\sin\theta_0\mathbf{e}_3' + \cos\theta_0\mathbf{e}_1') + mg(\cos\theta_0\mathbf{e}_3' - \sin\theta_0\mathbf{e}_1')\right]\\ &\quad = ma^2\left(-\Omega\omega_3'\sin\theta_0 - \frac{g}{a}\cos\theta_0\right)\mathbf{e}_2'.\end{aligned}$$

Using these with (4), we thus obtain the condition (note that $2I_1 = I_3 = \frac{1}{2}ma^2$)

$$\Omega^2 = \frac{4g\cos\theta_0}{a\cos\theta_0\sin\theta_0 + 6b\sin\theta_0} = \frac{4g\cot\theta_0}{a\cos\theta_0 + 6b},$$

which is the desired result.

12-5 (i) We here have to study the equation of motion for the CM of the sphere and the torque equation about the CM of the sphere, under the constraint provided by the rolling condition. Using cylindrical coordinates (ρ, φ, z), the position of the CM of the sphere can be represented as

$$\vec{R} = \rho(\cos\varphi\mathbf{i} + \sin\varphi\mathbf{j}) + z\mathbf{k}$$
$$\equiv \rho\hat{\rho} + z\mathbf{k}, \qquad (1)$$

where the coordinate z is related to ρ by (see the figure)

$$z = \frac{a + \rho\cos\alpha}{\sin\alpha}. \qquad (2)$$

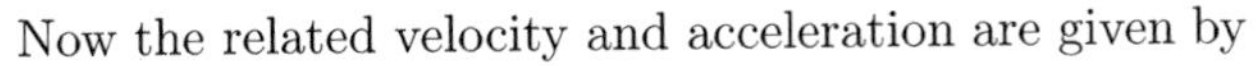

Now the related velocity and acceleration are given by

$$\frac{d\vec{R}}{dt} = \dot{\rho}\hat{\rho} + \rho\dot{\varphi}\hat{\varphi} + \dot{\rho}\cot\alpha\,\mathbf{k}, \qquad (3a)$$

$$\frac{d^2\vec{R}}{dt^2} = (\ddot{\rho} - \rho\dot{\varphi}^2)\hat{\rho} + (\rho\ddot{\varphi} + 2\dot{\rho}\dot{\varphi})\hat{\varphi} + \ddot{\rho}\cot\alpha\,\mathbf{k}, \qquad (3b)$$

where $\hat{\varphi} \equiv -\sin\varphi\mathbf{i} + \cos\varphi\mathbf{j}$. Introducing also the unit normal (to the surface of the cone) $\hat{n}$ and the unit tangent $\hat{t}$ according to

$$\hat{n} = -\cos\alpha\hat{\rho} + \sin\alpha\,\mathbf{k}, \quad \hat{t} = \sin\alpha\hat{\rho} + \cos\alpha\,\mathbf{k}, \qquad (4)$$

we can then write the equation of motion for the CM as

$$M\frac{d^2\vec{R}}{dt^2} = \vec{f} + N\hat{n} - Mg\mathbf{k}, \qquad (5)$$

where $\vec{f} = f_\varphi\hat{\varphi} + f_t\hat{t}$ represents the friction, and $N\hat{n}$ the normal force.

Denoting the angular velocity of the sphere by $\vec{\omega} = \omega_\rho\hat{\rho} + \omega_\varphi\hat{\varphi} + \omega_z\mathbf{k}$, we have the torque equation about the CM of the sphere

$$I\dot{\vec{\omega}} = (-a\hat{n}) \times \vec{f} = af_\varphi\hat{t} - af_t\hat{\varphi}, \qquad (6)$$

where $I = \frac{2}{5}Ma^2$. Using $\dot{\vec{\omega}} = (\dot{\omega}_\rho - \dot{\varphi}\,\omega_\varphi)\hat{\rho} + (\dot{\omega}_\varphi + \dot{\varphi}\,\omega_\rho)\hat{\varphi} + \dot{\omega}_z\mathbf{k}$, we thus find, from (6),

$$I(\dot{\omega}_\rho - \dot{\varphi}\,\omega_\varphi) = a\sin\alpha\, f_\varphi,$$
$$I(\dot{\omega}_\varphi + \dot{\varphi}\,\omega_\rho) = -af_t,$$
$$I\dot{\omega}_z = a\cos\alpha\, f_\varphi, \qquad (7)$$

and therefore the following relations:

$$\vec{f} = \frac{I\dot{\omega}_z}{a\cos\alpha}\hat{\varphi} - \frac{I(\dot{\omega}_\varphi + \dot{\varphi}\,\omega_\rho)}{a}\hat{t} \tag{8}$$

and

$$\dot{\omega}_z = \cot\alpha\,(\dot{\omega}_\rho - \dot{\varphi}\,\omega_\varphi). \tag{9}$$

We can use (8) to eliminate the friction $\vec{f}$ from our equation of motion (5) and also eliminate the normal force by taking the cross product of this equation with the unit normal $\vec{n}$. In view of (4) and (3b), we have

$$\begin{aligned}
\hat{n} \times M\frac{d^2\vec{R}}{dt^2} &= M\left(\frac{1}{\sin\alpha}\ddot{\rho} - \sin\alpha\rho\dot{\varphi}^2\right)\hat{\varphi} - M(\rho\ddot{\varphi} + 2\dot{\rho}\dot{\varphi})\hat{t},\\
\hat{n} \times \vec{f} &= -\frac{I\dot{\omega}_z}{a\cos\alpha}\hat{t} - \frac{I(\dot{\omega}_\varphi + \dot{\varphi}\,\omega_\rho)}{a}\hat{\varphi},\\
\hat{n} \times (-Mg\mathbf{k}) &= -Mg\cos\alpha\hat{\varphi},
\end{aligned} \tag{10}$$

and hence the resulting equation provides us with the following equations:

$$\frac{M}{\sin\alpha}\ddot{\rho} - M\sin\alpha\rho\dot{\varphi}^2 = -\frac{I}{a}(\dot{\omega}_\varphi + \dot{\varphi}\,\omega_\rho) - Mg\cos\alpha, \tag{11a}$$

$$M(\rho\ddot{\varphi} + 2\dot{\rho}\dot{\varphi}) = \frac{I\dot{\omega}_z}{a\cos\alpha}. \tag{11b}$$

We now incorporate the rolling constraint, requiring the instantaneous velocity of the contact point of the sphere to vanish. This is the case when

$$\begin{aligned}
\frac{d\vec{R}}{dt} &= a\vec{\omega} \times \vec{n}\\
&= a\omega_\varphi \sin\alpha\,\hat{\rho} - (a\omega_\rho\sin\alpha + a\omega_z\cos\alpha)\hat{\varphi} + a\omega_\varphi\cos\alpha\,\mathbf{k}.
\end{aligned} \tag{12}$$

Using (3a) with this condition yields the relations

$$\omega_\varphi = \frac{1}{a\sin\alpha}\dot{\rho}, \quad \omega_\rho = -\frac{\rho}{a\sin\alpha}\dot{\varphi} - \cot\alpha\,\omega_z, \tag{13}$$

dropping a redundant one. If we use these relations in (9), we are led to the following equation

$$\frac{a\dot{\omega}_z}{\cos\alpha} = -(\rho\ddot{\varphi} + 2\dot{\rho}\dot{\varphi}). \tag{14}$$

The two equations in (11b) and (14) can hold simultaneously only when $\dot{\omega}_z = (\rho\ddot{\varphi} + 2\dot{\rho}\dot{\varphi}) = 0$; so, both ω_z (or the z-component of the spin angular momentum) and $L = \rho^2\dot{\varphi}$ (or the z-component of the orbital angular momentum) are constants of motion. Also, using (13) in (11a) (and from the conserved nature of ω_z and $L = \rho^2\dot{\varphi}$), we obtain an equation involving only the cylindrical radial coordinate ρ:

$$\frac{7}{5}\ddot{\rho} - \left(\frac{2}{5} + \sin^2\alpha\right)\frac{L^2}{\rho^3} - \frac{2}{5}\cos\alpha\frac{La\omega_z}{\rho^2} + g\sin\alpha\cos\alpha = 0, \tag{15}$$

where we used $I = \frac{2}{5}Ma^2$. Now, integrating (15), we have the radial energy equation of the form

$$\frac{7}{10}\dot{\rho}^2 + V_{\text{eff}}(\rho) = \mathcal{E} \quad (\text{: const.}) \tag{16}$$

where $V_{\text{eff}}(\rho) = \frac{L^2}{2\rho^2}\left(\frac{2}{5} + \sin^2\alpha\right) + \frac{2La\omega_z}{5\rho}\cos\alpha + g\rho\sin\alpha\cos\alpha$. It is not difficult to show that (16) is in fact related to the statement of the conservation of total energy, i.e.,

$$\frac{1}{2}M\frac{d\vec{R}}{dt}\cdot\frac{d\vec{R}}{dt} + \frac{1}{5}Ma^2\vec{\omega}^2 + Mgz = E \quad (\text{: const.}). \tag{17}$$

(ii) From (15), we have circular orbits $\rho = \rho_0$ (: const.) if ρ_0 corresponds to the root of the equation

$$g\sin\alpha\cos\alpha\rho_0^3 - \frac{2}{5}(a\omega_z\cos\alpha)L\rho_0 - \left(\frac{2}{5} + \sin^2\alpha\right)L^2 = 0. \tag{18}$$

Using $L = \rho_0^2\dot{\varphi}$, this can be recast as a relation between the angular frequency $\dot{\varphi}$ and the radius ρ_0:

$$\left(\frac{2}{5} + \sin^2\alpha\right)\rho_0\dot{\varphi}^2 + \frac{2}{5}(a\omega_z\cos\alpha)\dot{\varphi} = g\sin\alpha\cos\alpha. \tag{19}$$

Here note that, for circular orbits, we have $\omega_\varphi = 0$ (see (13)); also, if β is the angle that the spin axis makes with the vertical (so that we can write $\omega_\rho = \omega_z\tan\beta$), we may use the second equation of (13) to set

$$a\omega_z = -\frac{\rho_0 \cos\beta}{\cos(\beta - \alpha)}\dot{\varphi}. \tag{20}$$

Using this relation, we may write (19) in the form

$$\rho_0\dot{\varphi}^2 \left[\left(\frac{2}{5} + \sin^2\alpha\right) - \frac{2}{5}\frac{\cos\alpha\cos\beta}{\cos(\beta - \alpha)}\right] = g\sin\alpha\cos\alpha. \tag{21}$$

The frequency of small oscillations about these circular orbits, Ω, can be found by inserting $\rho(t) = \rho_0 + \delta(t)$ in (15) or (16), according to the standard procedure. After some calculations, one will find that

$$\frac{7}{5}\Omega^2 = \dot{\varphi}^2\left(\frac{2}{5} + \sin^2\alpha\right) + \frac{2g}{\rho_0}\sin\alpha\cos\alpha \tag{22}$$

or, after using (21) to eliminate the ρ_0 and g dependences,

$$\Omega^2 = \frac{5}{7}\dot{\varphi}^2\left[3\left(\frac{2}{5} + \sin^2\alpha\right) - \frac{4}{5}\frac{\cos\alpha\cos\beta}{\cos(\beta - \alpha)}\right]. \tag{23}$$

12-6 Note that, in terms of the principal axes $(\mathbf{e}_1, \mathbf{e}_2, \mathbf{e}_3)$, we have the uniform gravity represented by (see footnote 5 in the text)

$$\vec{g} = -g\mathbf{k} = -g(-\sin\theta\cos\psi\mathbf{e}_1 + \sin\theta\sin\psi\mathbf{e}_2 + \cos\theta\mathbf{e}_3),$$

and hence the external torque by

$$\begin{aligned}\vec{\Gamma}^{(e)} &= R\mathbf{e}_3 \times (M\vec{g})\\ &= MgR\sin\theta\sin\psi\mathbf{e}_1 + MgR\sin\theta\cos\psi\mathbf{e}_2.\end{aligned}$$

Using this expression and the formulas for ω_1, ω_2 and ω_3 given in Sec. 12.1, we have Euler's equations (for $I_1 = I_2$)

$$\begin{aligned}I_1\frac{d}{dt}&(-\dot{\varphi}\sin\theta\cos\psi + \dot{\theta}\sin\psi) + (I_3 - I_1)\\ &\times(\dot{\varphi}\sin\theta\sin\psi + \dot{\theta}\cos\psi)(\dot{\varphi}\cos\theta + \dot{\psi})\\ &= MgR\sin\theta\sin\psi, \end{aligned}\tag{1}$$

$$\begin{aligned}I_1\frac{d}{dt}&(\dot{\varphi}\sin\theta\sin\psi + \dot{\theta}\cos\psi) - (I_3 - I_1)\\ &\times(-\dot{\varphi}\sin\theta\cos\psi + \dot{\theta}\sin\psi)(\dot{\varphi}\cos\theta + \dot{\psi})\end{aligned}$$

$$= MgR\sin\theta\cos\psi, \tag{2}$$

$$I_3\frac{d}{dt}(\dot\varphi\cos\theta + \dot\psi) = 0. \tag{3}$$

The last equation implies the conservation of p_ψ as in Sec. 12.2. On the other hand, if we multiply (1) by $\sin\psi$ and (2) by $\cos\psi$ and then add them, we obtain (after some straightforward algebra) the following equation

$$I_1\ddot\theta - I_1\dot\varphi^2\sin\theta\cos\theta + I_3(\dot\psi + \dot\varphi\cos\theta)\dot\varphi\sin\theta = MgR\sin\theta,$$

which coincides with the Lagrange equation for θ (see Sec. 12.2). Similarly, if we multiply (1) by $-\cos\psi$ and (2) by $\sin\psi$ and then add them, we are led to

$$I_1\ddot\varphi\sin\theta + 2I_1\dot\varphi\dot\theta\cos\theta - I_3(\dot\psi + \dot\varphi\cos\theta)\dot\theta = 0.$$

This is actually equivalent to the Lagrange equation for φ, which is

$$\begin{aligned} 0 &= \frac{d}{dt}\big[I_1\dot\varphi\sin^2\theta + I_3(\dot\psi + \dot\varphi\cos\theta)\cos\theta\big] \\ &= \sin\theta\big[I_1\ddot\varphi\sin\theta + 2I_1\dot\varphi\dot\theta\cos\theta - I_3(\dot\psi + \dot\varphi\cos\theta)\dot\theta\big], \end{aligned}$$

using (3) above.

12-7 If $\alpha \equiv \frac{2I_1MgR}{I_3^2\omega_3^2} \ll 1$, $\theta(t)$ takes value in a small region $\theta_1 \le \theta \le \theta_2$ where $\theta_2 = \theta_1 + 2a$ with $a \approx \frac{1}{2}\alpha\sin\theta_1$ (see (12.56) in the text). Hence, setting $\theta_1 = \theta_0 - a$ and $\theta(t) = \theta_0 + \delta(t)$, the given effective potential can be written as

$$\begin{aligned} V_{\text{eff}}(\theta) &= \frac{I_3^2\omega_3^2}{2I_1}\left\{\frac{[\cos(\theta_0 - a) - \cos(\theta_0 + \delta)]^2}{\sin^2(\theta_0 + \delta)} + \alpha\cos(\theta_0 + \delta)\right\} \\ &= \frac{I_3^2\omega_3^2}{2I_1}\left\{\frac{[\cancel{\cos\theta_0} + a\sin\theta_0 - \cancel{\cos\theta_0} + \delta\sin\theta_0]^2}{(\sin\theta_0 + \cos\theta_0\sin\delta)^2} + \alpha\cos\theta_0 \right. \\ &\qquad \left. - \alpha\delta\sin\theta_0 + O(\alpha^3)\right\} \\ &= \frac{I_3^2\omega_3^2}{2I_1}\{(a+\delta)^2 + \alpha\cos\theta_0 - \alpha\delta\sin\theta_0 + O(\alpha^3)\}. \end{aligned}$$

But, $(a+\delta)^2 = a^2 + (\alpha\sin\theta_0)\delta + \delta^2 + O(\alpha^3)$. Using this with the above expression, we find

$$V_{\text{eff}}(\theta) = \frac{I_3^2\omega_3^2}{2I_1}\left\{a^2 + \alpha\cos\theta_0 + \delta^2 + O(\alpha^3)\right\}.$$

Here note that

$$V_{\text{eff}}(\theta_0) = \frac{I_3^2\omega_3^2}{2I_1}\left\{\frac{[\cos(\theta_0 - a) - \cos(\theta_0)]^2}{\sin^2\theta_0} + \alpha\cos\theta_0\right\}$$
$$= \frac{I_3^2\omega_3^2}{2I_1}\Big[a^2 + \alpha\cos\theta_0 + O(\alpha^3)\Big].$$

Hence we have the harmonic approximation (note that $\delta \equiv \theta - \theta_0$)

$$V_{\text{eff}}(\theta) = V_{\text{eff}}(\theta_0) + \frac{1}{2}\frac{I_3^2}{I_1}\omega_3^2(\theta - \theta_0)^2.$$

12-8 (i) With the plane exerting only normal force, the CM of the top will move only vertically in a fixed line, which we take as our z-axis. Let $(\mathbf{e}_1, \mathbf{e}_2, \mathbf{e}_3)$ denote the principal axes of the top at its CM, $\mathbf{e}_3$ being the symmetric axis; associated principal moments of inertia are I_1, I_1, and I_3, and their orientations in space are specified in terms of three Euler's angles (θ, φ, ψ). The height of the CM above the plane is $R\cos\theta$. Hence the part of the kinetic energy due to the motion of the CM is $\frac{1}{2}MR^2\sin^2\theta\,\dot{\theta}^2$. So the Lagrangian for the system is

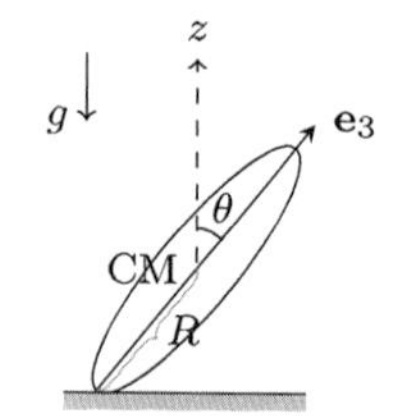

$$L = \frac{1}{2}MR^2\dot{\theta}^2\sin^2\theta + \frac{1}{2}I_1\dot{\theta}^2 + \frac{1}{2}I_1\dot{\varphi}^2\sin^2\theta + \frac{1}{2}I_3(\dot{\psi} + \dot{\varphi}\cos\theta)^2$$
$$- MgR\cos\theta.$$

We have two first integrals corresponding to cyclic coordinates φ and ψ:

$$I_1\dot{\varphi}\sin^2\theta + I_3(\dot{\psi} + \dot{\varphi}\cos\theta)\cos\theta = p_\varphi \quad (\text{: const.}), \tag{1}$$
$$I_3(\dot{\psi} + \dot{\varphi}\cos\theta) \equiv I_3\omega_3 = p_\psi \quad (\text{: const.}). \tag{2}$$

Also, as the Lagrangian does not depend on time explicitly, the energy

$$E = \frac{1}{2}(I_1 + MR^2\sin^2\theta)\dot{\theta}^2 + \frac{1}{2}I_1\dot{\varphi}^2\sin^2\theta + \frac{1}{2}I_3(\dot{\psi} + \dot{\varphi}\cos\theta)^2$$
$$+ MgR\cos\theta \tag{3}$$

is also a constant of motion. Using (1) and (2) in (3), we can also cast our energy conservation law by the form

$$\frac{1}{2}(I_1 + MR^2\sin^2\theta)\dot{\theta}^2 + V_{\text{eff}}(\theta) = E' \qquad \left(E' = E - \frac{p_\psi^2}{2I_3}\right) \quad (4)$$

with the effective potential

$$V_{\text{eff}}(\theta) = \frac{(p_\varphi - p_\psi\cos\theta)^2}{2I_1\sin^2\theta} + MgR\cos\theta.$$

(ii) If the top is released from $\theta = \theta_1$ with $\dot{\theta} = \dot{\varphi} = 0$ and $\dot{\psi} = \omega_3$ at $t = 0$, we have $p_\psi = I_3\omega_3$, $p_\varphi = I_3\omega_3\cos\theta_1$ and $E' = MgR\cos\theta_1$ so that

$$V_{\text{eff}}(\theta) = \frac{I_3^2\omega_3^2}{2I_1}\left[\frac{(\cos\theta - \cos\theta_1)^2}{\sin^2\theta} + \alpha\cos\theta\right] \quad \left(\alpha \equiv \frac{2I_1 MgR}{I_3^2\omega_3^2}\right).$$

This is identical with the effective potential expression in (12.53) in the text; so our discussion from (12.53) to (12.58), including the nutation pattern shown in the figure, also applies to this case. On the other hand, in the equations (12.59)–(12.61), one must now use the angular frequency $\omega_\theta = \frac{I_3\omega_3}{I_1 + MR^2\sin^2\theta_1}$ in view of (4).

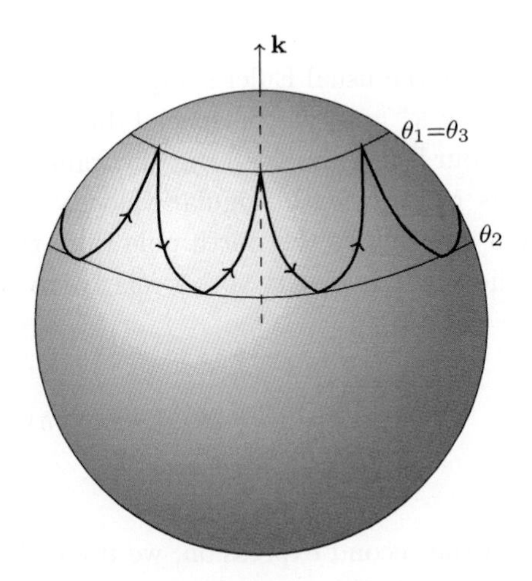

12-9 (i) The Lagrangian can be represented by

$$L = \frac{1}{2} I_3 \omega_3^2 + \frac{1}{2} I_1 (\omega_1^2 + \omega_2^2), \tag{1}$$

if $(\omega_1, \omega_2, \omega_3)$ denote the components of the angular velocity vector $\vec{\omega}$, as seen in an *inertial* frame, along the three principal axes of the gyrocompass $\mathbf{e}_1, \mathbf{e}_2, \mathbf{e}_3$). (The symmetry axis of the gyrocompass is $\mathbf{e}_3$.)

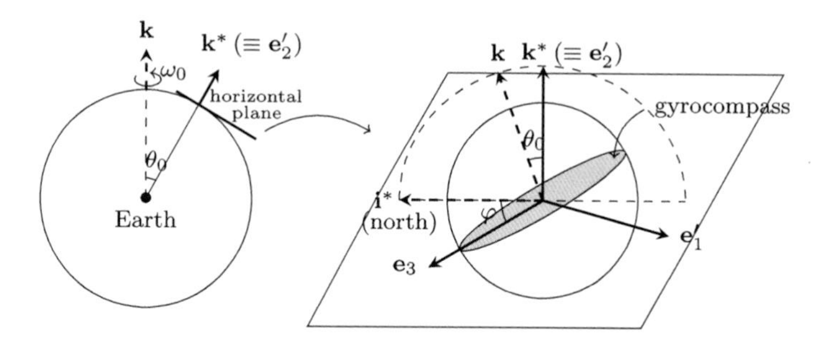

Let $\mathbf{k}^*(\equiv \mathbf{e}_2')$ denote the direction perpendicular to the horizontal plane in which the symmetry axis $\mathbf{e}_3$ moves, and let $\mathbf{e}_1' \equiv \mathbf{e}_2' \times \mathbf{e}_3$. (see the figure). Relative to the frame fixed to the Earth's surface, the angular velocity is then given by

$$\vec{\omega}^* = \dot{\varphi}\mathbf{e}_2' + \dot{\psi}\mathbf{e}_3, \tag{2}$$

as we choose the usual Euler's angles with θ fixed to the value $\frac{\pi}{2}$. (We took $\mathbf{k}^* = \mathbf{e}_2'$ as the special direction picked out when we define our Euler's angles.) But the frame fixed to the Earth's surface is rotating with angular velocity $\vec{\Omega} = \omega_0 \mathbf{k}$ ($\mathbf{k}$ is the direction pointing to the north pole from the Earth's center). Hence the angular velocity vector relative to an inertial frame is

$$\begin{aligned} \vec{\omega} &= \vec{\omega}^* + \omega_0 \mathbf{k} \\ &= -\omega_0 \sin\theta_0 \sin\varphi \mathbf{e}_1' + (\dot{\varphi} + \omega_0 \cos\theta_0)\mathbf{e}_2' \\ &\quad + (\dot{\psi} + \omega_0 \sin\theta_0 \cos\varphi)\mathbf{e}_3, \end{aligned} \tag{3}$$

where, in the second expression, we used (2) and the relation $\mathbf{k} = \cos\theta_0 \mathbf{e}_2' + \sin\theta_0(-\sin\varphi \mathbf{e}_1' + \cos\varphi \mathbf{e}_3)$ from the figure above.

Therefore, we find

$$\omega_3 = \dot{\psi} + \omega_0 \sin\theta_0 \cos\varphi \tag{4a}$$

$$\omega_1^2 + \omega_2^2 = \omega_1'^2 + \omega_2'^2 = (\dot{\varphi} + \omega_0 \cos\theta_0)^2 + \omega_0^2 \sin^2\theta_0 \sin^2\varphi. \tag{4b}$$

Inserting these in (1) gives the Lagrangian

$$L = \frac{1}{2} I_3 (\dot{\psi} + \omega_0 \sin\theta_0 \cos\varphi)^2 + \frac{1}{2} I_1 \left[(\dot{\varphi} + \omega_0 \cos\theta_0)^2 + \omega_0^2 \sin^2\theta_0 \sin^2\varphi \right]. \tag{5}$$

(ii) Considering the Lagrange equations with (5), we readily find

$$I_3(\dot{\psi} + \omega_0 \sin\theta_0 \cos\varphi) = I_3\omega_3 = \text{const.}, \tag{6a}$$

$$I_1\ddot{\varphi} = -I_3\omega_3\omega_0 \sin\theta_0 \sin\varphi + I_1\omega_0^2 \sin\theta_0^2 \sin\varphi\cos\varphi. \tag{6b}$$

For small φ, (6b) can be approximated by

$$I_1\ddot{\varphi} = -\omega_0 \sin\theta_0 [I_3\omega_3 - I_1\omega_0 \sin\theta_0]\varphi. \tag{7}$$

Hence, if $I_3\omega_3 - I_1\omega_0 \sin\theta_0 > 0$ or $\omega_3 > \left(\frac{I_1}{I_3}\right)\omega_0 \sin\theta_0$, the symmetry axis of the gyrocompass oscillates about the northern direction with the frequency of oscillation $\omega \approx \sqrt{\frac{I_3}{I_1}\omega_0\omega_3 \sin\theta_0}$ (to the leading order in ω_0).

12-10 We use as our generalized coordinates the X- and Y-coordinates of the center of the disk and the Euler's angles (θ, φ, ψ) which specify the orientations of three principal axes through the center. As in the solution to Problem 12-4, $\mathbf{e}_3' = \mathbf{e}_3 = \sin\theta(\cos\varphi\mathbf{i} + \sin\varphi\mathbf{j}) + \cos\theta\mathbf{k}$ denotes the direction of the symmetry axis of the disk; this also implies that the center of the disk is at distance $Z = a\sin\theta$ from the XY-plane. The Lagrangian, before taking the constraint conditions into account, reads

$$L = \frac{1}{2} m(\dot{X}^2 + \dot{Y}^2 + a^2\dot{\theta}^2 \cos^2\theta) + \frac{1}{2} I_1 (\dot{\theta}^2 + \dot{\varphi}^2 \sin^2\theta) + \frac{1}{2} I_3 (\dot{\psi} + \dot{\varphi}\cos\theta)^2 - mga\sin\theta, \tag{1}$$

where $I_1 = \frac{1}{4}ma^2$ and $I_3 = \frac{1}{2}ma^2$.

From the constraint that the disk rolls without slipping, the instantaneous velocity of the contact point of the disk should vanish;

using the notations employed in the solution to Problem 12-4, this can be expressed by the form

$$\{\dot{X}\mathbf{i} + \dot{Y}\mathbf{j} + a\dot{\theta}\cos\theta\mathbf{k}\} + \vec{\omega} \times (a\mathbf{e}'_1) = 0, \tag{2}$$

where $\mathbf{e}'_1 = \cos\theta(\cos\varphi\mathbf{i} + \sin\varphi\mathbf{j}) - \sin\theta\mathbf{k}$ is the unit vector directed from the center of the disk to the point of contact with the XY-plane. Using $\vec{\omega} = -\dot{\varphi}\sin\theta\mathbf{e}'_1 + \dot{\theta}\mathbf{e}'_2 + (\dot{\psi} + \dot{\varphi}\cos\theta)\mathbf{e}'_3$ (see our solution to the Problem 12-4) with (2), we then obtain the (nonholonomic) conditions

$$\dot{X} = a\dot{\theta}\sin\theta\cos\varphi + a(\dot{\psi} + \dot{\varphi}\cos\theta)\sin\varphi, \tag{3a}$$

$$\dot{Y} = a\dot{\theta}\sin\theta\sin\varphi - a(\dot{\psi} + \dot{\varphi}\cos\theta)\cos\varphi. \tag{3b}$$

For virtual displacements these give rise to the restrictions

$$\begin{aligned} &\delta X - a\sin\theta\cos\varphi\delta\theta - a\sin\varphi(\delta\psi + \cos\theta\delta\varphi) = 0, \\ &\delta Y - a\sin\theta\sin\varphi\delta\theta + a\cos\varphi(\delta\psi + \cos\theta\delta\varphi) = 0. \end{aligned} \tag{4}$$

Based on (1) and (4), we may now apply the D'Alembert–Lagrange method to write the equations of motion:

$$m\ddot{X} = \lambda_1, \tag{5a}$$

$$m\ddot{Y} = \lambda_2, \tag{5b}$$

$$\begin{aligned} I_1\ddot{\theta} + ma^2\frac{d}{dt}(\dot{\theta}\cos^2\theta) &= -ma^2\dot{\theta}^2\cos\theta\sin\theta + I_1\dot{\varphi}^2\cos\theta\sin\theta \\ &-I_3\omega_3\dot{\varphi}\sin\theta - mga\cos\theta - \lambda_1 a\sin\theta\cos\varphi - \lambda_2 a\sin\theta\sin\varphi, \end{aligned} \tag{5c}$$

$$\frac{d}{dt}(I_1\dot{\varphi}\sin^2\theta + I_3\omega_3\cos\theta) = -\lambda_1 a\sin\varphi\cos\theta + \lambda_2 a\cos\varphi\cos\theta, \tag{5d}$$

$$I_3\dot{\omega}_3 = -\lambda_1 a\sin\varphi + \lambda_2 a\cos\varphi, \tag{5e}$$

where $\omega_3 \equiv (\dot{\psi} + \dot{\varphi}\cos\theta)$, and λ_1, λ_2 are Lagrange multipliers introduced in association with two constraint conditions in (4). Then, from (3a,b)

$$\begin{aligned} \cos\varphi\dot{X} + \sin\varphi\dot{Y} &= a\dot{\theta}\sin\theta, \\ -\sin\varphi\dot{X} + \cos\varphi\dot{Y} &= -a\omega_3, \end{aligned} \tag{6}$$

and these may be combined with (5a,b) to obtain the relations

$$\begin{aligned}
\lambda_1 \cos\varphi + \lambda_2 \sin\varphi &= m\frac{d}{dt}(\cos\varphi\dot{X} + \sin\varphi\dot{Y}) \\
&\quad - m\dot{\varphi}(-\sin\varphi\dot{X} + \cos\varphi\dot{Y}) \\
&= ma\frac{d}{dt}(\dot{\theta}\sin\theta) + ma\dot{\varphi}\omega_3, \qquad (7a) \\
-\lambda_1 \sin\varphi + \lambda_2 \cos\varphi &= m\frac{d}{dt}(-\sin\varphi\dot{X} + \cos\varphi\dot{Y}) \\
&\quad + m\dot{\varphi}(\cos\varphi\dot{X} + \sin\varphi\dot{Y}) \\
&= -ma\dot{\omega}_3 + ma\dot{\varphi}\dot{\theta}\sin\theta. \qquad (7b)
\end{aligned}$$

We may now use (7a,b) with (5c–e) to obtain the following equations of motion (involving only the variables θ, φ and ω_3):

$$\begin{aligned}
&(I_1 + ma^2)\ddot{\theta} + (I_3 + ma^2)\sin\theta\omega_3\dot{\varphi} \\
&\qquad - I_1 \sin\theta\cos\theta\dot{\varphi}^2 + mga\cos\theta = 0, \qquad (8a) \\
&I_1 \sin\theta\ddot{\varphi} + 2I_1\cos\theta\dot{\varphi}\dot{\theta} - I_3\omega_3\dot{\theta} = 0, \qquad (8b) \\
&(I_3 + ma^2)\dot{\omega}_3 = ma^2\sin\theta\dot{\varphi}\dot{\theta}. \qquad (8c)
\end{aligned}$$

For steady motion, we put $\dot{\theta} = 0$ and then, from (8b,c), $\ddot{\varphi} = \dot{\omega}_3 = 0$. If we thus set $\theta = \theta_0$, $\dot{\varphi} = \Omega$ (: const.) and $\omega_3 = W$ (: const.) for this case, the following relation should hold between these three constants (thanks to (8a)):

$$(I_3 + ma^2)\sin\theta_0 W\Omega - I_1\sin\theta_0\cos\theta_0\Omega^2 + mga\cos\theta_0 = 0. \qquad (9)$$

Also, from (3a,b) we find that $\dot{X} = aW\sin\varphi$ and $\dot{Y} = -aW\cos\varphi$, with $\varphi = \Omega t + \text{const.}$: the center of the disk moves in a circle with a constant speed $V = a|W|$ and radius $R = a|W/\Omega|$. [A special case of steady motion is $\theta_0 = \frac{\pi}{2}$, corresponding to the plane of the disk being vertical. In this case, (9) requires that $W\Omega = 0$. If $\Omega = 0$ (and so $R \to \infty$), the disk rolls along a straight line and W is the rolling angular velocity; if $W = 0$ and hence $R = 0$, the disk spins in place about the vertical axis with angular velocity Ω.]

We now turn to the stability analysis of the above steady motion, setting

$$\theta(t) = \theta_0 + \bar{\theta}(t), \quad \dot{\varphi}(t) = \Omega + \overline{\Omega}(t), \quad \omega_3 = W + \bar{\omega}_3(t), \qquad (10)$$

where $\bar{\theta}$, $\overline{\Omega}$ and $\bar{\omega}_3$ represent small perturbations. Inserting these in the equations of motion (8a–c) and retaining only first-order terms, we find (with $I_1 = \frac{1}{4}ma^2$ and $I_3 = \frac{1}{2}ma^2$)

$$5\ddot{\bar{\theta}}(t) + \left(6\Omega W \cos\theta_0 - \Omega^2 \cos 2\theta_0 - 4\frac{g}{a}\sin\theta_0\right)\bar{\theta}(t) + (6W\sin\theta_0 - \Omega\sin 2\theta_0)\overline{\Omega}(t) + 6\Omega\sin\theta_0\bar{\omega}_3(t) = 0, \tag{11a}$$

$$\sin\theta_0\dot{\overline{\Omega}}(t) + 2\Omega\cos\theta_0\dot{\bar{\theta}}(t) - 2W\dot{\bar{\theta}}(t) = 0, \tag{11b}$$

$$3\dot{\bar{\omega}}_3(t) = 2\Omega\sin\theta_0\dot{\bar{\theta}}(t). \tag{11c}$$

We may then look for solutions in which $\bar{\theta}(t)$, $\overline{\Omega}(t)$ and $\bar{\omega}_3(t)$ oscillate with the same frequency γ, that is

$$\bar{\theta}(t) = A_1 e^{i\gamma t}, \quad \overline{\Omega}(t) = A_2 e^{i\gamma t}, \quad \bar{\omega}_3(t) = A_3 e^{i\gamma t}, \tag{12}$$

using complex notations. We then find that, from (11b),

$$A_2 = -\frac{2(\cos\theta_0\Omega - W)}{\sin\theta_0}A_1, \tag{13}$$

and, from (11c),

$$A_3 = \frac{2}{3}\Omega\sin\theta_0 A_1. \tag{14}$$

Then, if the forms in (12) are inserted into (11a) and the relations (13) and (14) are used, we obtain the following equation for γ^2:

$$5\gamma^2 = \Omega^2(3 + 2\sin^2\theta_0) - 10W\Omega\cos\theta_0 + 12W^2 - 4\frac{g}{a}\sin\theta_0. \tag{15}$$

Only when the given values of θ_0, Ω and W make the right hand side of (15) *positive*, the corresponding steady motion is stable. For the special case of a disk rolling in a straight line, i.e., $\theta_0 = \frac{\pi}{2}$ and $\Omega = 0$, (15) reduces to

$$5\gamma^2 = 12W^2 - 4\frac{g}{a}. \tag{16}$$

Hence the rolling is stable only if $W^2 > \frac{g}{3a}$. Another special case is that of a disk spinning about a vertical diameter, for which $\theta_0 = \frac{\pi}{2}$ and $W = 0$. Then, from the relation (15) which leads to

$5\gamma^2 = 5\Omega^2 - 4\frac{g}{a}$, the spinning is stable only for $\Omega > \sqrt{\frac{4g}{5a}}$. (This is the condition for "sleeping".)

12-11 See the solution to Problem 6-12, in which a closely related problem is considered. Note that, as shown in footnote 9 of Chapter 12 in the text, the quadrupole moment Q is equal to $-2(I_3 - I_1)$ in terms of related moments of inertia.

13 Hamiltonian Mechanics

13-1 To avoid possible confusion, let us write $\frac{\partial H}{\partial p_i} = \dot{\tilde{q}}_i$, so that we may have $\widetilde{L}(\mathbf{q}, \dot{\tilde{\mathbf{q}}}, t) = \sum_{i=1}^{n} \dot{\tilde{q}}_i p_i - H$, with $p_i = p_i(\mathbf{q}, \dot{\tilde{\mathbf{q}}}, t)$ derived using $\frac{\partial H}{\partial p_i} = \dot{\tilde{q}}_i$ $(i = 1, \ldots, n)$. Then, using $H = \sum_{i=1}^{n} p_i \dot{q}_i - L$,

$$\widetilde{L}(\mathbf{q}, \dot{\tilde{\mathbf{q}}}, t) = \sum_{i=1}^{n} (\dot{\tilde{q}}_i - \dot{q}_i) p_i + L.$$

Therefore, we find $\widetilde{L} = L$ if we can assert that $\dot{\tilde{q}}_i = \dot{q}_i$ $(i = 1, \ldots, n)$. But, note that, from $\dot{\tilde{q}}_i = \frac{\partial H}{\partial p_i}$ with $H = \sum_{i=1}^{n} p_i \dot{q}_i - L(\mathbf{q}, \dot{\mathbf{q}}, t)$,

$$\begin{aligned} \dot{\tilde{q}}_i &= \dot{q}_i + \sum_j p_j \frac{\partial \dot{q}_j}{\partial p_i}\bigg|_{\mathbf{q}} - \sum_j \frac{\partial \dot{q}_j}{\partial p_i}\bigg|_{\mathbf{q}} \frac{\partial L}{\partial \dot{q}_j} \\ &= \dot{q}_i \end{aligned}$$

where we used $\frac{\partial L}{\partial \dot{q}_j} = p_j$. Accordingly, $\widetilde{L} = L$.
[Remark: This result is just a consequence of performing successive Legendre transforms; equations of motion are not needed.]

13-2 (i) Given a Lagrangian $L(\mathbf{q}, \dot{\mathbf{q}}, t)$ which is a scalar under general coordinate transformations $q_i \to q_i' = q_i'(\mathbf{q}, t)$, we have

$$p_i' \equiv \frac{\partial L}{\partial \dot{q}_i'} = \frac{\partial L}{\partial \dot{q}_j} \frac{\partial \dot{q}_j}{\partial \dot{q}_i'} = \frac{\partial q_j}{\partial q_i'} p_j,$$

since $\dot{q}_i = \frac{\partial q_i}{\partial q_j'} \dot{q}_j' + \frac{\partial q_i}{\partial t}$ and so $\frac{\partial \dot{q}_i}{\partial \dot{q}_j'} = \frac{\partial q_i}{\partial q_j'}$.

(ii) Now, based on (i) and also $\dot{q}_i' = \frac{\partial q_i'}{\partial q_j} \dot{q}_j + \frac{\partial q_i'}{\partial t}$, the Hamiltonian using new phase-space coordinates becomes

$$H' \equiv \sum_{i=1}^{n} \dot{q}_i' p_i' - L$$

$$
\begin{aligned}
&= \sum_{i=1}^{n} \left(\frac{\partial q_i'}{\partial q_j} \dot{q}_j \frac{\partial q_k}{\partial q_i'} p_k + \frac{\partial q_i'}{\partial t} p_i' \right) - L \\
&= \left(\sum_{i=1}^{n} \dot{q}_i p_i - L \right) + \sum_{i=1}^{n} \frac{\partial q_i'}{\partial t} p_i' = H + \sum_{i=1}^{n} \frac{\partial q_i'}{\partial t} p_i'.
\end{aligned}
$$

(iii) With p_i' and H' given as in (i) and (ii), we have

$$
\begin{aligned}
\frac{\partial H'}{\partial p_i'} &= \frac{\partial}{\partial p_i'} \left(H + \sum_{j=1}^{n} \frac{\partial q_j'}{\partial t} p_j' \right) \\
&= \frac{\partial H}{\partial p_j} \frac{\partial p_j}{\partial p_i'} + \frac{\partial q_i'}{\partial t} \\
&= \dot{q}_j \frac{\partial q_i'}{\partial q_j} + \frac{\partial q_i'}{\partial t} = \dot{q}_i',
\end{aligned}
$$

where we used $\frac{\partial H}{\partial p_j} = \dot{q}_j$ and $\frac{\partial p_j}{\partial p_i'} = \frac{\partial q_i'}{\partial q_j}$. Similarly,

$$
\begin{aligned}
\frac{\partial H'}{\partial q_i'} &= \frac{\partial}{\partial q_i'} \left(H + \sum_{j=1}^{n} \frac{\partial q_j'}{\partial t} p_j' \right) \\
&= \frac{\partial H}{\partial q_j} \frac{\partial q_j}{\partial q_i'} + \frac{\partial H}{\partial p_j} \frac{\partial p_j}{\partial q_i'} + \frac{\partial q_k}{\partial q_i'} \frac{\partial^2 q_j'}{\partial q_k \, \partial t} p_j' \\
&= -\dot{p}_j \frac{\partial q_j}{\partial q_i'} + \dot{q}_j \frac{\partial p_j}{\partial q_i'} + \frac{\partial q_k}{\partial q_i'} \frac{\partial^2 q_j'}{\partial q_k \, \partial t} p_j' \\
&= -\left(\frac{\partial q_k'}{\partial q_j} \dot{p}_k' + \frac{\partial^2 q_k'}{\partial q_l \, \partial q_j} \dot{q}_l p_k' + \frac{\partial^2 q_k'}{\partial t \, \partial q_j} p_k' \right) \frac{\partial q_j}{\partial q_i'} \\
&\quad + \dot{q}_j \left(\frac{\partial q_l}{\partial q_i'} \frac{\partial^2 q_k'}{\partial q_l \, \partial q_j} p_k' \right) + \frac{\partial q_k}{\partial q_i'} \frac{\partial^2 q_j'}{\partial q_k \, \partial t} p_j' \\
&= -\dot{p}_i',
\end{aligned}
$$

where we used the facts that, from $p_j = \frac{\partial q_k'}{\partial q_j} p_k'$,

$$
\begin{aligned}
\dot{p}_j &= \frac{\partial q_k'}{\partial q_j} \dot{p}_k' + \frac{\partial^2 q_k'}{\partial q_l \, \partial q_j} \dot{q}_l p_k' + \frac{\partial^2 q_k'}{\partial t \, \partial q_j} p_k', \\
\frac{\partial p_j}{\partial q_i'} &= \frac{\partial q_l}{\partial q_i'} \frac{\partial^2 q_k'}{\partial q_l \, \partial q_j} p_i'.
\end{aligned}
$$

[Remark: More "elegant" derivation of these results, using the language of Poisson brackets, can be found in (13.71)–(13.78) of the text.]

13-3 (i)

$$H = \frac{1}{2m_1}\vec{p}_1^{\,2} + \frac{1}{2m_2}\vec{p}_2^{\,2} - \frac{1}{4\pi\epsilon_0}\frac{q^2}{|\vec{r}_1 - \vec{r}_2|} - qE_0(z_1 - z_2).$$

(ii) The canonical momenta conjugate to $\vec{R}$ and $\vec{r}$ are $\vec{P} = M\dot{\vec{R}}$ (with $M = m_1 + m_2$) and $\vec{p} = \mu\dot{\vec{r}}$ (with $\mu = \frac{m_1 m_2}{m_1+m_2}$), respectively. The Hamiltonian in terms of the variables $\vec{R}, \vec{r}, \vec{P}$ and $\vec{p}$ is given by

$$H = \frac{1}{2M}\vec{P}^{\,2} + \frac{1}{2\mu}\vec{p}^{\,2} - \frac{1}{4\pi\epsilon_0}\frac{q^2}{|\vec{r}|} - qE_0 z,$$

where $z = z_1 - z_2$.

(iii) With the spherical coordinates (r, θ, φ) used to represent the relative position $\vec{r} = (x, y, z)$, the related canonical momenta become $p_r = \mu\dot{r}$, $p_\theta = \mu r^2\dot{\theta}$ and $p_\varphi = \mu r^2 \sin^2\theta\dot{\varphi}$. Also, we have

$$\frac{1}{2\mu}\vec{p}^{\,2} = \frac{p_r^2}{2\mu} + \frac{p_\theta^2}{2\mu r^2} + \frac{p_\varphi^2}{2\mu r^2\sin^2\theta}$$

and hence the full Hamiltonian reads

$$H = \frac{1}{2M}\vec{P}^{\,2} + \frac{p_r^2}{2\mu} + \frac{p_\theta^2}{2\mu r^2} + \frac{p_\varphi^2}{2\mu r^2\sin^2\theta} - \frac{1}{4\pi\epsilon_0}\frac{q^2}{r} - qE_0 r\cos\theta.$$

(iv) As the coordinates $\vec{R}$ and φ are cyclic, we have the conservation laws

$$\vec{P} = m_1\dot{\vec{r}}_1 + m_2\dot{\vec{r}}_2 = \text{const.} \quad \left(\begin{array}{c}\text{: the total momentum} \\ \text{conservation}\end{array}\right),$$

$$p_\varphi = \mu r^2\sin^2\theta\dot{\varphi} = \text{const.} \quad \left(\begin{array}{l}\text{: the conservation of the} \\ \quad z\text{-component of the relative} \\ \quad \text{angular momentum}\end{array}\right).$$

Also, since the Hamiltonian has no explicit time dependence, we have the energy conservation law

$$H = E \quad (\text{: const.}).$$

13-4 Clearly, we have the conjugate momentum

$$\vec{p} = \frac{\partial L}{\partial \dot{\vec{r}}} = \frac{m_0 \dot{\vec{r}}}{\sqrt{1 - \dot{\vec{r}}^2/c^2}}$$

(which coincides with the relativistic mechanical momentum). We can invert the above equation in the form

$$\dot{\vec{r}} = \frac{c^2 \vec{p}}{\sqrt{m_0^2 c^4 + \vec{p}^2 c^2}}.$$

Hence the Hamiltonian is given by the expression

$$\begin{aligned} H &= [\dot{\vec{r}} \cdot \vec{p} - L]\Big|_{\dot{\vec{r}} = c^2 \vec{p}/\sqrt{m_0^2 c^4 + \vec{p}^2 c^2}} \\ &= \frac{c^2 \vec{p}^2}{\sqrt{m_0^2 c^4 + \vec{p}^2 c^2}} + m_0 c^2 \frac{m_0 c^2}{\sqrt{m_0^2 c^4 + \vec{p}^2 c^2}} + V(\vec{r}) \\ &= \sqrt{m_0^2 c^4 + \vec{p}^2 c^2} + V(\vec{r}), \end{aligned}$$

which has the same value as the relativistic energy (including the rest energy). Hamilton's equations are

$$\begin{aligned} \dot{\vec{r}} &= \frac{\partial H}{\partial \vec{p}} = \frac{c^2 \vec{p}}{\sqrt{m_0^2 c^4 + \vec{p}^2 c^2}}, \\ \dot{\vec{p}} &= -\frac{\partial H}{\partial \vec{r}} = -\vec{\nabla} V(\vec{r}). \end{aligned}$$

13-5 (i) From $\dot{\mathbf{q}} = \mathbf{A}^{-1}(\mathbf{q})\,\mathbf{p}$ and $\dot{\mathbf{q}}^{\mathrm{T}} = \mathbf{p}^{\mathrm{T}}\left(\mathbf{A}^{-1}(\mathbf{q})\right)^{\mathrm{T}} = \mathbf{p}^{\mathrm{T}}\left(\mathbf{A}^{\mathrm{T}}(\mathbf{q})\right)^{-1} = \mathbf{p}^{\mathrm{T}}\mathbf{A}^{-1}(\mathbf{q})$ (since $\mathbf{A}^{\mathrm{T}} = \mathbf{A}$), we obtain

$$\begin{aligned} H &= \sum_{i=1}^{n} p_i \dot{q}_i - L \\ &= \mathbf{p}^{\mathrm{T}} \mathbf{A}^{-1}(\mathbf{q})\,\mathbf{p} \\ &\quad - \left[\frac{1}{2}\mathbf{p}^{\mathrm{T}} \mathbf{A}^{-1}(\mathbf{q})\,\mathbf{A}(\mathbf{q})\mathbf{A}^{-1}(\mathbf{q})\,\mathbf{p} - U(\mathbf{q})\right] \\ &= \frac{1}{2}\mathbf{p}^{\mathrm{T}} \mathbf{A}^{-1}(\mathbf{q})\,\mathbf{p} + U(\mathbf{q}). \end{aligned}$$

(ii) With the identifications

$$\dot{\mathbf{q}} = \begin{pmatrix} \dot{\theta} \\ \dot{\varphi} \\ \dot{\psi} \end{pmatrix},$$

$$\mathbf{A} = \begin{pmatrix} I_1 \sin^2\psi + I_2 \cos^2\psi & (I_2 - I_1)\sin\theta\sin\psi\cos\psi & 0 \\ (I_2 - I_1)\sin\theta\sin\psi\cos\psi & \sin^2\theta(I_1\cos^2\psi + I_2\sin^2\psi) + I_3\cos^2\theta & I_3\cos\theta \\ 0 & I_3\cos\theta & I_3 \end{pmatrix},$$

the given Lagrangian assumes the form

$$L = \frac{1}{2}\mathbf{q}^{\mathrm{T}}\mathbf{A}\dot{\mathbf{q}} - MgR\cos\theta.$$

Here it is convenient to represent the above 3×3 matrix $\mathbf{A}$ by the form $\mathbf{A} = \mathbf{B}^{\mathrm{T}}\bar{\mathbf{I}}\,\mathbf{B}$, where

$$\mathbf{B} = \begin{pmatrix} -\sin\psi & \sin\theta\cos\psi & 0 \\ \cos\psi & \sin\theta\sin\psi & 0 \\ 0 & \cos\theta & 1 \end{pmatrix}, \quad \bar{\mathbf{I}} = \begin{pmatrix} I_1 & 0 & 0 \\ 0 & I_2 & 0 \\ 0 & 0 & I_3 \end{pmatrix}.$$

Then

$$\mathbf{B}^{-1} = \frac{1}{\sin\theta}\begin{pmatrix} -\sin\theta\sin\psi & \sin\theta\cos\psi & 0 \\ \cos\psi & \sin\psi & 0 \\ \cos\theta\cos\psi & \cos\theta\sin\psi & \sin\theta \end{pmatrix},$$

and so

$$\begin{aligned} \mathbf{A}^{-1} &= \mathbf{B}^{-1}\bar{\mathbf{I}}^{-1}\left(\mathbf{B}^{-1}\right)^{\mathrm{T}} \\ &= \frac{1}{\sin^2\theta}\begin{pmatrix} -\sin\theta\sin\psi & \sin\theta\cos\psi & 0 \\ \cos\psi & \sin\psi & 0 \\ \cos\theta\cos\psi & \cos\theta\sin\psi & \sin\theta \end{pmatrix}\begin{pmatrix} \frac{1}{I_1} & 0 & 0 \\ 0 & \frac{1}{I_2} & 0 \\ 0 & 0 & \frac{1}{I_3} \end{pmatrix} \\ &\quad \times \begin{pmatrix} -\sin\theta\sin\psi & \cos\psi & \cos\theta\cos\psi \\ \sin\theta\cos\psi & \sin\psi & \cos\theta\sin\psi \\ 0 & 0 & \sin\theta \end{pmatrix} \\ &= \frac{1}{\sin^2\theta}\times \end{aligned}$$

$$\begin{pmatrix} \sin^2\theta\left(\frac{\sin^2\psi}{I_1} + \frac{\cos^2\psi}{I_2}\right) & \frac{I_1 - I_2}{I_1 I_2}\sin\theta\sin\psi \times\cos\psi & \frac{I_1 - I_2}{I_1 I_2}\sin\theta\cos\theta \times\sin\psi\cos\psi \\ \frac{I_1 - I_2}{I_1 I_2}\sin\theta\sin\psi\cos\psi & \frac{\cos^2\psi}{I_1} + \frac{\sin^2\psi}{I_2} & \cos\theta\left(\frac{\cos^2\psi}{I_1} + \frac{\sin^2\psi}{I_2}\right) \\ \frac{I_1 - I_2}{I_1 I_2}\sin\theta\cos\theta \times\sin\psi\cos\psi & \cos\theta\left(\frac{\cos^2\psi}{I_1} + \frac{\sin^2\psi}{I_2}\right) & \cos^2\theta\left(\frac{\cos^2\psi}{I_1} + \frac{\sin^2\psi}{I_2}\right) + \frac{\sin^2\theta}{I_3} \end{pmatrix}.$$

Hence, we find the Hamiltonian

$$\begin{aligned}
H &= \frac{1}{2}\mathbf{p}^{\mathrm{T}}\mathbf{A}^{-1}\mathbf{p} + MgR\cos\theta \\
&= \frac{1}{2}p_\theta^2\left(\frac{\sin^2\psi}{I_1} + \frac{\cos^2\psi}{I_2}\right) + \frac{I_1 - I_2}{I_1 I_2}\frac{\sin\psi\cos\psi}{\sin\theta}p_\theta p_\varphi \\
&\quad + \frac{\cos\theta}{\sin\theta}\left(\frac{\cos^2\psi}{I_1} + \frac{\sin^2\psi}{I_2}\right)p_\theta p_\psi \\
&\quad + \frac{1}{2\sin^2\theta}p_\varphi^2\left(\frac{\cos^2\psi}{I_1} + \frac{\sin^2\psi}{I_2}\right) \\
&\quad + \frac{\cos\theta}{\sin^2\theta}\left(\frac{\cos^2\psi}{I_1} + \frac{\sin^2\psi}{I_2}\right)p_\varphi p_\psi \\
&\quad + \frac{\cos^2\theta}{2\sin^2\theta}\left(\frac{\cos^2\psi}{I_1} + \frac{\sin^2\psi}{I_2}\right)p_\psi^2 \\
&\quad + \frac{1}{2I_3}p_\psi^2 + MgR\cos\theta \\
&= \frac{1}{2}p_\theta^2\left(\frac{\sin^2\psi}{I_1} + \frac{\cos^2\psi}{I_2}\right) \\
&\quad + \frac{I_1 - I_2}{I_1 I_2}\frac{\sin\psi\cos\psi}{2\sin\theta}p_\theta(p_\varphi - p_\psi\cos\theta) \\
&\quad + \frac{1}{2\sin^2\theta}(p_\varphi - p_\psi\cos\theta)^2\left(\frac{\cos^2\psi}{I_1} + \frac{\sin^2\psi}{I_2}\right) \\
&\quad + \frac{1}{2I_3}p_\psi^2 + MgR\cos\theta,
\end{aligned}$$

where p_θ, p_φ, p_ψ are related to $\dot{\theta}$, $\dot{\varphi}$, $\dot{\psi}$ as (from $\mathbf{p} = \mathbf{A}(\mathbf{q})\dot{\mathbf{q}}$)

$$\begin{aligned}
p_\theta &= (I_1\sin^2\psi + I_2\cos^2\psi)\dot{\theta} - (I_1 - I_2)\sin\theta\sin\psi\cos\psi\dot{\varphi}, \\
p_\varphi &= (I_2 - I_1)\sin\theta\sin\psi\cos\psi\dot{\theta} \\
&\quad + \left[\sin^2\theta(I_1\cos^2\psi + I_2\sin^2\psi) + I_3\cos^2\theta\right]\dot{\varphi} + I_3\cos\theta\dot{\psi}, \\
p_\psi &= I_3\cos\theta\dot{\varphi} + I_3\dot{\psi}.
\end{aligned}$$

13-6 The Hamiltonian is given by

$$H = \frac{\vec{p}^2}{2m} - \vec{\omega}\cdot\vec{r}\times\vec{p} + V$$

with related Hamilton's equations of motion

$$\dot{\vec{r}} = \frac{\partial H}{\partial \vec{p}} = \frac{\vec{p}}{m} - \vec{\omega} \times \vec{r}$$
$$\dot{\vec{p}} = -\frac{\partial H}{\partial \vec{r}} = -\vec{\omega} \times \vec{p} - \vec{\nabla} V.$$

Use the relation $\vec{p} = m\dot{\vec{r}} + m\vec{\omega} \times \vec{r}$ (obtained from the first equation) in the second of Hamilton's equations, to obtain

$$\begin{aligned} m\ddot{\vec{r}} &= -m\vec{\omega} \times \dot{\vec{r}} - \vec{\omega} \times (m\dot{\vec{r}} + m\vec{\omega} \times \vec{r}) - \vec{\nabla} V \\ &= \underbrace{-2m\vec{\omega} \times \dot{\vec{r}}}_{\text{Coriolis force}} \underbrace{-m\vec{\omega} \times (\vec{\omega} \times \vec{r})}_{\text{centrifugal force}} - \vec{\nabla} V. \end{aligned}$$

(This agrees with the form in (9.17).)

13-7 To prove the assertion, let us calculate the related Jacobian

$$\begin{aligned} J &= \det\left(\frac{\partial \xi'_\alpha}{\partial \xi_\beta}\right) \\ &= \begin{vmatrix} \frac{\partial q'_1}{\partial q_1} & \cdots & \frac{\partial q'_1}{\partial q_n} & \cancel{\frac{\partial q'_1}{\partial p_1}} & \cdots & \cancel{\frac{\partial q'_1}{\partial p_n}} \\ \vdots & \vdots & \vdots & \vdots & \vdots & \vdots \\ \frac{\partial q'_n}{\partial q_1} & \cdots & \frac{\partial q'_n}{\partial q_n} & \cancel{\frac{\partial q'_n}{\partial p_1}} & \cdots & \cancel{\frac{\partial q'_n}{\partial p_n}} \\ \frac{\partial p'_1}{\partial q_1} & \cdots & \frac{\partial p'_1}{\partial q_n} & \frac{\partial p'_1}{\partial p_1} & \cdots & \frac{\partial p'_1}{\partial p_n} \\ \vdots & \vdots & \vdots & \vdots & \vdots & \vdots \\ \frac{\partial p'_n}{\partial q_1} & \cdots & \frac{\partial p'_n}{\partial q_n} & \frac{\partial p'_n}{\partial p_1} & \cdots & \frac{\partial p'_n}{\partial p_n} \end{vmatrix} \\ &= \det \mathbf{A} \cdot \det \mathbf{B} \end{aligned}$$

where $\mathbf{A}, \mathbf{B}$ denote two $n \times n$ matrices with $A_{ij} = \frac{\partial q'_i}{\partial q_j}$ and $B_{ij} = \frac{\partial p'_i}{\partial p_j} = \frac{\partial q_j}{\partial q'_i}$, respectively. Now observe that

$$J = \det \mathbf{A} \cdot \det \mathbf{B} = \det\left(\mathbf{A}\mathbf{B}^{\mathrm{T}}\right) = 1$$

since $\left(\mathbf{A}\mathbf{B}^{\mathrm{T}}\right)_{ij} = \sum_{k=1}^{n} A_{ik} B_{jk} = \sum_{k=1}^{n} \frac{\partial q'_i}{\partial q_k} \frac{\partial q_k}{\partial q'_j} = \delta_{ij}$. Hence the assertion follows.

13-8 (i) The initial phase space, defined by the conditions $p_1 \le p \le p_2$ and $E_1 \le \frac{1}{2m}p^2 + mgz \le E_2$, has the area given by (see the figure)

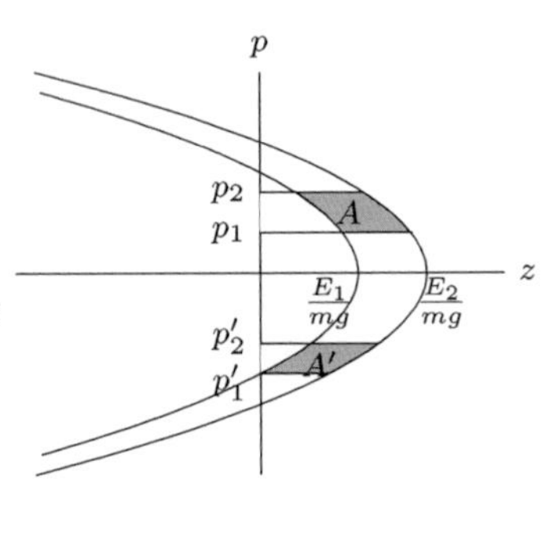

$$A = \int_{p_1}^{p_2} dp \int_{\frac{1}{mg}(E_1 - \frac{p^2}{2m})}^{\frac{1}{mg}(E_2 - \frac{p^2}{2m})} dz$$

$$= \frac{E_2 - E_1}{mg} \int_{p_1}^{p_2} dp$$

$$= \frac{E_2 - E_1}{mg}(p_2 - p_1).$$

(ii) Given initial values of p and z $(= \frac{1}{mg}(E - \frac{p^2}{2m}))$, the phase point $z' = z(t)$ and $p' = p(t)$ will be given by

$$p' = p - mgt, \quad z' = \frac{1}{mg}\left(E - \frac{p'^2}{2m}\right),$$

because of the conservation of energy. Hence the phase-space region at a later time t is determined by the conditions

$$p_1 - mgt \le p' \le p_2 - mgt,$$

$$E_1 \le E\left(= \frac{1}{2m}p'^2 + mgz'\right) \le E_2$$

and has the area (see the figure)

$$A' = \frac{E_2 - E_1}{mg}[(p_2 - mgt) - (p_1 - mgt)]$$

$$= \frac{E_2 - E_1}{mg}(p_2 - p_1) \quad (= A),$$

in agreement with Liouville's theorem.

13-9 (i) With the Boltzmann distribution of the form $\rho \propto e^{-\sum_{i=1}^{N} H^{(1)}(\vec{r}_n, \vec{p}_n)/k_BT} = \prod_{n=1}^{N} e^{-H^{(1)}(\vec{r}_n, \vec{p}_n)/k_BT}$, the probability that *any* given atom is to be found in the phase volume $(x, x+dx)$, $(y, y+dy)$, $(z, z+dz)$, (p_x, p_x+dp_x),

$(p_y, p_y + dp_y)$, $(p_z, p_z + dp_z)$ clearly should be given by

$$dP = \frac{e^{-H^{(1)}(\vec{r},\vec{p})/k_BT} d^3\vec{r}\, d^3\vec{p}}{\left[\int e^{-H^{(1)}(\vec{r},\vec{p})/k_BT} d^3\vec{r}\, d^3\vec{p}\right]} \qquad \left(\int dP = 1\right).$$

Then, in a gas of N atoms, the average number of atoms in the given phase cell, i.e., $\mathcal{n}(\vec{r}, \vec{p})\, d^3\vec{r}\, d^3\vec{p}$ can be equated to NdP; hence,

$$\mathcal{n}(\vec{r}, \vec{p}) = N \frac{e^{-H^{(1)}(\vec{r},\vec{p})/k_BT}}{\left[\int e^{-H^{(1)}(\vec{r},\vec{p})/k_BT} d^3\vec{r}\, d^3\vec{p}\right]}.$$

(ii) With the potential

$$V(\vec{r}) = \begin{cases} 0, & \text{for } \vec{r} \text{ in the range} \\ & (0 < x < a,\ 0 < y < b,\ 0 < z < c) \\ \infty, & \text{otherwise,} \end{cases}$$

the average number of atoms in the velocity range $(v_x, v_x + dv_x)$, $(v_y, v_y + dv_y)$, $(v_z, v_z + dv_z)$ should be given by (here $V = abc = \int e^{-V(\vec{r})/k_BT} d^3\vec{r}$)

$$\begin{aligned} &\left[\int \mathcal{n}(\vec{r}, \vec{p} = m\vec{v}) d^3\vec{r}\right] m^3 d^3\vec{v} \\ &\quad = N \frac{e^{-(\frac{1}{2}m\vec{v}^2)/k_BT} V m^3 d^3\vec{v}}{\left[V \int e^{-\vec{p}^2/2mk_BT} d^3\vec{p}\right]} \\ &\quad = N(2\pi m k_B T)^{-\frac{3}{2}} e^{-(\frac{1}{2}m\vec{v}^2)/k_BT} m^3 d^3\vec{v}, \end{aligned}$$

where we used $\int_{-\infty}^{\infty} e^{-\alpha x} dx = \sqrt{\frac{\pi}{\alpha}}$ $(\alpha > 0)$. This must coincide with $N\overline{\Phi}(\vec{v})\, d^3\vec{v}$ if $\overline{\Phi}(\vec{v})$ is the velocity-vector probability distribution. Hence,

$$\overline{\Phi}(\vec{v})\, d^3\vec{v} = \left(\frac{m}{2\pi k_B T}\right)^{3/2} e^{-(\frac{1}{2}m\vec{v}^2)/k_BT} d^3\vec{v}$$

which is the Maxwell–Boltzmann distribution.

(iii) With

$$V(\vec{r}) = \begin{cases} mgz, & \text{for } \vec{r} \text{ in the range} \\ & (0 < x < a,\, 0 < y < b,\, 0 < z < c) \\ \infty, & \text{otherwise,} \end{cases}$$

we find

$$\begin{aligned} &\int e^{-H^{(1)}(\vec{r},\vec{p})/k_BT} d^3\vec{r}\, d^3\vec{p} \\ &= ab \left[\int_0^c e^{-mgz/k_BT} dz\right] \left[\int e^{-\vec{p}^2/2mk_BT} d^3\vec{p}\right] \\ &= ab \left(\frac{k_BT}{mg}\right) \left[1 - e^{-mgc/k_BT}\right] \left(2\pi m k_BT\right)^{3/2}. \end{aligned}$$

Hence, for $\vec{r}$ in the range $(0 < x < a, 0 < y < b, 0 < z < c)$, we have

$$\begin{aligned} \mathcal{n}(\vec{r},\vec{p}) = \frac{N}{ab\left[1 - e^{-mgc/k_BT}\right]} \left(\frac{mg}{k_BT}\right) \left(2\pi m k_BT\right)^{-3/2} \\ \times e^{-(\frac{1}{2}m\vec{v}^2)/k_BT} e^{-mgz/k_BT}. \end{aligned}$$

According to this result, the distribution of velocities is the same at all height — it is Maxwellian, if the system is held at constant temperature T. But the gas becomes tenuous as we go up, i.e., the atomic number density $n(z)$ decreases with the height z exponentially. To understand this behavior, recall that an ideal gas satisfies the equation of state of the form $p = nk_BT$ where p is the (local) pressure. Here the pressure p should increase as we go down; if we apply the hydrostatic balance condition, $p = p(z)$ should satisfy the equation $\frac{dp}{dz} = -nmg$ (see Sec. 8.1). Inserting $p(z) = n(z)k_BT$ into this condition then yields $\frac{dn(z)}{dz} = -\frac{mg}{k_BT}n(z)$, which has the solution $n(z) = n_0 e^{-mgz/k_BT}$. This accounts for the z-dependence in $\mathcal{n}(\vec{r},\vec{p})$.

13-10 Based on the following representation of the Poisson bracket

$$[u,v] = -\left(J_{\alpha\beta}\frac{\partial u}{\partial \xi_\beta}\right)\frac{\partial}{\partial \xi_\alpha}v = \left(J_{\alpha\beta}\frac{\partial v}{\partial \xi_\beta}\right)\frac{\partial}{\partial \xi_\alpha}u,$$

it is a simple matter to verify the first relation:

$$
\begin{aligned}
[u, vw] &= -\left(J_{\alpha\beta}\frac{\partial u}{\partial \xi_\beta}\right)\frac{\partial}{\partial \xi_\alpha}(vw) \\
&= -J_{\alpha\beta}\frac{\partial u}{\partial \xi_\beta}\left(v\frac{\partial w}{\partial \xi_\alpha} + \frac{\partial v}{\partial \xi_\alpha}w\right) \\
&= v[u, w] + [u, v]w.
\end{aligned}
$$

On the other hand, for the second relation (which is the Jacobi identity), observe that we can write

$$
\begin{aligned}
&[u, [v, w]] + [v, [w, u]] + [w, [u, v]] \\
&\quad = J_{\alpha\beta}\frac{\partial u}{\partial \xi_\beta}\frac{\partial}{\partial \xi_\alpha}\left[\left(J_{\gamma\delta}\frac{\partial v}{\partial \xi_\delta}\right)\frac{\partial w}{\partial \xi_\gamma}\right] - J_{\alpha\beta}\frac{\partial v}{\partial \xi_\beta}\frac{\partial}{\partial \xi_\alpha}\left[\left(J_{\gamma\delta}\frac{\partial u}{\partial \xi_\delta}\right)\frac{\partial w}{\partial \xi_\gamma}\right] \\
&\qquad - \left(J_{\gamma\delta}\frac{\partial}{\partial \xi_\delta}\left[J_{\alpha\beta}\frac{\partial u}{\partial \xi_\beta}\frac{\partial v}{\partial \xi_\alpha}\right]\right)\frac{\partial w}{\partial \xi_\gamma} \\
&\quad = \left\{\cancel{J_{\alpha\beta}J_{\gamma\delta}\frac{\partial u}{\partial \xi_\beta}\frac{\partial^2 v}{\partial \xi_\alpha \partial \xi_\delta}} - \cancel{J_{\alpha\beta}J_{\gamma\delta}\frac{\partial v}{\partial \xi_\beta}\frac{\partial^2 u}{\partial \xi_\alpha \partial \xi_\delta}}\right. \\
&\qquad \left. - \cancel{J_{\gamma\delta}J_{\alpha\beta}\frac{\partial^2 u}{\partial \xi_\delta \partial \xi_\beta}\frac{\partial v}{\partial \xi_\alpha}} - \cancel{J_{\gamma\delta}J_{\alpha\beta}\frac{\partial u}{\partial \xi_\beta}\frac{\partial^2 v}{\partial \xi_\delta \partial \xi_\alpha}}\right\}\frac{\partial w}{\partial \xi_\gamma} \\
&\qquad + \left\{J_{\alpha\beta}J_{\gamma\delta}\frac{\partial u}{\partial \xi_\beta}\frac{\partial v}{\partial \xi_\delta} - J_{\alpha\beta}J_{\gamma\delta}\frac{\partial v}{\partial \xi_\beta}\frac{\partial u}{\partial \xi_\delta}\right\}\frac{\partial^2 w}{\partial \xi_\alpha \partial \xi_\gamma} \\
&\quad = 0.
\end{aligned}
$$

(Note that, in the last term, the expression inside the curly bracket is antisymmetric under the exchange $\alpha \leftrightarrow \gamma$; hence, upon multiplying the factor $\frac{\partial^2 w}{\partial \xi_\alpha \partial \xi_\gamma}$ which is symmetric under the exchange $\alpha \leftrightarrow \gamma$, we get zero.)

13-11 First, we may differentiate $p_i = \frac{\partial F_2(\mathbf{q}, \mathbf{P}, t)}{\partial q_i}$ with respect to q_j (for fixed $\mathbf{p}$) or with respect to p_j (for fixed $\mathbf{q}$) to obtain the relations

$$
0 = \frac{\partial^2 F_2}{\partial q_j \partial q_i} + \left.\frac{\partial P_k}{\partial q_j}\right|_{\mathbf{p}} \frac{\partial^2 F_2}{\partial P_k \partial q_i} = A_{ji} + \left.\frac{\partial P_k}{\partial q_j}\right|_{\mathbf{p}} B_{ki},
$$

$$
\delta_{ij} = \left.\frac{\partial P_k}{\partial p_j}\right|_{\mathbf{q}} \frac{\partial^2 F_2}{\partial P_k \partial q_i} = \left.\frac{\partial P_k}{\partial p_j}\right|_{\mathbf{q}} B_{ki}
$$

(see the hint for the definitions of the matrices $\mathbf{A}$ and $\mathbf{B}$), from which we conclude that

$$\left.\frac{\partial P_k}{\partial q_j}\right|_{\mathbf{p}} = -(\mathbf{A}\mathbf{B}^{-1})_{jk}, \qquad \left.\frac{\partial P_k}{\partial p_j}\right|_{\mathbf{q}} = (\mathbf{B}^{-1})_{jk}.$$

Similarly, by differentiating $Q_i = \frac{\partial F_2(\mathbf{q},\mathbf{P},t)}{\partial P_i}$ with respect to q_j or p_j, we have

$$\left.\frac{\partial Q_i}{\partial q_j}\right|_{\mathbf{p}} = \frac{\partial^2 F_2}{\partial q_j\,\partial P_i} + \left.\frac{\partial P_k}{\partial q_j}\right|_{\mathbf{p}} \frac{\partial^2 F_2}{\partial P_k\,\partial P_i} = B_{ij} - (\mathbf{A}\mathbf{B}^{-1}\mathbf{C})_{ji},$$

$$\left.\frac{\partial Q_i}{\partial p_j}\right|_{\mathbf{q}} = \left.\frac{\partial P_k}{\partial p_j}\right|_{\mathbf{q}} \frac{\partial^2 F_2}{\partial P_k\,\partial P_i} = (\mathbf{B}^{-1}\mathbf{C})_{ji}.$$

Using these results, we then find (note that $A_{kl} = A_{lk}$)

$$\begin{aligned}
[P_i, P_j]_{\mathbf{q},\mathbf{p}} &= \sum_k \left(\frac{\partial P_i}{\partial q_k}\frac{\partial P_j}{\partial p_k} - \frac{\partial P_i}{\partial p_k}\frac{\partial P_j}{\partial q_k}\right) \\
&= \sum_k \left(-(\mathbf{A}\mathbf{B}^{-1})_{ki}(\mathbf{B}^{-1})_{kj} + (\mathbf{B}^{-1})_{ki}(\mathbf{A}\mathbf{B}^{-1})_{kj}\right) \\
&= -\sum_{k,l} A_{kl}\left[(\mathbf{B}^{-1})_{li}(\mathbf{B}^{-1})_{kj} - (\mathbf{B}^{-1})_{ki}(\mathbf{B}^{-1})_{lj}\right] = 0, \\
[Q_i, Q_j]_{\mathbf{q},\mathbf{p}} &= \sum_k \left(\frac{\partial Q_i}{\partial q_k}\frac{\partial Q_j}{\partial p_k} - \frac{\partial Q_i}{\partial p_k}\frac{\partial Q_j}{\partial q_k}\right) \\
&= \sum_k \Big(\left[B_{ik} - (\mathbf{A}\mathbf{B}^{-1}\mathbf{C})_{ki}\right](\mathbf{B}^{-1}\mathbf{C})_{kj} \\
&\qquad - (\mathbf{B}^{-1}\mathbf{C})_{ki}\left[B_{jk} - (\mathbf{A}\mathbf{B}^{-1}\mathbf{C})_{kj}\right]\Big) \\
&= -\sum_{k,l} A_{kl}\left[(\mathbf{B}^{-1}\mathbf{C})_{li}(\mathbf{B}^{-1}\mathbf{C})_{kj} - (\mathbf{B}^{-1}\mathbf{C})_{ki}(\mathbf{B}^{-1}\mathbf{C})_{lj}\right] \\
&= 0, \\
[Q_i, P_j]_{\mathbf{q},\mathbf{p}} &= \sum_k \left(\frac{\partial Q_i}{\partial q_k}\frac{\partial P_j}{\partial p_k} - \frac{\partial Q_i}{\partial p_k}\frac{\partial P_j}{\partial q_k}\right)
\end{aligned}$$

$$= \sum_k \Big(\Big[B_{ik} - \left(\mathbf{A}\mathbf{B}^{-1}\mathbf{C}\right)_{ki} \Big] \left(\mathbf{B}^{-1}\right)_{kj}$$
$$+ \left(\mathbf{B}^{-1}\mathbf{C}\right)_{ki} \left(\mathbf{A}\mathbf{B}^{-1}\right)_{kj} \Big)$$
$$= \delta_{ij} - \sum_{k,l} A_{kl} \Big[\left(\mathbf{B}^{-1}\mathbf{C}\right)_{li} \left(\mathbf{B}^{-1}\right)_{kj} - \left(\mathbf{B}^{-1}\mathbf{C}\right)_{ki} \left(\mathbf{B}^{-1}\right)_{lj} \Big]$$
$$= \delta_{ij},$$

i.e., the fundamental Poisson-bracket relations are preserved. Hence the transformation $(\mathbf{q}, \mathbf{p}) \to (\mathbf{Q}, \mathbf{P})$ is canonical.

13-12 We may begin with the equations describing the time evolution of $P_i = P_i(\mathbf{q}, \mathbf{p}, t)$ and $Q_i = Q_i(\mathbf{q}, \mathbf{p}, t)$, i.e.,

$$\dot{P}_i = [P_i, H]_{\mathbf{q},\mathbf{p}} + \left.\frac{\partial P_i}{\partial t}\right|_{\mathbf{q},\mathbf{p}},$$
$$\dot{Q}_i = [Q_i, H]_{\mathbf{q},\mathbf{p}} + \left.\frac{\partial Q_i}{\partial t}\right|_{\mathbf{q},\mathbf{p}}. \quad (1)$$

Here, the transformation $(\mathbf{q}, \mathbf{p}) \to (\mathbf{Q}, \mathbf{P})$ being canonical, we have $[P_i, H]_{\mathbf{q},\mathbf{p}} = [P_i, H]_{\mathbf{Q},\mathbf{P}} = -\frac{\partial H}{\partial Q_i}$ and $[Q_i, H]_{\mathbf{q},\mathbf{p}} = [Q_i, H]_{\mathbf{Q},\mathbf{P}} = \frac{\partial H}{\partial P_i}$. To rewrite the term $\frac{\partial P_i}{\partial t}|_{\mathbf{q},\mathbf{p}}$ as an expression involving the generating function, consider taking the partial derivative $\frac{\partial}{\partial t}$ (for fixed $\mathbf{q}, \mathbf{p}$) with the equation $p_i = \frac{\partial F_2(\mathbf{q}, \mathbf{P}(\mathbf{q},\mathbf{p},t), t)}{\partial q_i}$; then, writing $\frac{\partial^2 F_2}{\partial P_j \partial q_i} \equiv B_{ji}$ (see Problem 13-11),

$$0 = \frac{\partial^2 F_2}{\partial t \partial q_i} + \left.\frac{\partial P_j}{\partial t}\right|_{\mathbf{q},\mathbf{p}} B_{ji}$$
$$\longrightarrow \quad \left.\frac{\partial P_j}{\partial t}\right|_{\mathbf{q},\mathbf{p}} = -\frac{\partial^2 F_2}{\partial t \partial q_i} (\mathbf{B}^{-1})_{ij} = -\frac{\partial^2 F_2}{\partial t \partial q_i} \frac{\partial P_j}{\partial p_i}, \quad (2)$$

since $\frac{\partial P_j}{\partial p_i} = (\mathbf{B}^{-1})_{ij}$. As for the term $\left.\frac{\partial Q_i}{\partial t}\right|_{\mathbf{q},\mathbf{p}}$, we use the equation $Q_i = \frac{\partial}{\partial P_i} F_2(\mathbf{q}, \mathbf{P}(\mathbf{q}, \mathbf{p}, t), t)$ to obtain (here, $\frac{\partial^2 F_2}{\partial P_j \partial P_i} \equiv C_{ji}$)

$$\left.\frac{\partial Q_i}{\partial t}\right|_{\mathbf{q},\mathbf{p}} = \frac{\partial^2 F_2}{\partial t \partial P_i} + \left.\frac{\partial P_j}{\partial t}\right|_{\mathbf{q},\mathbf{p}} C_{ji}$$
$$= \frac{\partial^2 F_2}{\partial t \partial P_i} - \frac{\partial^2 F_2}{\partial t \partial q_j} (\mathbf{B}^{-1}\mathbf{C})_{ji} = \frac{\partial^2 F_2}{\partial t \partial P_i} - \frac{\partial^2 F_2}{\partial t \partial q_j} \frac{\partial Q_i}{\partial p_j}, \quad (3)$$

where we made use of (2) and $\frac{\partial Q_i}{\partial p_j} = \left(\mathbf{B}^{-1}\mathbf{C}\right)_{ji}$. We then observe that

$$\begin{aligned}
\left[P_j, \frac{\partial F_2}{\partial t}\right]_{\mathbf{q},\mathbf{p}} &= \frac{\partial P_j}{\partial q_i} \left.\frac{\partial}{\partial p_i}\right|_{\mathbf{q},t} \frac{\partial}{\partial t} F_2(\mathbf{q}, \mathbf{P}(\mathbf{q}, \mathbf{p}, t), t) \\
&\quad - \frac{\partial P_j}{\partial p_i} \left.\frac{\partial}{\partial q_i}\right|_{\mathbf{p},t} \frac{\partial}{\partial t} F_2(\mathbf{q}, \mathbf{P}(\mathbf{q}, \mathbf{p}, t), t) \\
&= \frac{\partial P_j}{\partial q_i} \frac{\partial P_k}{\partial p_i} \frac{\partial^2 F_2}{\partial P_k\, \partial t} - \frac{\partial P_j}{\partial p_i} \left(\frac{\partial^2 F_2}{\partial q_i\, \partial t} + \frac{\partial P_k}{\partial q_i} \frac{\partial^2 F_2}{\partial P_k\, \partial t} \right) \\
&= -\frac{\partial P_j}{\partial p_i} \frac{\partial^2 F_2}{\partial q_i\, \partial t} + \cancel{[P_j, P_k]_{\mathbf{q},\mathbf{p}}} \frac{\partial^2 F_2}{\partial P_k\, \partial t} \\
&= \left.\frac{\partial P_j}{\partial t}\right|_{\mathbf{q},\mathbf{p}}
\end{aligned} \tag{4}$$

with the last identity following from $[P_j, P_k]_{\mathbf{q},\mathbf{p}} = 0$ and (2); similarly,

$$\begin{aligned}
\left[Q_i, \frac{\partial F_2}{\partial t}\right]_{\mathbf{q},\mathbf{p}} &= \frac{\partial Q_i}{\partial q_j} \left.\frac{\partial}{\partial p_j}\right|_{\mathbf{q},t} \frac{\partial}{\partial t} F_2(\mathbf{q}, \mathbf{P}(\mathbf{q}, \mathbf{p}, t), t) \\
&\quad - \frac{\partial Q_i}{\partial p_j} \left.\frac{\partial}{\partial q_j}\right|_{\mathbf{p},t} \frac{\partial}{\partial t} F_2(\mathbf{q}, \mathbf{P}(\mathbf{q}, \mathbf{p}, t), t) \\
&= \frac{\partial Q_i}{\partial q_j} \frac{\partial P_k}{\partial p_j} \frac{\partial^2 F_2}{\partial P_k\, \partial t} - \frac{\partial Q_i}{\partial p_j} \left(\frac{\partial^2 F_2}{\partial q_j\, \partial t} + \frac{\partial P_k}{\partial q_j} \frac{\partial^2 F_2}{\partial P_k\, \partial t} \right) \\
&= -\frac{\partial Q_i}{\partial p_j} \frac{\partial^2 F_2}{\partial q_j\, \partial t} + [Q_i, P_k]_{\mathbf{q},\mathbf{p}} \frac{\partial^2 F_2}{\partial P_k\, \partial t} \\
&= \left.\frac{\partial Q_i}{\partial t}\right|_{\mathbf{q},\mathbf{p}}
\end{aligned} \tag{5}$$

with the last identity following from $[Q_i, P_k]_{\mathbf{q},\mathbf{p}} = \delta_{ik}$ and (3). Therefore, it is possible to write

$$\dot{P}_i = \left[P_i, H + \frac{\partial F_2}{\partial t}\right]_{\mathbf{q},\mathbf{p}}, \quad \dot{Q}_i = \left[Q_i, H + \frac{\partial F_2}{\partial t}\right]_{\mathbf{q},\mathbf{p}},$$

or, defining $H' = H + \frac{\partial F_2}{\partial t}$,

$$\dot{P}_i = [P_i, H']_{\mathbf{q},\mathbf{p}} = [P_i, H']_{\mathbf{Q},\mathbf{P}} = -\frac{\partial H'}{\partial Q_i},$$
$$\dot{Q}_i = [Q_i, H']_{\mathbf{q},\mathbf{p}} = [Q_i, H']_{\mathbf{Q},\mathbf{P}} = \frac{\partial H'}{\partial P_i},$$

i.e., the standard Hamilton's equations of motion are secured.

13-13 For Q and P given by

$$Q = q\cos\alpha - p\sin\alpha, \qquad P = q\sin\alpha + p\cos\alpha,$$

we find

$$[Q, P]_{(q,p)} \equiv \frac{\partial Q}{\partial q}\frac{\partial P}{\partial p} - \frac{\partial Q}{\partial p}\frac{\partial P}{\partial q}$$
$$= \cos^2\alpha + \sin^2\alpha = 1,$$

and hence the above defines a canonical transformation. Also the above transformation implies that, for p and P,

$$p = \frac{1}{\sin\alpha}(q\cos\alpha - Q)$$
$$P = q\sin\alpha + \frac{\cos\alpha}{\sin\alpha}(q\cos\alpha - Q) = \frac{1}{\sin\alpha}(q - Q\cos\alpha),$$

while, for the function $F_1 = \frac{1}{2\sin\alpha}(q^2\cos\alpha - 2qQ + Q^2\cos\alpha)$, we find

$$\frac{\partial F_1}{\partial q} = \frac{1}{\sin\alpha}(q\cos\alpha - Q) = p,$$
$$-\frac{\partial F_1}{\partial Q} = \frac{1}{\sin\alpha}(q - Q\cos\alpha) = P.$$

Hence, F_1 is the appropriate generating function.

13-14 From the formulas

$$Q = \frac{\partial F_2}{\partial P} = \frac{m\omega^2 q^2}{2}\sec^2[\omega(t - P)],$$
$$p = \frac{\partial F_2}{\partial q} = -m\omega q\tan[\omega(t - P)],$$

we obtain

$$P = \frac{1}{\omega}\tan^{-1}\left(\frac{p}{m\omega q}\right) + t,$$
$$Q = \frac{p^2}{2m} + \frac{m}{2}\omega^2 q^2.$$

The Hamiltonian using new canonical variables is given by

$$H' = H + \frac{\partial F_2}{\partial t} = \frac{1}{2m}p^2 + \frac{1}{2}m\omega^2 q^2 - \frac{m\omega^2 q^2}{2}\sec^2[\omega(t-P)],$$
$$= \frac{1}{2m}p^2 + \frac{1}{2}m\omega^2 q^2 - \left(\frac{1}{2m}p^2 + \frac{m}{2}\omega^2 q^2\right)$$
$$= 0.$$

Hence, Hamilton's equations in new variables read

$$\dot{Q} = \frac{\partial H'}{\partial P} = 0 \quad \rightarrow \quad Q\left(= \frac{p^2}{2m} + \frac{m}{2}\omega^2 q^2\right) = \text{const.},$$
$$\dot{P} = -\frac{\partial H'}{\partial Q} = 0 \quad \rightarrow \quad P\left(= \frac{1}{\omega}\tan^{-1}\left(\frac{p}{m\omega q}\right) + t\right) = \text{const.}$$

13-15 From $G_3 = [G_1, G_2]$, we here have the equation

$$\frac{d}{dt}G_3 = [[G_1, G_2], H] + \frac{\partial}{\partial t}[G_1, G_2].$$

Now, use the Jacobi identity for Poisson brackets

$$[[G_1, G_2], H] = -[[G_2, H], G_1] + [[G_1, H], G_2]$$

and the obvious formula

$$\frac{\partial}{\partial t}[G_1, G_2] = \left[\frac{\partial G_1}{\partial t}, G_2\right] + \left[G_1, \frac{\partial G_2}{\partial t}\right]$$

in the right hand side of the above equation. We then obtain the desired relation

$$\frac{d}{dt}G_3 = \left[[G_1, H] + \frac{\partial G_1}{\partial t}, G_2\right] + \left[G_1, [G_2, H] + \frac{\partial G_2}{\partial t}\right]$$
$$= \left[\frac{dG_1}{dt}, G_2\right] + \left[G_1, \frac{dG_2}{dt}\right].$$

13-16 (i) Suppose that the equations of motion for the given system read

$$\frac{\partial H(\mathbf{q}, \mathbf{p}, t)}{\partial p_i} = \dot{q}_i, \qquad -\frac{\partial H(\mathbf{q}, \mathbf{p}, t)}{\partial q_i} = \dot{p}_i$$

in terms of canonical variables $(\mathbf{q}, \mathbf{p})$. Then consider making an infinitesimal canonical transformation $(\mathbf{q}, \mathbf{p}) \longrightarrow (\mathbf{Q} = \mathbf{q} + \delta\mathbf{q}, \mathbf{P} = \mathbf{p} + \delta\mathbf{p})$ with the help of the generating function $F_2(\mathbf{q}, \mathbf{P}, t) = \sum_i q_i P_i + \delta\lambda\, G(\mathbf{q}, \mathbf{P}, t)$, so that we have $\delta q_i = \frac{\partial G(\mathbf{q},\mathbf{p},t)}{\partial p_i}\delta\lambda$ and $\delta p_i = -\frac{\partial G(\mathbf{q},\mathbf{p},t)}{\partial q_i}\delta\lambda$, to first order in $\delta\lambda$.

In terms of these new canonical variables, the above equations of motion can be rewritten, to first order in $\delta\lambda$, in the form

$$\frac{\partial H'}{\partial P_i} = \dot{Q}_i, \qquad -\frac{\partial H'}{\partial Q_i} = \dot{P}_i,$$

where

$$\begin{aligned} H' &= H(\mathbf{q}, \mathbf{p}, t) + \frac{\partial F_2(\mathbf{q}, \mathbf{P}, t)}{\partial t} \\ &= H(\mathbf{q}, \mathbf{p}, t) + \frac{\partial G(\mathbf{q}, \mathbf{p}, t)}{\partial t}\delta\lambda + O\big((\delta\lambda)^2\big). \end{aligned} \tag{1}$$

The considered canonical transformation will be a symmetry of equations of motion only when we have, to first order in $\delta\lambda$,

$$H' = H(\mathbf{q}, \mathbf{p}, t)\Big|_{\mathbf{q}\to\mathbf{Q},\, \mathbf{p}\to\mathbf{P}} = H(\mathbf{q} + \delta\mathbf{q}, \mathbf{p} + \delta\mathbf{p}, t). \tag{2}$$

From (1) and (2), we may thus write the condition to be a symmetry of equations of motion by the form

$$\begin{aligned} \delta H &\equiv H(\mathbf{q} + \delta\mathbf{q}, \mathbf{p} + \delta\mathbf{p}, t) - H(\mathbf{q}, \mathbf{p}, t) \\ &= \frac{\partial G(\mathbf{q}, \mathbf{p}, t)}{\partial t}\delta\lambda + O\big((\delta\lambda)^2\big). \end{aligned} \tag{3}$$

(ii) Since we can write (see Sec. 13.4)

$$H(\mathbf{q} + \delta\mathbf{q}, \mathbf{p} + \delta\mathbf{p}, t) - H(\mathbf{q}, \mathbf{p}, t) = [H, G]\delta\lambda,$$

the condition (3) is equivalent to

$$[H, G] = \frac{\partial G(\mathbf{q}, \mathbf{p}, t)}{\partial t}.$$

This implies that, from Sec. 13.3,

$$\frac{dG(\mathbf{q}(t), \mathbf{p}(t), t)}{dt} = [G, H] + \frac{\partial G(\mathbf{q}, \mathbf{p}, t)}{\partial t} = 0,$$

i.e., $G(\mathbf{q}, \mathbf{p}, t)$ is conserved.

(iii) For the infinitesimal Galilean transformations

$$\delta\vec{r}_i = t\hat{u}\delta\lambda, \quad \delta\vec{p}_i = m_i\hat{u}\delta\lambda \quad (i = 1, \ldots, N), \tag{5}$$

we can take the generator

$$G = \sum_i \vec{p}_i \cdot \hat{u}t - \sum_i m_i \vec{r}_i \cdot \hat{u} = \vec{P} \cdot \hat{u}t - M\vec{R} \cdot \hat{u},$$

where $\vec{P} \equiv \sum_i \vec{p}_i$ is the total momentum, and $\vec{R} = \frac{1}{M}\sum_i m_i \vec{r}_i$ the center of mass. Since $\hat{u}$ can assume any direction, we can thus say that the vector

$$\vec{G} = \vec{P}t - M\vec{R} \tag{6}$$

is a constant of motion. Assuming that the total momentum is conserved, this gives simply $\vec{P} = M\dot{\vec{R}}$, the relation connecting the total momentum to center-of-mass velocity.

Under the above transformations, the change in the Hamiltonian should be given by (see (3) above)

$$\delta H = \hat{u} \cdot \frac{\partial \vec{G}}{\partial t} \delta\lambda = \hat{u} \cdot \vec{P} \delta\lambda. \tag{7}$$

If we write $H = T + V$ where $T = \sum_{i=1}^{N} \frac{\vec{p}_i^{\,2}}{2m_i}$, we find from (5) that $\delta T = \sum_i \vec{p}_i \cdot \hat{u}\delta\lambda = \vec{P} \cdot \hat{u}\delta\lambda$. Thus the change in the kinetic energy is exactly what is demanded by (7), and so this relation reduces to $\delta V = 0$. If V is a scalar function involving the particle positions only through the differences $\vec{r}_{ij} = \vec{r}_i - \vec{r}_j$, the requirement $\delta V = 0$ imposes no further restriction since $\delta\vec{r}_{ij} = 0$ under the transformation (5). But, if V is allowed to contain terms depending on momenta, it must contain them only through the combinations $\frac{\vec{p}_i}{m_i} - \frac{\vec{p}_j}{m_j}$ $(= \vec{v}_{ij})$, to satisfy the condition $\delta V = 0$. (Thus the most general form of interaction in an N-particle system which is invariant under the Galilean transformation may be described by a scalar potential energy function of the relative position vectors and the relative velocity vectors.)

13-17 Since $\vec{r}(t) \cdot \vec{L} = \vec{M} \cdot \vec{L} = 0$, the motion takes place in a plane perpendicular to $\vec{L}$ in which both $\vec{r}(t)$ and $\vec{M}$ lie. Then, choosing an appropriate coordinate system in the plane of motion, we may write

$$\vec{M} = |\vec{M}|(\cos\theta_0 \mathbf{i} + \sin\theta_0 \mathbf{j})$$

and

$$\vec{r}(t) = r(t)(\cos\theta(t)\,\mathbf{i} + \sin\theta(t)\,\mathbf{j}).$$

Now we find

$$\begin{aligned}\vec{r}\cdot\vec{M} &= |\vec{M}|r\cos(\theta-\theta_0)\\ &= \frac{1}{m}\vec{L}^2 - \kappa r,\end{aligned}$$

or, for $r = r(\theta)$,

$$\frac{1}{r(\theta)} = \frac{\kappa m}{\vec{L}^2}\left[1 + \frac{|\vec{M}|}{\kappa}\cos(\theta-\theta_0)\right].$$

Note that

$$|\vec{M}| = \sqrt{\frac{2}{m}E\vec{L}^2 + \kappa^2} = \kappa e,$$

where $E = \frac{m}{2}\dot{\vec{r}}^2 - \frac{\kappa}{r}$ is the total energy, and $e = \sqrt{1 + \frac{2EL^2}{m\kappa^2}}$ the eccentricity of the orbit. Hence this agrees with the orbit given in Sec. 4.4.

The nature of the orbit, of course, depends on E. If $E < 0$, then the orbit is closed and is an ellipse or a circle. If $E \geq 0$, the orbit is open and is a parabola or a hyperbola.

13-18 (i) From the Poisson brackets

$$\begin{aligned}\left[\frac{1}{2m}p_kp_k, A_{ij}\right] &= -\frac{1}{4m^2}\frac{\partial(p_kp_k)}{\partial p_l}\frac{\partial}{\partial x_l}(\cancel{p_ip_j} + m^2\omega^2x_ix_j)\\ &= -\frac{\omega^2}{2}\delta_{kl}p_k(\delta_{li}x_j + \delta_{lj}x_i)\\ &= -\frac{\omega^2}{2}(p_ix_j + p_jx_i),\\ \left[\frac{m}{2}\omega^2x_kx_k, A_{ij}\right] &= \frac{\omega^2}{4}\frac{\partial(x_kx_k)}{\partial x_l}\frac{\partial}{\partial p_l}(p_ip_j + \cancel{m^2\omega^2x_ix_j})\\ &= \frac{\omega^2}{2}\delta_{kl}x_k(\delta_{li}p_j + \delta_{lj}p_i) = \frac{\omega^2}{2}(p_jx_i + p_ix_j),\end{aligned}$$

we have that

$$[H, A_{ij}] = 0,$$

i.e., A_{ij} $(i, j = 1, 2)$ are conserved.

(ii) The orbit equation $f(x, y) = 0$ can be found if one inserts the relations $p_x^2 = 2mA_{11} - m^2\omega^2 x^2$ and $p_y^2 = 2mA_{22} - m^2\omega^2 y^2$ into $p_x^2 p_y^2 = (2mA_{12} - m^2\omega^2 xy)^2$. The result is

$$A_{22}x^2 - 2A_{12}xy + A_{11}y^2 = \frac{2}{m\omega^2}(A_{11}A_{22} - A_{12}^2), \qquad (1)$$

where $A_{11}A_{22} - A_{12}^2 = \frac{1}{4}\omega^2(xp_y - yp_x)^2 = \frac{1}{4}\omega^2 L^2 > 0$; this corresponds to a closed orbit for any initial conditions — an ellipse having the center at $x = y = 0$. [Note that, introducing new coordinates $\bar{x}$, $\bar{y}$ defined relative to appropriately rotated axes, the left hand side of (1) can be changed to a form $C\bar{x}^2 + D\bar{y}^2$ with some positive constants C and D.] Also, upon writing $x = r\cos\theta$ and $y = r\sin\theta$, the orbit equation (1) can be written in the form given in Sec. 4.3; this equation shows that a unique r value corresponds to every polar angle θ modulo $2n\pi$ ($n = 0, 1, 2, \ldots$), as should be the case for a closed orbit.

(iii) The given Poisson bracket relations are obtained as

$$\begin{aligned}
[S_1, S_2] &= \frac{1}{8m^2\omega^2}[p_x p_y + m^2\omega^2 xy,\, p_y^2 - p_x^2 + m^2\omega^2(y^2 - x^2)] \\
&= \frac{1}{8m^2\omega^2}\{2m^2\omega^2(xp_y - yp_x) + 2m^2\omega^2(-p_x y + p_y x)\} \\
&= \frac{1}{2}(xp_y - yp_x) = S_3, \\
[S_3, S_1] &= \frac{1}{4m\omega}[xp_y - yp_x,\, p_x p_y + m^2\omega^2 xy] \\
&= \frac{1}{4m\omega}\{p_y^2 - m^2\omega^2 x^2 - p_x^2 + m^2\omega^2 y^2\} \\
&= S_2,
\end{aligned}$$

etc.

13-19 The relation $[L_i, M_j] = \epsilon_{ijk}M_k$ (i.e., $\vec{M}$ is a vector observable as discussed in Sec. 13.4.6 in the text) can be shown as follows. From $[L_i, x_j] = \epsilon_{ijk}x_k$, $[L_i, p_j] = \epsilon_{ijk}p_k$ and $p_i L_i = 0$,

$$\begin{aligned}
[L_i, M_j] &= \left[L_i, \frac{1}{m}\epsilon_{jkl}p_k L_l - \frac{\kappa}{r}x_j\right] \\
&= \frac{1}{m}\epsilon_{jkl}\{[L_i, p_k]L_l + p_k[L_i, L_l]\} - \frac{\kappa}{r}[L_i, x_j] \\
&= \frac{1}{m}\epsilon_{jkl}\{\epsilon_{ikm}p_m L_l + p_k\epsilon_{ilm}L_m\} - \frac{\kappa}{r}\epsilon_{ijk}x_k
\end{aligned}$$

$$
\begin{aligned}
&= \frac{1}{m}\{(\cancel{\delta_{ji}\delta_{lm}} - \delta_{jm}\delta_{li})p_m L_l + (\cancel{-\delta_{ji}\delta_{km}} + \delta_{jm}\delta_{ki})p_k L_m\} \\
&\quad - \frac{\kappa}{r}\epsilon_{ijk}x_k \\
&= \frac{1}{m}\epsilon_{ijk}\epsilon_{klm}p_l L_m - \frac{\kappa}{r}\epsilon_{ijk}x_k \\
&= \epsilon_{ijk}M_k.
\end{aligned}
$$

For the other relation, observe that

$$
\begin{aligned}
[M_i, M_j] &= \left[\frac{1}{m}\epsilon_{ikl}p_k L_l - \frac{\kappa}{r}x_i, \frac{1}{m}\epsilon_{jmn}p_m L_n - \frac{\kappa}{r}x_j\right] \\
&= \frac{1}{m^2}\epsilon_{ikl}\epsilon_{jmn}\{[p_k, p_m L_n]L_l + p_k[L_l, p_m L_n]\} \\
&\quad - \frac{\kappa}{m}\epsilon_{ikl}\left\{p_k\left[L_l, \frac{1}{r}x_j\right] + \left[p_k, \frac{x_j}{r}\right]L_l\right\} \\
&\quad + \frac{\kappa}{m}\epsilon_{jkl}\left\{p_k\left[L_l, \frac{1}{r}x_i\right] + \left[p_k, \frac{x_i}{r}\right]L_l\right\} \\
&= \frac{1}{m^2}\epsilon_{ikl}\epsilon_{jmn}\Big(-p_m\epsilon_{nk\alpha}p_\alpha L_l + p_k p_m\epsilon_{ln\alpha}L_\alpha \\
&\quad + p_k\epsilon_{lm\alpha}p_\alpha L_n\Big) \\
&\quad - \frac{\kappa}{m}\epsilon_{ikl}\left(p_k\epsilon_{lj\alpha}\frac{x_\alpha}{r} - \partial_k\left(\frac{x_j}{r}\right)L_l\right) \\
&\quad + \frac{\kappa}{m}\epsilon_{jkl}\left(p_k\epsilon_{li\alpha}\frac{x_\alpha}{r} - \partial_k\left(\frac{x_i}{r}\right)L_l\right).
\end{aligned}
$$

The first term on the right can be simplified as

$$
\begin{aligned}
&\frac{1}{m^2}\epsilon_{jmn}\Big[-(\delta_{in}\delta_{l\alpha} - \delta_{i\alpha}\delta_{ln})p_m p_\alpha L_l + (\delta_{in}\delta_{k\alpha} - \delta_{i\alpha}\delta_{kn})p_k p_m L_\alpha \\
&\qquad + (\delta_{im}\delta_{k\alpha} - \delta_{i\alpha}\delta_{km})p_k p_\alpha L_n\Big] \\
&= \frac{1}{m^2}\cancel{\epsilon_{jmn}p_m p_i L_n} + \frac{1}{m^2}\epsilon_{jin}\vec{p}^{\,2}L_n - \frac{1}{m^2}\cancel{\epsilon_{jmn}p_m p_i L_n} \\
&= \frac{1}{m^2}\epsilon_{jin}\vec{p}^{\,2}L_n.
\end{aligned}
$$

Using this result in the above expression, we then find

$$
[M_i, M_j] = \frac{1}{m^2}\epsilon_{jin}\vec{p}^{\,2}L_n - \frac{\kappa}{m}\epsilon_{ikl}\left[p_k\epsilon_{lj\alpha}\frac{x_\alpha}{r} - \left(\frac{1}{r}\delta_{kj} - \frac{x_k x_j}{r^3}\right)L_l\right]
$$

$$
\begin{aligned}
&+\frac{\kappa}{m}\epsilon_{jkl}\left[p_k\epsilon_{li\alpha}\frac{x_\alpha}{r}-\left(\frac{1}{r}\delta_{ki}-\frac{x_kx_i}{r^3}\right)L_l\right]\\
&=-\frac{2}{m}\left(\frac{1}{2m}\vec{p}^2-\frac{\kappa}{r}\right)\epsilon_{ijk}L_k\\
&+\frac{\kappa}{m}\left[\frac{1}{r}p_jx_i-\frac{1}{r^3}(x_ix_j\vec{r}\cdot\vec{p}-r^2x_jp_i)\right.\\
&\left.-\frac{1}{r}p_ix_j+\frac{1}{r^3}(x_ix_j\vec{r}\cdot\vec{p}-r^2x_ip_j)\right]\\
&=-\frac{2}{m}\left(\frac{1}{2m}\vec{p}^2-\frac{\kappa}{r}\right)\epsilon_{ijk}L_k.
\end{aligned}
$$

13-20 (i) Based on the formula in Sec. 12.1, one will find

$$
\begin{pmatrix}\mathbf{e}_1\\ \mathbf{e}_2\\ \mathbf{e}_3\end{pmatrix}=\begin{pmatrix}
\cos\theta\cos\varphi\cos\psi-\sin\varphi\sin\psi & \cos\theta\sin\varphi\cos\psi+\cos\varphi\sin\psi & -\sin\theta\cos\psi\\
-\cos\theta\cos\varphi\sin\psi-\sin\varphi\cos\psi & -\cos\theta\sin\varphi\sin\psi+\cos\varphi\cos\psi & \sin\theta\sin\psi\\
\sin\theta\cos\varphi & \sin\theta\sin\varphi & \cos\theta
\end{pmatrix}
\times\begin{pmatrix}\mathbf{e}_1^{(0)}\\ \mathbf{e}_2^{(0)}\\ \mathbf{e}_3^{(0)}\end{pmatrix}
$$

or

$$
\begin{pmatrix}\mathbf{e}_1^{(0)}\\ \mathbf{e}_2^{(0)}\\ \mathbf{e}_3^{(0)}\end{pmatrix}=\begin{pmatrix}
\cos\theta\cos\varphi\cos\psi-\sin\varphi\sin\psi & -\cos\theta\cos\varphi\sin\psi-\sin\varphi\cos\psi & \sin\theta\cos\varphi\\
\cos\theta\sin\varphi\cos\psi+\cos\varphi\sin\psi & -\cos\theta\sin\varphi\sin\psi+\cos\varphi\cos\psi & \sin\theta\sin\varphi\\
-\sin\theta\cos\psi & \sin\theta\sin\psi & \cos\theta
\end{pmatrix}
\times\begin{pmatrix}\mathbf{e}_1\\ \mathbf{e}_2\\ \mathbf{e}_3\end{pmatrix}.
$$

Analogous relations can be used to relate L_1, L_2, L_3 to $L_1^{(0)}, L_2^{(0)}$ and $L_3^{(0)}$. Also, using the expressions of $\omega_1, \omega_2, \omega_3$ in Sec. 12.1, we can represent the components of $\vec{L}$ relative to the body-fixed basis $\{\mathbf{e}_i\}$ as

$$
\begin{aligned}
L_1&=I_1\omega_1=I_1(-\dot{\varphi}\sin\theta\cos\psi+\dot{\theta}\sin\psi),\\
L_2&=I_2\omega_2=I_2(\dot{\varphi}\sin\theta\sin\psi+\dot{\theta}\cos\psi),\\
L_3&=I_3\omega_3=I_3(\dot{\varphi}\cos\theta+\dot{\psi}),
\end{aligned}
$$

and the rotational kinetic energy is

$$
\begin{aligned}
T &= \frac{1}{2}I_1\omega_1^2 + \frac{1}{2}I_2\omega_2^2 + \frac{1}{2}I_3\omega_3^2 \\
&= \frac{1}{2}I_1(-\dot{\varphi}\sin\theta\cos\psi + \dot{\theta}\sin\psi)^2 \\
&\quad + \frac{1}{2}I_2(\dot{\varphi}\sin\theta\sin\psi + \dot{\theta}\cos\psi)^2 \\
&\quad + \frac{1}{2}I_3(\dot{\varphi}\cos\theta + \dot{\psi})^2.
\end{aligned}
$$

Then, for the canonical momenta conjugate to ψ, φ and θ, one obtains the following expressions:

$$
\begin{aligned}
p_\psi \equiv \frac{\partial T}{\partial \dot{\psi}} &= I_3(\dot{\varphi}\cos\theta + \dot{\psi}) = L_3 \\
&= L_1^{(0)}\sin\theta\cos\varphi + L_2^{(0)}\sin\theta\sin\varphi + L_3^{(0)}\cos\theta, \\
p_\varphi \equiv \frac{\partial T}{\partial \dot{\varphi}} &= -I_1\sin\theta\cos\psi\omega_1 + I_2\sin\theta\sin\psi\omega_2 + I_3\cos\theta\omega_3 \\
&= -L_1\sin\theta\cos\psi + L_2\sin\theta\sin\psi + L_3\cos\theta = L_3^{(0)}, \\
p_\theta \equiv \frac{\partial T}{\partial \dot{\theta}} &= I_1\sin\psi\omega_1 + I_2\cos\psi\omega_2 \\
&= L_1\sin\psi + L_2\cos\psi = -L_1^{(0)}\sin\varphi + L_2^{(0)}\cos\varphi.
\end{aligned}
$$

From these results, we can thus make the identifications

$$
\begin{aligned}
L_1 &= -\frac{\cos\psi}{\sin\theta}(p_\varphi - p_\psi\cos\theta) + p_\theta\sin\psi, \\
L_2 &= \frac{\sin\psi}{\sin\theta}(p_\varphi - p_\psi\cos\theta) + p_\theta\cos\psi, \\
L_3 &= p_\psi,
\end{aligned}
$$

and

$$
\begin{aligned}
L_1^{(0)} &= \frac{\cos\varphi}{\sin\theta}(p_\psi - p_\varphi\cos\theta) - p_\theta\sin\varphi, \\
L_2^{(0)} &= \frac{\sin\varphi}{\sin\theta}(p_\psi - p_\varphi\cos\theta) + p_\theta\cos\varphi, \\
L_3^{(0)} &= p_\varphi.
\end{aligned}
$$

(ii) Using the representations of (L_1, L_2, L_3) and $(L_1^{(0)}, L_2^{(0)}, L_3^{(0)})$ in terms of canonical variables $(\theta, \varphi, \psi, p_\theta, p_\varphi, p_\psi)$ (see (i)), it is straightforward to verify the given Poisson-bracket relations. For instance,

$$\begin{aligned}
[L_3^{(0)}, L_1^{(0)}] &= \left[p_\varphi, \frac{\cos\varphi}{\sin\theta}(p_\psi - p_\varphi\cos\theta) - p_\theta\sin\varphi\right] \\
&= \frac{\sin\varphi}{\sin\theta}(p_\psi - p_\varphi\cos\theta) + p_\theta\cos\varphi = L_2^{(0)}, \\
[L_3, L_1] &= \left[p_\psi, -\frac{\cos\psi}{\sin\theta}(p_\varphi - p_\psi\cos\theta) + p_\theta\sin\psi\right] \\
&= -\frac{\sin\psi}{\sin\theta}(p_\varphi - p_\psi\cos\theta) - p_\theta\cos\psi = -L_2, \\
[L_3^{(0)}, L_1] &= \left[p_\varphi, -\frac{\cos\psi}{\sin\theta}(p_\varphi - p_\psi\cos\theta) + p_\theta\sin\psi\right] = 0.
\end{aligned}$$

Here, $(L_1^{(0)}, L_2^{(0)}, L_3^{(0)})$ being the components of the angular momentum vector relative to space-fixed axes $\{\mathbf{e}_i^{(0)}\}$, they satisfy the usual Poisson-bracket relations corresponding to rotation generators, i.e., $[L_i^{(0)}, L_j^{(0)}] = \epsilon_{ijk}L_k^{(0)}$; also, as $\mathbf{e}_a = \sum_i(\mathbf{e}_a)_i\mathbf{e}_i^{(0)}$ for each $a = 1, 2, 3$ defines a vector observable, we should find

$$[L_i^{(0)}, (\mathbf{e}_a)_j] = \epsilon_{ijk}(\mathbf{e}_a)_k \quad \text{(for each } a\text{)}.$$

This can be checked explicitly using the representations of $L_1^{(0)}$ in (i) and $(\mathbf{e}_1)_1 = \cos\theta\cos\varphi\cos\psi - \sin\varphi\sin\psi$, $(\mathbf{e}_1)_2 = \cos\theta\sin\varphi\cos\psi + \cos\varphi\sin\psi$, $(\mathbf{e}_1)_3 = -\sin\theta\cos\psi$ (if one wishes to verify $[L_i^{(0)}, (\mathbf{e}_1)_j] = \epsilon_{ijk}(\mathbf{e}_1)_k$). Then, $L_a = \vec{L}\cdot\mathbf{e}_a = L_i^{(0)}(\mathbf{e}_a)_i$ and so $[L_i^{(0)}, L_a] = 0$ (i.e., L_a for each a defines a scalar observable). Now

$$\begin{aligned}
[L_a, L_b] &= [L_i^{(0)}(\mathbf{e}_a)_i, L_j^{(0)}(\mathbf{e}_b)_j] \\
&= L_i^{(0)}[(\mathbf{e}_a)_i, L_j^{(0)}(\mathbf{e}_b)_j] \\
&= -L_i^{(0)}\epsilon_{jik}(\mathbf{e}_a)_k(\mathbf{e}_b)_j \\
&= -L_i^{(0)}(\mathbf{e}_a\times\mathbf{e}_b)_i \\
&= -L_i^{(0)}\epsilon_{abc}(\mathbf{e}_c)_i = -\epsilon_{abc}L_c,
\end{aligned}$$

or $[\tilde{L}_a, \tilde{L}_b] = \epsilon_{abc}\tilde{L}_c$ if we define $\tilde{L}_a \equiv -L_a$. [We then note that $[\tilde{L}_a, (\mathbf{e}_b)_i] = -[L_j^{(0)}(\mathbf{e}_a)_j, (\mathbf{e}_b)_i] = -\epsilon_{jik}(\mathbf{e}_b)_k(\mathbf{e}_a)_j = (\mathbf{e}_a\times\mathbf{e}_b)_i = \epsilon_{abc}(\mathbf{e}_c)_i$ for each i.]

(iii) For a free rigid body with $H = \frac{1}{2}I_1\omega_1^2 + \frac{1}{2}I_2\omega_2^2 + \frac{1}{2}I_3\omega_3^2 = \sum_{i=1}^{3}\frac{1}{2I_i}L_i^2$, we have from the time evolution equation

$$\begin{aligned}\frac{dL_i}{dt} &= [L_i, H] \\ &= -\sum_{j,k}\frac{1}{I_j}\epsilon_{ijk}L_jL_k \\ &= -\sum_{j,k}\epsilon_{ijk}\omega_j\omega_k I_k,\end{aligned}$$

where we used $[L_i, L_j] = -\epsilon_{ijk}L_k$. This gives

$$I_1\dot{\omega}_1 = -\omega_2\omega_3 I_3 + \omega_3\omega_2 I_2 \quad \text{or} \quad I_1\dot{\omega}_1 + (I_3 - I_2)\omega_2\omega_3 = 0,$$

and, similarly,

$$\begin{aligned}I_2\dot{\omega}_2 + (I_1 - I_3)\omega_3\omega_1 = 0, \\ I_3\dot{\omega}_3 + (I_2 - I_1)\omega_1\omega_2 = 0,\end{aligned}$$

i.e., we obtain Euler's equations in Sec. 12.1 (with $\vec{\Gamma}^{(e)} = 0$).